Energy Performance and Indoor Climate Analysis in Buildings

Energy Performance and Indoor Climate Analysis in Buildings

Special Issue Editors

Jarek Kurnitski
Andrea Ferrantelli
Martin Thalfeldt

MDPI • Basel • Beijing • Wuhan • Barcelona • Belgrade

MDPI

Special Issue Editors
Jarek Kurnitski
Tallinn University of Technology
Estonia

Andrea Ferrantelli
Tallinn University of Technology
Estonia

Martin Thalfeldt
Norwegian University of Science and Technology
Estonia

Editorial Office
MDPI
St. Alban-Anlage 66
4052 Basel, Switzerland

This is a reprint of articles from the Special Issue published online in the open access journal *Energies* (ISSN 1996-1073) from 2018 to 2019 (available at: https://www.mdpi.com/journal/energies/special_issues/Energy_Performance_Indoor_Climate_Analysis_Buildings)

For citation purposes, cite each article independently as indicated on the article page online and as indicated below:

LastName, A.A.; LastName, B.B.; LastName, C.C. Article Title. *Journal Name* **Year**, *Article Number,* Page Range.

ISBN 978-3-03921-379-5 (Pbk)
ISBN 978-3-03921-380-1 (PDF)

Cover image courtesy of Jarek Kurnitski, Martin Thalfeldt and Andrea Ferrantelli.

Contents

About the Special Issue Editors

Jarek Kurnitski is Professor at Tallinn University of Technology and Aalto University, and Vice President of REHVA, Federation of European Heating and Air-Conditioning Associations, a non-profit organization representing more than 120,000 HVAC engineers and energy experts. He is Leader of the Estonian Center of Excellence in Research ZEBE, Zero Energy and Resource Efficient Smart Buildings, and Leader of the Nearly Zero Energy Buildings NZEB research group, which today operates at both universities. He is internationally known for the preparation of technical definitions for nearly zero energy buildings through many activities in REHVA Technology and Research Committee and contributions to European standards. Recently, he chaired a task force for preparing the European residential ventilation guidebook. He has been deeply involved in work towards improving the energy efficiency of the built environment in Estonia and Finland, with major contributions to the development of energy calculation frames for present energy performance regulations.

Andrea Ferrantelli is a Postdoctoral Researcher at Tallinn University of Technology. He obtained his MSc in theoretical physics at Turin University (Italy) and his PhD in theoretical particle cosmology at the University of Helsinki (Finland). He is interested in the physical modeling of HVAC and in energy efficient buildings.

Martin Thalfeldt recently completed his postdoc at Norwegian University of Science and Technology and returned to Tallinn University of Technology as Professor of Building Services (HVAC). His research interest is especially in advanced modeling of heating and ventilation systems for energy analyses and sizing purposes. He is actively involved in REHVA TRC, where he is currently chair of the office buildings' common EPC (Energy Performance Certificate) Task Force.

Preface to "Energy Performance and Indoor Climate Analysis in Buildings"

This Special Issue is dedicated to HVAC systems, load shifting, indoor climate, and energy and ventilation performance analyses in buildings. All of these topics are important for improving the energy performance of new and renovated buildings within the roadmap towards low energy and nearly zero energy buildings (NZEBs). On the background of this development, stringent NZEB requirements have recently been established and will soon be used in all EU Member States, as well as similar long-term zero energy building targets in Japan, the US, and other countries. To improve energy performance and occupant comfort and wellbeing at the same time, new technical solutions are evidently required. The research covered in this Special Issue provides evidence and examples of how such new technical solutions have worked, in practice, in new or renovated buildings, and also discusses potential problems and how solutions should be further developed. Energy performance and indoor climate improvements are also a challenge for calculation and sizing methods. More detailed approaches are needed in order to be able to design and size dedicated systems correctly, and to be capable for accurate quantification of energy saving effects. To avoid common performance gaps between calculated and measured performance, occupant behavior and building operation have to be adequately addressed. This demonstrates the challenge of high performance buildings as, in the end, comfortable buildings with good indoor climate which are easy and cheap to operate and maintain are expected by end customers. Ventilation performance, heating and cooling, sizing, energy predictions and optimization, load shifting, and field studies are some of the key topics of the articles in this Special Issue, contributing to the future of high performance buildings with reliable operation.

Jarek Kurnitski, Andrea Ferrantelli, Martin Thalfeldt
Special Issue Editors

Article

The Impact of Air Pressure Conditions on the Performance of Single Room Ventilation Units in Multi-Story Buildings

Alo Mikola *, Raimo Simson and Jarek Kurnitski

Nearly Zero Energy Research Group, Tallinn University of Technology, Tallinn 19086, Estonia
* Correspondence: alo.mikola@taltech.ee; Tel.: +372-566-47-035

Received: 30 April 2019; Accepted: 27 June 2019; Published: 9 July 2019

Abstract: Single room ventilation units with heat recovery is one of the ventilation solutions that have been used in renovated residential buildings in Estonia. In multi-story buildings, especially in a cold climate, the performance of units is affected by the stack effect and wind-induced pressure differences between the indoor and the outdoor air. Renovation of the building envelope improves air tightness and the impact of the pressure conditions is amplified. The aim of this study was to predict the air pressure conditions in typical renovated multi-story apartment buildings and to analyze the performance of room-based ventilation units. The field measurements of air pressure differences in a renovated 5-story apartment building during the winter season were conducted and the results were used to simulate whole-year pressure conditions with IDA-ICE software. Performance of two types of single room ventilation units were measured in the laboratory and their suitability as ventilation renovation solutions was assessed with simulations. The results show that one unit stopped its operation as a heat recovery ventilator. In order to ensure satisfactory indoor climate and heat recovery using wall mounted units the pressure difference values were determined and proposed for correct design.

Keywords: single room ventilation unit; building pressure condition; stack effect; wind pressure; ventilation renovation; decentralized ventilation unit

---

1. Introduction

In Estonia, multi-story apartment buildings constitute about 60% of the whole dwelling stock, and the majority (75%) of the buildings were built primarily in 1961–1990 [1]. Due to the increase in the price of energy, the energy policies of the European Union [2], the age, construction quality, and poor thermal insulation of the buildings, as well as both morally and technically outdated, obsolete heating and ventilation systems, there is an increasing need for retrofitting [3–7]. Part of the building stock built before the 1990s has already been renovated but for many apartment buildings this process is yet to start [6,8].

Typical multi-story apartment buildings have been built with natural ventilation, where fresh outdoor air enters through leaks or openings of the windows and doors, mixes with the warm room air, and leaves the building through shafts in the bathroom and kitchen. With retrofitting the building envelope, in order to achieve necessary thermal insulation for reducing the energy consumption for space heating, the air tightness of the building increases and the air flow through cracks and leaks is reduced, which makes the air change with natural ventilation very poor and does not provide the required air change rate [9]. Several analyses on the performance of ventilation in old Estonian dwellings [3,4,8] show that average indoor air CO_2 in occupied period is 1225 ppm which means the air change rate is too low to ensure good indoor air quality. As concluded in previous studies [10–16], there is a strong correlation between ventilation and health. With the renovation of old apartment buildings, the improvement of ventilation is unavoidable in order to provide healthy indoor environment for the occupants [17].

During the period 2010 to 2014 a total number of 663 apartment buildings were renovated using the renovation grant scheme [18]. The main principle of this grant schemes was to improve indoor air climate and energy efficiency of Estonian apartment buildings. There were 3 different grant levels, but in order to qualify for the highest financial support of 35% of the renovation costs provided by the state, the designed ventilation system was required to include heat recovery. Few solutions used in new buildings are suited for retrofitting purposes, mainly for construction-technological reasons. Other factors that affect the choice of suitable system are the cost of the system, the volume of construction work, aesthetics, adjustability and the costs of maintenance and operation. The impact of ventilation on the energy use of buildings can be between 30–60% for new and retrofitted buildings [5,8,19], thus heat recovery from the exhaust air is inevitable. Depending on the type of the heat exchanger (HEX) used in the air handling unit (AHU), it is possible to recover either sensible and latent heat or only sensible heat from the exhaust air [20,21].

The need for electricity to move the air increases at higher ventilation rates, becoming in some cases the main factor of increase in the final energy demand [22–24]. Ductless systems with room-based air handling units tend to have the lowest construction and operation costs, and to be simplest in design and most aesthetic [25]. The lack of ducts is a clear advantage since the most common problems are caused by the poor installation quality of ducts and inadequate project design [4]. It is also essential for the ventilation unit to have a low electric power consumption, suitable acoustic properties [26] and sufficient energy saving performance, which is strongly related to outdoor climatic conditions, the enthalpy efficiency, fan power consumption and necessary fresh air change rate [27].

One way to save energy from grid-connected electrical appliances would also be a real-time control strategy based on Model Predictive Control for the energy scheduling [28]. Chen et al. have presented the development of a model predictive control strategy for the hybrid ventilation solution [29]. As this model is still a prototype, it needs more testing to analyze the detailed possibilities of Model Predictive Control strategies.

The two most commonly used types of room-based devices used to renovate ventilation systems of apartment buildings during the retrofits 2010–2014 are: unit with recuperative plate HEX and centrifugal fans (Figure 1a) and unit with regenerative ceramic HEX and an axial fan (Figure 1b). The single-fan-based unit works in cycles, switching between the supply and exhaust mode every 60–70 s. During the exhaust cycle, the heat from the warm exhaust air is accumulated in the ceramic comb-like HEX and is then used to heat up the cold outdoor air during the supply cycle.

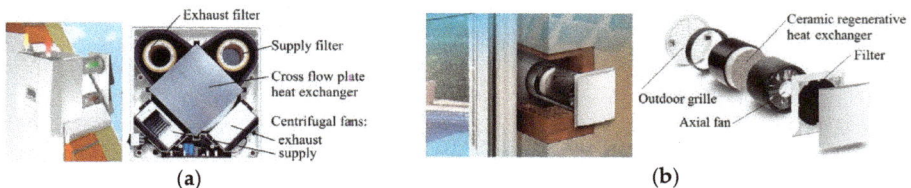

Figure 1. Types of room-based ventilation units used in the renovation of old apartment buildings: (**a**) with recuperative cross-flow plate HEX and (**b**) with regenerative ceramic heat exchanger (HEX).

Since the ventilation units are mounted inside the exterior wall of the building, the performance of the units is directly affected by the pressure differences between indoor and outdoor air across the building envelope. During the ventilation renovation in 2010–2014, the natural exhaust ventilation system was often not replaced. It means that the room-based ventilation units had to operate together with the natural ventilation system. The pressure difference in buildings with natural ventilation is caused mainly by the wind and the stack effect. There are numerous studies on both the stack effect [30–34] and the wind-induced pressure [35–37] in different building types with variable height, geometry and location. Wind conditions depend on the location and surroundings of the building. The stack effect depends on the height of the building and the temperature difference between indoor and outdoor air. The temperature differences of the air cause density differences that induce buoyancy force; the warm indoor air rises and is replaced by the colder outdoor air through the building envelope during the heating season. Studies indicate larger air pressure difference over the building envelope in more airtight buildings [19,38]. Shafts, staircases and other vertical openings, but also leaks through the cracks in floors, walls and ceilings can contribute considerably to the stack effect [39].

Indoor air quality measurements in Estonian renovated apartment buildings have shown that room-based ventilation units are not ensuring the necessary air change rate [18,40]. To secure the success of the renovation work, it is necessary to find out the reasons why the air change rate is below the designed values. Based on the described practical need, the main aim of this study is to analyze the performance of room-based ventilation units in typical renovated multi-story apartment buildings. Mikola et al. [40] have measured the air change rate in apartment buildings with room-based ventilation units and pointed out that problem may be caused by the high indoor and outdoor pressure difference and incorrect dimensioning of the fans. Thus, present study allows a detailed examination of these hypotheses. The results of the study can provide an innovative overview of the performance of the room-based ventilation units in renovated apartment buildings.

2. Methods

The performance of the exterior wall mounted single room ventilation units with regenerative and recuperative HEX were studied. Firstly, the on-site measurements were made in a renovated five-story apartment building. In next step, the measurements of units with regenerative HEX and recuperative HEX were performed in laboratory conditions. The results of field and laboratory measurements were used to compile a simulation model of the studied renovated apartment building with regenerative ventilation units. The next step was to calibrate the simulation model according to the measured indoor-and outdoor pressure differences, indoor temperatures, airflows, and outdoor climate data. Then the simulations of indoor and outdoor pressure differences, fan performance curves and heat recovery were performed. Lastly, the simulation and measurement results of room-based AHUs were analyzed and the conclusions on the performance of room-based ventilation units in apartment buildings were outlined. The flow chart of the main methods of the study is described in Figure 2.

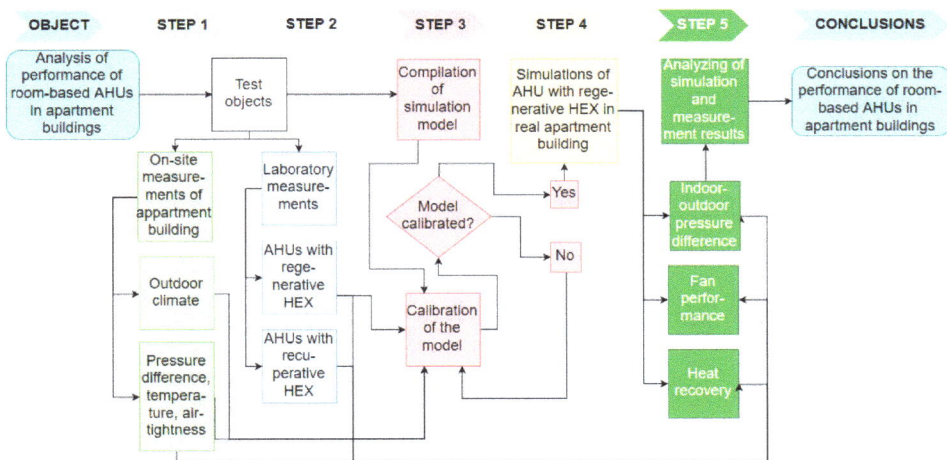

Figure 2. The flow chart of the performed studies. AHU: air handling unit.

2.1. Measurements

The pressure differences between the indoor and outdoor air across the exterior wall were measured during a 3-month period of the heating season in renovated 5-story apartment building. The fan performance and the temperature efficiency of room-based ventilation unit with regenerative and recuperative HEX were studied in TalTech technological facility.

2.1.1. The Studied Building

The studied building was a typical precast large concrete panel 5-story apartment building located in an urban area built in 1975 with 30 apartments, 2 staircases and a full cellar. The building is connected on both sides with two buildings of the same type. The height of stories is 2.7 m and the height of rooms is 2.5 m. The building is heated with water radiators by district heating. The floor plan of the building is shown in Figure 3a and cross-section in Figure 3b.

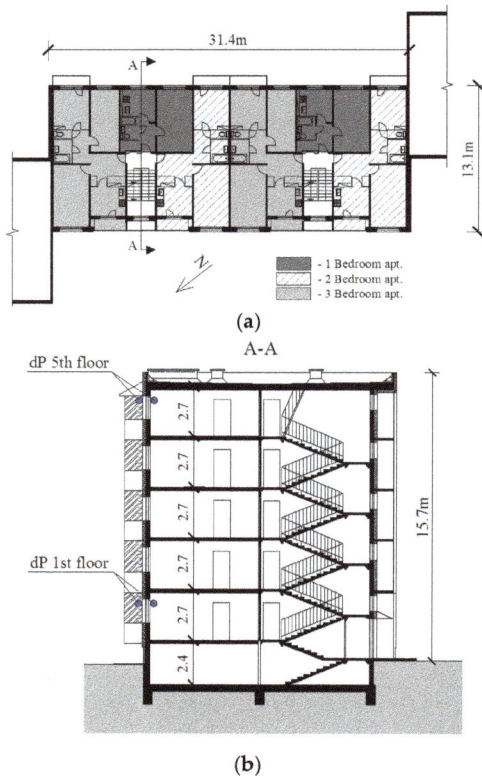

Figure 3. (a) Floor plan and (b) cross-section with pressure difference measurement point locations of the studied building.

The building was renovated in the years 2003 and 2012: the exterior walls, roof, and balconies were insulated, the windows were replaced; and the heating system was reconstructed. The thermal transmittances of the envelope before and after the retrofitting are presented in the Table 1. For ventilation, wall mounted regenerative HEX ventilation units were installed in the bedrooms and living rooms, in the bathrooms and kitchens natural exhaust ventilation was used (Figure 4). The diameter of the ventilation shafts with round cross-sections is 140 mm and height varies between

0.7 m and 12.6 m, depending on the story. All the units with regenerative HEX were controlled from the control center. It means that all the units worked in same speed and in same working cycle.

Table 1. Thermal transmittances of the envelope before and after retrofitting.

Part of the Thermal Envelope	Thermal Transmittance, W/(m²·K)	
	Before	After
External walls	1.05	0.22
Roof	0.45	0.15
Doors	2.0	1.2
Windows	2.9	1.4

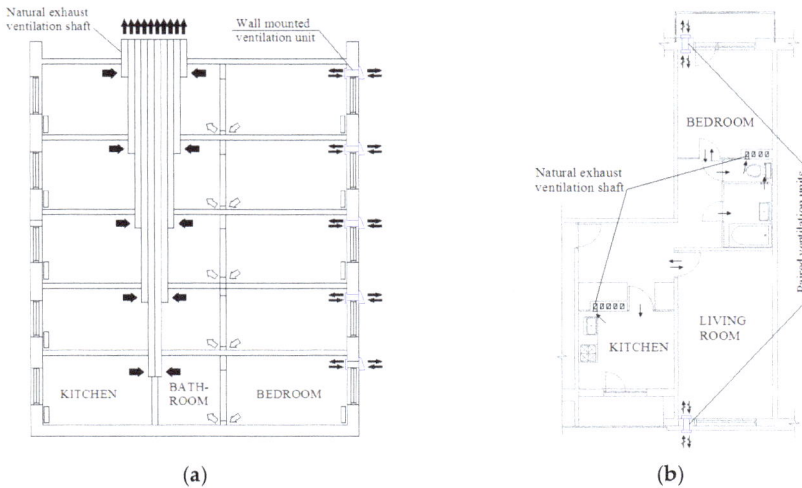

(a) (b)

Figure 4. Principle solution of the ventilation system with room-based units: (**a**) cross-section and (**b**) floor plan of a typical apartment.

2.1.2. Field Measurements

Pressure differences between the indoor and outdoor air across the exterior wall were measured during a 3-month period of the heating season (December 2013–February 2014) in 4 apartments located on different floors. Measurements were taken at the height of 2 m from the floor level (Figure 3b). For the outdoor pressure component, a plastic tube of 4 mm in diameter was planted through the window seal with one end connected to the pressure transducer. Diaphragm-type pressure transducers were used: Dwyer Magnesense MS-221 and Onset T-VER-PXU-L both with the measuring range of −50 to +50 Pa (output 0–10 VDC) and accuracy of 1% full scale output. Readings were taken in every 1 min and an average of the readings was saved with a 10-min interval using Onset Hobo U12 and Squirrel Q2010 data loggers. For environment measurements and data logging Hobo U12 devices were used with the temperature measuring range of −20 to +70 °C with accuracy ±0.35 °C and relative humidity 5% to 95% with accuracy ±2.5% of full-scale output.

2.1.3. Laboratory Measurements

In the laboratory-controlled environment, performance and efficiency of two types of room-based ventilation units were studied. The core elements of the first unit are cross-flow plate HEX, centrifugal fan, supply and exhaust filters, transfer ducts and outdoor hoods (see Figure 1a). The second unit has a ceramic HEX, axial fan, mounting tube, outdoor hood, inner cover and air filter (see Figure 1b).

The first device is a constant flow ventilation unit compared to the latter, which works in cycles, switching with 70-s intervals between the supply and exhaust mode. During the exhaust cycle, heat from the warm exhaust air is accumulated in the ceramic comb and is then used to heat up the cold outdoor air during the supply cycle. Both devices are installed without the heating coil.

The setup of the experiment of recuperative units is shown in Figure 5a and setup of the regenerative HEX is shown in Figure 5b. In the case of the unit with regenerative HEX, the temperature sensor was placed in the center of the airflow behind the air distributor. To measure the room and exhaust air temperature, another temperature sensor was placed to the top of the room 0.3 m away from the ventilation unit. The outdoor air temperature was measured close to the fresh air grille. The measuring cone was placed over the inner cap of the unit and the air speed was measured inside the cone. To measure the outside pressure, the pressure sensor was installed through the window to outside. The inside and outside pressure were both measured at a height of two meters. In case of unit with recuperative HEX, the temperature sensors were installed on the top of the unit inside the supply and exhaust airflow. The airflow, fresh air temperature, and pressure difference were measured in the same way as described in case of unit with regenerative HEX.

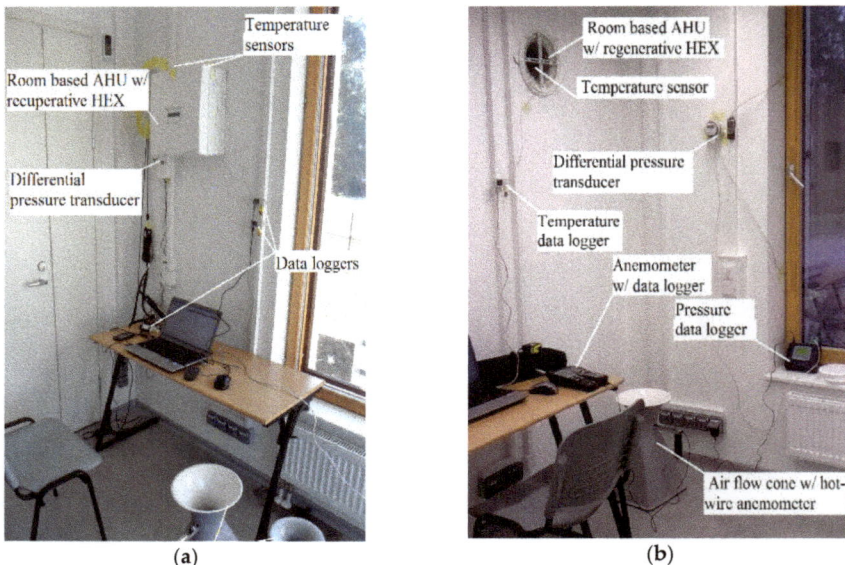

Figure 5. Experimental setup with the studied room-based ventilation units: (**a**) with recuperative HEX and (**b**) regenerative HEX.

During the experiments, the pressure difference between the indoor and outdoor air, outdoor temperature, supply/exhaust air temperature and air speed inside the measuring cone were measured and logged in every second. The same temperature and pressure sensors and loggers were used as in field measurements. The measured air speed was used to calculate the volumetric air flow through the unit. Airflow measurements were carried out using Testo 435-4 measuring instrument/data logger with hot-wire anemometry probe (measuring range from 0 to 20 m/s, with accuracy 0.01 m/s and +4% of reading). The pressure conditions in the test room were achieved using a central air handling unit and by adjusting the supply/exhaust valves of the ventilation system.

2.1.4. Temperature Efficiency of Room-Based Units

The temperature efficiency was used to quantify the effect of heat recovery of the studied ventilation system. As the main purpose of the study is to evaluate the performance of the room-based ventilation units, then the temperature ratio (efficiency) η_{temp} is defined as [25,26]:

$$\eta_{temp} = \frac{\overline{t_{sup}} - t_{out}}{t_{exh} - t_{out}},$$ (1)

where t_{out} is the outdoor air temperature and t_{exh} is the exhaust air temperature. The time-averaged value of the supply air temperature of ventilation units with the regenerative HEX, has to be used which is given by [26]:

$$\overline{t_{sup}} = \frac{1}{\tau} \int_{t=0}^{t=\tau} t_{sup}(t) \cdot dt$$ (2)

where t is the time and τ is the semi-period, which means the duration of the supply or extract process. In the case of the recuperative HEX, the process is in a steady state [26]:

$$t_{sup}(t) = \text{const.}$$ (3)

2.2. Computational Model

2.2.1. Description of Simulation Model

A model of the building was created and simulated using IDA Indoor Climate and Energy (IDA ICE) software version 4.6 developed by Equa Simulation AB (Figure 6). Each room of the composed building model is the separate zone. As there is common natural ventilation exhaust channel for the bathroom and toilet, these rooms were composed as a one zone. The building model was calibrated using the measured data from field studies. A custom climate file with hourly wind data, outdoor temperature and relative humidity of the measurement period from the local weather station located ~1 km from the site was used for the validation process. The orientation of the building is presented in Figure 3a.

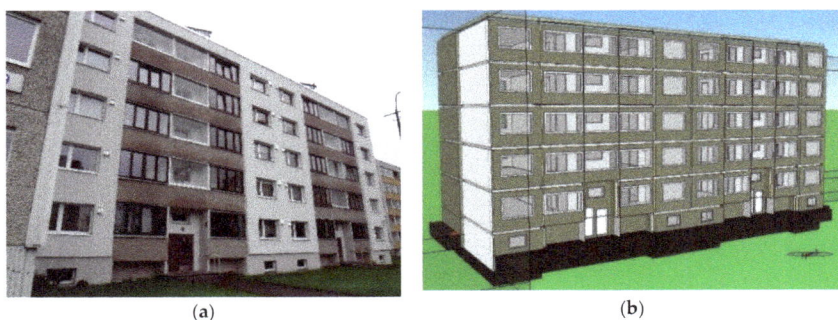

Figure 6. The studied (a) apartment building and (b) the simulation model.

For whole-year simulation, weather data from Estonian Test Reference Year (TRY) were used. The TRY is constructed using selected months from a number of calendar years, and may be used for many applications, such as indoor climate and energy simulations, HVAC system performance, or simulation of active or passive solar energy systems [41]. The air tightness of the building was defined with air leakage rate per envelope area at 50 Pa of pressure difference (q_{50}). A value of 3.0 $m^3/(h \cdot m^2)$ was used in the calibration process, as also achievable with the renovation of building envelope for pre-fabricated large-panel buildings [42].

The ventilation units were inserted according to the standard renovation solution that means 1 pair-wise ventilation system, which consist of 2 separate units, was added in every living room and bedroom. The natural exhaust ventilation system was not renovated and it continued the work as before. To modulate the natural ventilation system, the "chimney" component was used. Chimney takes into account the height and length of ventilation channels but also friction and minor pressure losses. Chimney elements were added to the kitchens and toilets or bathrooms. Two chimney components were added per each apartment. The airflows of natural exhaust systems were measured using hot wire anemometer with a cone. During the air flow measurements, the specific indoor and outdoor parameters were also measured and for the model calibration average values of airflow measurements were used.

The studied ventilation unit with regenerative HEX was modelled using IDA-ICE advanced modelling interface. The exterior wall leak module was used to calculate the differential pressure across the building exterior wall, which was used as an input for supply and exhaust air flow control accordingly to the laboratory measurements results. The main principles of the model are described in Figure 7. The pressure-airflow dependencies were inserted to the linear segment controller and connected to the respective air terminal. To model room-based units, some simplifications were made. Firstly, the standard ventilation unit macro was used and the control signal to the HEX was removed. The working cycle of the unit is 60 s in supply mode and after that the unit is turned off and the pairing device is switched on for 60 s in exhaust mode. Switching the units between supply and exhaust mode is achieved using the "gain" component. Regulating the supply and exhaust airflow was performed according to the differential pressure variable (DPA_S) of the exterior wall in "leak" component.

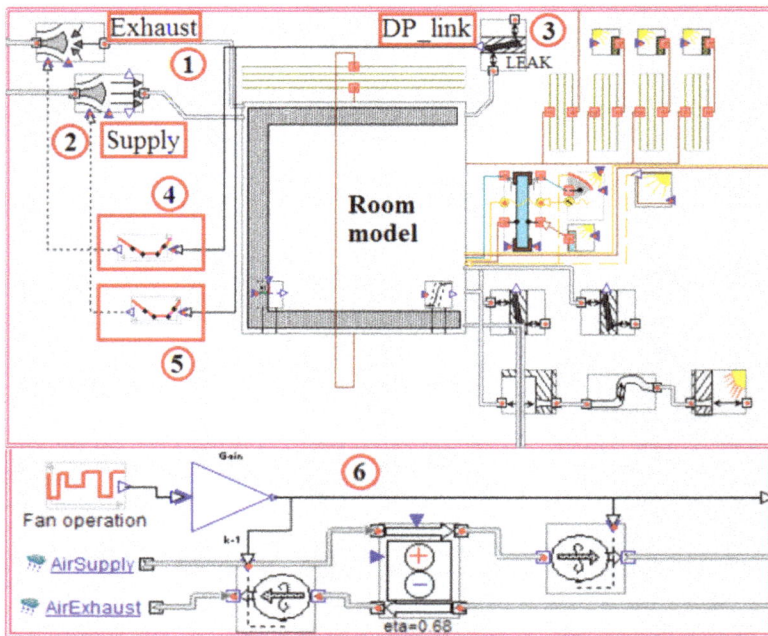

Figure 7. Schematics of the studied ventilation units modelling in IDA Indoor Climate and Energy (IDA ICE): (1) exhaust air terminal; (2) supply air terminal; (3) exterior wall leak module; (4) and (5) linear segment controllers; (6) ventilation unit module.

2.2.2. Air Pressure Calculations

The simulation model was calibrated according to the outdoor climate, indoor temperature, air change rate and air tightness measurements. During the calibration, the values of measured pressure difference were compared to the simulated data. The wind pressure distribution around the house is composed in way the wind flow is horizontal and an atmospheric boundary layer is neutral without vertical airflow [19]. The static wind pressure p_{wind} (Pa) outside the building facades is given by [19,43]:

$$p_{wind} = C_p \cdot \varrho_a \cdot U^2/2, \tag{4}$$

where C_p (dimensionless) is the pressure coefficient, ϱ_a is the air density (kg/m^3) and U (m/s) is the local wind velocity.

Pressure coefficients are empirically derived parameters determined either experimentally in a wind tunnel [44,45] or numerically using computational fluid dynamics [46,47]. In studied building model the wind-induced pressure conditions were simulated using constant wind-pressure coefficients defined at 45° intervals of a wind direction. Approximate values of wind pressure coefficients were used on external boundaries based on the exposure of the building. The pre-coded values of the "semi-exposed" option founded accurate enough (see Table 2).

Table 2. Facade average wind pressure coefficients used in the building simulation.

Facade	Orientation	Wind Angle (°)							
		0	45	90	135	180	225	270	315
Exterior wall	NE	0.4	0.2	−0.6	−0.5	−0.3	−0.5	−0.6	0.2
Exterior wall	SE	0.25	0.06	−0.35	−0.6	−0.5	−0.6	−0.35	0.06
Exterior wall	SW	0.4	0.2	−0.6	−0.5	−0.3	−0.5	−0.6	0.2
Exterior wall	NW	0.4	0.2	−0.6	−0.5	−0.3	−0.5	−0.6	0.2
Roof		−0.8	−0.8	−0.8	−0.8	−0.8	−0.8	−0.8	−0.8

The local wind velocity is calculated according to the simplified method for combining weather information with air tightness to calculate residential air infiltration (LBL method) recommended by American Society of Heating, Refrigerating and Air-Conditioning Engineers (ASHRAE) [43]. The local wind velocity at height h U(h) (m/s) is calculated by the equation:

$$U(h) = U_m \cdot k \cdot (h/h_m)^a, \tag{5}$$

where U_m (m/s) is the measured wind speed at the weather station (at a height of 10 m), h (m) the height from the surface of the ground, h_m (m) the height of the measurement equipment and constants k and a are the terrain coefficients. For the terrain coefficients k and a ASHRAE [43] recommended values for suburban terrain of 0.67 and 0.25 respectively were used. The LBL method, that is used in simulation model, has been proposed by Sherman and Grimsrud [48] and Modera et al. [49]. Modera et al. [49] have pointed out the typical values of terrain parameters for the standard terrain classes. The IV class is described as urban, industrial or forest areas and fitted the best with the conditions of tested apartment building. Sherman and Grimsrud [48] have pointed out that this method can also be used when the wind speed was not measured on-site.

The airflow Q (kg/s) through the bi-directional leakage opening is simulated in the building model with the empirical power law equation [19]:

$$Q = C \cdot \Delta P^n, \tag{6}$$

where C (dimensionless) is a flow coefficient (related to the opening), ΔP (Pa) is the pressure difference over the opening and n is a flow exponent which is characterizing the flow regime. The infiltration air flow is calculated for the facade of every zone [19]. The leakage openings in model are distributed over the building model according to the total infiltration airflow.

3. Results

3.1. Field Measurements

The results of field measurements showed that the pressure difference across the building envelope was negative during the entire measurement period in the first floor apartment and mostly negative in the fifth floor apartment (see Figure 8a). The occasional peaks toward zero-pressure difference are most likely caused by using the cooker hood, opening the windows or external doors to the balcony or staircase, the peaks and periods toward greater difference indicate the wind-induced effect. Pressure difference caused by wind can be dominant also for longer periods. The results indicate that the pressure difference is mostly caused by the stack effect being strongly dependent on the outdoor temperature in the bottom floor apartment, whereas on the top floor the dependence is weak due to the smaller height of the shaft (see Figure 8a). The measured indoor temperature during the measurement period in both apartments was roughly between 20 and 22 °C. The dependence between the indoor and outdoor pressure and temperature is shown in Figure 8b. In the first-floor apartment the value of linear correlation coefficient R^2 is 0.7483 and in fifth floor 0.0281.

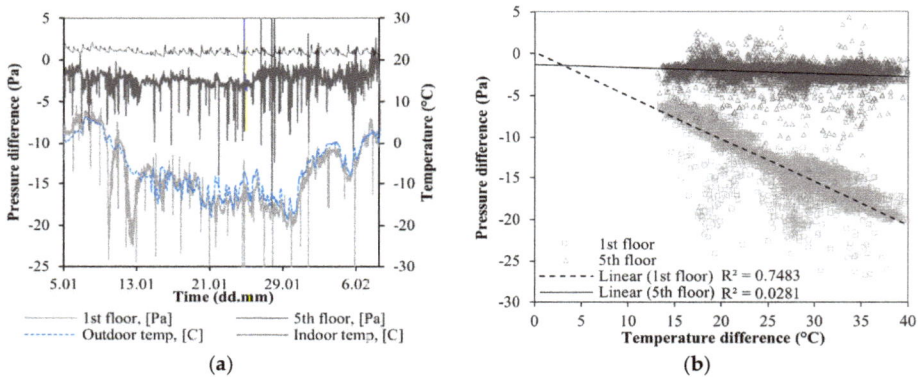

Figure 8. (a) The measured indoor and outdoor pressure differences in the first and fifth floor apartments, indoor and outdoor temperature in the heating season during a one-month period; (b) The dependence of pressure conditions on the indoor–outdoor temperature difference.

3.2. The Laboratory Measurements

Based on the results of laboratory measurements the fan performance and HEX temperature efficiency of observed AHUs were studied. The measurement results of the performance of the unit with ceramic HEX are shown in Figure 9a,b. In the beginning of the tests the value of underpressure in room was −2 Pa which means that supply and extract airflows were equal. After the pressure difference increased the extract air flow decreased and supply airflow increased at the same time. As results indicate, the supply and extract airflows are equal only at very low pressure differences. The greater the difference, the more the air flows differ. It can be seen that in case of 75% fan power, with differential pressure over −20 Pa the extract airflow is close to zero and the supply airflow around 60 m³/h (Figure 9b). The supply–exhaust cycles, which are presented in Figure 9b, show quick drop of the supply air temperature after the cycle change. During the tests, the outdoor air temperature was close to −5 °C. If the supply and extract airflows are equal, the supply air temperature was about 7 °C but if the pressure difference was increased from 0 Pa–20 Pa in test room, the supply air temperature at the end of the supply working cycle was about −2 °C.

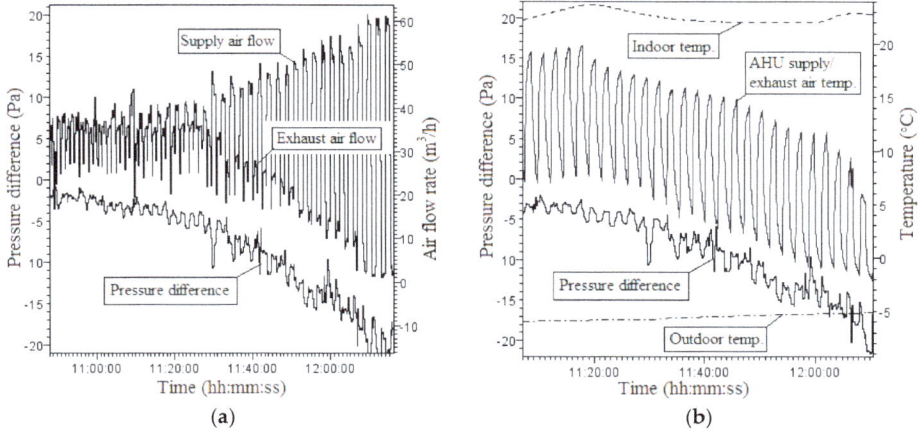

Figure 9. The results of the laboratory measurements of performance of the room unit with regenerative HEX at 75% fan power: (**a**) Pressure difference and air flows; (**b**) pressure difference and supply/exhaust air temperature.

The fan performance curves were constructed for the fan speed levels of 25%, 50%, 75%, and 100% (see Figure 10a). The fan performance curves show how the supply and extract airflows of the ventilation units are related to the in-and outdoor pressure difference. It is also possible to present how the pressure difference is related to the temperature efficiency of studied ventilation units (see Figure 10b). The results indicate that if the pressure difference rises then the temperature efficiency decreases. The same trend appears for all tested fan speeds. For example, in case the 50% speed level, the temperature efficiency is over 0.5 if the pressure difference is smaller than 4 Pa.

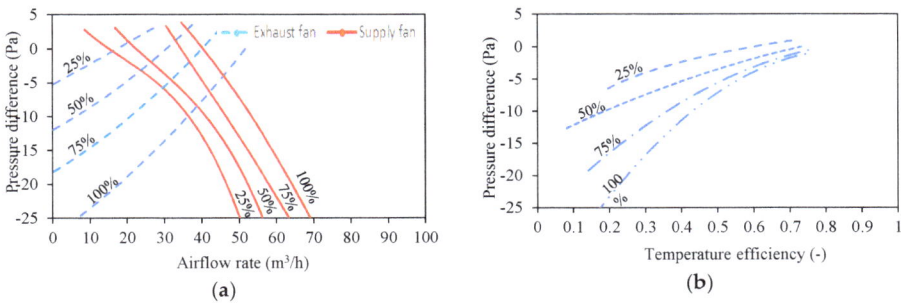

Figure 10. (**a**) Measurement based fan performance curves of room-based ventilation units with ceramic regenerative HEX; (**b**) measurement based temperature efficiencies of room-based ventilation units with ceramic regenerative HEX.

The fan performance curves and temperature efficiency graphs were also constructed for the ventilation units with recuperative HEX (see Figure 11a,b). The fan performance curves were constructed for the fan speed levels 25%, 50%, 75%, and 100%. Compared to the ventilation units with regenerative HEX, the units with recuperative can perform effectively in case of higher pressure differences between indoor and outdoor air. At the same time, if the pressure difference is −20 Pa at fan speed level 50%, the supply airflow is about 15% higher than exhaust airflow. The temperature efficiency of ventilation units with recuperative HEX is presented in Figure 11b. Compared to units with regenerative HEX, the temperature efficiency of studied ventilation units is significantly better at higher pressure difference conditions. The pressure difference influences the temperature efficiency the most in lower fan speed levels.

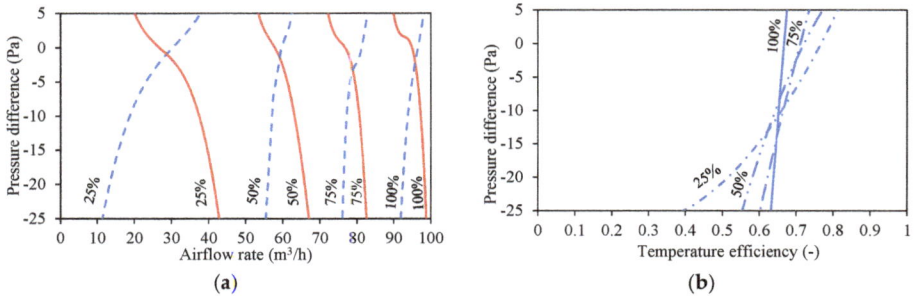

Figure 11. (**a**) Measurement based fan performance curves of room-based ventilation units with recuperative cross-flow plate HEX; (**b**) Measurement based temperature efficiencies of room-based ventilation units with recuperative cross-flow plate HEX.

3.3. Simulation Results

The simulation results achieved in the validation process are in good concordance with the field measurements (see Figure 12a), considering the fact that approximate wind pressure coefficients, performance of natural exhaust ventilation and wind data from an off-site weather station was used. The whole-year simulation results are presented in Figure 12b. The results show that the whole building is under negative pressure for 63% of the year (5521 h per year). In the first-floor apartments, the pressure difference is below −10 Pa for 22% (1927 h per year) and lower than −20 Pa for 2% of the year (180 h per year). The pressure difference across the exterior wall during the heating season in the 5-story building can be as high as −30 Pa on the first floor, −20 Pa on the third floor and −15 Pa on the fifth floor. Although the performed whole-year simulations has been done according to only one building and some simplifications have been done during the simulation process, the results confirm the fact that room based ventilation systems in 5-story buildings have to cope with the pressure difference which is more than −20 Pa.

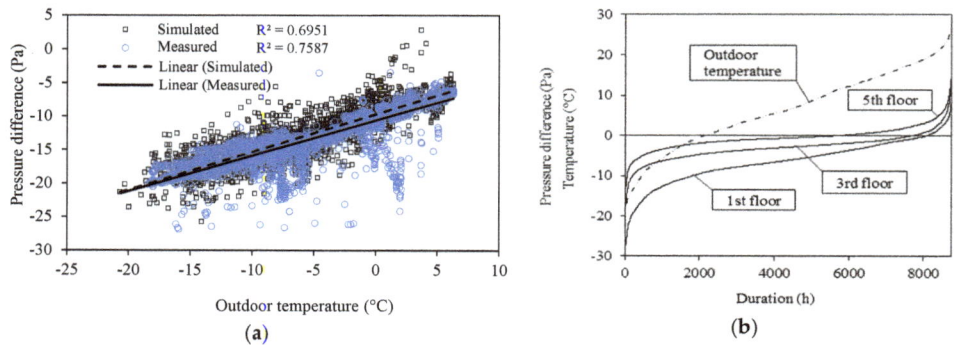

Figure 12. (**a**) Dependence between measured and simulated indoor and outdoor pressure differences on the indoor–outdoor temperature difference in the first-floor apartment. (**b**) The outdoor temperature and simulated whole-year pressure difference in the top, middle and bottom floor apartments of the 5-story apartment building.

The simulations of the single room ventilation units with the regenerative ceramic HEX were made using the same calibrated model of 5-story apartment building which was used in whole-year pressure difference simulations. The performance data of the fans and HEX has been taken from the results of laboratory tests that are described in pt. 3.2. As the studied ventilation units have to ensure the low noise level in living room and bedroom, the unit can only work in 30–50% speed level. An example of supply temperature and airflow rates simulation results of the ventilation unit with ceramic HEX, located in first floor, are shown as duration curves in Figure 13. During heating season, supply air temperature is relatively close to the outdoor temperature (Figure 13a) and that supply airflow rate is much higher than exhaust airflow rate (Figure 13b).

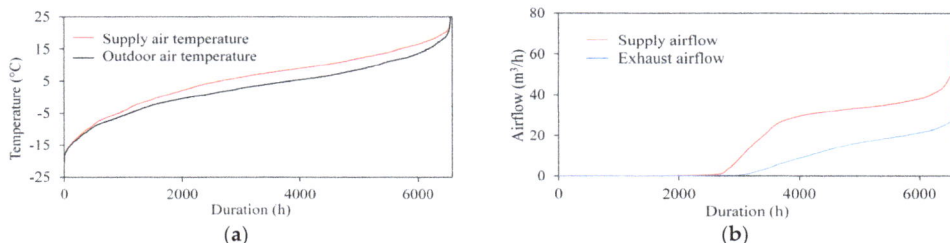

Figure 13. (**a**) Simulated supply air temperature and (**b**) airflow rates duration curves for ventilation unit with regenerative HEX during heating period in 1st floor apartment (September–May).

4. Discussion

The field measurements show that the pressure difference between the indoor and outdoor air in the bottom floor apartment depends heavily on the outdoor temperature, indicating the influence of the stack effect, whereas on the top floor, due to the smaller height of the exhaust ventilation shaft, the dependence is weak. Similar results have been also shown in other studies [50,51]. Kiviste and Vinha [50] have studied different Finish buildings and found that in some cases there can be large underpressure conditions (<−15 Pa) between in-and outdoor environments. Kalamees et al. [28] showed that in most critical cases, the air pressure across the building envelope may rise up to 30 Pa and the main reason for that is airtight building envelope with unbalanced ventilation. This study confirms that in renovated 5-stories apartment buildings with natural exhaust ventilation, the pressure difference across the building envelope can rise up to 20–30 Pa. Analyzing the fan performance of room-based ventilation units with regenerative HEX, it can be concluded that the pressure differences over the envelope were caused by the natural ventilation and density differences between the indoor and outdoor air. The room-based ventilation units itself did not play a significant role to increase the pressure drop across the building envelope.

During the analysis of the pressure difference measurement results and calibrating the model, the occasional peaks toward zero-pressure difference (see Figure 8a). These peaks are most likely caused by opening the windows or external doors to the balcony or staircase, the peaks and periods toward greater difference indicate the wind-induced effect. As the kitchen hoods and exhaust fans were installed in some apartments of studied building, the peaks can also be caused by these components of mechanical ventilation. Pressure difference caused by wind can be dominant also for longer periods. Although the wind-induced component varies in a wide range and depends on multiple variables, its contribution to the pressure conditions can be considerable, and thus special attention needs to be paid to buildings in wind exposed locations. Kalamees et al. [28] found that wind primarily influence the peak air pressure values and comparing the average values, the influence of wind is small. Despite the wind effect and other uncertainties during the pressure measuring and model compilation, we can see that dependence between measured and simulated indoor and outdoor pressure differences is quite strong (see Figure 12a). In the first floor the value of linear correlation coefficient of simulated data is 0.6951 and the correlation coefficient of measured data is 0.7587. The values of correlation coefficients, together with the similarity in results of simulation and measurement indicates that the calibration of the model was successful.

The laboratory measurement results of the studied room-based ventilation units show that supply and extract airflows are equal only at very low pressure differences. The greater the gap, the more the airflows differ. It can be seen, that the unit with plate HEX is performing considerably better, mainly due to centrifugal fans, but also because of the constant airflows. In case of the unit with regenerative HEX, the pressure difference is causing low-pressure axial fan to perform poorly: smaller volumes of exhaust air to flow through the unit during exhaust cycle, lowering the heat quantity accumulated in the HEX and leaving the outdoor air heating insufficient. The effect escalates with lower outdoor temperatures and higher pressure differences, in which case the heat transfer in the unit degrades and larger volumes of cold air are entering the ventilated space. It means that the pressure difference across the building envelope is closely related to the temperature efficiency of studied ventilation systems.

The performance of room-based ventilation units have also been monitored in previous studies. Smith et al. have developed the ventilation units with plastic rotary HEX [25,44]. They have made airflow measurements of the tested units using the tracer gas method. The main conclusion of this study was that the temperature efficiency of studied unit on equal supply and exhaust airflows is about 0.83–0.84 [25]. As these tests were not performed in different indoor and outdoor pressure conditions, it is not possible to make the conclusion, how these units would perform in apartment buildings in cold climate region.

Based on the simulation results, the pressure difference across the exterior wall during the heating season in a five-story building can be as high as −30 Pa on the first floor, −20 Pa on the third floor, and −15 Pa on the fifth floor. Comparing the simulated pressure conditions and the measured characteristics of the ventilation unit, poor performance of the unit can be expected. The simulation results of room-based units with regenerative HEX show that during heating season, supply air temperature is relatively close to the outdoor temperature and that supply airflow rate is much higher than exhaust airflow rate. At the same time the unit with recuperative HEX can ensure the temperature efficiency of the unit over 0.5 even under negative pressure as high as −25 Pa, making it possible to use the device in first floor apartments.

Several studies have shown that room-based ventilation units in Estonian apartment buildings are not ensuring the necessary air change rate [3,7,18]. Mikola et al. [3] point out that as these room units generate high sound power level, people switched ventilation units to the 30% of the maximum airflow. If tested room units with regenerative HEX work at 30% speed level and the pressure difference across the building envelope is −8 Pa then the supply airflow is 2 times higher than the exhaust airflow from the room. Mikola et al. [3] also measured the air change rate in apartments with regenerative room-based units. According to the measurements results, the average air exchange rate was $0.18\ h^{-1}$ and the average airflow per surface area of an apartment was 0.12 L/(s*m^2). According to the indoor climate category III, the general ventilation airflow in old apartments should be at least 0.35 L/(s*m^2) or $0.5\ h^{-1}$ and airflow in living rooms and bedrooms should be at least 0.6 L/(s*m^2) or 4 L/(s*person). The indoor climate category III requirements for the air change rate were also a minimum requirement to apply for the renovation grant. The air exchange rate met the requirements of III class in 6% of the apartments with room-based ventilation units [3]. In rest of the apartments, the minimum requirement of renovation grant scheme, was not ensured. It can be concluded, that the measured room-based ventilation renovation solution in apartment buildings does not ensure the necessary air change rate.

In case of both studied room-based ventilation units, the only possible to protect the HEX from freezing in cold climate is to reduce the supply airflow. As proven in this study, the exhaust airflow can be very small if the air pressure difference across the building envelope is high. That is the reason why there is high risk of ice formation in HEX, which complicates using these units in rooms with high humidity. The freezing process of HEXs of room-based units is not analyzed in detail during this study, so it would be worth to study this in future.

Based on previous studies [1,8,40] in Estonian climate, the following ventilation renovation solutions have performed better than room ventilation units:

- Apartment based supply and exhaust ventilation system with heat recovery, with unit located in apartment, in corridor or staircase.
- Centralized supply and exhaust ventilation system with heat recovery, with unit located on roof, pipes located in external wall or in apartment.
- Centralized exhaust ventilation system with fresh air radiators and heat pump heat recovery.

5. Conclusions

In this study, field measurements of pressure difference across the building envelope were carried out during a three-month period of the heating season in a fully renovated five-story apartment building. The results were used to validate the IDA-ICE whole building simulation model allowing to simulate hourly whole-year pressure conditions and airflows. Considering the measured and simulated pressure conditions, the performance of two different single room ventilation units was studied: one of the units was a device with a recuperative cross-flow plate HEX and two centrifugal fans and another with a regenerative ceramic HEX and an axial fan. The units were tested in TalTech technological facility, where supply and exhaust temperatures and airflow rates were measured under changing pressure conditions and different fan speeds. Fan and heat recovery efficiency curves were created and modelled in IDA-ICE for whole-year performance assessment.

In both cases of the studied ventilation units, pressure differences generated large differences in the supply and exhaust air flow rates. Because of the higher pressure rise, the airflow balance difference was much smaller in case of the unit with centrifugal fans compared to the unit with axial fan. This resulted in the smaller change in heat recovery efficiency of the recuperative HEX, compared to the regenerative HEX case which practically lost its heat recovery because of dominating stack effect pressure.

The simulation results show, that in cold periods, apartments in the first floor can be under negative pressure as high as −20 Pa for longer periods of time. In ventilation system planning, values of −10 Pa in fifth floor, −15 Pa in third floor and −20 Pa in first floor apartments can be recommended to be used as design values for ventilation units. The simulation results of single room units with regenerative HEX show that during heating season, supply air temperature was close to the outdoor temperature and that supply airflow rate was much higher than exhaust airflow rate, showing that the unit operated as air intake. Due to the differences in supply and exhaust airflows, there is a risk for freezing the heat exchanger, which excludes using studied ventilation units in rooms with high humidity.

The laboratory measurement results confirmed, that the axial fan used in the ventilation unit was not capable to work in typical pressure conditions occurring in multi-story building in cold periods, in order to achieve sufficient air change rate, heat recovery and supply air temperature, with noise levels under acceptable limits. In the case of the unit with recuperative HEX, under the same circumstances, the temperature efficiency of the unit remained higher than 0.5 even under negative pressure as high as−25 Pa, making it possible to use the device in first floor apartments.

Author Contributions: The laboratory and field measurements were performed by A.M. and R.S. Analyses of the measured data was carried out by A.M. and R.S. The simulation model was calibrated by R.S. The research principles and methods of the study were developed by A.M., R.S. and J.K.

Funding: This research was supported by the Estonian Centre of Excellence in Zero Energy and Resource Efficient Smart Buildings and Districts, ZEBE, grant 2014-2020.4.01.15-0016 funded by the European Regional Development Fund.

Acknowledgments: Authors would also like to thank Fund Kredex for cooperation of our research work.

Conflicts of Interest: The authors declare no conflict of interest.

References

1. Kuusk, K.; Kalamees, T.; Maivel, M. Cost effectiveness of energy performance improvements in Estonian brick apartment buildings. *Energy Build.* **2014**, *77*, 313–322. [CrossRef]
2. Directive (EU) 2018/844 of the European Parliament and of the Council of 30 May 2018 Amending Directive 2010/31/EU on the Energy Performance of Buildings and Directive 2012/27/EU on Energy Efficiency. Available online: https://eur-lex.europa.eu/legal-content/EN/TXT/?uri=OJ:L:2018:156:TOC (accessed on 30 April 2019).
3. Mikola, A.; Kõiv, T.-A.; Tennokese, K. Improving the indoor climate and energy saving in renovated apartment buildings in Estonia. In Proceedings of the 12th REHVA World Congress CLIMA 2016, Aalborg, Denmark, 22–25 May 2016.
4. Mikola, A.; Kõiv, T.-A.; Kalamees, T. Quality of ventilation systems in residential buildings: Status and perspectives in Estonia. In Proceedings of the International Workshop Securing the Quality of Ventilation Systems in Residential Buildings: Status and Perspectives, Brussels, Belgium, 18–19 March 2013.
5. Kalamees, T.; Kuusk, K.; Arumägi, E.; Ilomets, S.; Maivel, M. The analysis of measured energy consumption in apartment buildings in Estonia. In Proceedings of the IEA Annex 55 (RAP-RETRO), Working Meeting, Vienna, Austria, 23–25 April 2012.
6. Kurnitski, J.; Kuusk, K.; Tark, T.; Uutar, A.; Kalamees, T.; Pikas, E. Energy and investment intensity of integrated renovation and 2030 cost optimal savings. *Energy Build.* **2014**, *75*, 51–59. [CrossRef]
7. Hamburg, A.; Kalamees, T. The Influence of Energy Renovation on the Change of Indoor Temperature and Energy Use. *Energies* **2018**, *11*, 3179. [CrossRef]
8. Kuusk, K.; Kalamees, T. Estonian Grant Scheme for Renovating Apartment Buildings. *Energy Procedia* **2016**, *96*, 628–637. [CrossRef]
9. Paap, L.; Mikola, A.; Kõiv, T.-A.; Kalamees, T. Airtightness and ventilation of new Estonian apartments constructed 2001-2010. In Proceedings of the Joint Conference 33rd AIVC Conference and 2nd TightVent Conference Optimising Ventilative Cooling and Airtightness for [Nearly] Zero-Energy Buildings, IAQ and Comfort, Copenhagen, Denmark, 10–11 October 2012.
10. Armstrong, P.; Dirks, J.; Klevgard, L.; Matrosov, Y.; Olkinuora, J. Infiltration and Ventilation in Russian Multi-Family Buildings. In Proceedings of the ACEEE Summer Study on Energy Efficiency in Buildings, Residential Buildings: Technologies, Design, and Performance Analysis, Pacific Grove, CA, USA, 25–31 August 1996.
11. Wargocki, P.; Sundell, J.; Bischof, W.; Brundrett, G.; Fanger, P.; Gyntelberg, F.; Hanssen, S.; Harrison, P.; Pickering, A.; Seppanen, O.; et al. Ventilation and health in non-industrial indoor environments: Report from a European Multidisciplinary Scientific Consensus Meeting (EUROVEN). *Indoor Air* **2002**, *12*, 113–128. [CrossRef] [PubMed]
12. Iversen, M.; Bach, E.; Lundqvist, G. Health and comfort changes among tenants after retrofitting of their housing. *Environ. Int.* **1986**, *12*, 161–166. [CrossRef]
13. Menzies, R.; Tamblyn, R.; Farant, J.; Hanley, J.; Nunes, F.; Tamblyn, R. The Effect of Varying Levels of Outdoor-Air Supply on the Symptoms of Sick Building Syndrome. *N. Engl. J. Med.* **1993**, *328*, 821–827. [CrossRef] [PubMed]
14. Seppanen, O.; Fisk, W.; Mendell, M. Association of Ventilation Rates and CO_2 Concentrations with Health andOther Responses in Commercial and Institutional Buildings. *Indoor Air* **1999**, *9*, 226–252. [CrossRef]
15. Ucci, M.; Ridley, I.; Pretlove, S.; Davies, M.; Mumovic, D.; Oreszczyn, T.; McCarthy, M.; Singh, J. Ventilation rates and moisture-related allergens in UK dwellings. In Proceedings of the 2nd WHO International Housing and Health Symposium, Vilnius, Lithuania, 29 September–1 October 2004.
16. Ruotsalainen, R.; Jaakkola1, J.; Ronnberg, R.; Majanen, A.; Seppanen, O. Symptoms and Perceived Indoor Air Quality among Occupants of Houses and Apartments with Different Ventilation Systems. *Indoor Air* **1991**, *1*, 428–438. [CrossRef]
17. Noris, F.; Adamkiewicz, G.; Delp, W.; Hotchi, T.; Russell, M.; Singer, B.; Spears, M.; Vermeer, K.; Fisk, W. Indoor environmental quality benefits of apartment energy retrofits. *Build. Environ.* **2013**, *68*, 170–178. [CrossRef]

18. Hamburg, A.; Kalamees, T. Improving the indoor climate and energy saving in renovated apartment buildings in Estonia. In Proceedings of the 9th International Cold Climate HVAC 2018, Kiruna, Sweden, 12–15 March 2018.

19. Jokisalo, J.; Kurnitski, J.; Korpi, M.; Kalamees, T.; Vinha, J. Building leakage, infiltration, and energy performance analyses for Finnish detached houses. *Build. Environ.* **2009**, *44*, 377–387. [CrossRef]

20. Papakostas, K.; Kiosis, G. Heat recovery in an air-conditioning system with air-to-air HEX. *Int. J. Sustain. Energy* **2014**, *34*, 221–231. [CrossRef]

21. El Fouih, Y.; Stabat, P.; Rivière, P.; Hoang, P.; Archambault, V. Adequacy of air-to-air heat recovery ventilation system applied in low energy buildings. *Energy Build.* **2012**, *54*, 29–39. [CrossRef]

22. Merzkirch, A.; Maas, S.; Scholzen, F.; Waldmann, D. Field tests of centralized and decentralized ventilation units in residential buildings—Specific fan power, heat recovery efficiency, shortcuts and volume flow unbalances. *Energy Build.* **2016**, *116*, 376–383. [CrossRef]

23. Kim, M.; Baldini, L. Energy analysis of a decentralized ventilation system compared with centralized ventilation systems in European climates: Based on review of analyses. *Energy Build.* **2016**, *111*, 424–433. [CrossRef]

24. Santos, H.; Leal, V. Energy vs. ventilation rate in buildings: A comprehensive scenario-based assessment in the European context. *Energy Build.* **2012**, *54*, 111–121. [CrossRef]

25. Smith, K.; Svendsen, S. Development of a plastic rotary HEX for room-based ventilation in existing apartments. *Energy Build.* **2015**, *107*, 1–10. [CrossRef]

26. Manz, H.; Huber, H.; Schälin, A.; Weber, A.; Ferrazzini, M.; Studer, M. Performance of single room ventilation units with recuperative or regenerative heat recovery. *Energy Build.* **2000**, *31*, 37–47. [CrossRef]

27. Liu, J.; Li, W.; Liu, J.; Wang, B. Efficiency of energy recovery ventilator with various weathers and its energy saving performance in a residential apartment. *Energy Build.* **2010**, *42*, 43–49. [CrossRef]

28. Hosseini, S.M.; Carli, R.; Dotoli, M. Model Predictive Control for Real-Time Residential Energy Scheduling under Uncertainties. In Proceedings of the IEEE International Conference on Systems, Man, and Cybernetics (SMC), Miyazaki, Japan, 7–10 October 2018; pp. 1386–1391.

29. Chen, J.; Augenbroe, G.; Song, X. Lighted-weighted model predictive control for hybrid ventilation operation based on clusters of neural network models. *Autom. Constr.* **2018**, *89*, 250–265. [CrossRef]

30. Kalamees, T.; Kurnitski, J.; Jokisalo, J.; Eskola, L.; Jokiranta, K.; Vinha, J. Measured and simulated air pressure conditions in Finnish residential buildings. *Build. Serv. Eng. Res. Technol.* **2010**, *31*, 177–190. [CrossRef]

31. Khoukhi, M.; Al-Maqbali, A. Stack Pressure and Airflow Movement in High and Medium Rise buildings. *Energy Procedia* **2011**, *6*, 422–431. [CrossRef]

32. Khoukhi, M.; Yoshino, H.; Liu, J. The effect of the wind speed velocity on the stack pressure in medium-rise buildings in cold region of China. *Build. Environ.* **2007**, *42*, 1081–1088. [CrossRef]

33. Acred, A.; Hunt, G. Multiple Flow Regimes in Stack Ventilation of Multi-Story Atrium Buildings. *Int. J. Vent.* **2013**, *12*, 31–40. [CrossRef]

34. Tovar, T.; Garrido, C. Stack-Driven Ventilation in Two Interconnected Rooms Sharing a Single Opening and Connected to the Exterior by a Lower Vent. *Int. J. Vent.* **2010**, *9*, 211–226. [CrossRef]

35. Kalamees, T.; Arumägi, E.; Tähiste, M. Air tightness of apartment buildings in Estonia. In Proceedings of the 4th International Symposium on Building and Ductwork Air Tightness (BUILDAIR)/30th AIVC Conference: Trends in High Performance Buildings and the Role of Ventilation, Berlin, Germany, 1–2 October 2009.

36. Gładyszewska-Fiedoruk, K.; Gajewski, A. Effect of wind on stack ventilation performance. *Energy Build.* **2012**, *51*, 242–247. [CrossRef]

37. Etheridge, D. Wind Turbulence and Multiple Solutions for Opposing Wind and Buoyancy. *Int. J. Vent.* **2009**, *7*, 309–320. [CrossRef]

38. Kalamees, T.; Korpi, M.; Eskola, L.; Kurnitski, J.; Vinha, J. The distribution of the air leakage places and thermal bridges in Finnish detached houses and apartment buildings. In Proceedings of the 8th Symposium on Building Physics in the Nordic Countries: Dept. of Civil Engineering, Copenhagen, Denmark, 16–18 June 2008.

39. Jo, J.; Lim, J.; Song, S.; Yeo, M.; Kim, K. Characteristics of pressure distribution and solution to the problems caused by stack effect in high-rise residential buildings. *Build. Environ.* **2007**, *42*, 263–277. [CrossRef]

40. Mikola, A.; Kalamees, T.; Kõiv, T.-A. Performance of ventilation in Estonian apartment buildings. *Energy Procedia* **2017**, *132*, 963–968. [CrossRef]

41. Kalamees, T.; Kurnitski, J. Estonian Test Reference Year for Energy Calculations. *Proc. Est. Acad. Sci. Eng.* **2006**, *12*, 40–58.
42. Kuusk, K.; Kalamees, T.; Link, S.; Ilomets, S.; Mikola, A. Case-study analysis of concrete large-panel apartment building at pre- and post low-budget energy-renovation. *J. Civ. Eng. Manag.* **2016**, *23*, 67–75. [CrossRef]
43. ASHRAE. Airflow around buildings. In *Book 1993 ASHRAE Handbook-Fundamentals*; American Society of Heating, Refrigerating and Air Conditioning Engineers: Atlanta, GA, USA, 1993.
44. Smith, K.; Svendsen, S. The effect of a rotary HEX in room-based ventilation on indoor humidity in existing apartments in temperate climates. *Energy Build.* **2016**, *116*, 349–361. [CrossRef]
45. Peng, X.; Yang, L.; Gavanski, E.; Gurley, K.; Prevatt, D. A comparison of methods to estimate peak wind loads on buildings. *J. Wind Eng. Ind. Aerodyn.* **2014**, *126*, 11–23. [CrossRef]
46. Montazeri, H.; Blocken, B. CFD simulation of wind-induced pressure coefficients on buildings with and without balconies: Validation and sensitivity analysis. *Build. Environ.* **2013**, *60*, 137–149. [CrossRef]
47. Grosso, M. Wind pressure distribution around buildings: A parametrical model. *Energy Build.* **1992**, *18*, 101–131. [CrossRef]
48. Sherman, M.H.; Grimsrud, D. Infiltration-pressurization correlation: Simplified physical modeling. *ASHRAE Trans.* **1980**, *86*, 778–807.
49. Modera, M.P.; Sherman, M.H.; Grimsrud, D.T. A Predictive Air Infiltration Model-Long-Term Field Test Validation. In Proceedings of the Semi-annual meeting of the American Society of Heating, Refrigerating, and Air Conditioning Engineers, Houston, TX, USA, 24 January 1982; Available online: https://www.osti.gov/biblio/6285265 (accessed on 4 March 2019).
50. Kiviste, M.; Vinha, J. Air pressure difference measurements in Finnish municipal service buildings. *Energy Procedia* **2017**, *132*, 879–884. [CrossRef]
51. Leivo, V.; Kiviste, M.; Aaltonen, A.; Turunen, M.; Haverinen-Shaughnessy, U. Air Pressure Difference between Indoor and Outdoor or Staircase in Multi-family Buildings with Exhaust Ventilation System in Finland. *Energy Procedia* **2015**, *78*, 1218–1223. [CrossRef]

energies

MDPI

Article

Supporting the Smart Readiness Indicator—A Methodology to Integrate A Quantitative Assessment of the Load Shifting Potential of Smart Buildings

Thomas Märzinger [1] **and Doris Österreicher** [2,*]

1　Department of Material Sciences and Process Engineering, Institute for Chemical and Energy Engineering, University of Natural Resources and Life Sciences, 1190 Vienna, Austria; thomas.maerzinger@boku.ac.at
2　Department of Landscape, Spatial and Infrastructure Sciences, Institute of Spatial Planning, Environmental Planning and Land Rearrangement, University of Natural Resources and Life Sciences, 1190 Vienna, Austria
*　Correspondence: doris.oesterreicher@boku.ac.at; Tel.: +43-1-47654-85515

Received: 23 April 2019; Accepted: 17 May 2019; Published: 22 May 2019

Abstract: With the third revision of the Energy Performance of Buildings Directive (EPBD) issued in July 2018, the assessment of buildings now has to include a Smart Readiness Indicator (SRI) to consider the fact that buildings must play an active role within the context of an intelligent energy system. In order to support the development of the SRI, this article describes a methodology for a simplified quantitative assessment of the load shifting potential of buildings. The aim of the methodology is to provide a numerical, model-based approach, which allows buildings to be categorized based on their energy storage capacity, load shifting potential and their subsequent interaction with the grid. A key aspect is the applicability within the Energy Performance Certificate (EPC) in order to provide an easy to use calculation, which is applied in addition to the already established energy efficiency, building services and renewable energy assessments. The developed methodology is being applied to theoretical use cases to validate the approach. The results show that a simplified model can provide an adequate framework for a quantitative assessment for the Smart Readiness Indicator.

Keywords: smart buildings; smart readiness indicator; energy efficiency; energy performance of buildings directive; energy flexibility; load shifting; demand response

1. Introduction

By 2050 the EU must reduce its greenhouse gas emissions to 80% below 1990-based levels [1]. Considering that the building sector currently contributes up to 40% to the overall emission levels, the European Commission is recognizing that energy efficiency in buildings can add significantly to this reduction. For 2030 a new binding energy efficiency target of 32.5% has been defined in a political agreement between the European Commission, the European Parliament and the European Council [2], which should subsequently trigger legislative measures and financial incentives on the respective national levels. Energy efficiency contributes in this context to the reduction of costs for consumers, reduction of the import dependency in Europe and redirecting investments towards smart and sustainable infrastructure [3]. This also highlights, that higher energy efficiency must be achieved by innovative measures without increasing the financial burden on the consumers.

Innovation plays one of the key roles to facilitate the transformation towards an energy efficient building stock in Europe. In order to boost clean energy innovation, the EU has defined several fundamental initiatives: The 'Strategy on accelerating clean energy innovation' [3], together with the 'Accelerating clean energy in buildings' initiative [4] define a comprehensive plan for the main policy levers. Thus in regard to energy, the focus of the Horizon 2020 funding—the core of the research and innovation pillar—is on the decarbonization of the EU building stock [5]. The funding schemes should ensure that industry and research are strongly working together towards innovative systems, products and business models to accelerate this transformation. Another key aspect to increase efficiency in the building sector is the legislative framework on a European level, which directly influences the national building energy performance standards. Understanding the economic, political and social drivers in reducing energy and greenhouse gas emission in the buildings industry are crucial in order to propose new policies related to climate change. Studies that outline the impact of energy policies contribute significantly to the adequate framing of future policies [6].

The related key European legislative framework documents are the Energy Efficiency Directive [7], the Renewable Energy Directive [8] and Energy Performance of Buildings Directive (EPBD) [9]. Each of these directives provides targets and framework conditions for the implementation into national regulations and standards. The EPBD sets out the conditions for the national and regional building regulations, which subsequently define the quality of the European building stock. Currently the EPBD states that an Energy Performance Certificate (EPC), which includes an assessment of the energy efficiency as well as the building services and renewable energy systems, has to be submitted to the local building authorities for approval. With the latest revision of the EPBD in July 2018 [10] the assessment now has to include a Smart Readiness Indicator (SRI) to take into account the capacity of the building to manage and optimize itself and to interact with occupants and the grid. A study, subsequently commissioned by the EC, consisting of an assessment methodology for the SRI has been issued as a guiding document for the member states. The provided methodology is focusing on a wide range of qualitative aspects in a rather complex matrix approach, covering a series of impact criteria, domains and domain services [11]. However the approach proposed in this study is highly qualitative. An expert judgment, with appraisals to be carried out by certified assessors, dominates the application of the methodology. Even though this approach addresses relevant aspects related to so-called Smart Buildings, it does not include a calculation of the actual load shifting potential or "grid friendliness" of a building, which is one of the key objectives of the SRI. Since an assessment of the building to adapt to user and grid cannot be solely based on a qualitative appraisal, the expert judgments should be complemented by a quantitative and replicable assessment.

Therefore in this article we propose to support the existing SRI approach with a methodology based on a quantitative approach to assess the load shifting potential of buildings. The aim of the proposed methodology is to provide a numerical, model-based approach, which allows buildings to be categorized based on their energy storage capacity, load shifting potential and their subsequent interaction with the grid. One of the key aspects of the proposed methodology is to provide an easy to use calculation within the EPC, which is applied in addition to the already established energy efficiency, building services and renewable energy assessments.

In the next sections the legislative and content related framework conditions related to the EPBD and SRI are outlined. This is followed by a description of state-of-the-art research in the context of demand response, load shifting and smart grid in relation to buildings. A separate sub-section is dedicated to the development of Smart Buildings. Subsequently Section 3 outlines the main principles behind the proposed methodology, followed by the results in Section 4, where the mathematical models are being tested on theoretical use cases in order to validate the approach. Finally, the discussion and conclusions deliver a review of the approach and provide an outlook on how this methodology might be implemented in the revisions of the national and regional building regulations the member states have to undertake following the latest update of the legally binding EPBD.

2. Background

In 2002 the European Parliament has released the first version of the EPBD in order to promote energy efficiency in buildings with the aim of fulfilling the climate targets as specified in the Kyoto protocol [9,12]. By adhering to the guidelines set out in the EPBD the member states acknowledge the fact, that within the EU, buildings are responsible for approximately 40% of energy consumption and 36% of CO_2 emissions [2]. Stricter guidelines, including the commitment to Nearly Zero Energy

Buildings (NZEBs) have been added in the 2010 revision of the EPBD [13], stipulating that all new buildings must be built to this standard as of 1st January 2021. The NZEBs characterize a significant opportunity to provide future-proof buildings with minimal energy consumption. Other aspects include the mandatory issuance of energy performance certificates in all advertisements and whenever a building is sold or rented. In addition, member states had to set cost-optimal levels for minimum energy performance requirements for new buildings and renovations and had to draw up a list of financial measures to improve the energy efficiency in buildings [13]. In regards to aspects relating to governments' own building stock the member states must ensure that at least 3% of the total floor area of buildings owned and occupied by central government are renovated, that only highly efficient buildings are purchased by the governments and that long-term national building renovation strategies are drawn up to be included in their National Energy Efficiency Action Plans [2]. Judging the impact of this legislation, one can look at overall emission figures: Between 1990 and 2016, greenhouse gas emissions in the EU have been reduced by 22% [14]. In parts, this can also be attributed to the stricter efficiency guidelines for the European building stock.

In the third revision of the EPBD issued in July 2018 [10] further improvements have been made in order to accelerate the reduction of greenhouse gas emissions associated with the building sector. With the 2018 amendment, EU member states will have to establish stronger long-term renovation strategies with sound financial components, aiming at decarbonizing the national building stock by 2050. With around 40% of Europe's housing stock dating pre-1960 [15], renovation becomes ever more important as climate targets are getting raised. Smart technologies will have to be promoted e.g., by adding conditions related to control engineering and building automation systems. Other measures include the increased support of e-mobility by adding requirements on respective infrastructure for buildings as well as aspects on ventilation and air quality to increase the health and well-being for the occupants [2,10]. Finally, one of the key new requirements includes a provision for the assessment of buildings with the definition of a Smart Readiness Indicator (SRI) to give justice to the fact, that buildings must play an active role within the context of an intelligent energy system.

2.1. Smart Readiness Indicator (SRI)

As the EPBD is providing the framework for the national building regulations, a key factor for including additional parameters in mandatory assessments is the applicability to a national or regional context and the usability within a formal assessment procedure. Based on the legislation, the SRI should ideally be a combination of a simple indicator whilst at the same time factoring in a series of highly complex aspects related to the management and interaction capability of the building with its occupants and the grid:

> "... The smart readiness indicator should be used to measure the capacity of buildings to use information and communication technologies and electronic systems to adapt the operation of buildings to the needs of the occupants and the grid and to improve the energy efficiency and overall performance of buildings. The smart readiness indicator should raise awareness amongst building owners and occupants of the value behind building automation and electronic monitoring of technical building systems and should give confidence to occupants about the actual savings of those new enhanced-functionalities ... " [10]

Even though there is no general definition of the term Smart Building within the revised EPBD the smartness of the building is defined by having the technical capability of (a) managing itself efficiently (b) being able to interact with and respond to its occupants (c) being able to actively and passively interact with the grid. This approach has been applied in the legislative document that the:

> "... methodology shall rely on three key functionalities relating to the building and its technical building systems: (a) the ability to maintain energy performance and operation of the building through the adaptation of energy consumption for example through use of energy from renewable sources; (b) the ability to adapt its operation mode in response to

the needs of the occupant while paying due attention to the availability of user-friendliness, maintaining healthy indoor climate conditions and the ability to report on energy use; and (c) the flexibility of a building's overall electricity demand, including its ability to enable participation in active and passive as well as implicit and explicit demand response, in relation to the grid, for example through flexibility and load shifting capacities … ". [10]

In order to support the member states with a definition of the SRI, a technical study has been commissioned by the European Commission DG Energy. The aim of the study was to propose a methodological framework for the SRI and the smart services that the indicator builds on. The study also provided a first EU wide impact assessment of the newly developed indicator [11]. The proposed SRI-calculation methodology builds on a single score system that classifies the building's smart readiness. The score is based on eight impact criteria: energy, flexibility for the grid, self-generation, comfort, convenience, wellbeing & health, maintenance & fault prediction and information to occupants. Each impact criterion is expressed as a percentage of the maximum score that can be achieved for the evaluated building. Each impact criterion is in turn the weighted average of 10 domain scores, which constitute of heating, cooling, domestic hot water, controlled ventilation, lighting, dynamic building envelope, on site renewable energy generation, demand side management, electric vehicle charging and monitoring and control. For each domain various functionality levels are defined, with higher functionality levels reflecting a smarter implementation of the respective service. The overall methodology has been tested on two use cases and is thoroughly described in the Final Report of the study. The study concludes that the developed methodology follows the principles as outlined in the EPBD 2018 whilst being able to be practically implemented. Even though the methodology is not binding for the member states, it provides a template that is stated to be flexible enough to be adapted to local framework conditions [11].

Since the above described report [11] has been finalized in August 2018 and the implementation of the amended EPBD has not yet been included within the national laws of the member states, so far there is no empirical data on the validity of the methodology or the usability available. However, based on the case study assessment undertaken within the above study, one can conclude that: (a) the appraisal heavily relies on a (potentially subjective) expert judgment, (b) the assessment needs to be carried out within an operational building and might not be applicable for the use of the building design and (c) the assessment can be time consuming as there are a high number of criterions and domains that need to be covered. It is also understood, that the assessment is undertaken on site by a qualified consultant, which inherently necessitates an accreditation and certification structure. Whilst this is a valid way forward to address the increasing management capabilities of buildings, it is questionable if the strong focus on building and information system technology adequately addresses the requirement for an indicator that still should have energy and resource efficiency at its core. The question remains as to who defines *smartness* and how it can be measured.

2.2. Demand Response and Load Shifting in Buildings

Buildings have been inherently energy autarkic in early settlements. With increasing energy demand, it was advantageous to move from individual energy supply to a more centralized supply and thus physically separate consumption and generation. For this purpose, the energy supply networks were built, which commonly encompass large scale centralized energy generation and wide distribution networks surpassing regional and national borders. The Liberalization of the energy market at the end of the 20th century together with the understanding of the correlation between CO_2 emissions and climate change have resulted in a transformation process towards a less fossil fuel dependent and more sustainable system. With the advance of renewable energy generation, the distribution grids become ever more complex as the grids need to respond to heavily fluctuating supply. The transition of the current centralized market towards a smart market in an intelligent smart grid has been exemplary documented by Aichele et al. [16]. The increasing demands on this structure make it imperative to switch from a centralized to a decentralized and interactive system. For this purpose, numerous concepts and methodologies have already been developed and tested, with some relevant studies as briefly outlined below. In summary, these concepts are termed Smart Grids and Smart Markets.

Assessing the heat dynamics of buildings Madsen et al. have already proposed in 1995 to apply building performance data and statistical methods instead of simply relying on the physical characteristics of the building and building systems [17]. At a time when building energy modeling was not yet an inherent part of advanced building design processes, this approach presented an alternative route to assessing the thermal behavior of buildings. The methodology has been further improved by applying the use of stochastic differential equations to assess heat dynamics [18] and by including a methodology for the identification of suitable models in this context [19].

In order to actively interact with the grid, buildings are moving from consumers to producers, which resulted in the terminology of the prosumer (producer + consumer). Using buildings to produce energy by integrating renewable energy systems constitutes one of the key principles of our changing energy system. Another key aspect is to use the storage capacity of buildings in order to shift loads over time. Buildings with a high thermal mass can thus activate their thermal capacities by storing thermal energy (adding energy for later use) or conserving thermal energy (using the previously stored energy). Le Dreau and Heiselberg have evaluated the dynamic behavior of buildings by applying thermal storage solutions as opposed to water or battery storage solutions. The study highlights the importance of high efficiency buildings, as poorly insulated buildings have a short autonomy and thus thermal capacity over a longer time constant compared to buildings with a low heating and cooling energy demand [20]. Other storage devices in buildings can include e.g., ice storage units or water tanks as well as batteries. Selecting the best storage type for a certain use depends on a series of factors. Using optimization models for an appropriate selection process can determine the most advantageous system as outlined by Xu et al. [21]. However, it is imperative to consider the current market situation as prices for building systems can significantly vary over time. In an analysis carried out for batteries used in the automotive industry a cost reduction averaging from over USD300/kWh to under USD100/kWh is projected for 2020 to 2025. The automotive market can thus significantly contribute to the wider application (and second life) of electrical batteries for the use in buildings [22]. In a study undertaken to assess the application of all electric storage systems in buildings, the results showed that electrical batteries could successfully perform load shifting and peak-shaving in order to take advantage of price differences of a smart market [23].

A surplus of energy generation can also be efficiently managed by using power-to-heat or power-to-gas storage. Whilst power-to-gas is currently mainly applied in large-scale projects, it provides a suitable technology to convert surplus electrical energy into hydrogen to be used directly as a final energy carrier or to be converted into e.g., methane or liquid fuels [24]. Power-to-heat on the other hand can be used directly in buildings by exploiting the entire thermal mass available for activation. Converting surplus electrical energy in times of high electrical renewable energy generation through wind and solar into locally stored heat constitutes one of the great potential buildings offer for short-term demand response [25]. Whilst smart grids are mostly associated with electricity grids, the use of smart thermal grids opens up the thermal networks for the integration of decentralized thermal energy generation. In a study undertaken in the framework of the Austrian Climate and Energy Fund an in depth analysis of the potential of thermal prosumers concludes that whilst the technical capabilities are manageable, business models must still be developed in order to provide the necessary incentives for the thermal grid operators to allow a large integration of decentralized thermal energy integration [26]. Lund et al. also argue, that the challenge of thermal networks, the 4th Generation District heating (4GDH), will be in the utilization of low temperature heat sources (such as ambient and waste heat from cooling) and the interaction with zero and low energy buildings [27].

Whilst the management of the grids is focused on the optimization for the grid operator, the building side needs to respond to the requirements of the grid whilst maintaining optimal comfort and efficiency levels for the building. Building management systems (BMS) already contribute to the optimization of the building and its building services systems. A BMS can control any devices from passive architectural elements, such as shading or opening elements to active systems such as heating, cooling, ventilation and lighting systems. In combination with sensors that measure a range of system parameters such as room temperature, humidity, lux levels, CO_2 or any set point temperatures to name just a few, the BMS can actively contribute to increased comfort levels and reduction of the overall energy demand. Using the BMS with distributed energy resources (DERs) in order to reduce energy costs and overall CO_2 emissions is a way forward in managing buildings on a system level. Optimization models, which work with different pricing and energy demand scenarios support the understanding of how buildings can be actively integrated into the energy grids [28]. Other studies have shown that user comfort optimization and energy use reduction can be achieved by applying Smart-Context-Awareness Management (Smart-CAM) by using smart building ontology and context awareness mechanisms [29]. Clustering buildings in order to further increase the load shifting potential and thus enhance the energy flexibility of whole regions is the consequential result from the optimization of the single entity [30]. Accompanying policy measures to increase the use of building energy management systems play a crucial role in this context [31]. Real-time optimization models are increasingly necessary to adjust demand and supply in the smart grids. Genetic algorithms can support the assessment of load shifting towards improving the application of buildings as distributed thermal storage within a smart grid [32].

Within the Horizon 2020 research, there is an uptake of research and innovation projects, which have smart buildings and associated sectors as a part of their projects plans. Based on an analysis from 2017, 42 projects with a total funding of 304.1 million Euros between 2014 and 2017 have directly or indirectly addressed the subject of smart buildings within their research. 15 out of the 42 analyzed projects explicitly deal with demand response, 19 each with the control of appliances and individual buildings, 18 with the subject of full automation and 29 with user interfaces [33]. Even though this data represents only a snapshot of projects running at that time, it nevertheless highlights that this area is likely to progress further with the new legislation of the EPBD being transformed into national law in the next years in the member states.

2.3. Smart Buildings

In order to give justice to the terminology of smart, one has to consider how this term has been attributed to the field of construction over time. In the European Commission's Strategic Energy Technology (SET) Plan 2011 [34] smart cities were defined within their own thematic field. This has been the first time that in addition to individual technologies (such as wind or geothermal power) a systemic topic has been included in the SET Plan. It has thus been acknowledged, that—in addition to the incremental improvement of individual technologies—only the added focus on the overall system will bring a paradigm shift in the energy system and substantial changes in energy and resource efficiency. The optimization of the overall system at the interface of urban planning, mobility, industry, buildings and energy thus opened up a new focus.

The potential for optimization on this system level can be attributed to two main factors: (1) Based on the UN World Urbanization Prospects [35] by the middle of 2009, the number of people living in urban areas had surpassed the number living in rural areas. This trend is strongly continuing with virtually all of the expected growth in the world population concentrated in the urban areas of the less developed regions. The second factor can be attributed to the advance of (2) Information and communication technology (ICT), which has improved exponentially over the last two decades. The collection, storage and analysis of large quantities of data have facilitated new potentials such as the Internet of things (IoT) and machine-to-machine (M2M) communication. In conclusion the world is becoming more urban than rural and new systemic approaches made possible by the advance of ICT can support cities and regions in becoming more sustainable whilst ensuring a high quality of life.

Smart Buildings have been cited along Smart Cities and Smart Grids as the new way forward towards a transformation of our current fossil-fuel dependent energy structure into an energy and resource efficient system. Even though the smartness of buildings has been roughly defined in the latest version of the EPBD as outlined in the chapters above, smart can and should encompass much more than only interaction of the building system with the building itself, the occupants or the grid. Looking back at the actual need for buildings, one has to acknowledge that the fundamental aim of constructing buildings was always to provide shelter from the climate and security from potential enemies. Thus, autochthonous architecture is characterized by the provision of a comfortable, secure indoor environment under the given climatic conditions and with the available resources. Transferring this fundamental approach to the present, building smart is thus building according to the local climate by exploiting passive design measures to increase energy efficiency in an integrated design. After exploiting passive design measures, the next step is the adequate and resource-oriented integration of technical building systems and renewable energy systems. Only then can we optimize systems by exploiting synergies through connection and load shifting, which opens up new potentials to increase efficiency of bigger systemic entities such as districts or cities.

A methodology for energy-efficient and sustainable building design must consecutively focus on a structured approach starting with energy-demand reduction (Step 1), followed by energy-efficient equipment (Step 2), renewable energy systems (Step 3) and as a last step the system optimization and interaction with the grid (Step 4) as outlined in Figure 1 The approach should ensure that passive measures are always fully exploited before any active technology—and system-intense measures are considered. The architecture plays a key role in defining the least possible energy demand. Thus demand-side measures are optimized before distribution and supply measures are even considered, resulting in efficient use of resources and adequate application of technical and renewable building energy systems [36]. Whilst Step 1, 2 and 3 are accounted for in the EPCs by calculation the relevant indicators, the last Step 4 encompassing the interaction with the grid has yet to be included in any EPC assessment. Adding to the already existing indicators of heating and cooling energy demand (based on Step 1) as well as the primary energy demand and CO_2 emissions (consecutively resulting from Steps 2 and 3), the SRI as outlined in the last amendment of the EPBD is in this context the key indicator for this Step 4 and should in the future constitute a mandatory part in the EPC assessment.

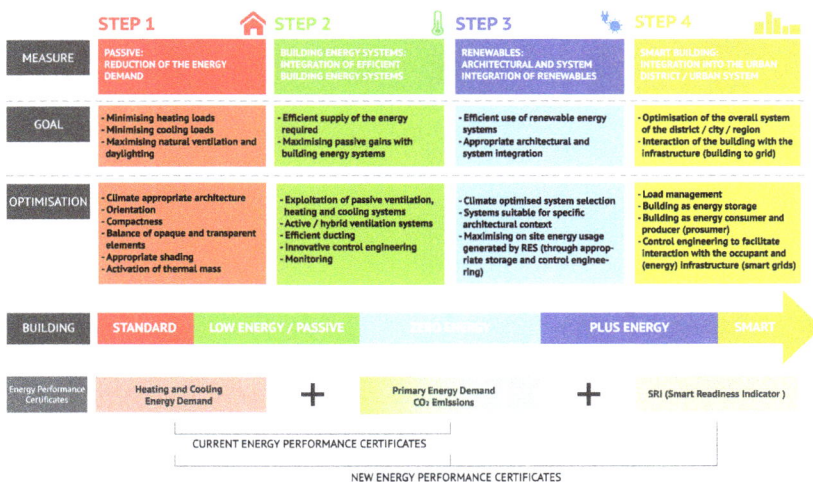

Figure 1. General methodology: development of building-energy concepts in a 4-step approach, adapted from authors' graphic from [36].

An indicator for smartness can thus include a variety of aspects such as: building to high efficiency standards, responding automatically to external conditions (e.g., weather, grid), managing building systems automatically and in response to the user and the grid, supplying large amounts of renewable energy on site and feeding energy back into the grid. This can subsequently be measured by, for example: summing up the number of responsive appliances in buildings, assessing how many technical items can be controlled with a building management system, assessing if and how the building and its systems support energy and grid management and assessing how many systems contribute to the internal comfort of the occupants.

This however would constitute a rather qualitative way forward, where an assessor has to decide, which systems contribute to what extent to the smartness of the building. An example of such a qualitative but nonetheless thorough approach has been proposed within the study commissioned by DG Energy [11] as a methodology for the SRI as outlined above. The question however remains, if this type of assessment favors the increased use of smart appliances without adequately addressing the underlying aim of the SRI, which is inherently to reduce resource and energy consumption by the management and connection of systems. Another disadvantage of this qualitative approach lies in the timing of the implementation of the respective systems. Whilst most energy related systems (heating, cooling, ventilation) form an inherent part of the overall building design, additional smart systems might not necessarily be included in the planning of the building as appliances and other gadgets are more likely to be installed by the final user rather than the building developer.

Focusing again on this fundamental principle of resource and energy efficiency, one can conclude that:

" ... A smart building is offering maximal quality of living with minimal consumption of resources by (1) exploiting architectural passive design measures to increase energy efficiency, by (2) joining internal and external systems and integrating renewable energy infrastructure and by (3) optimizing load management and load shifting to increase energy efficiency on all levels of the building, district or city ... ". [37]

Extracting this for the above stated question on how we can measure the smartness of a building:

" ... We can assess the load shifting potential of a building by calculating how much of the thermal and electrical energy can be loaded from the grid, stored and unloaded into the grid over a certain period of time whilst maintaining required comfort levels ... ". [37]

3. Methodology

Whilst there exist numerous approaches to analyze the thermal and electrical behavior of the building as outlined in the introduction section, there is still a need to address the actual load shifting potential of buildings by an assessment, which can be carried out as part of the SRI within the EPC. The aim of this article is thus to provide a methodology for a simplified quantitative assessment for an SRI with a numerical, model-based approach. Following the requirements of the current EPBD as outlined above, the methodology covers the flexibility of interaction with a grid and thus grid-friendliness of a building. The proposed approach focuses on this aspect only, as it can be assumed that the ability of a building to adapt to energy performance requirements as well as the ability to respond to the needs of the user are covered by the integration of a BMS, which in turn is a pre-requisite for the building systems to interact with the grid. With the proposed numerical approach, buildings should be able to be rated and subsequently categorized with a single indicator per energy type based on their energy storage capacity, load shifting potential and subsequent interaction with the grid.

3.1. Description of the Model

A key aspect of the proposed methodology is the applicability within the scope of the EPBD and subsequently in the EPC in order to provide an easy to use calculation, which can be carried out in addition to the already established calculations as outlined above. Therefore, the use of input data, which is already being applied in the current EPC as well as other data, which is easily accessible at an early design stage, constitutes a prerequisite for the development of the methodology. Another crucial requirement is that the calculation can be both carried out during the design phase as well as during operation, i.e., covering both new and existing buildings.

Currently for the calculation of the EPC all relevant aspects related to the building as well as its technical building systems need to be considered. Thus the characteristics of the building (e.g., physical and thermal properties) and technical systems (e.g., type, efficiency), including the respective energy sources, are already part of the assessment. As a result the overall energy demand in terms of heating, cooling ventilation and power as well as the primary energy demand and associated CO_2 emissions are calculated. The source of the final energy as well as the grid connection is therefore already included in the input data. At this point it is thus already clear, if the building will be connected to a grid (electrical, thermal, gas) or if the building is independent from a grid. In order to calculate the SRI as proposed in this article, additional characteristics related to the storage potential of the building and its systems need to be included in a future assessment. Factors that should be considered within the model are the physical structural storage capacity of the building (e.g., activation of thermal mass), the physical technical storage capacity (e.g., chemical, electrical, thermal) and the potential of thermal/electrical conversion systems (e.g., heat pumps). One of the key input parameters is also the time component as it is highly relevant how long a certain system can obtain, store and dispatch energy. The model foresees, that an indicator is calculated for each energy source provided via an energy network. The calculation itself is dependent on the type of energy source, e.g., electricity, thermal (heating, cooling) or gas. In Figure 2 the SIR calculation development is outlined to describe the input data for the proposed SRI methodology.

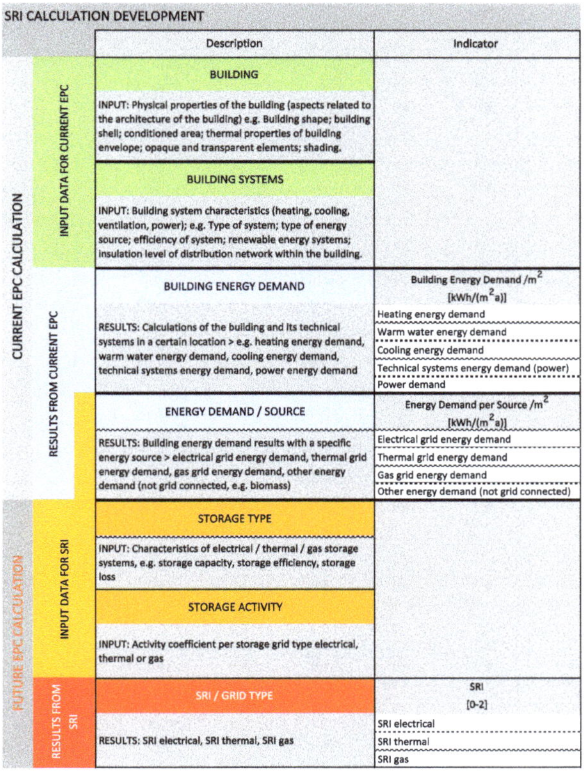

Figure 2. Smart Readiness Indicator (SRI) calculation development.

The input data already included in the current EPC (building, building systems) are needed for the calculation of the current EPC (building energy demand, energy demand/source). These results from the current EPC (energy demand/source) provide together with the additional input data for the SRI (storage type, storage activity) the basis for the future EPC calculations. As a result, one SRI is subsequently calculated per grid type. The most relevant grid types are considered to be electrical, thermal and gas, however additional grid types can be added if necessary.

3.1.1. System Boundaries

The calculated smart readiness indicator (SRI) should represent the ability of a building for the meaningful participation in smart grid concepts. For this reason, two key aspects are considered in the first step:

- Is there a grid with appropriate energy sources available?
- To what extent is the building able to communicate with the corresponding grid?

The building is examined with regard to: (1) qualitative and (2) quantitative criteria. In the qualitative criteria (1), a distinction is made in terms of the interaction potential and activity with the grid: no possibility to shift energy to and from the grid, the possibility to shift the energy demand according to the needs of the grid or the possibility to fully participate in smart grid solutions. The possibility of interaction is in this context independent of the actual building energy demand. The interaction is based on the usefulness for the networks, i.e., is it useful to take energy from the grid or is it useful to give energy back to the grid. The quantitative criteria (2) of the building are defined as independent of the used technology. For the active interaction of a building with a smart grid concept, the temporal flexibility of a building is essential with regard to its characteristics relating to energy consumption, energy production or both. To take this fact into account, the building is simplified and acts as energy storage in the model (see Figure 2).

To determine the quantitative values, the building energy requirements are based on the relevant standards per m^2 or, if calculations have already been carried out, on the actual building energy demand. Since smart-grid concepts require periods other than one year, the building energy requirement must be adapted to the selected period. This value is used as a reference and the building storage is normalized in this regard. To counteract any oversupply of stored energy, an upper bound was added in the model function. Energy costs or associated costs for auxiliary energy to load or unload the storage have been explicitly excluded from the model. Following the EPCs currently applied in the member states, the focus lies on the energy rather than the costs aspects.

The proposed SRI does not assess the efficiency of the building nor of the technical building systems, as this is already part of the current Energy Performance Certificate calculations. In addition, it does not take into account the capability of the interaction of the building with the user, i.e., the building management systems that allow the user to communicate with the respective technology items (often referred to as "smart home technology") as this is assumed to be covered with a BMS.

The proposed SRI provides a quantitative assessment of the interaction capabilities of the building with the networks. It can answer the question of:

"What is the potential of the building to take energy from the grid, store it over a certain period of time and again dispatch it back to the grid?"

Within this context the methodology does not differentiate to where the energy is stored (e.g., in thermal mass, water storage tank or electrical battery). It only assesses the connection of the respective grid to the building (electrical, thermal, gas) and the loading and unloading capabilities in this context. If the energy source is not grid connected (e.g., biomass, coal or other energy source) then it is not taken into account. The BMS is considered to be a prerequisite in order for the building systems to interact with the relevant networks. The SRI calculation system boundaries are subsequently defined by the energy storage types and the activity with the respective grids (See Figure 3). For example, if the main thermal energy source of the building is biomass, then there might not be a thermal grid connection nor a gas connection. In this case only the electrical grid connection can be considered for the SRI calculation. Therefore one SRI must be calculated separately for each grid connection (electrical, thermal, gas). If only one grid connection is available, then the two others do not apply. It should be noted that purposefully there is no distinction made between district heating, low-temperature district heating, district cooling or other thermal grids such as anergy-networks, as it is assumed that only one thermal grid connection per building is realized. However, within the methodology other grid types can be added if necessary.

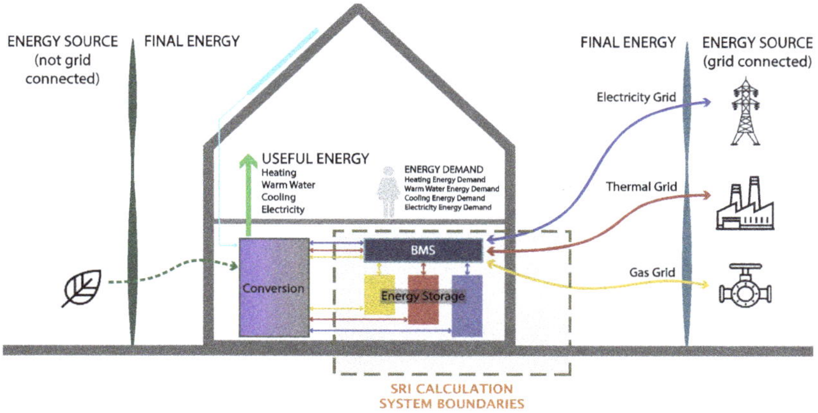

Figure 3. System boundaries of SRI calculation.

3.1.2. Description of the Indicator

To describe the ability of a building to interact with the utility grids, four different options are distinguished, depending on the activity of the building. From these options the following possible values are derived:

- No grid available n/a.
- No interaction with the grid; activity coefficient 0.
- Passive interaction with the grid; activity coefficient 1.
- Active interaction with the grid; activity coefficient 2.

In this context "no interaction with the grid" means that no storage or load shifting potential is available, the building is a simple consumer. A "passive interaction with the grid" requires the building to offer storage and/or load shifting potential to the grid. The load shifting is however only one-directional from the grid to the building. The "active interaction with the grid" stands for an energy flexible building that provides storage and/or load shifting capabilities and offers bi-directional load shifting from the grid to the building as well as from the building to the grid. This building would be able to produce as well as consume energy and consequently be a prosumer.

3.1.3. Description of the Model Functions

The basis for the selection of the function was to connect two barriers with a smooth curve. Consequently, the following sigmoid function was chosen:

$$Y(x) = L + \frac{U - L}{\left(1 + Ce^{-G(x-M)}\right)^{\frac{1}{v}}} \tag{1}$$

With the individual coefficients of $Y(x)$ as follows:

L Lower asymptote.
U Upper asymptote.
G Growth rate.
$v > 0$ Affects the growth near the asymptote.
C Coefficient influencing $Y(0)$.
M Move the curve to the right.

From this basic equation a normalized curve, which leads from 0 to 1 and goes through the point $(1, 0.5)$, has been created. Subsequently the chosen coefficients are as follows:

$L = 0$
$U = 1$
$G = 6$
$v = 1$
$C = 1$
$M = 1$

From the selected curve the subsequent function can be derived:

$$Y(x) = \frac{1}{\left(1 + e^{-6(x-1)}\right)} \tag{2}$$

For the calculation of the proposed SRI the following variables are necessary:

ED Energy demand of the building per energy source for the selected time period τ.
SC Storage capacity of the respective storage in the building.
η_{SC} Efficiency factor of the storage capacity. (Here the efficiency for loading as well as unloading the storage must be considered). $\eta_{SC} := \eta_C \cdot \eta_D$
η_C Efficiency factor of the storage capacity for charge.
η_D Efficiency factor of the storage capacity for discharge.
ζ_{SC} Storage loss during the selected period in full storage (e.g., self-discharge, heat losses)
AC Activity coefficient for the building.

Following Equation (2) and the variables as outlined above, the following equation can be defined as a result for the SRI:

$$SRI = AC \cdot Y\left(\frac{SC}{ED} \cdot \eta_{SC} \cdot (1 - \zeta_{SC})\right) = \frac{AC}{\left(1 + e^{-6\left(\left(\frac{SC}{ED} \cdot \eta_{SC} \cdot (1 - \zeta_{SC})\right) - 1\right)}\right)} \tag{3}$$

Based on this equation the required characteristics for the SRI methodology can be achieved. The figures below depict the SRI curves based on the Equation (3) for the various activity coefficients. Figure 4 shows the SRI curves with the activity coefficient 1 and Figure 5 shows the SRI curves with the activity coefficient 2. The SRI with an activity coefficient of 0 would result in all values to be equally 0. With an activity coefficient 1, the resulting SRI is between 0 and 1 and with an activity coefficient of 2, the SRI falls between 0 and 2.

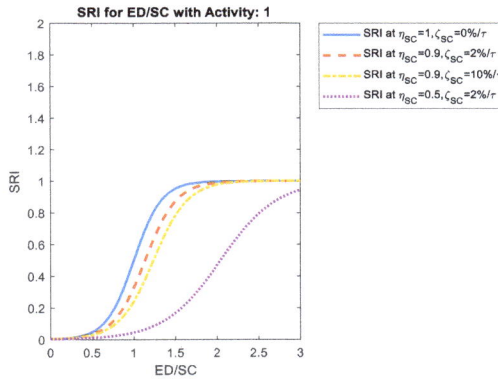

Figure 4. Example for SRI with activity coefficient 1.

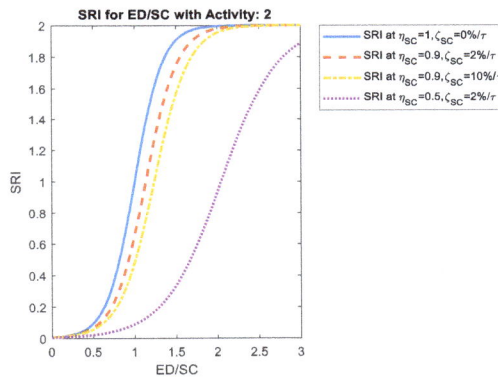

Figure 5. Example for SRI with activity coefficient 2.

3.2. Characteristics of the SRI Methdology

The described SRI methodology has been chosen based on a series of requirements that form an essential part of the methodology. These requirements can be described as follows:

- There is a convergence of the SRI curve in respect to the upper and lower limit.
- The upper and lower limit can be freely selected.
- The period length τ can be freely selected: e.g., one day for a 24 h load profile or one year for a seasonal load profile, depending on the time period; the energy demand (ED) needs to be adjusted accordingly (e.g., for thermal loads the total heating energy demand would need to be divided by the heating days whereas for the total warm water energy demand this would need to be divided by 365 days if only a 24 h period is selected).
- Inclusion of system losses
- Inclusion of system efficiency
- The storage capacity is given in relation to the amount of energy, which is taken from the grid within a defined period.
- Query for the extent of interaction between the building and the grid. The query asks, whether the system can actively, passively or not at all interact with the grid (i.e., no grid available n/a; no interaction with the grid; activity coefficient 0; passive interaction with the grid; activity coefficient (1); active interaction with the grid; activity coefficient 2.

For the load profile of the charge and discharge of any storage in the building the following model has been assumed, which is based on the below equations:

$$E(t) = \int_0^t P(s) - P_L(s)ds \tag{4}$$

$$E(0) = E(\tau) = 0 \tag{5}$$

$$E(t_1) = SC \tag{6}$$

$$E(t_2) = SC - \int_{t_1}^{t_2} P_L(s)ds \triangleq (1 - \zeta_{SC}) \cdot SC \tag{7}$$

$$P(t) = \begin{cases} \eta_C \cdot P_N & t \in]0, t_1] \\ 0 & t \in]t_1, t_2] \\ -\eta_D \cdot P_N & t \in]t_2, \tau] \end{cases} \tag{8}$$

$$P_L(t) = \zeta_{SC} \cdot \frac{SC}{\tau} \tag{9}$$

With the individual coefficients as follows:

$E(t)$ Energy in the storage at the time t in kWh.
$P(t)$ Load Power in the storage at the time t in kW.
$P_L(t)$ System losses kW.
η_C Efficiency factor at charge.
η_D Efficiency factor at discharge.
ζ_{SC} Loss coefficient of the system.
P_N Rated power of the grid connection of the building in kW.
SC Storage capacity in kWh.

This load profile is of particular relevance, if the building generates its own energy e.g., by means of renewable energy. For each grid type (electrical, thermal, gas) a separate load profile must be calculated. Figure 6 shows the graphical representation of the load profile of the charge and discharge of any storage in the building.

If there are multiple storage system within the same energy type (e.g., two or more thermal storage systems) then the SRI can either be depicted as a max (SRI) or as a weighted average. Therefore, the proposal is, that for the SRI used in the EPCs, SRIs for all storage systems of one energy type are calculated and only the maximum is considered for the certificate. To assess a combined SRI with multiple storage systems per energy type, the following Equations (10)–(15) can be applied:

$$SC = \sum_{i=1}^N SC_i \tag{10}$$

$$\omega_i = \frac{SC_i}{SC} \tag{11}$$

$$\eta_C = \sum_{i=1}^N \omega_i \cdot \eta_{Ci} \tag{12}$$

$$\eta_D = \sum_{i=1}^N \omega_i \cdot \eta_{Di} \tag{13}$$

$$\eta_{SC} = \eta_C \cdot \eta_D \tag{14}$$

$$\zeta_{SC} = \sum_{i=1}^{N} \omega_i \cdot \zeta_{SCi} \qquad (15)$$

With the individual coefficients as follows:

N Number of storage systems per energy type.
ω_i Weighting factor for i
SC_i Storage capacity of the respective storage i in the building.
η_{Ci} Efficiency factor of the storage capacity for charge of storage i.
η_{Di} Efficiency factor of the storage capacity for discharge of storage i.
ζ_{SC_i} Storage loss during the selected period in full storage of storage i.

LOAD PROFILE FOR STORAGE CHARGE AND DISCHARGE FUNCTION

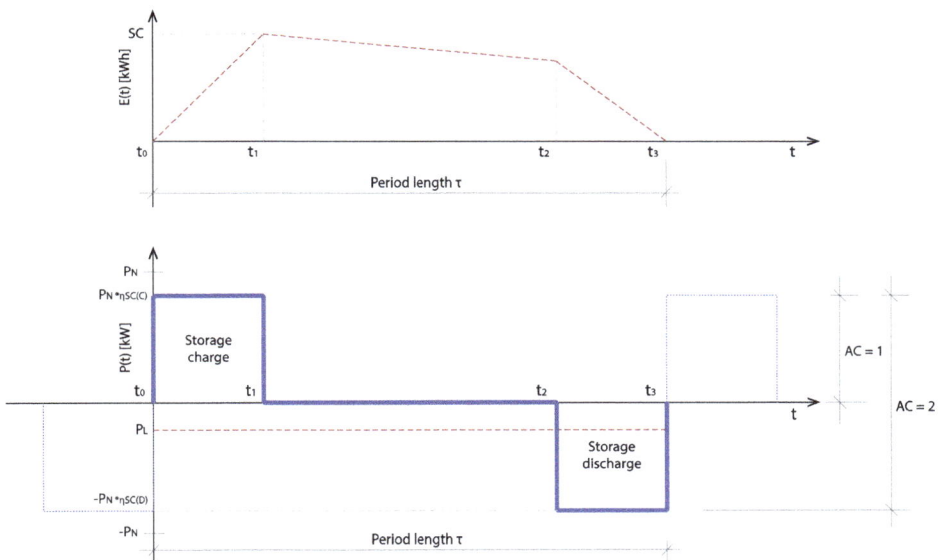

Figure 6. Load profile for the assumed function of the storage charge and discharge model.

4. Application of the Methodology in Theoretical Use Cases

To assess whether the methodology can be used as a standard assessment within the EPCs, four use cases, which are considered to be typically covering a wide range of building stock, have been selected. The main characteristics for two residential and two office buildings have been chosen in order to test a range of potential applications. The use cases cover existing buildings, considered to have little interaction with the grid as well as newer buildings, which already have an active and bi-directional interaction with the grid. In the following the uses case are explained in more detail and subsequently the results for the SRI are presented and discussed.

4.1. Description of Use Cases

In Figure 7 the SRI use case matrix with the relevant input data for the four selected uses cases is shown. In addition to the characteristics of the building (type, year of construction, area, number of people) and energy related data (heating energy demand, warm water energy demand, cooling energy demand, technical systems energy demand and power demand) the thermal and electrical storage capacities as well as the activity coefficient with the thermal and electrical grids differ within the four use cases. The option for *technical storage gas* has also been included as in the future the potential to generate, store and unload (bio)gas might be a viable option for certain building types. The proposed SRI methodology can thus equally cover this option. However, for the purpose of the selected use cases gas grids have not been considered.

- Use case 1: Single family building built in 1990; the building has not yet been refurbished, the external envelope is therefore poor compared to state of the art standards; the heating system is based on a typical floor heating system connected to district heating; no cooling; the building is not able to store thermal energy from the grid (activity coefficient 0 electrical; activity coefficient 0 thermal).
- Use case 2: Multi-storey residential building built in 2010; the quality of the external envelope is better than use case 1 with a resulting heating energy demand half of that of use case 1; the heating is equal to use case 1 with floor heating connected to district heating; no cooling; an electrical battery has been added; the building is able to store thermal energy from the grid in a water storage tank and to load and unload electrical energy from the grid (activity coefficient 2 electrical; activity coefficient 1 thermal;).
- Use case 3: office building built in 2010; external envelope to a high standard based on building regulations as of 2010; heating and cooling energy is generated by ground-coupled heat pumps; the building is connected to district heating; an electrical battery has been added; the building is able to load and unload electrical energy from the grid (activity coefficient 2 electrical; activity coefficient 1 thermal).
- Use case 4: office building to be built in 2020; external envelope twice as high as use case 3; the heating and cooling is equal to use case 3; in addition the system is connected to a district heating grid; a building integrated photovoltaic system is installed; an electrical battery has been added; the building is able to load and unload electrical as well as thermal energy to and from the grid (activity coefficient 2 electrical; activity coefficient 2 thermal).

SRI CALCULATION INPUT MATRIX

INPUT DATA FOR CURRENT EPC

USE CASE	USE CASE 1			USE CASE 2			USE CASE 3			USE CASE 4		
BUILDING	Residential / Non-Res.	Area GFA [m²]	People No.	Residential / Non-Res.	Area GFA [m²]	People No.	Residential / Non-Res.	Area GFA [m²]	People No.	Residential / Non-Res.	Area GFA [m²]	People No.
Key data	Residential	150	4	Residential	800	25	Non-Res.	5000	400	Non-Res.	5000	400
Description	single family building; 1990			multi-storey residential building, 2010			office building, 2010			future office building, 2020		

BUILDING SYSTEMS	Type of system	Type of system	Type of system	Type of system
Heating system	District heating	District heating	Heat pump	Heat pump; District heating
Warm water system	District heating	District heating	Heat pump	Heat pump; District heating
Cooling system	n/a	n/a	Heat pump	Heat pump; District heating
Power system	Grid	Grid	Grid	Grid; PV

RESULTS FROM CURRENT EPC

BUILDING ENERGY DEMAND	Demand /m² [kWh/(m²a)]	Total Demand [kWh/a]	Demand /m² [kWh/(m²a)]	Total Demand [kWh/a]	Demand /m² [kWh/(m²a)]	Total Demand [kWh/a]	Demand /m² [kWh/(m²a)]	Total Demand [kWh/a]
Heating energy demand	100	15.000	50	40.000	30	150.000	15	75.000
Warm water energy demand	13	1.950	13	10.400	5	25.000	5	25.000
Cooling energy demand	n/a	n/a	n/a	n/a	30	n/a	20	n/a
Technical systems energy demand	5	750	3	2.400	5	25.000	5	25.000
Power demand	30	4.500	20	16.000	20	100.000	15	75.000

ENERGY DEMAND / SOURCE	Source /m² [kWh/(m²a)]	BED [kWh/a]	Source /m² [kWh/(m²a)]	BED [kWh/a]	Source /m² [kWh/(m²a)]	BED [kWh/a]	Source /m² [kWh/(m²a)]	BED [kWh/a]
Electrical grid energy demand	35	5.250	23	18.400	25	125.000	20	100.000
Thermal grid energy demand	113	16.950	63	50.400	65	325.000	40	200.000
Gas grid energy demand	n/a	n/a	n/a	n/a	n/a	n/a	n/a	n/a
Other (not grid connected)	n/a	n/a	n/a	n/a	n/a	n/a	n/a	n/a

INPUT DATA FOR SRI

STORAGE TYPE	Capacity SC [kWh]	Efficiency η_{SCQ} [%]	η_{SCD} [%]	Loss ζ_c [%]	Capacity SC [kWh]	Efficiency η_{SCQ} [%]	η_{SCD} [%]	Loss ζ_c [%]	Capacity SC [kWh]	Efficiency η_{SCQ} [%]	η_{SCD} [%]	Loss ζ_c [%]	Capacity SC [kWh]	Efficiency η_{SCQ} [%]	η_{SCD} [%]	Loss ζ_c [%]
Storage electrical	7	93	93	0,1	80	93	93	0,1	400	93	93	0,1	700	93	93	0,1
Storage thermal	0	90	90	1	200	90	90	1	1.000	90	90	1	800	90	90	1
Storage gas	n/a	n/a	n/a	n/a	n/a	n/a	n/a	n/a	n/a	n/a	n/a	n/a	n/a	n/a	n/a	n/a

STORAGE ACTIVITY	Activity coefficient k [n/a, 0,1,2]	Time [d]	Activity coefficient k [n/a, 0,1,2]	Time [d]	Activity coefficient k [n/a, 0,1,2]	Time [d]	Activity coefficient k [n/a, 0,1,2]	Time [d]
Storage grid type electrical	0	1	2	1	2	1	2	1
Storage grid type thermal	0	1	1	1	1	1	2	1
Storage grid type gas	n/a	n/a	n/a	n/a	n/a	n/a	n/a	n/a

Figure 7. SRI use case matrix showing input data for use cases 1, 2, 3 and 4.

4.2. Results of Use Cases

The results highlight that, based on the activity coefficient, the SRI figures are in the range of 0 to 1 for activity coefficient 1 and in the range of 0 to 2 for activity coefficient 2 (Figure 8). It is evident, that a higher SRI stands for a higher flexibility of the building to store and subsequently dispatch energy back to the grid on demand.

- Results of use case 1: The activity coefficient for all grids is 0, thus the SRI is similarly 0 as the building is not able to actively engage with the energy networks. The building is a simple consumer.
- Results of use case 2: Since the building can store energy from the thermal grid in a water tank but cannot actively discharge energy back to the thermal grid, the SRI thermal is below 1 and due to the limited storage capacity relatively low. The SRI electrical is with 1.81 high, since the electrical battery can actively store surplus energy from the grid and dispatch it back into the electrical network.
- Results of use case 3: The first example for the office building has similarly to use case 2 an activity coefficient of 2 for the electrical storage and an activity coefficient of 1 for the thermal storage system. This results in a SRI below 1 for the thermal connection and a just above 1 SRI for the electrical coefficient. The latter is due to the relatively small storage capacity of the electrical battery.
- Results of use case 4: This use case is the only one with an activity coefficient of 2 for both the thermal as well as the electrical connection. The resulting SRI of 2 shows that the electrical storage is at its limit for the charge and discharge of the electrical energy. The SRI thermal is with 1.91 relatively high due to the high thermal storage capacity of the system. It can be seen that compared to use case 3, the SRI is with a lower thermal storage capacity higher, due to the fact that in this use case, the system can charge as well as discharge thermal energy from the grid and is thus bi-directional.

SRI CALCULATION RESULTS

		USE CASE	USE CASE 1	USE CASE 2	USE CASE 3	USE CASE 4
INPUT DATA FOR SRI		STORAGE ACTIVITY	Activity coefficient lk [n/a, 0,1,2]	Activity coefficient lk [n/a, 0,1,2]	Activity coefficient lk [n/a, 0,1,2]	Activity coefficient lk [n/a, 0,1,2]
		Storage grid type electrical	0	2	2	2
		Storage grid type thermal	0	1	1	2
		Storage grid type gas	n/a	n/a	n/a	n/a
RESULTS FROM SRI		SRI / GRID TYPE	SRI / grid type [0-2]	SRI / grid type [0-2]	SRI / grid type [0-2]	SRI / grid type [0-2]
		SRI electrical	0.00	1.81	1.03	2.00
		SRI thermal	0.00	0.17	0.54	1.91
		SRI gas	n/a	n/a	n/a	n/a

Figure 8. SRI results showing activity coefficients and subsequent results for use cases 1, 2, 3 and 4.

5. Discussion

The outcome of the uses cases show that the calculated SRI provides a clear result in regard to the load shifting potential of a building and its interaction with the grid. It thus delivers one single number per grid type, which can be used to depict the potential interaction of the various energy carriers of the building with the respective energy networks (and subsequently with other buildings). It can also be used to assess the potential of the building to interact with the energy markets. The SRI thus provides a quantitative assessment, highlighting that the building either does not at all interact with the grid (SRI 0), that the building can shift its own energy demand based on the requirements of the grid and thus take energy from the grid on demand (SRI near 1) or that the building can charge and discharge its own energy demand and that of a similar building thus taking energy from the grid and shifting energy back to the grid (SRI near 2).

The methodology could similarly be applied to any other building use case, with the prerequisite that the building has one or more active storage systems (otherwise the SRI is by default 0) and that these storage systems can be actively managed by a BMS. If the building generates its own energy (e.g., by means of renewable energy) and that energy can be stored and shifted to the grid, then the energy sources can be assessed based on the load profiles as outlined in Equations (4)–(9) and Figure 8. The SRI always needs to be calculated per energy grid (i.e., electrical, thermal, gas). Whilst any additional grid can be added, it is not useful to combine the SRI across the various grids, as this would not provide a meaningful answer. The fact that one SRI is calculated per energy type is an inherent advantage, as buildings with no connection to the grid (for example buildings with no connection to a thermal network) are not disadvantaged (the SRI would then show not available). However, if there is a grid available and the building is not actively engaging with the grid, then the SRI is 0.

The time period could theoretically be determined by the respective regulations, i.e., to define if a specific storage is required for one day or one season. This could also be viewed differently for different energy sources. Thus a specific time period could be defined for a specific energy source (e.g., 1 day for electricity; 1 week for thermal energy and 1 year for gas) making the SRI comparable.

In order to assess the relationship of the storage capacity to the energy demand, the function could be inversed as shown in Figure 9 and Equation (16). This could be useful for an assessment of a whole building quarter or district as it provides the view from the system perspective of the energy provider.

$$SP = \min\left(\frac{5}{2}, \max\left(0, -\frac{\ln\left(\frac{2}{SRI} - 1\right)}{6} + 1\right)\right) \tag{16}$$

With the individual coefficients as follows:

SP   Storage potential of the building (Relationship of *SC/EB*; dimensionless)

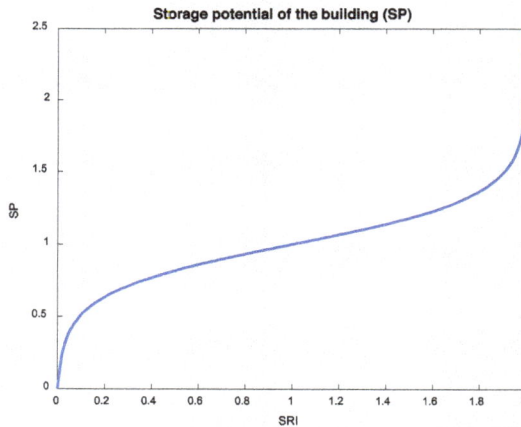

Figure 9. Use of the SRI to assess the load shifting potential of buildings.

Overall, there are several key advantages of providing a qualitative assessment for the SRI as proposed in this article: One of the most important aspects is that the methodology builds on already existing and well-established assessments and provides only an add-on, as existing quantitative information is used to calculate the SRI (see Figure 2). The calculation can thus be included in the existing EPC systems of the member states and planners can calculate the SRI already during the design phase of the building in addition to other energy related indicators that already need to be provided when applying for a building permit. The input data for the storage characteristics that would be required at this stage could be supplied by the technology providers. This would avoid having potentially costly assessments, based on a series of qualitative measures, which would need to be carried out by accredited assessors once the building is operational as currently proposed in the template for the SRI assessments [11]. The SRI calculation as proposed in this article can be used to serve as a planning aid for other applications e.g., to assess Smart Grid and Smart Market strategies if the indicator is made available for whole districts or cities. Furthermore, it does not reward an unnatural proportion of the storage capabilities of the building towards the grids. It does however consider the efficiency of the energy transfer and the increase of the SRI is within a meaningful range a maximum. One of the key advantages of the proposed SRI compared to a qualitative assessment is that there is no specific state of the art in terms of technical systems required for the methodology to be applied. This means that (a) it can be avoided that the SRI is becoming meaningless once the technology advances and (b) that the indicator cannot be used to evaluate how many services are sold, if certain technologies are installed or how many smart Apps are included. Thus, the SRI is limited to the bare essentials and is not at risk of reaching the highest category only with paid services. The proposed indicator can also be used within the framework of future research proposals, as the potential pilot projects can be rated based on their "grid friendliness" in addition to impacts on energy efficiency and renewable energy integration.

Even though the methodology offers a series of advantages, there are several limits, that need to be taken into account: For one, the methodology does not consider the efficiency of the smart grid as it only considers the interaction of the building with the network. It must also be noted, that neither the energy efficiency of the building nor the building systems are assessed as part of the SRI, as this is already an inherent part of the current EPCs. If the building offers storage systems for the same type of energy within the same building, then those must be considered in a greatly simplified manner. In addition, the methodology only considers the fact, that a BMS is available in the building and it assumes that a user, energy and/or market-oriented function is defined within the BMS. This subsequently means that essential information (e.g., energy use of certain systems) related to the user is considered as standard in this model. One of the key drawbacks of the methodology is however the fact that currently temporary storage capabilities cannot be assessed (e.g., batteries in e-mobility systems such as cars or bikes that only serve as temporary storage for the building). This could be an opportunity for a further development of the SRI. Another aspect that could be further developed would be a cost assessment, which is currently not part of the proposed methodology.

6. Conclusions

The Smart Readiness Indicator (SRI) as outlined in the latest amendment of the Energy Performance of Buildings Directive (EPBD) is considered to be a key indicator to:

> " ... measure the capacity of buildings to use information and communication technologies and electronic systems to adapt the operation of buildings to the needs of the occupants and the grid and to improve the energy efficiency and overall performance of buildings ... ". [10]

With the European legislation coming into force in the next years, a sound methodology for the calculation of the SRI must be developed. Whilst a template for a rather qualitative procedure already exists [11], this article outlines a quantitative approach based on the load shifting potential and subsequent active interaction of a building with a grid.

One of the key advantages of this approach is the usability and applicability with data that is already being calculated as part of the EPCs, which are mandatory in the member states. The proposed SRI does not include the efficiency of the building nor of the technical building systems, as this is already part of the assessments currently in place. Whilst the capability of the interaction of the building with the user as well as the capability of efficient energy management is not explicitly addressed, these aspects are inherently covered as the active interaction of any building with the grids necessitates the integration of a BMS. The methodology follows in logic the input data already applied in the current EPC and only needs additional input data related to the storage system and the potential to interact with the respective grids. The application of the proposed methodology in a series of use cases outlines that it can be applied to a wide variety of building types. The methodology is future proof in a sense that it is equally applicable for current as well as for innovative technologies that are yet to be implemented. In addition it can also be used to assess the load shifting capabilities of building quarters or districts to support energy providers in their grid developments.

It is shown that the methodology can provide an adequate framework for a quantitative assessment of the load shifting potential of buildings, which can be integrated into the calculations for the current EPC. The proposed approach thus complements and supports the development and implementation of the Smart Readiness Indicator in the Member States.

Author Contributions: Author D.Ö. contributed mainly to the conceptualization, methodology, writing of the original draft, review and editing. Author T.M. contributed mainly to the development of the methodology, the mathematical equations, the analysis and the writing of the original draft.

Funding: This research received no external funding.

Acknowledgments: Supported by the University of Natural Resources and Life Sciences, Vienna Open Access Publishing Fund.

Conflicts of Interest: The authors declare no conflict of interest.

References

1. European Commission. *COM(2011) 112 Final. A Roadmap for Moving to a Competitive Low Carbon Economy in 2050*; European Commission: Brussels, Belgium; Luxembourg, 2011.
2. European Commission. Policies, Information and Services, Energy/Topics/Buildings/Energy Efficiency/Buildings. Available online: https://ec.europa.eu/energy/en/topics/energy-efficiency/buildings (accessed on 10 March 2019).
3. European Commission. *COM(2016) 763 Final. Accelerating Clean Energy Innovation*; European Commission: Brussels, Belgium; Luxembourg, 2016.
4. European Commission. *COM(2016) 860 Final, Annex 1. Accelerating Clean Energy in Buildings*; European Commission: Brussels, Belgium; Luxembourg, 2016.
5. European Commission—Fact Sheet; Towards Reaching the 20% Energy Efficiency Target for 2020, and Beyond. European Commission Press Release. 1 February 2017. Available online: http://europa.eu/rapid/press-release_MEMO-17-162_en.htm (accessed on 10 March 2019).
6. Allouhi, A.; El Fouih, Y.; Kousksou, T.; Jamil, A.; Zeraouli, Y.; Mourad, Y. Energy consumption and efficiency in buildings: Current status and future trends. *J. Clean. Prod.* **2015**, *109*, 118–130. [CrossRef]
7. *Directive 2012/27/EU of the European Parliament and of the Council of 25 October 2012 on Energy Efficiency*; L315/1-56; Official Journal of the European Union: Brussels, Belgium, 2012.
8. *Directive 2009/28/EC of the European Parliament and of the Council of 23 April 2009 on the Promotion of the Use of Energy from Renewable Sources*; L140/16-62; Official Journal of the European Union: Brussels, Belgium, 2009.
9. *Directive 2002/91/EU of the European Parliament and of the Council of 16 December 2002 on the Energy Performance of Buildings*; L1/65; Official Journal of the European Union: Brussels, Belgium, 2003.
10. *Directive (EU) 2018/844 of the European Parliament and of the Council of 30 May 2018 Amending Directive 2010/31/EU on the Energy Performance of Buildings Directive 2012/27/EU on Energy Efficiency*; L156/75; Official Journal of the European Union: Brussels, Belgium, 2018.
11. Verbeke, S.; Ma, Y.; Van Tichelen, P.; Bogaert, S.; Gómez Oñate, V.; Waide, P.; Bettgenhäuser, K.; Ashok, J.; Hermelink, A.; Offermann, M.; et al. *Support for Setting Up a Smart Readiness Indicator for Buildings and Related Impact Assessment, Final Report*; Study Accomplished under the Authority of the European Commission DG Energy 2017/SEB/R/1610684; Vito NV: Mol, Belgium, 2018.
12. United Nations Climate Change. 'Kyoto Protocol—Targets for the First Commitment Period'. Available online: https://unfccc.int/process/the-kyoto-protocol (accessed on 5 October 2018).
13. *Directive 2010/31/EU of the European Parliament and of the Council of 19 May 2010 on the Energy Performance of Buildings (Recast)*; L153/13; Official Journal of the European Union: Brussels, Belgium, 2010.

14. Eurostat. 'Greenhouse Gas Emissions Statistics—Emission Inventories'. Data from June 2018. Available online: https://ec.europa.eu/eurostat/statistics-explained/index.php/Greenhouse_gas_emission_statistics (accessed on 5 October 2018).

15. Atanasiu, B.; Despret, C.; Economidou, M.; Maio, J.; Nolte, I.; Rapf, O. *Europe's Buildings under the Microscope, A Country-by-Country Review of the Energy Performance of Buildings*; Buildings Performance Institute Europe (BPIE): Bruxelles, Belgium, 2011; ISBN 9789491143014.

16. Aichele, C.; Doleski, O.D. *Smart Markt, Vom Smart Grid Zum Intelligenten Energiemarkt*; Springer Vieweg: Wiesbaden, Germany, 2014. [CrossRef]

17. Madsen, H.; Holst, J. Estimation of continuous-time models for the heat dynamics of a building. *Energy Build.* **1995**, *22*, 67–79. [CrossRef]

18. Andersen, K.K.; Madsen, H.; Hansen, L.H. Modelling the heat dynamics of a building using stochastic differential equations. *Energy Build.* **2000**, *31*, 13–24. [CrossRef]

19. Bacher, P.; Madsen, H. Identifying suitable models for the heat dynamics of buildings. *Energy Build.* **2011**, *43*, 1511–1522. [CrossRef]

20. Le Dréau, J.; Heiselberg, P. Energy flexibility of residential buildings using short term heat storage in the thermal mass. *Energy* **2016**, *111*, 991–1002. [CrossRef]

21. Xu, Z.B.; Guan, X.H.; Jia, Q.S.; Wu, J.; Wang, D.; Chen, S.Y. Performance Analysis and Comparison on Energy Storage Devices for Smart Building Energy Management. *IEEE Trans. Smart Grid* **2012**, *3*, 2136–2147. [CrossRef]

22. Berckmans, G.; Messagie, M.; Smekens, J.; Omar, N.; Vanhaverbeke, L.; Van Mierlo, J. Cost Projection of State of the Art Lithium-Ion Batteries for Electric Vehicles up to 2030. *Energies* **2017**, *10*, 1314. [CrossRef]

23. Georgakarakos, A.D.; Mayfield, M.; Hathway, E.A. Battery Storage Systems in Smart Grid Optimised Buildings. *Energy Procedia* **2018**, *151*, 23–30. [CrossRef]

24. Wulf, C.; Linßen, J.; Zapp, P. Review of Power-to-Gas Projects in Europe. *Energy Procedia* **2018**, *155*, 367–378. [CrossRef]

25. Kohlhepp, P.; Hagenmeyer, V. Technical Potential of Buildings in Germany as Flexible Power-to-Heat Storage for Smart-Grid Operation. *Energy Technol.* **2017**, *5*, 1084–1104. [CrossRef]

26. Lichtenegger, K.; Moser, A.; Muschick, D.; Reiterer, D.; Wöss, D.; Leitner, A. Einbindung von dezentralen Einspeisern in Wärmenetze, der Prosumer am Wärmemarkt. AEE Arbeitsgemeinschaft Erneuerbare Energie NÖ-Wien. Wien. 2018. Available online: http://www.aee-now.at/cms/fileadmin/downloads/projekte/bine2/Publikationen/Bine2plus%20Leitfaden%20final.pdf (accessed on 10 March 2019).

27. Lund, H.; Ostergaard, P.A.; Chang, M.; Werner, S.; Svendsen, S.; Sorknaes, P.; Thorsen, J.E.; Hvelplund, F.; Mortensen, B.O.G.; Mathiesen, B.V.; et al. The status of 4th generation district heating: Research and results. *Energy* **2018**, *164*, 147–159. [CrossRef]

28. Pooranian, Z.; Abawajy, J.H.; Vinod, P.; Conti, M. Scheduling Distributed Energy Resource Operation and Daily Power Consumption for a Smart Building to Optimize Economic and Environmental Parameters. *Energies* **2018**, *11*, 1348. [CrossRef]

29. Degha, H.E.; Laallam, F.Z.; Said, B. Intelligent context-awareness system for energy efficiency in smart building based on ontology. *Sustain. Comput. Inform. Syst.* **2019**, *21*, 212–233. [CrossRef]

30. Vigna, I.; Pernetti, R.; Pasut, W.; Lollini, R. New domain for promoting energy efficiency: Energy Flexible Building Cluster. *Sustain. Cities Soc.* **2018**, *38*, 526–533. [CrossRef]

31. Rocha, P.; Siddiqui, A.; Stadler, M. Improving energy efficiency via smart building energy management systems: A comparison with policy measures. *Energy Build.* **2015**, *88*, 203–213. [CrossRef]

32. Kolokotsa, D. The role of smart grids in the building sector. *Energy Build.* **2016**, *116*, 703–708. [CrossRef]

33. Moseley, P. EU Support for Innovation and Market Uptake in Smart Buildings under the Horizon 2020 Framework Programme. *Buildings* **2017**, *7*, 105. [CrossRef]

34. SETIS Strategic Energy Technologies Information System, 5th SET Plan Conference 2011. Available online: https://setis.ec.europa.eu/set-plan-conferences/5th-set-plan-conference-2011 (accessed on 10 March 2019).

35. United Nations, Department of Economic and Social Affairs, Population Division. *World Urbanization Prospects, the 2009 Revision*; ESA/P/WP/215; United Nations: New York, NY, USA, 2010.

36. Österreicher, D. A Methodology for Integrated Refurbishment Actions in School Buildings. *Buildings* **2018**, *8*, 42. [CrossRef]

Energies **2019**, *12*, 1955

37. Österreicher, D. The Role of Smart Buildings for Energy Efficiency. Smart Buildings for a Greener Europe: Emerging Policy and Practice. Concerted Action Energy Performance of Buildings. La Valetta, Malta. 14.02.2017. Available online: https://www.epbd-ca.eu/wp-content/uploads/2017/02/2.-Doris-the-role-of-smart-buildings_170214.pdf (accessed on 26 September 2018).

Article

Modeling an Alternate Operational Ground Source Heat Pump for Combined Space Heating and Domestic Hot Water Power Sizing

Kaiser Ahmed [1],*, Jevgeni Fadejev [2] and Jarek Kurnitski [1,2]

[1] Department of Civil Engineering, Aalto University, 02150 Espoo, Finland; jarek.kurnitski@ttu.ee
[2] Department of Civil Engineering and Architecture, Tallinn University of Technology, 19086 Tallinn, Estonia; jevgeni.fadejev@ttu.ee
* Correspondence: kaiser.ahmed@aalto.fi

Received: 25 March 2019; Accepted: 27 May 2019; Published: 3 June 2019

Abstract: This study developed an alternate operational control system for ground source heat pumps (GSHP), which was applied to determine combined space heating and domestic hot water (DHW) power equations at design temperature. A domestic GSHP with an alternate control system was implemented in a whole building simulation model following the heat deficiency for space heating based on degree minute counting. A simulated GSHP system with 200 L storage tank resulted in 13%–26% power reduction compared to the calculation of the same system with existing European standards, which required separate space heating and DHW power calculation. The periodic operation utilized the thermal mass of the building with the same effect in the case of light and heavy-weight building because of the very short cycle of 30 min. Room temperatures dropped during the DHW heating cycle but kept within comfort range. The developed equations predict the total power as a function of occupancy, peak and average DHW consumption with variations of 0%–2.2% compared to the simulated results. DHW heating added the total power in modern low energy buildings by 21%–41% and 13%–26% at design temperatures of −15 °C and −26 °C, respectively. Internal heat gains reduced the power so that the reduction effect compensated the effect of DHW heating in the case of a house occupied by three people. The equations could be used for power sizing of any heat pump types, which has alternate operation principle and hydronic heating system.

Keywords: ground source heat pump; heating power; sizing; DHW heating; space heating; alternate operation

1. Introduction

Energy performance of buildings has been continuously improved by imposing the new energy regulations. These reduced energy use in buildings for space conditioning, lighting, and appliances. However, energy use for DHW heating has kept the same contribution as before, which is considered as the most dominating one in buildings where the space heating (SH) need is low. Energy use for DHW heating seemed unpredictable, which is mainly caused by different DHW usages at occupant and apartment levels [1–3]. The detailed hourly DHW usages profile affects the power sizing of a heating system; however, an accurate sizing method does not exist yet for a combined space heating and DHW generation. In this context, the DHW hourly usages profile was used during the development of DHW heating with a GSHP for an apartment building [4]. Various sizes of storage tanks, an operational principle with corresponding power needs of HP, were analyzed for an apartment building, which was occupied by 75 people. The hourly profile of DHW was represented for a large number of occupants, which dampened the peaks of hot water usages and it cannot be used to estimate the DHW heating need for single-family houses, because the hourly DHW profile of a smaller occupant group has more strong peaks in morning and evening time [5,6]. Additionally, high delivered temperatures for DHW heating may have critical issues in a system's sizing. A two-stage heat pump system was introduced in Reference [7], which consisted of two HPs and two storage tanks. HPs were connected in a series and two tanks stored thermal masses for space and DHW heating. The first HP operated between the ground source and a low temperature storage tank, which acted like a heat storage system for space heating. Another HP operated between a low temperature storage tank to a high temperature storage tank, which acted like the heat storage system for DHW heating. This system found 31% of electricity savings; however, it was not justified that how this system was economically feasible for single-family houses [7].

Many studies have discussed the optimal sizing power of HP, i.e., how many percentages of the load are covered by the HP. The optimal power of HP was calculated in cooperation with a heat recovery exchanger that based on a coefficient of performance (COP), investment cost, system operation hours, and lifetime of the system [8]. The sizing was done in order to provide the sanitary hot water (SHW) only, not for the total heating need of both space and DHW heating. Similarly, the sizing details of an air source heat pump (ASHP) with a gas-boiler system for dwellings were explored in Reference [9]. This hybrid system introduced the shifting of the percentage of loads from the HP to the gas-boiler system in order to make the system more cost-efficient without compromising the occupants' thermal comfort level. In another context, the performance of the on-off and modulating air to water HP were compared in Reference [10]. The sizing effects of HP on the annual energy performance were discussed. The on-off HP with the storage system met about 95%–98% of the annual heating demand.

GSHP was found as the best possible alternative, which could supply heat for space and DHW heating for new-detached buildings, new apartment buildings, and existing apartment buildings that were built in the 1960s [11]. Besides, GSHP technology was reviewed in term of investment cost for each kW of installed power, running cost, financial savings, payback period, user benefits, community benefits and benefits to utilities, which ranked it as the most beneficial heating solution [12]. Rivoire et al. (2018) assessed the performance of GSHP in residential, hotel, and office buildings from six European cities [13]. A dynamic energy simulation tool was used to evaluate the performance, which also considered the effect of different envelope solutions on buildings' energy use. The results highlighted the annual energy demand and installation power of GSHP. The study emphasized the introduction of an additional heating system that enabled to compensate for the peak demand of heating need and reduced the installed power of GSHP [13]. Heating demand in a low energy single-family house seems low, which can be covered by a GSHP system only with minimum power that available in the market. In a similar context, the indoor thermal condition was studied in a low energy office building where GSHP was used as the primary heating source [14]. The study examined one-year precise monitoring data of indoor thermal conditions and found that rooms' temperatures were in between 20 and 23 °C during working hours of the heating season [14].

The control strategy affects both energy savings and sizing of heating power. An experimental study was performed in order to validate the numerical model that included the control strategy, aiming to account space and DHW heating demands in residential buildings [15]. Results highlighted the effect of control sensors on overall GSHP's ON/OFF numbers, duration of a cycle as well as assessing the impact on an overall COP [15]. In a similar context, Dong and Lam (2014) introduced a control system that integrated local weather conditions with a predicted occupant behavior pattern to show the energy savings [16]. The control system reduced the heating and cooling energy use by 30.1% and 17.8%, respectively while maintaining a specified thermal comfort range [16]. Such findings showed the importance of real demand-oriented control system modeling that could reduce the sizing power and energy use in buildings. Similarly, Arabzadeh et al. (2018) developed a cost-optimal operation schedule of GSHP for a Finnish detached house in cooperation with a demand response control system that based on building heating needs and dynamics of electricity price. Results showed that the proposed control system reduced the total heating energy cost by around 12% [17]. In the similar context, a control strategy of water to air HP was introduced in Reference [18], which enabled the control of the speed of the compressor for different operational modes such as SH, space cooling (SC), DHW heating, and simultaneous SC and DHW heating. This control system gave priority of DHW heating over SH, which could fail to fulfil the SH demand in case of large draw-off volumes of DHW and a longer draw-off duration. Besides, a neural predictive controller was developed in Reference [19], which could predict the GSHP heating power. The controller operated the GSHP compressor either to switch on and off according to the SH need [19]. However, this controller is not suitable in such a case where GSHP is the only source of providing heat for both SH and DHW heating.

To summarize the existing studies in the literature body, it may be concluded that the power sizing of HP at design temperature has not been discussed. In addition, there are no equations that calculate the HP power directly, which account for both space and DHW heating. Moreover, there is no study showing that how stable room temperatures have been provided by an alternate HP's control system with a priority of DHW heating, where low indoor temperatures may be expected during the DHW heating.

Precise knowledge is required for the power estimation of GSHP compared to the gas-boiler (GB) and district heating (DH) systems. The boiler power is not strictly limited, and most often the system is oversized. DH system also needs more accurate sizing than the GB system due to the sizing of the heat exchanger and water flow based pricing. However, the DH system has two separate heat exchangers (HX) for space and DHW heating that simplifies the problem, and there is no choice of periodic operation. EN standards estimate the heat load power for space and DHW heating separately [20,21]. Besides, the energy calculation of GSHP systems has well explained in EN standards [22]; however, these standards cannot be applied for the power sizing of GSHP. In a similar context, Finnish building regulations estimate the GSHP power from simultaneous space and DHW heating. However, there is no guidance in Reference [23] regarding how to calculate the simultaneous DHW power. Therefore, in practice, DHW power is calculated at a constant flow rate (assuming enough large DHW accumulation tank) or even neglected, if considerably smaller than space and supply air-heating power. Moreover, it is not well defined how sizing needs to be done in order to combine the space and DHW heating load in low energy buildings. The absence of sufficient guidelines in current building standards, limited knowledge of GSHP's sizing for a single-family house, a limited system capacity of GSHP, as well as considerable investment cost for producing of each kW of power, encourages one to find out the exact GSHP's sizing solutions for single-family houses. We focused on the sizing of most common domestic GSHP, which operated in a way that it switched from one heating mode to another (SH mode, DHW heating mode) for a given time interval. During this interval, the room temperature, as well as delivered DHW temperature, should stay in acceptable limits [24].

The working principle of common domestic GSHPs is shown in Figure 1. The system consists of a heat pump, heat source, heat sink, electrical module, circulation pumps, and a control system. GSHP is connected to the brine and heating medium circuits, which takes up the heat at a certain temperature and delivers the heat at a higher temperature. The performance of GSHP depends on heating curve set points, the heating system in a room, control strategy, and so on. Low-temperature radiator heating system can be used in low energy buildings and the heating curve might be higher than 5.0 °C compared to the floor heating system [25].

The objective of this study was to develop the alternate control model, allowing to determine the theoretical heating power need of a monovalent HP system at a designed outdoor temperature as a function of space and DHW heating needs. The hourly DHW usage profiles with two visible peaks, typical for single-family houses (Figure 2), were used for the power sizing of a GSHP. Besides, occupancy variations from three to six people in single-family houses, effects of internal heat gains were also considered in order to develop the sizing power equations for GSHP. Based on the determined heating power, a heat pump can be selected so that it covers 100% of the power or less of the power, that question was let out of the scope of this paper. The outlines of this study are the following:

- an alternate operational control system for GSHP was implemented in the IDA indoor climate and energy (IDA-ICE) plant model by developing a new control module, which was tested against measured data;
- maximum power was studied at different scenarios, i.e., occupant number, internal heat gains, building structural types, different design outdoor temperatures, DHW usage variations, variations of DHW usage peaks;
- equations that calculated the GSHP power at outdoor design temperature were developed and tested;
- the accuracy of equations was tested to calculate the GSHP's power for old buildings (followed the Finnish 1976 building regulations), building in different climate regions, variations of DHW usages, and so on. These could show the suitability of the control system as well as the total power estimation of GSHP for buildings in central European countries and buildings that may require major renovations.

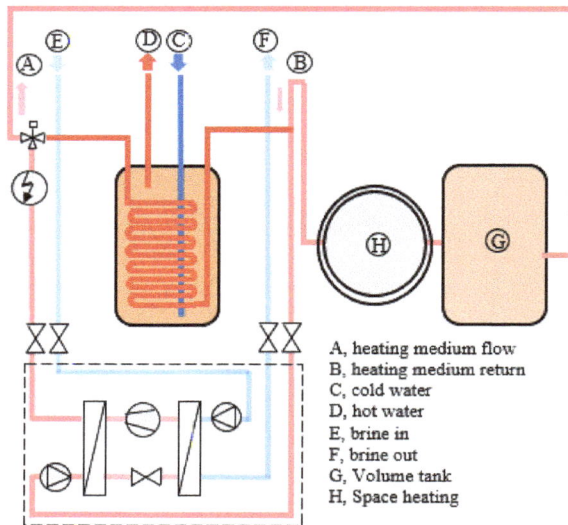

A, heating medium flow
B, heating medium return
C, cold water
D, hot water
E, brine in
F, brine out
G, Volume tank
H, Space heating

Figure 1. Typical domestic ground source heat pump (GSHP) with an alternate operation principle [26].

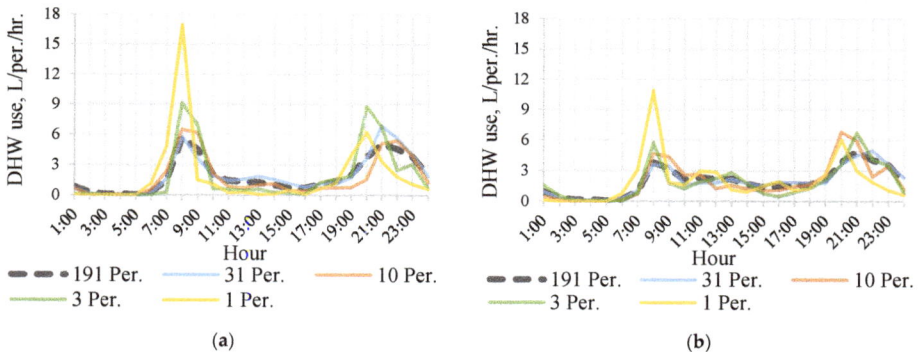

Figure 2. Proposed profiles of five groups for the month of November (a) weekday, (b) total day (all days that include weekdays and weekends).

The findings of this study were expressed as the power equations at design temperature, allowing to determine 100% of the total heating for space and DHW heating for a single-family house. These equations can be applied for any types of HP power sizing, which has an alternative operational principle and hydronic heating system.

2. Method

2.1. Space Heating Load

Heat losses from a building can be calculated from Equation (1). Heat transfer coefficients of building components (such as façade, roof, floor, glaze), ventilation rate, air change rate, thermal bridges, and infiltration rate were estimated according to the Finnish building code [27].

$$\Phi_{space} = \Phi_T + \Phi_I + \Phi_V \tag{1}$$

$$q_v, = \frac{q_{50} \, A_{env}}{3600 \, x} \tag{2}$$

where, Φ_{space}, heating power for space (W), Φ_T, sum of transmission heat losses through building envelope (external wall, ground, roof, windows, doors, thermal bridges, etc.) (W), Φ_I, heat losses caused by infiltration (W), Φ_V, heat losses caused by ventilation (W), q_v, infiltration rate ($\frac{m^3}{s}$), q_{50}, air leakage rate of building envelope ($\frac{m^3}{h.m^2}$), A_{env}, envelope area (m²), x, factor that based on the building height. The values for single, double, 3–4 storied, 5 or more were 35, 24, 20, and 15, respectively [27].

The same approach was used to estimate the heat load for a heated space in FprEN12831-1 [20].

$$\Phi_{HL,i} = \Phi_{T,i} + \Phi_{V,i} + \Phi_{hu,i} \tag{3}$$

where, $\Phi_{HL,i}$, design heat losses of the heated space of i (W), $\Phi_{T,i}$, sum of transmission heat losses through building envelope of i (external wall, ground, roof, windows, doors, thermal bridges, infiltration, etc.) (W), $\Phi_{V,i}$, design heat losses caused by ventilation of i (W), $\Phi_{hu,i}$, additional heating up power for the heated space of i (W).

We used dynamic simulation in order to determine the power need for space heating, which also accounted for the infiltration rate, as shown in Equation (2). According to the Finnish building code, the design outdoor temperature and indoor heating set points were −26 °C and 21 °C, respectively [23]. Besides, we considered the outdoor design temperature of −15 °C (for Strasbourg) could show the suitability of GSHP's power sizing equations. In addition, no solar heat gain was used for computational analysis. Furthermore, occupancy rate, unit load for lighting and appliance, usages profiles for single-family houses were summarized from References [28,29].

2.2. DHW Heating Load

We used the daily average DHW use of 43 L/person/day for Finnish residential buildings [30]. This average DHW use per day was multiplied with monthly consumption factor [30], which further distributed according to the hourly factors during 24 h of each a day [5], as shown in Figure 2. As GSHP's power was sized for a single-family house, the occupant number was varied from three to six. Thus, the obtained daily DHW use for a single-family house was calculated by using the following equation:

$$V_{T,d} = n \, F_m V_d \tag{4}$$

where, $V_{T,d}$, total DHW use by a single-family house ($\frac{L}{day}$), n, number of occupant (dimensionless), F_m, monthly DHW usage factor (dimensionless), V_d, DHW use by each occupant ($\frac{L}{person.day}$).

The $V_{T,d}$ was distributed in hourly basis according to the DHW hourly profile of three people [5]. Besides, the three people's DHW Weekday profile was used, because of having higher peaks compared to the three people weekend profile and three people total profile (all days that include weekdays and weekends) [5].

In addition to the dynamic simulation, the sizing power for a DHW heating was done according to the FprEN 12831-3:2016 standard [21]. The sizing steps for a DHW system are well defined in Reference [21], including the storage losses and DHW loading factor, compared to the Finnish standard [23]. According to the EN standard, power needs curve for DHW (needs curve) and power supply from the system (supply curve) was used. The need curve was determined by measuring the volume flow rate based on the load profile. Besides, temperatures of hot and cold water were considered. In order to develop the supply curve, available heat source and corresponding data such as storage thermal losses, distribution losses, power, and, system efficiency were also considered.

$$\Phi_{eff} = \Phi_N \left[1 - \frac{T_{Sto,m}(t) - T_c}{T_{ch,HG}(t) - T_c} \right] - \Phi_{w,Sto} - \Phi_{w,dis} \tag{5}$$

$$\Phi_{eff} = \rho_w \, C_w V_D \left(T_{W,draw} - T_c \right) \tag{6}$$

$$T_{Sto,m}(t) = T_{Sto,m,t0} + \left(T_{ch,HG} - T_{Sto,m,t0} \right) \left(1 - e^{-\frac{t}{\tau}} \right) \tag{7}$$

$$\tau = \frac{m_{Sto} \, c_w}{U_{HE} \, A_{HE}} * 0.06 \tag{8}$$

where, Φ_{eff}, effective power of water heating system (kW), Φ_N, nominal power of heat generator (kW), $T_{Sto,m}(t)$, mean water temperature in storage tank at time t (°C), T_c, cold water temperature (°C), $T_{ch,HG}(t)$, supply temperature from heat generators at time t (°C), $\Phi_{w,Sto}$, heat losses from storage at time t (kW), $\Phi_{w,dis}$, distribution heat losses (kW), ρ_w, water density ($\frac{kg}{l}$), C_w, water heat capacity ($\frac{kJ}{kgK}$), V_D, design flow rate ($\frac{l}{s}$), $T_{W,draw}$, temperature of water withdrawn (°C), $T_{Sto,m,t0}$, mean water temperature of the storage tank at the time reheating is switched on (°C), τ, time constant of storage tank during loading (min), t, time (min), m_{Sto}, water mass in the storage volume (kg), c_w, heat capacity of water in the storage volume ($\frac{kJ}{kgK}$), U_{HE}, thermal transmittance of the heat exchanger ($\frac{W}{m^2 K}$), A_{HE}, effective surface of the heat exchanger (m²).

2.3. Simulation Model

The IDA indoor climate and energy (IDA-ICE, version 4.7.1) simulation tool was used for computational analysis. The model was developed in an IDA-ICE advanced level interface, where users could connect system components, connection editing, output data logging, and so on. The model was developed in the early stage building optimization (ESBO) plant model, which is available in the tools library. IDA-ICE are well-established energy simulation tools, and many studies have been validated the reliability, performance, and accuracy compared to the field data [4,31–34].

The reference single-family house with a heated net floor area of 171.1 m² was used, as shown in Figure 3. It was assumed that the building model could use as a typical representative single-family house in Finland. Building input parameters were selected according to two building regulations, i.e., modern low energy (passive) building, old building regulation of 1976. The effects of heavyweight and lightweight construction concepts on indoor thermal condition were also discussed. The detailed building information is given in Table 1.

The external wall (from inside to outside) of a lightweight building consisted of gypsum board, air gap and EPS insulation. The heavyweight building used LECA block additionally. The external wall followed the old building regulation of 1976 and used the concrete wall instead of an LECA block. Roof structure (from inside to outside) consisted of gypsum board, air gap, and EPS insulation, wooden frame, and metal sheet. Similarly, the floor structure kept the same configuration (from inside to outside) for all cases, i.e., parquet, concrete, and EPS insulation. For the internal floor, a wooden floor slab was used in lightweight building cases whereas hollow concrete slab was used in heavyweight building cases. The internal floor of an old building followed the old building regulation of 1976 that used concrete floor. For the low energy building case, the model considered a balanced ventilation system with a heat recovery unit, Table 1. Thus, supply air was heated by recovered heat from the heat exchanger and heating coil. A low-temperature underfloor heating system (35/28 °C) with a PI controller was considered. The design power for underfloor heating system was 35 W/m² at a maximum temperature drop of 7 °C. The pipe located at 0.035 m below from the floor surface top and heat transfer coefficient was 30 W/m²K.

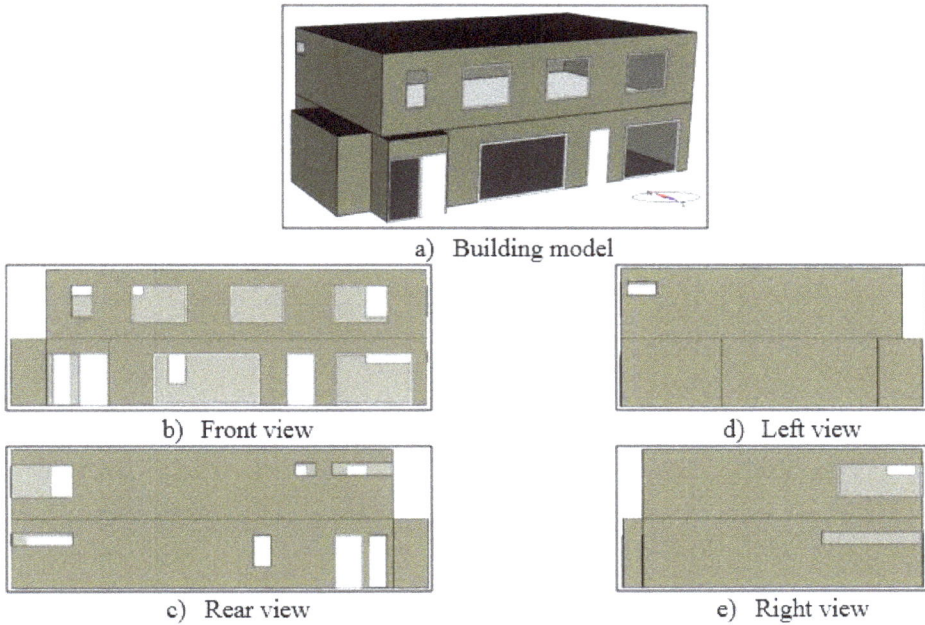

a) Building model

b) Front view

d) Left view

c) Rear view

e) Right view

Figure 3. (**a**) Building model and views of (**b**) Front, (**c**) Rear, (**d**) Left, (**e**) Right.

For building model according to old building regulation of 1976, there was no air-handling unit. The model was equipped with a mechanical exhaust ventilation system only. The heating need was provided by a radiator system (55/45 °C) with a PI controller. A detailed radiator model (available in the IDA-ICE library) was considered, and total heat generation was calculated by using the following equation:

$$P_r = KLdT^{n_r} \tag{9}$$

where, P_r, heat flux from water (W), n_r, coefficient of power law, which depends on the radiator height and width ($\frac{W}{m°C}$), dT, difference between the mean surface temperature of radiator and the air temperature, L, radiator length (m), K, radiator constant (dimensionless).

A stratified tank model with a capacity of 200 L was used. The tank model had five different layers, and it was highly insulated. This model also considered the storage heat losses and distribution heat losses.

Table 1. A detailed description of input data for a reference building model.

	Low Energy Building Regulation	Old Building Regulation of 1976
External wall Area—156.5 m^2		
U value	$0.1 \times \frac{W}{m^2 K}$ [35]	$0.4 \times \frac{W}{m^2 K}$
Roof Area—83.05 m^2		
U value	$0.06 \times \frac{W}{m^2 K}$ [35]	$0.35 \times \frac{W}{m^2 K}$
Ground floor Area—88.25 m^2		
U value	$0.06 \times \frac{W}{m^2 K}$ [35]	$0.4 \times \frac{W}{m^2 K}$
Window Area—33.22 m^2		
U value	$0.6 \times \frac{W}{m^2 K}$ [35]	$2.1 \times \frac{W}{m^2 K}$
G value	0.46	0.46
Frame to glaze ratio	0.1	
Leakage rate, q_{50}	$1.0 \times \frac{m^3}{h\,m^2}$	
Thermal bridge		
Ext. wall to internal slab,	0.0574 W/K/(m joint) [36]	0.0689 W/K/(m joint)
Ext. wall to ext. wall,	0.0336 W/K/(m joint) [36]	0.0403 W/K/(m joint)
Ext. window parameter,	0.02395 W/K/(m joint) [36]	0.0287 W/K/(m joint)
External door perimeter	0.0 W/K/(m perimeter) [36]	0 W/K/(m perimeter)
Roof to ext. wall,	0.0519 W/K/(m joint) [36]	0.0623 W/K/(m joint)
Ext. slab to ext. walls	0.0515 W/K/(m joint) [36]	0.0618 W/K/(m joint)
Ventilation rate	$62 \times \frac{1}{s}$	
Fan power	$1.5 \times \frac{kW}{\frac{m^3}{s}}$, always on [35]	
Temperature ratio (Heat recovery efficiency)	85%, always on [35]	-
Supply air temperature	[1] 17 °C	-
Minimum exhaust temperature	[3] 0 °C	
Heating set point	21 °C	
Occupancy rate	[2] $42 \times \frac{m^2}{person}$ [28]. Usages profile [28], $118.3 \times \frac{W}{person}$ [29]	
Lighting	$8 \times \frac{W}{m^2}$ [28], Usages profile [28]	
Appliance	$2.4 \times \frac{W}{m^2}$ [28], Usages profile [28]	

[1] Supply air temperature after the heating coil was set to 17 °C. Fan induced the air temperature of 1 °C. [2] Occupancy rate was distributed according to occupant number (i.e., 3, 4, 5, and 6) in 171.1 m^2. [3] Limits the heat exchanger capacity to 45% at a design temperature of −26 °C.

3. Results and Discussion

3.1. GSHP and Control System Model

An existing GSHP alternate control strategy was followed, enabling to supply the heat either for space or DHW heating. In order to illustrate the operation principle of such type of HP, the measured data of GSHP during 24 h at one cold day are shown in Figure 4. This is an example, how the modeled control principle works in reality. Heat energy for space heating was supplied according to the heating curve. The corresponding deficits of heat were counted as degree minutes (DM) value. Similar to the degree day (DD) method, the DM value is a cumulative number of the difference between actual flow temperature (T_a) and setpoint temperature of flow (T_s) for given elapsed times in minutes. The default DM value with a couple of degree temperatures (as a dead band) were assigned in the control system setting. If the estimated DM value was fallen beyond the default value (in this case, −60 °C·min), the system would start immediately. In some extreme cases, if GSHP failed to supply sufficient heat, an electrical top-up heater switched on according to the default steps and fulfilled the need at the demand side. In this case, the top-up heater switched on in six steps based on the DM value. For instance, top up heater operated for step 1 (1 kW, −180 < DM < −120), step 2 (2 kW, −240 < DM < −180), step 3 (3 kW, −300 < DM < −240), step 4 (4 kW, −360 < DM < −300), step 5 (5 kW, −420 < DM < −360), and step 6 (6 kW, DM < −420).

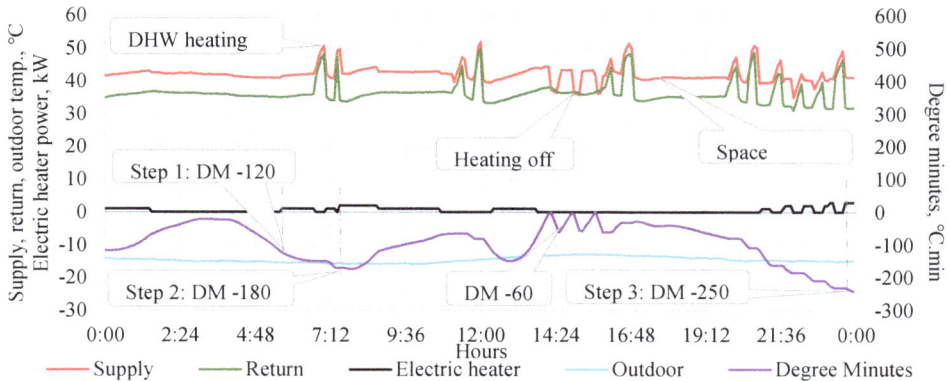

Figure 4. An example of measured GSHP operation: supply temperature, return temperature, outdoor temperature, electric top-up heater power and degree minutes of heat deficit.

The red line represents the supply water of temperatures for space and DHW heating. When GSHP ran in space heating mode, the supply water temperatures were in between 40 and 45 °C. Besides, if GSHP ran in a DHW heating mode, the supply water temperatures were higher than 45 °C (peak shown in Figure 4). In some periods the top-up electric heater switched on in a couple of steps in order to compensate the heat deficit. GSHP was in heating off mode if the DM value was higher than zero (heating off in Figure 4).

In this study, a monovalent ON/OFF brine to water GSHP (driven by electricity) was used to produce heat energy for space and DHW heating. The developed control system switched the heat generator either to DHW or to space heating (SH) mode, and it was forced to run the GSHP at the maximum load capacity during the operation period.

The GSHP control system accounted for the deficit of heat for space and DHW on a priority basis. The time interval was implemented in the plant model. The time interval means the maximum time to operate in one heating (space heating or DHW) mode if demand existed and after that time the HP pump was switched to another mode. The time interval was a user-defined value such as 15 min, 20 min, 30 min, etc.

The supply of heat was regulated according to the heating curve, and the set point temperature of delivered DHW. The control system calculated the deficits of heat as the DM value. Based on the DM value and delivered water temperature of DHW, the HP compressor was in ON/OFF mode. The differences were counted as DM value for each minute if $T_a \leq T_s$ and the cumulative number was obtained for a given elapsed minutes. The following equation was used to calculate the DM value:

$$DM = \sum_{i=1}^{t_n} (T_a - T_s)\, t_n \tag{10}$$

where, DM, degree minutes (°C·min), T_s, heating curve setpoint temperature (°C), T_a, flow temperature at actual condition (°C), t_n, elapsed time (minutes).

The schematic diagram of the model is shown in Figure 5. For modeling of the periodic HP operation, two HPs were used in the model instead of one HP in reality. The model was designed in such a way that only one HP operated at a time or both HPs were switched off. The supervising control system was placed between the two HPs, which decided the active state of one HP among two HPs and ensured only one HP operation at the same time. One HP and one storage tank were operated for the space heating facility. Similarly, another HP and another storage tank were operated for DHW heating. The properties of both HPs' and storage tanks were similar. The control system generated signals based on the heat deficits of space or DHW heating and then switched to SH mode, or DHW mode or OFF mode. In SH mode, only one HP for space heating was in operation, and another HP for DHW heating was switched off. Similarly, only one HP for DHW heating was in operation in DHW mode, and another HP for space heating was switched off. If there was no heat demand for space and DHW heating, both HPs were switched-off. Heating of each mode could be possible for a given time interval, which was a user-defined parameter. It could be possible to choose either 15 min, 20 min, 30 min interval and so on based on the demand side. This study considered the maximum time interval of 30 min after that HP switched to another mode. The control system follows the following steps, which is also illustrated in Figure 6.

- Step 1: DHW mode—started with DHW heating mode (DHW heating priority) for a maximum of 30 min. If continuous heating was required for DHW, the DHW mode would continue for a maximum period of 30 min. However, if no was heating required for DHW after some time interval, for instance, 15 min or 20 min of DHW heating period, the mode switched to SH mode (step 2) or OFF mode (step 3).
- Step 2: Space heating (SH) mode—continued for the next 30 min and afterwards retained back to step 1 if heat was required for DHW heating. If there were no heat demands for DHW heating but space heating, the SH mode would continue for the next 30 min period. If there were no heat demands for space and DHW, then it switched to OFF mode.
- Step 3: OFF mode—referred to as the state of GSHP where there was no heat demand for space and DHW heating that could allow the GSHP switched off.

It was a continuous process, which could able to provide heat for space and DHW heating. The best thing about this control system is that it is also suitable for extreme cases. For instance, at night time when the outdoor temperatures were very low, and the house had a less DHW consumption, the control system operated the GSHP at SH mode. It was able to continue at SH mode (because there were constant heat demands for SH and no heat demand for DHW), it would keep a stable indoor thermal condition. Similarly, at morning (for instance 7:00–8:00 h) when the peak demand of DHW was high, the control system operated the GSHP at both modes with a maximum interval of 30 min for each mode. It could keep the DHW supply temperature at an acceptable level.

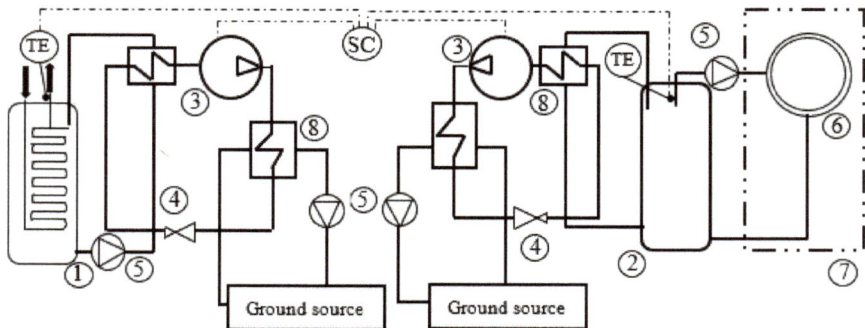

1– DHW storage tank, 2 – SH storage tank, 3 – compressor, 4 – expansion valve, 5 – circulation pump, 6 – radiator network, 7 – zone room, 8 – heat exchanger TE – temperature sensor, SC – supervising controller

Figure 5. Schematic diagram of modeled typical domestic GSHP with an alternate operation.

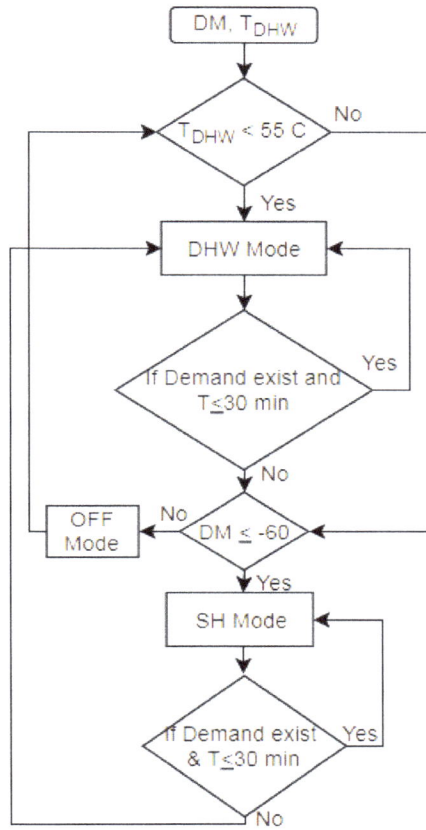

Figure 6. Operation of the control system in a flowchart.

This schematic diagram of a model as shown in Figure 5 was implemented in IDA-ICE (Figure 7). The condenser side of GSHP was connected to the stratified tank to meet the heating demand side. No additional top-up heater was needed, and the heating curve was controlled according to the outdoor air temperature. GSHP also met the performance maps of manufacturer specific product. GSHP operated with maximum power whenever the temperature at demand side dropped below the heating setpoint temperature. The operation process was continued until it reached the desired setpoint temperature. Moreover, the model implemented in IDA-ICE, with two GSHPs and two storage tanks are shown in Figure 7. Besides, the detailed of heat pump's control system is shown in Figure 8.

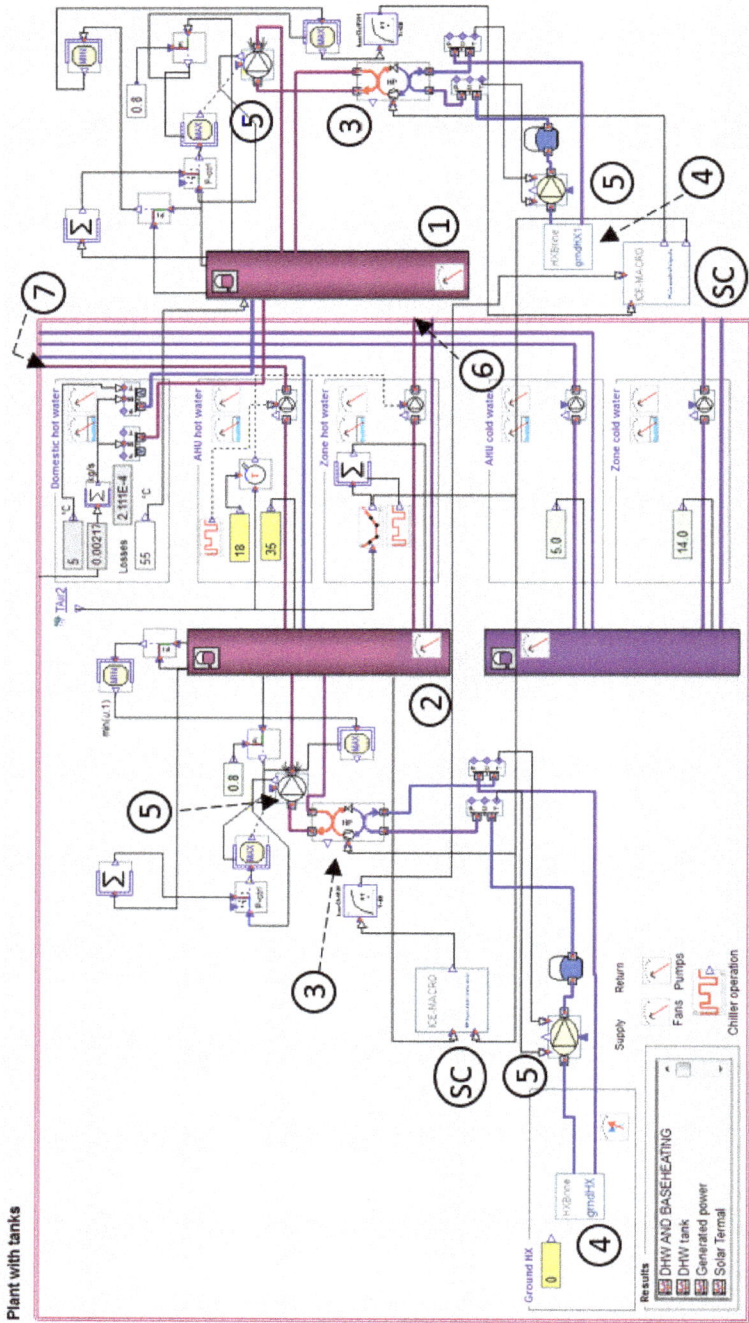

Figure 7. Implemented in the plant model.

1– DHW storage tank, 2 – SH storage tank, 3 – HP, 4 – ground source, 5 – circulation pump, 6 – radiator or underfloor heating network, 7 – AHU hot water, SC – supervising controller

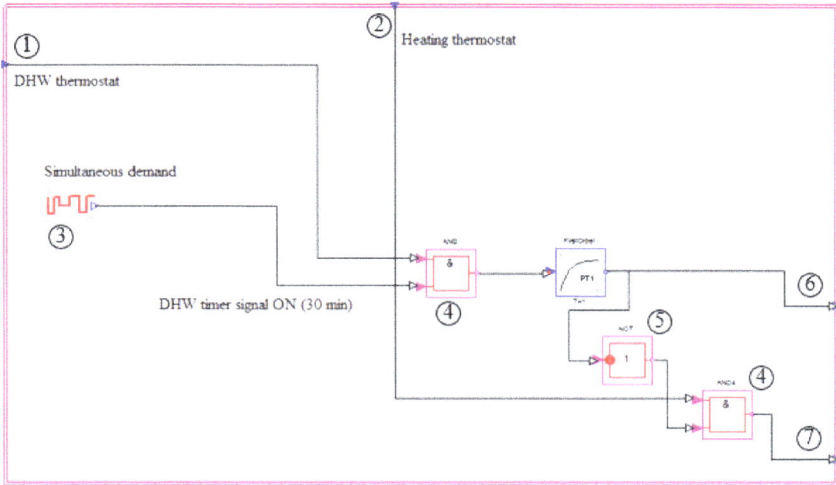

1– DHW demand signal, 2 – SH demand signal, 3 – simultaneous demand signal, 4 – AND logic gate, 5 – NOT logic gate, 6 – DHW heating signal, 7 – SH signal

(a)

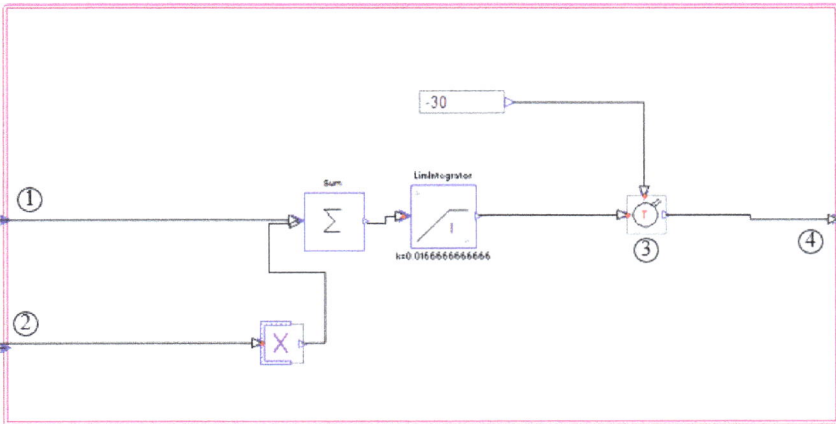

1– Tank actual temperature, 2 – heating curve temperature, 3 – temperature thermostat, 4 – Heat pump control signal

(b)

Figure 8. (a) Heat pump degree minute control system, (b) deliver control signal to the heat pump.

3.2. Operation Principle of the Control System

This section shows the operation of a control system according to the DM value. The control system recorded the delivered water temperature for space and DHW heating. If the delivered water temperature for space heating was less than the given set point temperature, the control system counted it as a heat demand signal for SH. Similarly, if DHW delivered temperature was less than 55 °C than the control system identified it as heat demand signal for DHW heating. Afterwards, the DM value was compared to the default setting value (user-defined DM value). The default setting value was considered as −60 °C·min (could be chosen between 0 to −120 °C·min). The GSHP compressor was in ON status and accelerated the heating production if the DM value was below than −60 °C·min. It also kept a similar status until the DM value reached zero. When DM was higher than zero, the GSHP compressor was switched off (OFF mode). During the on status, it worked according to the mentioned steps.

The dynamic simulation was performed during the heating season, i.e., 1st of December to 28th of February, 2018 and the results of indoor temperature, DHW outlet temperature, GSHP control signal, DHW signal, and heating signal for four people occupied a single-family house during 24 h are reported in Figure 9.

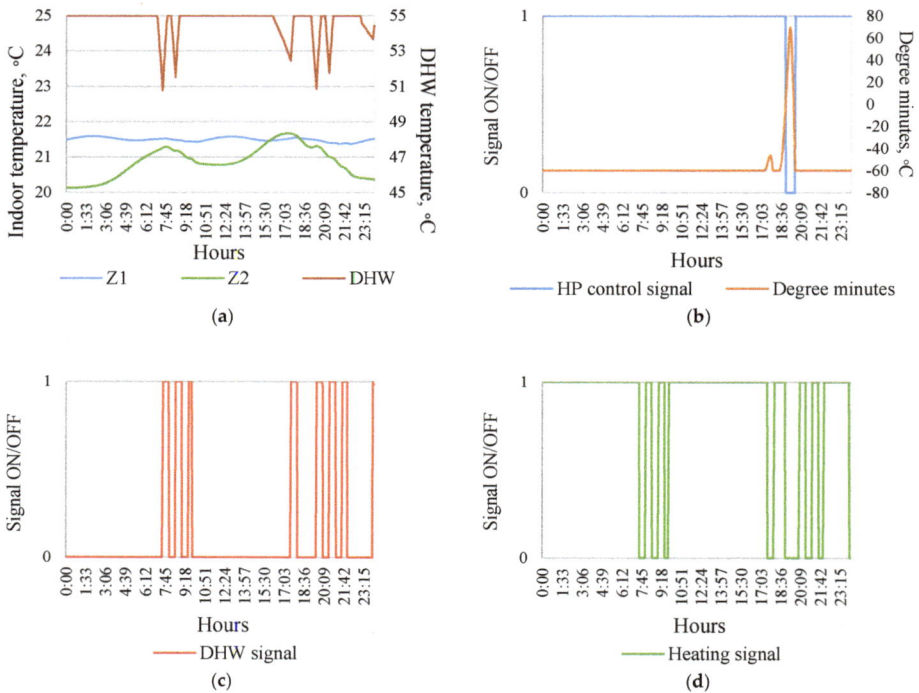

Figure 9. (a) Indoor temperature and domestic hot water (DHW) supply temperature, (b) GSHP control and degree minute (DM) value (c) only DHW signal, (d) only heating signal for a single-family house occupied by four people.

The control system determined the operational state of GSHP, i.e., ON, or OFF. The GSHP was in ON status for all hours in a day except 18:57 h (DM was 5.0 °C·min) to 19:39 h (DM was −42.8 °C·min), as shown in Figure 9b. During OFF mode, the DM was between 5.0 and −42.8 °C·min. However, the OFF mode happened at different times in different days because of the dynamic process. The indoor temperatures were higher than 21 °C for Zone 1 (Ground floor), and Zone 2 (First floor) and DHW outlet temperatures were not below than 51 °C (Figure 9a). Total heat deficits were estimated as the DM value, which was limited to −60 °C·min in this case. The GSHP started if DM was ≤−60 °C·min and it ran in DHW heating mode. GSHP ran in DHW heating mode for a maximum of 30 min, and alternatively, it switched to SH mode, if heating needs exist for SH.

The duration of DHW heating was short (9:33 to 9:48 h) due to the fulfilment of demand, as shown in Figure 9c. However, GSHP ran with SH mode for a minimum of 30 min duration, and it remained the same state if there would no heating demand for DHW, and continuous heating demand for space heating (Figure 9d). Besides, if the GSHP ran with SH mode and there was a continuous withdrawal of DHW from the tank, i.e., 7:06 to 7:27 h (Figure 9a); apparently the DHW outlet temperature was below the setpoint temperature. In this case, the DHW mode would not start immediately. It would start only after SH mode (minimum duration of SH mode was 30 min).

Energy use for DHW varied due to the different DHW usages volume, which mainly depended on the occupants' number and hourly DHW profile. Heat demand for DHW heating had significant effects on system sizing compared to the constant space heating need. Indoor thermal condition, DHW outlet temperature, and GSHP operational signal for six people occupied single-family house are reported in Figure 10. The indoor temperatures were in the acceptable limit of Category II, according to EN 15251 [24]. During peak hours of DHW use, the outlet temperatures fell nearly 45 °C, which might be considered as the worst cases. These scenarios only happened if six people used the DHW at the same time in a single-family house.

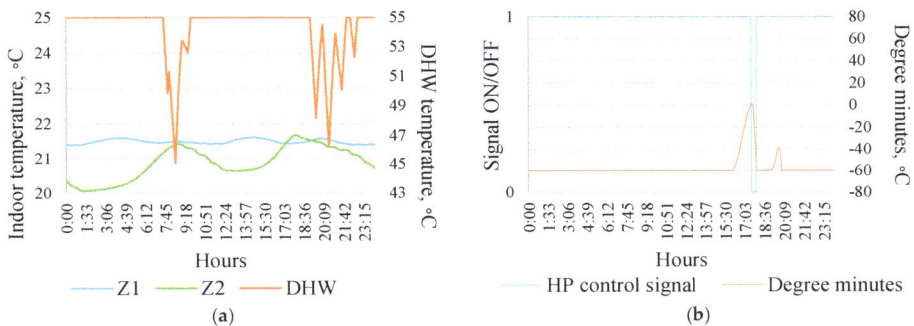

Figure 10. (**a**) Indoor temperature and DHW supply temperature, (**b**) GSHP control and DM value signal for a single-family house occupied by six.

3.3. GSHP Power Comparison

The total GSHP powers for a single-family house with different occupant groups were obtained by hand calculation and simulation approach, as shown in Figure 11. The hand calculation approach considered the steady state condition of DHW use with a uniform DHW profile and a three people DHW profile, following the Finnish building code [27]. Besides, the effects of storage tank were not accounted for. Daily DHW use was distributed equally into 24 h of a day and power of DHW heating was calculated for any hour (uniform 0 L). In the three people DHW profiles, the daily average DHW use was distributed according to the hourly usages factor, Figure 2. The maximum DHW usage volume was found at 8:00, which was used to calculate the maximum power required for DHW heating (EN 0 L). According to the Finnish building code [23], the heating power is the sum of simultaneous space heating, supply air heating, and DHW powers. Uniform 0 L case might not offer a realistic solution for single-family houses due to neglecting the hourly DHW consumption peaks. In addition, DHW and space heating powers were summed according to EN 0 L case, which led to overestimated GSHP power.

In the simulation approach, developed in this study, storage tank effects and DHW profile were considered. This Simulated 200 L case was compared with EN 200 L calculated according to the standard. In these both cases, a storage tank of 200 L was added. The Simulated 200 L case resulted in a 14%–44% power reduction compared to EN 0 L and EN 200 L cases, and the powers were 4%–14% higher compared to the non-realistic uniform 0 L case.

The hourly profile of DHW was not suitable for the'EN 0 L' case because peaks of DHW usage require a storage tank, which can compensate for the required thermal mass at the minute level.

For 'uniform 0 L', DHW usages were distributed evenly, and no peaks occurred. These two cases demonstrated the manual calculation method in order to show how Finnish and EN regulations estimated the power at static condition. For all realistic cases, a 'simulated 200 L' water tank was used as shown in Figure 5.

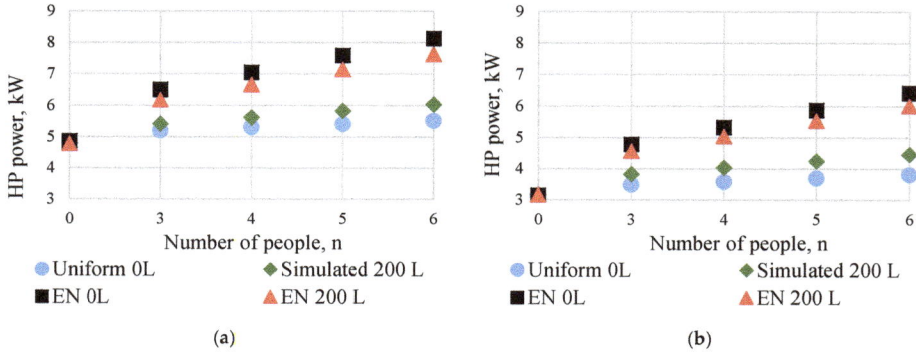

Figure 11. GSHP powers for design outdoor temperature of (**a**) −26 °C and (**b**) −15 °C.

3.4. Effect of Internal Heat Gains

The Finnish building code does not take into account the internal heat gains in order to estimate the design heat load [23]. However, EN standards account for the internal heat gain as optional heat load to estimate the design heat load [21]. DHW use and internal heat gains tend to appear simultaneously in a single-family house. Effect of internal heat gains was studied because it has a significant impact on system sizing, especially in low energy buildings. Moreover, solar heat gains were not accounted for system sizing due to the absence of solar heat during the night. The GSHP's control system also performed while accounting internal heat gains. The indoor temperature, DHW outlet temperature, DHW signal, and space heating signal for four and six people's occupied single-family house while considering internal heat gains are reported in Figure 12. Indoor temperatures for both zones (ground and first floor) were found in the acceptable limit (Category II of EN15251 standard) during the simulated period. The DHW outlet temperatures fell below the setpoint temperature at peak consumption hours, which was recovered when the GSHP ran with DHW heating mode (Figure 12a,c). Besides, GSHP ran with full power either for DHW heating mode or for SH mode (Figure 12b,d). The reduced HP sizing power with considering internal heat gains was calculated from Equation (12), Section 3.8.

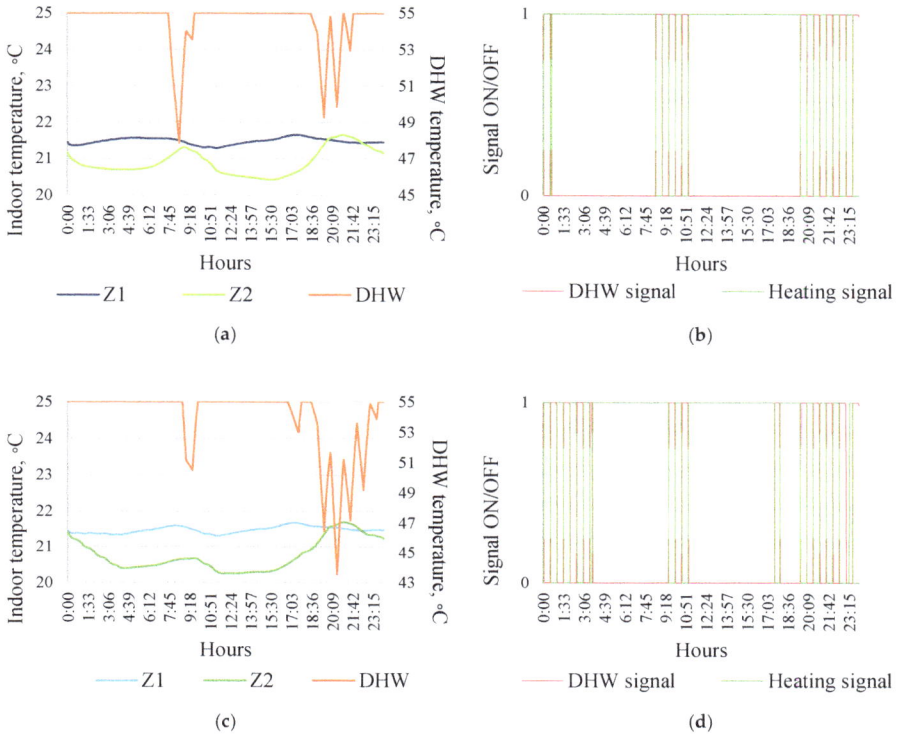

Figure 12. Internal gains are considered (**a**) Indoor temperature and DHW supply temperature for a single-family house occupied by four people, (**b**) heating and DHW signals for a single-family house occupied by four people, (**c**) indoor temperature and DHW supply temperature for a single-family house occupied by six people, (**d**) heating and DHW signals a single-family house occupied by six people.

3.5. Operational Performance of Control System at Different Scenarios

This section shows the suitability of this control system while sizing the heating power for old buildings (building parameters followed the old building regulation of 1976) or renovated buildings. The space heating need was higher compared to the low energy buildings and internal heat gains from occupant, lighting, and appliance did not seem so significant as low energy buildings. Heat losses from old buildings were quite fast so that the long duration heating mode could fail to serve the demand side. Duration of heating mode changed from 30 min to 15 min in order to keep the room temperature in an acceptable limit. According to the EN 15251 standard, the acceptable limit of indoor temperature must not be less than 18 °C (Category III) for old buildings [24]. GSHP operated with SH mode due to continuous space heating demand signals during the GSHP operation period. High GSHP power reduced the DHW heating mode duration, which allowed switching back to SH mode. Indoor temperature, DHW outlet temperature, DHW signal, and heating signal for houses occupied by four and six people are shown in Figure 13.

The effects of heating the mode duration on indoor temperature and DHW outlet temperature are reported in Figure 14. Effects were observed for those buildings, which required higher space and DHW heating. However, fewer effects were found in low energy buildings and building that occupied with a small group of occupants.

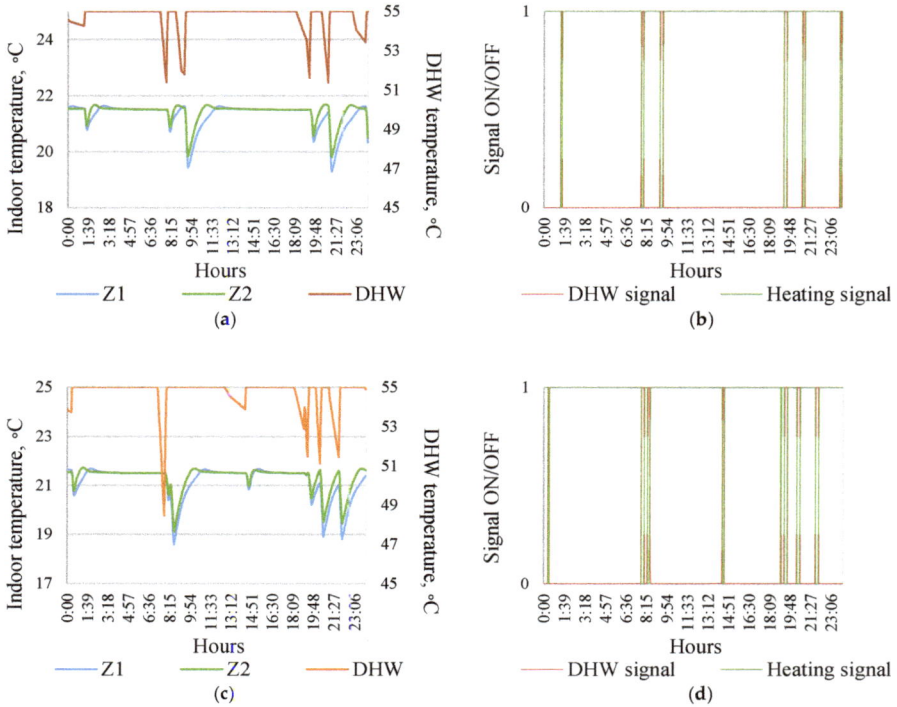

Figure 13. Building parameters followed the old building regulation of 1976 (**a**) indoor temperature and DHW supply temperature for an old single-family house occupied by four people, (**b**) heating and DHW signals for an old single-family house occupied by four people, (**c**) indoor temperature and DHW supply temperature an old single-family house occupied by six people, (**d**) heating and DHW signals for an old single-family house occupied by six people.

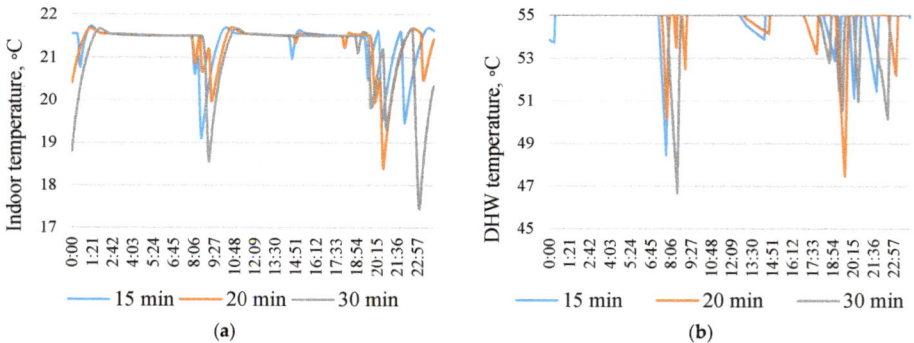

Figure 14. Effect of heating mode duration on (**a**) indoor temperature and (**b**) DHW outlet temperature for an old single-family house occupied by six people.

3.6. Simulation with Design Outdoor Temperature of −15 °C

Simulations were repeated for another design temperature of −15 °C in order to validate the GSHP power equations in Section 3.8. In order to keep the same comparison platform, 30 min heating mode duration and internal gains also considered. Indoor temperature and DHW outlet temperature for four and six people's occupied single-family house are shown in Figure 15. DHW outlet temperatures seemed critical compared to the indoor temperature. The GSHP power was low at a design temperature of −15 °C compared to the extreme Finnish case, i.e., −26 °C (Section 3.8). This low GSHP power resulted in the lower DHW outlet temperatures. This problem could overlook by picking the short heating mode duration, i.e., 20 min instead of 30 min.

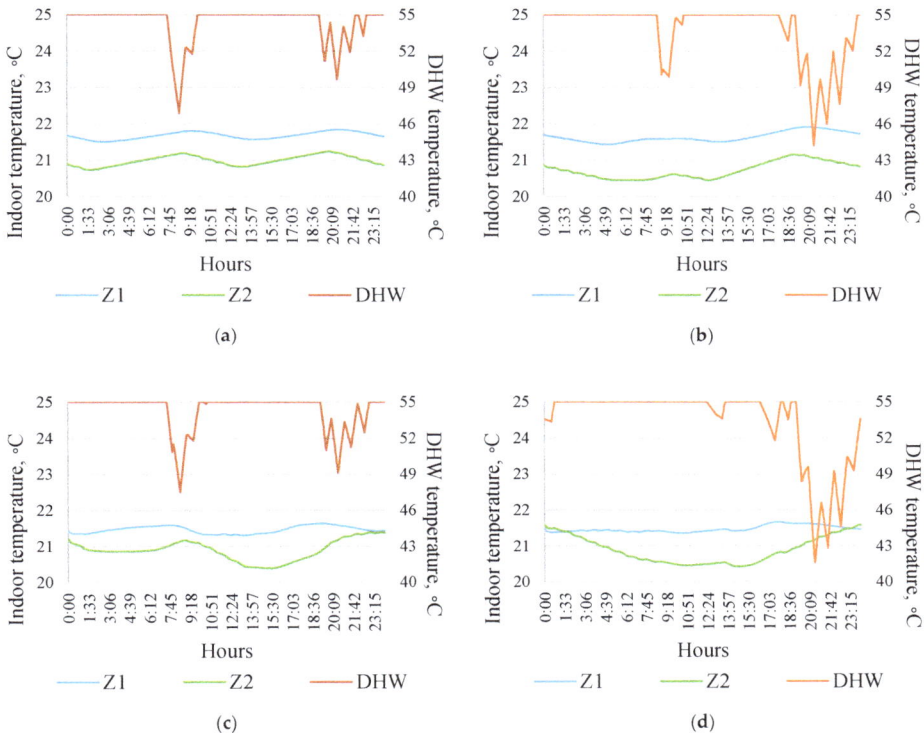

Figure 15. Indoor temperature and DHW outlet temperature for (**a**) a single-family house occupied by 4 people, (**b**) a single-family house occupied by six people, and indoor temperature and DHW supply temperature when internal gains were considered for (**c**) a single-family house occupied by 4 people, (**d**) a single-family house occupied by six people.

3.7. Thermal Mass and Heating System Effects

The effects of thermal mass and different space heating systems on indoor air temperature were evaluated in modern low energy passive building. Simulation with the LW, HW, radiator, and underfloor heating allowed us to analyze the effect of thermal mass and indoor heating solutions. The effects of the thermal mass were not significant, and are reported in Figure 16. No significant difference in indoor temperature was found due to the structural differences, i.e., HW, LW (Figure 16a,b). In addition, radiator heating was simulated, and results showed no significant difference in indoor air temperature between underfloor heating and radiant floor heating (Figure 16). The difference likely could not be seen because the HP periodic operation interval was so small (30 min) so that only the first centimeters of the structure were active. Active thermal masses were very limited because of the 30 min interval of heating.

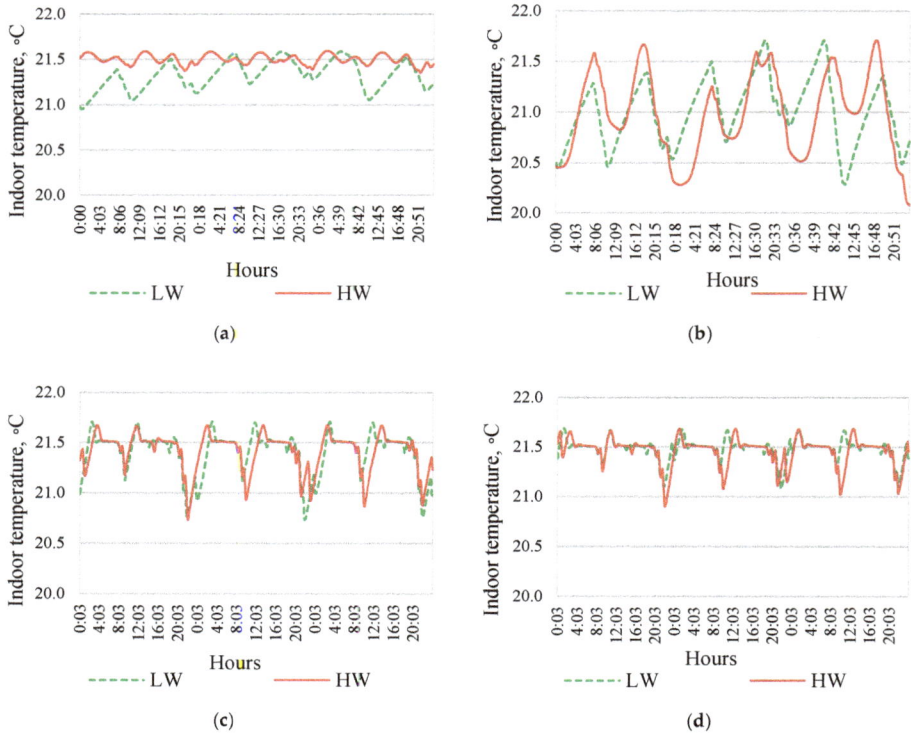

Figure 16. Indoor temperature at (**a**) the ground floor with floor heating, (**b**) the first floor with floor heating, (**c**) the ground floor with radiator heating, and (**d**) the first floor with radiator heating.

3.8. Equations for GSHP Power

The simulated results were used to set up the heating power equations. The simulated GSHP powers were obtained for different occupant groups, alongside with internal heat gains, building that were built according to new and old building regulations of 1976, and at different design temperatures. Simulated power refers to the dynamic simulation results. The empirical equations for GSHP power estimation were developed based on simulated results, which were tested for different scenarios. Total design power of GSHP can be estimated with Equation (11) for the case of without internal heat gains. However, internal heat gains cut-off a good percentage of space heating demand, which may be taken into account for the system's power sizing in low energy buildings, Equation (12). The simulated and predicted GSHP powers are shown in Figure 17.

$$P = \varnothing + nC \tag{11}$$

$$P = \varnothing + nC - 1.05 \left(A\, L_C\, L_u + A\, A_C\, A_u + n\, O_T\, O_u \right) \tag{12}$$

where, P, GSHP power (kW), $\varnothing$, total heat losses from building (kW), C, constant ($C = 0.21$ for case of 3 people DHW usage profile, where daily consumption rate of 47 $\frac{L}{person.day}$), n, number of people in house ($n = 3, 4, 5, 6$), A, floor area (m²), L_u, usages rate of lighting (dimensionless), L_C, unit load for lighting ($\frac{W}{m^2}$), A_u, usages rate of appliances (dimensionless), A_C, unit load for appliances ($\frac{W}{m^2}$), O_u, usages rate of occupancy (dimensionless), O_T, body heat loss ($\frac{W}{person}$).

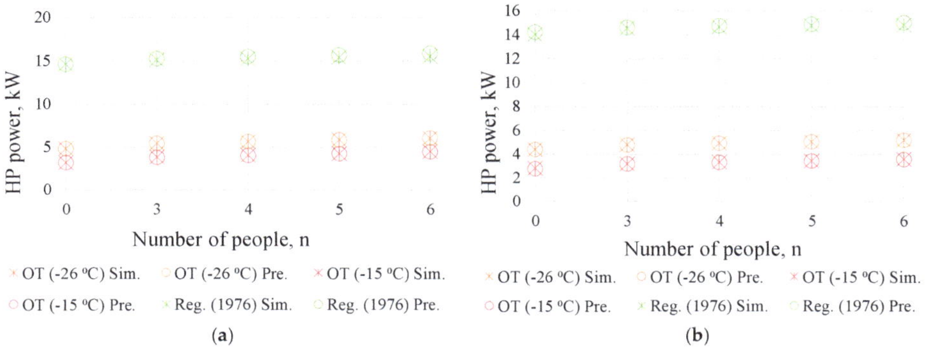

Figure 17. Simulated and predicted GSHP powers (**a**) for space and DHW heating without consideration of internal gains, (**b**) for space and DHW heating with consideration of internal gains. (OT (−26 °C) Sim. refers to simulated results with an outdoor temperature of −26 °C, OT (−26 °C) Pre. refers to predicted results with an outdoor temperature of −26 °C, Reg. (1976) Sim. refers to simulated results for a building that built according to old building regulations of 1976).

The predicted GSHP powers, obtained from empirical Equations (11) and (12), showed a good agreement with the simulated results. The predicted results varied 0%–2.2% compared to the simulated values for both cases (Figure 17a,b). The required GSHP power, with internal heat gains, reduced the total heating powers by 3%–19%. Higher effect of internal heat gains was found in new modern buildings compared to old buildings. Besides, the effects were more visible for buildings in Strasbourg (design temperature of −15 °C) compared to the Finnish buildings (design temperature of −26 °C). In the case of a house occupied by three people, the internal heat gain effect (the reduction of 0.63 kW) exactly compensated for the effect of the DHW (an increase of 0.62 kW). Furthermore, DHW heating power accounted for 21%–41% of the total heating power in the modern building with a design temperature of −15 °C and 13%–26% with a design temperature of −26 °C. The contributions of DHW heating in the old building were only in between 3% and 9%.

The performance of equations was tested for cases of different daily, and peak DHW, uses. The GSHP power for different daily DHW use and hourly peaks can be calculated by using Equations (13) and (14), respectively.

$$\frac{C_1}{Q_1} = \frac{C_2}{Q_2} \tag{13}$$

$$\frac{C_1}{P_1} = \frac{C_2}{P_2} \tag{14}$$

where, Q_1, DHW usage for given reference case ($\frac{L}{person.day}$), C_1, constant for given reference case (dimensionless), Q_2, new DHW's usage rate ($\frac{L}{person.day}$), C_2, constant (dimensionless) (constant obtained for case of new DHW's usage rate or new DHW's peak demand), P_1, peak demand for given reference case ($\frac{L}{person.hour}$), P_2, new DHW's peak demand ($\frac{L}{person.hour}$).

With this GSHP control system, C_1 was equivalent to 0.21 for a case of three people DHW profiles, where reference case had a daily usage rate of 47 L/person/day [30] and a peak demand of 4.77 L/person/h [5]. The obtained value of C_2 was used in Equations (11) and (12) in order to estimate the GSHP power. This study obtained the GSHP powers for daily usage rates of 40, 47, 55 L/person and for daily peaks of 4.77, 8.91 L/person/hour, as shown in Figure 18.

The predicted GSHP powers had shown a good agreement with the simulated results. For the case of different daily consumptions, the variation was found to be 0%–2% compared to the simulated results. Besides, the developed equations predicted the GSHP powers more precise for the case of different usage peaks. The impact of internal heat gains is well visible in Figure 18. The cutoff powers of GSHP were estimated at 11%–14% for cases with considering the internal heat gains.

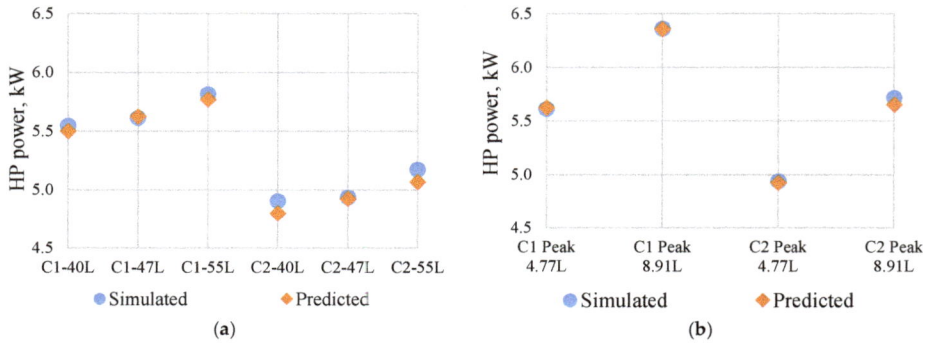

Figure 18. Simulated and predicted heat pump powers (**a**) for different DHW usage rate (L/person/day) (**b**) for different DHW usage peaks based on different DHW hourly profiles (C1 case refers without considering internal heat gains, C2 case refers with considering internal heat gains).

4. Conclusions

This study developed an alternate operational control system for GSHP, which was applied to determine the total space and DHW heating power equations at design temperature for single-family houses. Data obtained from previous studies such as a daily average of DHW use, hourly DHW profile, occupancy profile, appliances profiles, lighting profiles were used in the whole building simulation model. The performance of the developed control model was tested against measured data. Reference building properties, different DHW usage rates and design temperatures of −26 °C and −15 °C were used in order to size the total design power of GSHP for modern and old single-family houses. The following conclusions can be drawn:

- The alternate operational principle of GSHP was implemented in IDA-ICE whole building simulation model. The GSHP's control system was modeled with DHW priority switching from SH to DHW heating mode based on demand-side signals and a given duration of the heating period.
- The control of heat deficiency was modeled based on DM accounting. The default DM value (Section 3.1) and heating mode duration were user-defined, which enabled us to fit the GSHP operation for new and old buildings in different climates.
- Simulated GSHP system with a 200 L storage tank resulted in a 13%–26% power reduction compared to the calculation of the same system with EN standards, which required separate space heating and DHW power calculation.
- The periodic operation utilized the thermal mass of the building, and the sensitivity analyses with light and heavyweight buildings revealed the same effect because of the very short heating cycle of 30 min. With room a temperature setpoint of 21 °C, room temperatures only slightly decreased, so that indoor thermal comfort of Category II in a modern house and Category III in an old house were achieved. Besides, the DHW outlet temperature was in an acceptable limit.
- The contribution of DHW heating formed 2%1–41% of total heating power in a modern building with a design temperature of −15 °C and 13%–26% with a design temperature of −26 °C.
- Internal heat gains reduced the GSHP powers by 3%–19% when taken into account in the simulation. In the case of a house occupied by three people, the internal heat gain effect (the reduction of 0.63 kW) exactly compensated for the effect of the DHW (an increase of 0.62 kW). It brings

a question, should the effect of internal heat gains be taken into account for heat pump sizing in low energy buildings?

- The total heating power equation was obtained in the form of the space heating power at a design temperature plus 0.21 kW of the DHW heating power per person. The equation was tested in modern and old single-family houses with design temperatures of −15 °C and −26 °C where the deviations remained between 0%–2.2%. It was demonstrated that the equations could be easily modified for different DHW daily usages or peak usages. For that purpose, other equations were provided.

Author Contributions: K.A. wrote the paper and conceived, designed, and performed the simulations. J.F. conceived and developed the initial control model. J.K. was the principal investigator responsible for the study design.

Funding: This research received no external funding.

Acknowledgments: This research was supported by the K.V. Lindholms Stiftelse foundation and by Estonian Centre of Excellence in Zero Energy and Resource Efficient Smart Buildings and Districts, ZEBE, grant 2014-2020.4.01.15-0016 funded by the European Regional Development Fund.

Conflicts of Interest: The authors declare no conflict of interest.

Abbreviation

A	Floor area: m^2
A_C	Unit load for appliances, $\frac{W}{m^2}$
A_{env}	Envelope area, m^2
A_{HE}	Effective surface area of the heat exchanger, m^2
A_u	Usages rate of appliances, dimensionless
C	Constant, dimensionless
C_w	Water heat capacity, $\frac{kJ}{kgK}$
C_1	Constant for given reference case, dimensionless
C_2	Constant, dimensionless
DM	degree minutes, °C·min
dT	Difference between the mean surface temperature of a radiator and the room's air temperature, °C
F_m	Monthly DHW consumption factor, dimensionless
K	Radiator constant, dimensionless
L	Radiator length, m
L_C	Unit load for lighting, $\frac{W}{m^2}$
L_u	Usages rate of lighting, dimensionless
m_{Sto}	Water mass in the storage volume, kg
n	Number of occupants, dimensionless
n_r	Power law coefficient, which depends on the radiator type and width, $\frac{W}{m \cdot °C}$
O_u	Usages rate of occupancy, dimensionless
O_T	Body heat losses, $\frac{W}{person}$
P	GSHP's power, kW
P_r	Heat flux from the water, W
P_1	Peak demand for a given reference case, $\frac{L}{person.\ hour}$
P_2	New DHW peak demand, $\frac{L}{person.\ hour}$
Q_1	DHW consumption for a given reference case, $\frac{L}{person.\ day}$
Q_2	New DHW consumption rate, $\frac{L}{person.\ day}$
q_v	Infiltration rate, $\frac{m^3}{s}$
q_{50}	Air leakage rate of building's envelope, $\frac{m^3}{h.m^2}$
t	Time, minutes
t_n	Elapsed time, minutes
T_a	Actual flow temperature, °C

T_c	Cold water temperature, $°C$
$T_{ch,HG}$ (t)	Supply temperature from a heat generator at time t, $°C$
T_s	Flow temperature's set point, $°C$
$T_{Sto,m}$ (t)	Mean water temperature in a storage tank at time t, $°C$
$T_{Sto,m,t0}$	Mean water temperature of the storage tank at the time the reheating is switched on, $°C$
$T_{W, draw}$	Water withdrawn temperature, $°C$
U	Thermal transmittance, $\frac{W}{m^2 K}$
U_{HE}	Thermal transmittance of the heat exchanger, $\frac{W}{m^2 K}$
V_D	Design flow rate, $\frac{l}{s}$
V_d	DHW consumption by each occupant, $\frac{L}{person. \, day}$
$V_{T,d}$	Total DHW consumption by a single-family house, $\frac{L}{day}$
x	Factor that based on the building height, dimensionless
$\varnothing$	Total heat losses from the building, kW
Φ_{eff}	Effective power for water heating, kW
$\Phi_{HL,i}$	Design heat losses of the heated space of i, W
$\Phi_{hu,i}$	Additional heating up power for the heated space of i, W
Φ_I	Heat losses caused by infiltration, W
Φ_N	Nominal power for a heat generator, kW
Φ_{space}	Space heating power, W
Φ_T	Sum of transmission heat losses through building's envelope, W
$\Phi_{T,i}$	Sum of design transmission heat losses through building's envelope of i, W
Φ_V	Heat losses caused by ventilation, W
$\Phi_{V,i}$	Design heat losses caused by ventilation of i, W
$\Phi_{w, dis}$	Distribution heat losses, kW
$\Phi_{w, Sto}$	Heat losses from a storage tank at time t, kW
ρ_w	Water density, $\frac{kg}{l}$
τ	Time constant of storage tank during the loading period, minute

References

1. Ahmed, K.; Pylsy, P.; Kurnitski, J. Domestic hot water profiles for energy calculation in Finnish residential buildings. In Proceedings of the Young Researcher International Conference on Energy Efficiency, Linz, Austria, 25–26 February 2015.
2. Ahmed, K.; Pylsy, P.; Kurnitski, J. Hourly consumption profiles of domestic hot water for Finnish apartment buildings. In Proceedings of the CLIMA 2016 12th REHVA World Congress, Aalborg, Denmark, 22–25 May 2016.
3. Ferrantelli, A.; Ahmed, K.; Pylsy, P.; Kurnitski, J. Analytical modelling and prediction formulas for domestic hot water consumption in residential Finnish apartments. *Energy Build.* **2017**, *143*, 53–60. [CrossRef]
4. Niemela, T.; Manner, M.; Laitinen, A.; Sivula, T.M.; Jokisalo, J.; Kosonen, R. Computational and experimental performance analysis of a novel method for heating of domestic hot water with a ground source heat pump system. *Energy Build.* **2018**, *161*, 22–40. [CrossRef]
5. Ahmed, K.; Pylsy, P.; Kurnitski, J. Hourly consumption profiles of domestic hot water for different occupant groups in dwellings. *Sol. Energy* **2016**, *137*, 516–530. [CrossRef]
6. Rodriguez-Hidalgo, M.C.; Rodriguez-Aumente, P.A.; Lecuona, A.; Legrand, M.; Ventas, R. Domestic hot water consumption vs. solar thermal energy storage: The optimum size of the storage tank. *Appl. Energy* **2012**, *97*, 897–906. [CrossRef]
7. Yrjölä, J.; Laaksonen, E. Domestic hot water production with ground source heat pump in apartment buildings. *Energies* **2015**, *8*, 8447–8466. [CrossRef]
8. Hervas-Blasco, E.; Pitarch, M.; Navarro-Peris, E.; Corberan, J.M. Optimal sizing of a heat pump booster for sanitary hot water production to maximize benefit for the substitution of gas boilers. *Energy* **2017**, *127*, 558–570. [CrossRef]
9. Stafford, A. An exploration of load-shifting potential in real in-situ heat-pump/gas-boiler hybrid systems. *Build. Serv. Eng. Res. Technol.* **2017**, *38*, 450–460. [CrossRef]

10. Bagarella, G.; Lazzarin, R.; Noro, M. Sizing strategy of on–off and modulating heat pump systems based on annual energy analysis. *Int. J. Refrigerat.* **2016**, *65*, 183–193. [CrossRef]

11. Hakamies, S.; Hirvonen, J.; Jokisalo, J.; Knuuti, A.; Kosonen, R.; Niemela, T.; Paiho, S.; Pulakka, S. Heat Pumps in Energy and Cost Efficient Nearly Zero Energy Buildings in Finland. Teknologian tutkimuskeskus VTT Oy, 2015. Available online: https://www.vtt.fi/inf/pdf/technology/2015/T235.pdf (accessed on 2 June 2019).

12. Rawlings, R.H.D.; Sykulski, J.R. Ground source heat pumps: A technology review. *Build. Serv. Eng. Res. Technol.* **1999**, *20*, 119–129. [CrossRef]

13. Rivoire, M.; Casasso, A.; Piga, B.; Sethi, R. Assessment of energetic, economic and environmental performance of ground-coupled heat pumps. *Energies* **2018**, *11*, 1941. [CrossRef]

14. Li, H.; Xu, W.; Yu, Z.; Wu, J.; Sun, Z. Application analyze of a ground source heat pump system in a nearly zero energy building in China. *Energy* **2017**, *125*, 140–151. [CrossRef]

15. El-Baz, W.; Tzscheutschler, P.; Wagner, U. Experimental study and modeling of ground source heat pumps with combi-storage in buildings. *Energies* **2018**, *11*, 1174. [CrossRef]

16. Dong, B.; Lam, P.K. A real-time model predictive control for building heating and cooling systems based on the occupancy behavior pattern detection and local weather forecasting. *Build. Simul.* **2014**, *7*, 89–106. [CrossRef]

17. Arabzadeh, V.; Alimohammadisagvand, B.; Jokisalo, J.; Siren, K. A novel cost-optimizing demand response control for a heat pump heated residential building. *Build. Simul.* **2018**, *11*, 533–547. [CrossRef]

18. Bouheret, S.; Bernier, M. Modelling of a water-to-air variable capacity ground-source heat pump. *Taylor Francis* **2017**, *11*, 283–293. [CrossRef]

19. Salque, T.; Marchio, D.; Riederer, P. Neural predictive control for single-speed ground source heat pumps connected to a floor heating system for typical French dwelling. *Build. Serv. Eng. Res. Technol.* **2014**, *35*, 182–197. [CrossRef]

20. FprEN 12831-1. *Energy Performance of Buildings—Method for Calculation of the Design Heat Load—Part 1: Space Heating Load, Module M3-3*; EPB Center: Chattanooga, TN, USA, 2016.

21. FprEN 12831-3. *Energy Performance of Buildings—Method for Calculation of the Design Heat Load—Part 3: Domestic Hot Water Systems Heat Load and Characterisation of Needs, Module M8-2, M8-3*; EPB Center: Chattanooga, TN, USA, 2016.

22. FprEN 15316-4-2. *Energy Performance of Buildings—Method for Calculation of System Energy Requirements and System Efficiencies—Part 4-2: Space Heating Generation Systems, Heat Pump Systems, Module M3-8-2, M8-8-2*; EPB Center: Chattanooga, TN, USA, 2016.

23. Energiatehokkuus. Rakennuksen Energiankulutuksen ja Lammitystehontarpeen Laskenta. 2018. Available online: http://energiatodistus.motiva.fi/midcom-serveattachmentguid-1e842192ba36dce421911e8bb94c9f17026c02fc02f/ohje-rakennuksen_energiankulutuksen_ja_la--mmitystehontarpeen_laskenta_20-12-2017.pdf (accessed on 2 June 2019).

24. EN 15251. *Indoor Environmental Input Parameters for Design and Assessment of Energy Performance of Buildings Addressing Indoor Air Quality, Thermal Environment, Lighting and Acoustics*; European Commission: Brussels, Belgium, 2007.

25. Maivel, M.; Kurnitski, J. Heating system return temperature effect on heat pump performance. *Energy Build.* **2015**, *94*, 71–79. [CrossRef]

26. NIBE F1245. *Ground Source Heat Pump with Integrated Water Heater*; NIBE: San Francisco, CA, USA, 2018.

27. Finnish Building Code. Ymparistominsterion asetus uuden rakennuksen energiatehokkuudesta. 2017. Available online: https://www.finlex.fi/fi/laki/alkup/2017/20171010 (accessed on 2 June 2019).

28. Ahmed, K.; Akhondzada, A.; Kurnitski, J.; Olesen, B. Occupancy schedules for energy simulation in New prEN16798-1 and ISO/FDIS 17772-1 standards. *Sustain. Cities Soc.* **2017**, *35*, 134–144. [CrossRef]

29. Ahmed, K.; Kurnitski, J.; Olesen, B. Data for occupancy internal heat gain calculation in main building categories. *Data Brief.* **2017**, *15*, 1030–1034. [CrossRef] [PubMed]

30. Ahmed, K.; Pylsy, P.; Kurnitski, J. Monthly domestic hot water profiles for energy calculation in Finnish apartment buildings. *Energy Build.* **2015**, *97*, 77–85. [CrossRef]

31. Achermann, M.; Zweifel, G. *Radtest Radiant Heating and Cooling Test Cases a Report of Task 22, Subtask C, Building Energy Analysis Tools Comparative Evaluation Tests*; Technical Report; International Energy Agency: Luzern, Switzerland, April 2003.

32. Ahmed, K.; Sistonen, E.; Simson, R.; Kurnitski, J.; Kesti, J.; Lautso, P. Radiant panel and air heating performance in large industrial building. *Build. Simul.* **2018**, *11*, 293–303. [CrossRef]

33. Kurnitski, J.; Ahmed, K.; Simson, R.; Sistonen, E. Temperature distribution and ventilation in large industrial halls. In Proceedings of the 9th Windsor Conference: Making Comfort Relevant, Cumberland Lodge, UK, 7–10 April 2016; pp. 340–348.

34. Travesi, J.; Maxwell, G.; Klaassen, C.; Holtz, M. *Empirical Validation of Iowa Energy Resource Station Building Energy Analysis Simulation Models, Report of Task 22, Subtask Building Energy Analysis Tools*; International Energy Agency—Solar Heating and Cooling Programme, Technical Report; International Energy Agency: Luzern, Switzerland, June 2001.

35. Kurnitski, J.; Saari, A.; Kalamees, T.; Vuolle, M.; Niemelä, J.; Tark, T. Cost optimal and nearly zero (nZEB) energy performance calculations for residential buildings with REHVA definition for nZEB national implementation. *Energy Build.* **2011**, *43*, 3279–3288. [CrossRef]

36. Ivanov, J. *Metalli TN 3 Büroohoone, Lisainvesteeringu ja Energiatõhususe Analüüs*; Tallinn University of Technology: Tallinn, Estonia, 2016.

energies

MDPI

Article

Development of a Space Heating Model Suitable for the Automated Model Generation of Existing Multifamily Buildings—A Case Study in Nordic Climate

Lukas Lundström [1,2,*], Jan Akander [3] and Jesús Zambrano [1]

[1] School of Business, Society and Engineering, Mälardalen University, 72123 Västerås, Sweden; jesus.zambrano@mdh.se
[2] Eskilstuna Kommunfastighet AB, 63005 Eskilstuna, Sweden
[3] Division of Building, Energy and Environment Technology, Department of Technology and Environment, University of Gävle, 80176 Gävle, Sweden; jan.akander@hig.se
* Correspondence: lukas.lundstrom@mdh.se

Received: 29 November 2018; Accepted: 30 January 2019; Published: 2 February 2019

Abstract: Building energy performance modeling is essential for energy planning, management, and efficiency. This paper presents a space heating model suitable for auto-generating baseline models of existing multifamily buildings. Required data and parameter input are kept within such a level of detail that baseline models can be auto-generated from, and calibrated by, publicly accessible data sources. The proposed modeling framework consists of a thermal network, a typical hydronic radiator heating system, a simulation procedure, and data handling procedures. The thermal network is a lumped and simplified version of the ISO 52016-1:2017 standard. The data handling consists of procedures to acquire and make use of satellite-based solar radiation data, meteorological reanalysis data (air temperature, ground temperature, wind, albedo, and thermal radiation), and pre-processing procedures of boundary conditions to account for impact from shading objects, window blinds, wind- and stack-driven air leakage, and variable exterior surface heat transfer coefficients. The proposed model was compared with simulations conducted with the detailed building energy simulation software IDA ICE. The results show that the proposed model is able to accurately reproduce hourly energy use for space heating, indoor temperature, and operative temperature patterns obtained from the IDA ICE simulations. Thus, the proposed model can be expected to be able to model space heating, provided by hydronic heating systems, of existing buildings to a similar degree of confidence as established simulation software. Compared to IDA ICE, the developed model required one-thousandth of computation time for a full-year simulation of building model consisting of a single thermal zone. The fast computation time enables the use of the developed model for computation time sensitive applications, such as Monte-Carlo-based calibration methods.

Keywords: energy performance modeling; gray box; satellite-based solar radiation data; meteorological reanalysis data; ISO 52016-1

1. Introduction

Building energy performance modeling is essential for energy planning, management, and efficiency. To improve energy efficiency of building stocks, knowledge on how each building is currently performing is needed, often referred to as baselines. These baselines can for example be used to compare and rank building performance, estimate energy and cost saving potential of energy efficiency measures, and detect faults when systems are malfunctioning. Energy use of a building depends on aspects such as physical and thermal properties, climate, control and operational schemes, occupants' presence, and behavior. Information regarding these aspects is often uncertain and scattered, sometimes missing altogether, and establishing the energy performance of buildings can therefore be a tedious task [1,2].

Building energy modeling is often categorized into two major approaches: law-driven (forward) and data-driven (inverse) [1,3]. Law-driven models apply a given set of physical laws that govern the system. Conversely, data-driven models use system behavior as a predictor for system performance. An advantage of data-driven models is that they generally require less input, while law-driven models can offer more flexibility regarding the fine adjustment of a building's subsystems and components [1]. Examples of building energy simulation software that utilize a law-driven approach include EnergyPlus [4] and IDA ICE (https://www.equa.se/en/ida-ice). These types of software calculate building energy consumption based on detailed building and environmental information such as climate, operation schedules, building geometrics, and the shading/sheltering impact of the surroundings [1,2]. Examples of using law-driven simulation software for parameter identification can be found in Kang and Krarti [5], Tian et al. [6]. Data-driven modeling approaches have gained much research attention in recent years. The review article by Amasyali and El-Gohary [7] identified more than 50 articles using data-driven approaches. Weaknesses with data-driven approaches are that these may not perform well outside their training range and that statistical models yield limited knowledge of under-laying aspects that govern energy use. The main weaknesses of law-driven approaches are that they require detailed input and tend to model how the building is designed to operate and not how it actually operates.

Gray-box modeling combines law-driven and data-driven approaches, thereby leveraging the advantages and minimizing the disadvantages of both approaches, see for example Amasyali and El-Gohary [7], Bacher and Madsen [8], Boodi et al. [9]. In gray-box models, internal parameter and equations are physically interpretable and thus provide knowledge about the modeled building. Gray-box models are calibrated to better match a building's actual operation performance, but can also provide physically reasonable results outside the training data range. The authors in Boodi et al. [9] recognized a need for more research efforts regarding model structures for gray box models for residential buildings.

The ISO 52016-1:2017 [10] standard, which replaces the older ISO 13790:2008 [11] standard, presents a set of calculation methods for a building energy needs and internal air temperatures. The hourly method described in the standard proposes a system of linear equations that model heat transfer through opaque and transparent components of the envelope and air exchange between the internal and external environments. The calculation result is hourly internal air and component temperatures and heating and cooling loads. Each construction component (e.g., roof, windows, and walls) is modeled as serially connected RC (resistance and capacitance) thermal networks. Compared to the old ISO 13790:2008 [11] standard, the new hourly method is a much closer representation of algorithms used in whole building simulation software, such as IDA ICE or EnergyPlus. The simplistic hourly method of the now deprecated ISO 13790:2008 [11] has been employed and explored in several scientific works, for example [12–15]. To the best of our knowledge, there has not yet been any scientific works employing the newer ISO 52016-1:2017 [10] standard.

Law-driven approaches require detailed input regarding the physical properties of the building. Most existing buildings were built before the era of modern BIM (building information modeling) and thus lack such information in digitized form. However, 3D shape representation of existing building can be achieved through the use of aerial images or LiDAR (light detection and ranging) or a combination of both [16]. Such 3D shape representations allow for the determination of buildings' footprint, height, orientation, and surface areas. The Swedish Land Survey (Lantmäteriet) has collected LiDAR data for almost all of Sweden [17]. Parameters describing buildings' thermal properties and technical system are, however, usually not readily accessible in a digitized format. Gray-box modeling approaches have proven successful for estimating such uncertain or unknown parameters [7–9].

Energy meter readings of high resolution are required for gray-box modeling approaches. In Sweden, since 2015, energy bills have been required to be based on actual energy consumption [18]. Therefore, most Swedish district heating system operators have automatic meter reading systems installed that gather hourly or sub-hourly readings to centralized databases. Most Swedish cities have extensive district heating networks, and district heating operators have an approximately 90% share of the heating market for multifamily buildings [19]. Approximately 94% of Swedish multifamily buildings have hydronic heating systems, and approximately 94% of these have thermostatic control valves installed [20].

From a Swedish multifamily building portfolio owner's perspective (as energy strategist at Eskilstuna Kommunfastighet AB and corresponding author responsible for the energy performance of a 7000 apartment portfolio), it would be beneficial to have an energy performance model of each building—a model revealing both current and potential energy performance of the building. Such a model needs to be law-driven and detailed enough so that its parameters reveal meaningful and actionable information. However, the model also needs to be simple enough, so that basic models of existing buildings can be auto-generated from readily available data sources. It also needs to have a structure that allows calibration with actual meter readings, to ensure it represents the actual performance of the building and not only the intended/designed performance. A practical model fulfilling the above-mentioned requirements is currently missing: detailed simulation software (e.g., IDA ICE or EnergyPlus) requires too detailed input to be readily auto-generated and are too slow and complex to be readily calibrated, while the data-driven and gray-box models proposed in the literature are not detailed enough and too far from physics to reveal meaningful interpretable information. This paper presents a space heating model aiming at filling the identified gap.

Methodology and Outline

The ISO 52016-1:2017 [10] standard employs an RC network to calculate hourly internal air temperatures, hourly heating and cooling loads, and space heating/cooling needs by summing the loads over certain time periods (e.g., monthly). This article presents the development of a thermal RC network that is based on the ISO 52016-1:2017 standard. However, our implementation is further simplified by lumping building elements by type and decreasing the number of temperature nodes in opaque building elements. Since the ISO 52016-1:2017 omits the conditions of use, the standard calculates the minimum energy need when assuming perfect temperature control and no system losses. To enable the calculation of actual energy use for space heating, comparable to metered values, a typical Swedish hydronic radiator heating system model [21,22] was implemented. Moreover, the following boundary condition procedures are proposed: a procedure to acquire satellite-based solar radiation data and conversion to a suitable format; a procedure to obtain and utilize meteorological reanalysis data (air temperature, ground temperature, wind, and thermal radiation and surroundings); procedures to account for impact from shading objects, window blinds, and wind- and stack-driven air infiltration, and variable exterior surface heat transfer coefficients. The developed model (including the RC network, heating system, and pre-processing steps) is from here on called ISO14N, where "ISO" denotes that it is based on the ISO 52016-1:2017 [10] standard and "14N" denotes the 14 temperature nodes (and 14 system equations). Figure 1 shows an overview of the proposed ISO14N modeling framework: from data input to pre-processing the boundary conditions data and constructing the RC network and to the simulation procedure.

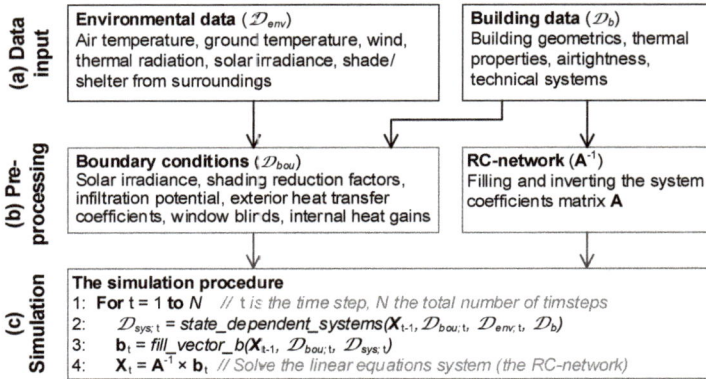

Figure 1. Process from data input (**a**) to pre-processing (**b**) and to the simulation procedure (**c**).

As a validation, the simulation results were compared with results obtained with the well established and validated (https://www.equa.se/en/ida-ice/validation-certifications) simulation software IDA ICE. The main motivation behind using IDA ICE instead of the verification case provided by the ISO 52016-1:2017 [10] was that it also enabled testing the procedures that are not derived from the standard. Although the proposed RC network is generic, the use of a continuously operating hydronic radiator heating system and a constant air flow ventilation system limits the application of the proposed model to mainly multifamily buildings in Nordic climates. Therefore, the work is delimited to energy use for space heating for continuously heated multifamily buildings. Domestic hot water use, internal heat gains, and variable air flow ventilation are left for further work. Parallel work developing the calibration procedure, beyond the scope of this article, has been conducted and has influenced the design choices. The construction of the linear equation system of the RC network, the heating system, and the simulation procedure was implemented using the probabilistic programming language Stan [23], which parses to C++ code before compiling. The pre-processing procedures and the data input handling were implemented in the R statistical language based package dplyr [24]. Code can be requested from the corresponding author. The ISO 52010:2017 [25] standard (solar irradiance at an arbitrarily tilted and oriented surface) calculation procedure was implemented in C++ and published as an open source R-package (https://github.com/lukas-rokka/solarCalcISO52010).

The main features of the proposed ISO14N model are introduced in Section 2, while the pre-processing of the boundary conditions data is presented in Section 3. Section 4 describes the required data input and used data sources. Sections 5.1–5.6, 5.8, and 5.9 compare simulation results of the developed ISO14N model with results obtained with the IDA ICE. Section 5.7 show results from the proposed shading reduction. Details on constructing the linear equation system of the RC network are given in Appendix A.

2. The ISO14N Space Heating Model

In this paper, a one-zone model representation for the whole building is illustrated. Figure 2 displays the proposed one-zone RC network. All external walls *ew* are modeled as a lumped 3-node element, all window glazing *gl* are modeled as a lumped 2-node element, and the external roof *rf* and the ground floor *gf* are modeled with 3-node elements. The remaining internal mass *im* (internal walls, intermediate floors, and adiabatic external walls) are represented with a 2-node element. Thus, the thermal network consists of 14 unknown temperature nodes (including the internal air node θ_{int}). The thermal model can be represented as a system of linear equations:

$$\mathbf{A} \times \mathbf{X} = \mathbf{b} \tag{1}$$

where **A** is a 14×14 square matrix holding system coefficients, **X** is the state vector holding the 14 node temperatures, and **b** is a vector holding determined terms (node temperatures from the previous time step and known boundary condition values). Appendix A shows the details of the model matrices. For every time step t, the system of equations needs to be solved to compute the new states:

$$X_t = A/b_t = A^{-1} \times b_t. \tag{2}$$

Matrix inversion is computationally intensive. However, if matrix **A** only consists of constants, this computation step is only required once before the iterative simulation procedure. Vector **b** holds terms that can be treated as pre-processed data input (e.g., solar irradiance) and terms that need to be calculated at each simulation step due to dependence on the state of the node temperatures (e.g., thermal storage in nodes). Figure 1c shows the simulation procedure in pseudo code.

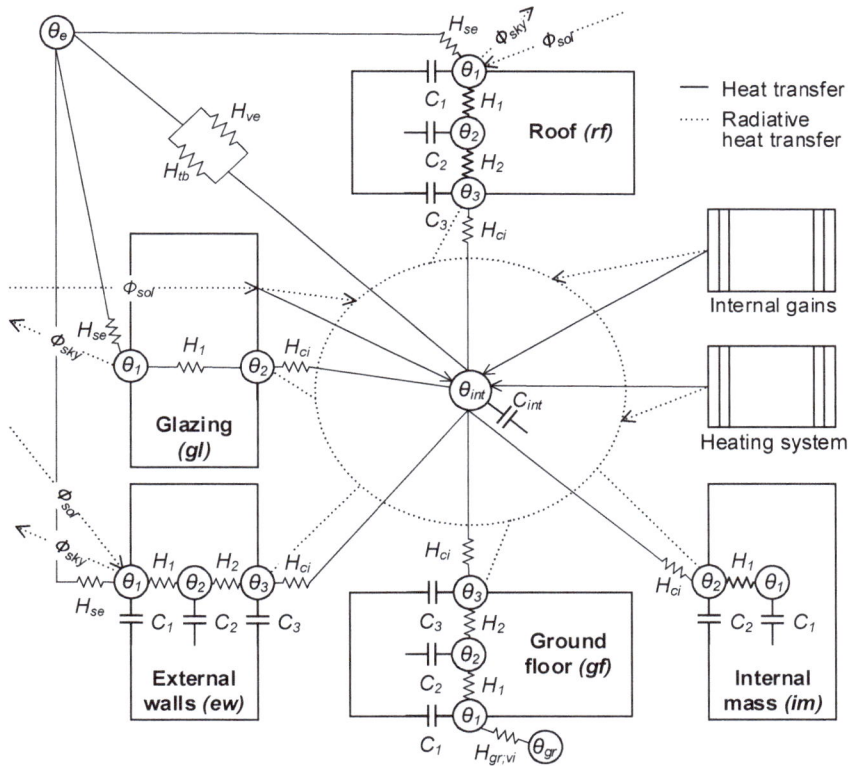

Figure 2. An RC network representation of the proposed ISO14N model.

2.1. Geometrical and Thermal Properties

In this paper, the calculation is conducted normalized to per square meter floor (a deviation from the standard). One benefit is that geometrical data can be given as ratios (r) between the total interior surface area of each building element type and the total floor area (unit m_s/m_{fl}). These ratios can be seen as weighting factors describing how large an impact one type of building element has on the average thermal balance of the whole building. It is straightforward to estimate these ratios from readily available building data:

$$r_{rf} = r_{gf} = 1/N_{fl}$$
$$r_{ev} = P \cdot H_{fl} \cdot N_{fl}/A_{fl} - r_{gl} \tag{3}$$
$$r_{im} = (2 - 2/N_{fl}) + 1.5$$

where N_{fl} is the number of floors of the building, A_{fl} is the total floor area of the building, P is the building perimeter, and H_{fl} is the average internal floor height. The term $(2 - 2/N_{fl})$ in r_{im} calculates the surface area of internal floors and ceilings per total floor area and will result in a value between 0 and 2, depending on the number of floors. The additional 1.5 constant represents the internal walls, and it is based on the old ISO 13790:2008 [11] standard (where total interior surface area per total floor area is given as 4.5). The glazing-to-floor area ratio r_{gl} is a property that is often regulated in standards and building codes, typically ranging between 0.1 and 0.2 for Swedish multi-family buildings.

Weighted surface heat transfer coefficients for each building element type are as follows: $el = [rf, ew, gl, im, gf]$, which are abbreviations of [roof, external walls, glazing, internal mass, ground floor]:

$$
\begin{aligned}
H_{se;rf;t} &= r_{rf} \cdot (h_{ce;rf;t} + (1 - F_{sky;hor}) \cdot 4.14), & H_{ci;rf} &= r_{rf} \cdot 0.7, & H_{ri;rf} &= r_{rf} \cdot 5.13 \\
H_{se;ew;t} &= r_{ew} (\cdot h_{se;ew;t} + (1 - F_{sky;ver}) \cdot 4.14), & H_{ci;ew} &= r_{ew} \cdot 2.5, & H_{ri;ew} &= r_{ew} \cdot 5.13 \\
H_{se;gl;t} &= r_{gl} \cdot (h_{se;gl;t} + (1 - F_{sky;ver}) \cdot 4.14), & H_{ci;gl} &= r_{gl} \cdot 2.5, & H_{ri;gl} &= r_{gl} \cdot 5.13 \\
& & H_{ci;im} &= r_{im} \cdot 2.5, & H_{ri;im} &= r_{im} \cdot 5.13 \\
H_{se;gf} &= r_{gf} \cdot h_{gr;vi}, & H_{ci;gf} &= r_{gf} \cdot 5.0, & H_{ri;gf} &= r_{gf} \cdot 5.13
\end{aligned}
\tag{4}
$$

where $H_{se;el;t}$ [W/(K $\cdot$ m$_{fl}^2$)] represents normalized variable heat transfer coefficients (convective + radiative) between the exterior surface of the elements and external air at time step t, $h_{ce;el;t}$ [W/(K $\cdot$ m$_s^2$)] represents exterior surface convective heat transfer coefficients at time step t (see Section 3.1), $F_{sky;hor}$ and $F_{sky;ver}$ are the sky view factors for horizontal and vertical elements respectively, h_{re} is the surface exterior radiative heat transfer coefficient, $h_{gr;vi}$ is the heat transfer coefficient towards a virtual ground layer (see Equation (8)), $H_{ci;el}$ represents weighted convective heat transfer coefficients between the interior surface side of the elements and the internal air node, and $H_{ri;el}$ represents weighted radiative heat transfer coefficients of interior surface nodes. The constants are taken from ISO 52016-1:2017 [10] (Table 25) and describe conventional convective and radiative surface heat transfer coefficients for interior or exterior surfaces oriented upwards, horizontally, or downward.

In ISO 52016-1:2017 [10], opaque building elements (walls, roof, etc.) are by default described as 5-node and window glazing as 2-node elements. The distribution of the thermal resistances are slightly weighted towards the center of the building elements, while heat capacity nodes are weighted depending on chosen class (mass concentrated externally, internally, inside, or equally). In this paper, 3-node elements are used for opaque elements; therefore, there is less flexibility in distributing the thermal properties over the elements. The thermal resistance is equally distributed:

$$
\begin{aligned}
H_{1;rf} &= H_{2;rf} = 2/R_{rf}, & H_{1;gl} &= 1/R_{gl} \\
H_{1;ew} &= H_{2;ew} = 2/R_{ew}, & H_{1;im} &= 1/R_{im}
\end{aligned}
\tag{5}
$$

where the floor area normalized thermal resistance R [m$_{fl}^2$ $\cdot$ °C/W] is calculated as

$$R_{el} = 1/(r_{el} \cdot U_{el}) - 1/(H_{ci;el} + H_{ri;el}) - 1/H_{se;el} \tag{6}$$

where U_{el} [W/(°C $\cdot$ m$_s^2$)] is the thermal transmittance of building element *el*. The heat capacity of 3-node building elements roof *rf* and external walls *ew* is distributed according to Table 1. The table is based on the ISO 52016-1:2017 [10] but altered to suit 3-node elements. Table B.14 of ISO 52016-1:2017 [10] specifies κ_m values according to five classes (from very light to very heavy, 14–70 Wh/(°C $\cdot$ m$_s^2$)). The simulation procedure is conducted with normalized values, which are acquired by multiplying with surface ratios: $C_{el} = r_{el} \cdot \kappa_{m;el}$.

Table 1. Node heat capacity distribution for the the 3-node building elements roof rf and external walls ew. Class I: concentrated on interior side; Class E: concentrated on exterior side; Class IE: divided on interior and exterior side; Class D: equally distributed; Class M: inside/centered.

Node	Class I	Class E	Class IE	Class D	Class M
1	0.10	0.50	0.40	0.33	0.10
2	0.40	0.40	0.20	0.33	0.80
3	0.50	0.10	0.40	0.33	0.10

The internal mass im building element is configured for 24 h thermal admittance cycles:

$$H_{1;im} = 1/1.0, \qquad \kappa_{1;im} = 0.85 \cdot \kappa_{m;im}, \qquad \kappa_{2;im} = 0.15 \cdot \kappa_{m;im}. \tag{7}$$

The ground floor element needs to account for effects of the ground, and its thermal resistance and heat capacity are distributed as follows:

$$H_{2;gf} = 2/R_{gf}, \qquad H_{1;gf} = 1/(R_{gf}/2 + R_{gr}), \qquad H_{se;gf} = r_{gr} \cdot h_{gr;vi}$$
$$\kappa_{3;gf} = \kappa_{m;gf}/2, \qquad \kappa_{2;gf} = \kappa_{m;gf}/2, \qquad \kappa_{1;gf} = \kappa_{gr} \tag{8}$$

where R_{gr} and κ_{gr} represent the thermal resistance and heat capacity of a 0.5 m thick soil layer (0.25 m$_s^2 \cdot$ K/W and 280 Wh/(K $\cdot$ m$_s^2$) are used as default values), R_{gf} is the thermal resistance of the ground floor including the effects of the ground, and $h_{gr;vi}$ is the thermal transmittance of a virtual ground layer. R_{gf} and $h_{gr;vi}$ are calculated according to ISO 13370:2017 [26].

2.2. Hydronic Heating System

Heat dissipation from, in Sweden typically, a hydronic heating system with radiator panels, at time step t, can be approximated with the following non-linear equation [21,22]:

$$\phi_{hyd;t} = H_{hyd} \cdot (\theta_{hyd;lmtd;t})^{n_{hyd}} \cdot u_{trv;t} \tag{9}$$

where H_{hyd} is the per floor square meter normalized radiator constant, n_{hyd} is the radiator exponent (this depends on the design and type, but a value of 1.3 is commonly used for typical Swedish radiator panels [21,22]), $u_{trv;t}$ is the control signal from the local thermostatic radiator valve(s) (TRV), and $\theta_{hyd;lmtd;t}$ is the logarithmic mean temperature difference between the radiator and the internal air temperature, which is calculated as

$$\theta_{hyd;lmtd;t} = \frac{\theta_{hyd;sup;t} - \theta_{hyd;ret;t}}{\ln\left(\theta_{hyd;sup;t} - \theta_{int;t-1}\right) - \ln\left(\theta_{hyd;ret;t} - \theta_{int;t-1}\right)} \tag{10}$$

where $\theta_{hyd;sup;t}$ and $\theta_{hyd;ret;t}$ are the supply/return temperatures to/from the radiators, $\theta_{int;t-1}$ is the internal air temperature from the previous time step (internal air temperature is unknown at time step t until the linear equations system has been solved). The supply temperature $\theta_{hyd;sup}$ is usually centrally controlled. For Swedish multifamily buildings, $\theta_{hyd;sup}$ is usually controlled by linear interpolation from a look-up table. Such a look-up table consists of value pairs specifying supply temperature set-points $\theta_{hyd;sup;set}$ at certain external air temperature θ_e. an example.

For a hydronic heating system without local TRVs, the following empirical equation was derived to estimate the return temperature:

$$\theta_{hyd;ret;t} = \theta_{hyd;sup;t} - b \cdot (\theta_{hyd;sup;t} - \theta_{int;t-1})^a$$
$$a = n_{hyd} - \Delta\theta_{hyd;d}/200, \qquad b = \frac{\Delta\theta_{hyd;d}}{(\theta_{hyd;sup;d} - \theta_{int;set})^a} \tag{11}$$

where $\Delta\theta_{hyd;d}$ is the temperature drop between supply and return temperatures at design power output, $\theta_{hyd;sup;d}$ is the supply temperature at design power output, and $\theta_{int;set}$ is the internal air set-point. Equation (11) was derived empirically and chosen to suit common Swedish radiator configurations. Validation is provided as Supplementary Materials. The TRV(s) are modeled as proportional controllers (P-controllers):

$$u_{trv;t} = 0.5 \cdot u_{trv;t-1} + 0.5 \cdot \max(0, \min(1, (\theta_{int;set} - \theta_{int;t-1})/\theta_{trv;pb})) \tag{12}$$

where $\theta_{int;set}$ is the set-point for desired indoor temperature, $\theta_{trv;pb}$ is the proportional band of the TRV (typically ranging between 0.5–2.0 °C). Taking the average of current step and previous time step dampens oscillations in the control signal.

2.3. Ventilation

Heat transfer due to active ventilation is modeled as

$$\phi_{ve;t} = \kappa\rho_a \cdot Q_{ve;t} \cdot (1 - \eta_{ve}) \cdot (\theta_{int;t-1} - \theta_{e;t}) \tag{13}$$

where $\kappa\rho_a$ is the heat capacity of air per volume 1.21 $W \cdot s/(l \cdot K)$, $Q_{ve;t}$ is the specific air flow rate $[l/(s \cdot m_{fl}^2)]$, η_{ve} is the temperature transfer efficiency of the heat recovery unit, $\theta_{int;t-1}$ is the internal air temperature from the previous time step t, and $\theta_{e;t}$ is the external air temperature at current time step t.

2.4. Infiltration

Infiltration is driven by pressure differences across the building envelope caused by unbalanced mechanical ventilation, wind, and air temperature difference between internal and external air (known as the stack or buoyancy effect). One of the most established single zone models for estimating air infiltration rates is the Alberta air infiltration model (AIM-2), originally presented by Walker and Wilson [27] and referred to as the "enhanced model" in the ASHRAE Handbook—Fundamentals [3]. The AIM-2 model was developed for dwellings and has proven to be fairly accurate at estimating air infiltration rates for dwellings [28]. Hayati et al. [29] successfully used and evaluated the AIM-2 model for Swedish churches, modeled as large single zone models. The proposed thermal model of this paper is implemented as a single zone model, but this single zone is a representation of the average of a whole building (that could be a height of several floors). Thus, the AIM-2 model is implemented with modifications as to make the model more suitable for multi-floor buildings. Compared to earlier representations [3,27–29], equations are presented such that parameters are separated from parts that can be treated as pre-calculable data input. Thermal power losses due to infiltration are calculated as

$$\phi_{inf;t} = C_{inf} \cdot Q_{inf;t}^* \cdot \kappa\rho_a \cdot (\theta_{int;t-1} - \theta_{e;t}) \tag{14}$$

where C_{inf} $[l/(s \cdot Pa^n \cdot m_{fl}^2)]$ is the infiltration coefficient (often referred to as the flow coefficient), n is the flow exponent, $\kappa\rho_a$ is the heat capacity of air per volume [1.21 $Ws/(l \cdot °C)$], and Q_{inf}^* is the potential specific air flow rate $[Pa^n]$ due to air infiltration. Q_{inf}^* is pre-calculated (see Section 3.2), the flow exponent n is treated as a constant ($n = 0.73$ [30]), while C_{inf} needs to be provided as an input parameter.

2.5. Radiant Heat Gains Exterior Surfaces

Net radiant heat gains on exterior surfaces due to thermal radiation exchange with the sky dome and absorption of solar irradiance are expressed as

$$\phi_{re;rf;t} = r_{rf} \cdot (\alpha_{sol} \cdot I_{tot;hor;sh;t} - \phi_{sky;rf;t})$$
$$\phi_{re;ew;t} = r_{ew} \cdot (\alpha_{sol} \cdot I_{tot;ver;sh;t} - \phi_{sky;ew;t}) \qquad (15)$$
$$\phi_{re;gl;t} = r_{gl} \cdot (-\phi_{sky;gl;t})$$

where r is the weighting ratio of the element, α_{sol} is the solar absorption coefficient of the exterior surface (0.5 is used as default), I_{tot} is the calculated total hemispherical solar irradiance (see Section 3.3), ϕ_{sky} is the extra thermal radiation to the sky, subscript *ver* denotes vertical surfaces, subscript *hor* denotes horizontal surfaces, and subscript *sh* denotes shading. Thermal radiation exchange with the surrounding environment is accounted for in Section 3.1. The extra thermal radiation to the sky variable follows the definition used in [10] (positive values denote heat loss; negative values denote heat gains). However, an alternative calculation procedure is proposed so that it can be derived from climate model data sources:

$$\phi_{sky;el;t} = F_{sky;el} \cdot \epsilon_{el} \cdot (\sigma \cdot (\theta_{1;el;t-1} + 273.15)^4 - \phi_{strd}) \qquad (16)$$

where $F_{sky;el}$ is the sky view factor for element *el* (0.5 used for vertical elements external walls and glazing and 1.0 used for the horizontal roof element), ϵ is the emissivity of the surfaces (0.9 used as default), $\theta_{1;el}$ is the exterior surface temperature of the element *el*, σ is the Stefan–Boltzmann constant (5.67×10^{-8} W/(m$^2 \cdot$ K^4)), 273.15 is the Kelvin to Celsius conversion constant, and ϕ_{strd} is the "surface thermal radiation downwards" variable retrieved from the climate reanalysis dataset ERA5 (see Section 4.2).

2.6. Radiant Heat Gains Interior Surfaces

Interior surfaces are subjected to solar and thermal radiation, expressed as

$$\phi_{ri;el;t} = f_{el} \cdot (f_{r;sol} \cdot \phi_{sol;gl;t} + f_{r;int} \cdot \phi_{int;t} + f_{r;hyd} \cdot \phi_{hyd;t}) \qquad (17)$$

where f_{el} is the surface area fraction of the element *el*, and f_r is the fraction of radiative heat ($f_r = 1 - f_c$). Default convective fractions from the ISO 52016-1:2017 [10] standard: $f_{c;sol} = 0.1$, $f_{c;int} = 0.4$, and $f_{c;hyd} = 0.4$. The surface area fractions f_{el} are simply calculated by dividing the ratio (r) of each building element with the sum of all ratios of the five building element types.

2.7. Operative Temperature

The ISO 52016-1:2017 [10] calculates operative temperature as follows:

$$\theta_{op;t} = 0.5 \cdot (\theta_{int;t} + \theta_{int;r;mn;t}) \qquad (18)$$

where $\theta_{int;r;mn}$ is the mean radiant temperature, calculated as the area weighted mean of interior surface temperatures of all building elements. However, Equation (18) omits the impact of radiative heating from the hydronic heating system. With radiator panels of the known surface area and surface temperatures, their contribution to the operative temperature can be area-weighted in the same manner as interior walls. An equation that does not require such detailed knowledge of the heating system is proposed:

$$\theta_{op;t} = 0.5 \cdot (\theta_{int;t} + \theta_{int;r;mn;t} + C_{hyd;r;mn} \cdot f_{r;hyd} \cdot \phi_{hyd;t}) \qquad (19)$$

where $C_{hyd;r;mn}$ is a weighting and conversion coefficient expressing the impact of the radiative heating part of the hydronic heating system on the operative temperature. The unit of the $C_{hyd;r;mn}$ parameter is the same as that of a floor area normalized thermal resistance, m$^2_{fl} \cdot$ °C/W.

3. Pre-Processing of Boundary Conditions Data

3.1. Convective Heat Transfer Coefficients for Weather-Exposed Exterior Surfaces

The ISO 52016-1:2017 [10] standard provides constant conventional surface heat transfer coefficients for exterior surfaces (h_{ce}). However, h_{ce} varies strongly depending on weather conditions [31,32]. Here, we implement the correlation equations by Liu et al. [32], derived by computational fluid dynamic calculations. Compared to earlier works [31], Liu et al. [32] accounts for wind sheltering effects, and separate correlations factors are provided for vertical walls and the roof. Additionally, our implementation accounts for surface roughness by adopting parts of the DOE-2 convection model [4] method.

$$h_{ce;rf;t} = h_{c;n} + R_{f;rf} \cdot \left(\sqrt{h_{c;n}^2 + h_{c;f;hor;t}^2} - h_{c;n} \right)$$
$$h_{ce;ew;t} = h_{c;n} + R_{f;ew} \cdot \left(\sqrt{h_{c;n}^2 + h_{c;f;ver;t}^2} - h_{c;n} \right) \tag{20}$$
$$h_{ce;gl;t} = \sqrt{h_{c;n}^2 + h_{c;f;ver;t}^2}$$

where $h_{c;n}$ is the natural convective heat transfer coefficient, $h_{c;f}$ is the forced (wind effects) convective heat transfer coefficient, and R_f is the surface roughness multiplier. The surface roughness multiplier is set as smooth $R_{f;rf} = 1.11$ for the roof and as medium rough $R_{f;ew} = 1.52$ for external walls [4]. The authors in Liu et al. [32] calculated the natural convective heat transfer coefficient as depending on temperature difference between external air and the exterior surface of the building element: $h_{c;n;t} = 1.52 \cdot \Delta\theta_t^{0.36}$ W/(K · m$_s^2$). Here we assume a constant temperature difference of $\Delta\theta = 5$ °C, which results in a constant $h_{c;n} = 2.7$ W/(K · m$_s^2$). Assuming $h_{c;n}$ as a constant allows treating Equation (20) as pre-calculable data input. Wind has the strongest impact on the exterior surface convective heat transfer, and the forced heat transfer coefficients are calculated as

$$h_{c;f;ver;t} = F_{ww} \cdot (3.39 - 5.03 \cdot \lambda_p) \cdot U_{loc;ver}^{0.94} + (1 - F_{ww}) \cdot (1.15 + 0.82 \cdot \lambda_p) \cdot U_{loc;ver}^{0.94}$$
$$h_{c;f;hor;t} = (3.57 + 1.72 \cdot \lambda_p) \cdot U_{loc;hor}^{0.84} \tag{21}$$

where F_{ww} is the fraction of windward-oriented exterior surfaces, λ_p is the plan area density, and U_{loc} is the local wind speed. F_{ww} is assumed as a constant of 0.5 but could also be calculated as a variable depending on wind orientation using an approach similar to that taken for solar irradiance in Equation (29). The plan area density λ_p represents the projected built area viewed from above to the total area in consideration [32]. Conceptually it relates to the wind sheltering factor used for the wind infiltration calculation (see Equation (26)), and here we present them in the same table, Table 2, and categorize the λ_p values used in [32] into the sheltering classes defined in Table 5, Chapter 16 ASHRAE [3]. The local wind speeds are calculated according to Equation (22), where the height H is set as the building height for horizontal surfaces and half the building height for vertical surfaces.

$$U_{loc} = U_{10m} \cdot 1.59 \cdot \left(\frac{H_{loc}}{\delta} \right)^\alpha \tag{22}$$

where U_{10m} is the meteorological wind speed at 10 m height, the constant 1.59 describes the terrain conditions of the meteorological wind speed measurement site, H_{loc} is the height above ground for local wind calculation point, and δ and α are atmospheric boundary layer coefficients for different terrain categories, provided, e.g., in Table 1 of Chapter 24 of the ASHRAE Handbook—Fundamentals [3] ($\delta = 370$ and $\alpha = 0.22$ for urban and suburban terrain).

Table 2. Plan area densities λ_p [32], wind sheltering factors λ_w [3], and solar shielding factors λ_{sol} for five shelter classes: ranging from (1) no shelter to (5) a highly dense urban area (neighboring buildings closer than one building height).

Shelter Class	1	2	3	4	5
λ_p	0.00	0.04	0.11	0.25	0.44
λ_w	1.00	0.90	0.70	0.50	0.30
λ_{sol}	3.00	2.50	1.90	1.30	0.70

3.2. Infiltration Potential

Potential air infiltration from wind-driven and stack-effect-driven infiltration is not additive, which in the AIM-2 [3] is modeled with the following superposition formula:

$$Q^*_{inf;t} = \left((Q^*_{s;t})^{\frac{1}{n}} + (Q^*_{w;t})^{\frac{1}{n}} - 0.33 \cdot (Q^*_{s;t} \cdot Q^*_{w;t})^{\frac{1}{2n}} \right)^n \tag{23}$$

where the flow exponent n is treated as a constant ($n = 0.73$ [30]), and Q^*_s and Q^*_w are the contributions of the stack and the wind effects. Compared to the original formula [3,27], Equation (23) is modified such that $Q^* = Q/C_{inf}$ [Pa^n]. The contribution due to stack effect is calculated as

$$Q^*_{s;t} = C_s \cdot \left(\rho_a \cdot g_0 \cdot H^* \cdot \frac{|\theta_{int;set} - \theta_{e;t}|}{\theta_{int;set} + 273.15} \right)^n \tag{24}$$

where C_s [(Pa/K)^n] is the stack coefficient, ρ_a [kg/m^3] is the density of air, g_0 [m/s^2] is the acceleration of gravity, H^* [m] is the modified building's ceiling height, $\theta_{int;set}$ is the internal temperature set point, and $\theta_{e;t}$ is the external air temperature at time step t. The set point temperature is used instead of the actual internal air temperature so that Equation (24) can be treated as a pre-calculable data variable. The AIM-2 model was originally designed as a single zone model for residential building with up to three floor levels, assuming free air contact between the floor levels. Using the model as such for high rise buildings would probably overestimate the stack effect. Thus, a modified building height parameter H^* is introduced:

$$H^* = \begin{cases} N_{fl} \cdot 0.5 \cdot H_{fl} & \text{if } N_{fl} < 3 \\ N_{fl} \cdot 0.5 \cdot H_{fl}/3 & \text{if } N_{fl} \geq 3 \end{cases} \tag{25}$$

where N_{fl} is the number of floor levels, and $0.5 \cdot H_{fl}$ is half the average floor height. The rationale behind dividing by 3 for non-dwelling buildings with three or more floor levels is that the stack effect of such buildings will be a mix of the per zone/floor level height-induced stack effect and the whole building stack stack effect due to elevator shafts and stairways. The contribution due to wind effect is calculated as

$$Q^*_{w;t} = C_w \cdot \left(\frac{\rho_a \cdot (\lambda_w \cdot U_{loc})^2}{2} \right)^n \tag{26}$$

where the C_w [(Pa · s^2/m^2)^n] is the wind coefficient, λ_w is the wind sheltering factor given in Table 2, and U_{loc} is the local wind speed at the building's uppermost eaves height calculated with Equation (22).

In the AIM-2 air infiltration model [27], the stack and wind coefficients (C_s and C_w) can be calculated as functions of leakage distribution, occurrence of flue, and the type of foundation (crawlspace, basement, or slab-on-grade) [27]. Assuming evenly distributed leakage, no flue, and slab-on-grade foundation, $C_s = 0.25$[(Pa/K)^n] and $C_w = 0.22$[(Pa · s^2/m^2)^n]. The infiltration coefficient C_{inf} is given in liters instead of cubic meters and normalized to square meter floor. Otherwise, it is the same coefficient as described in [3,27]. Thus, it can be derived from building air leakage databases or from measurements (such as blower door pressurization tests).

3.3. Solar Heat Gains

Solar heat gains transmitted through windows glazing (gl) are calculated as

$$\phi_{sol;gl;t} = r_{gl} \cdot g_{gl} \cdot g_{bl} \cdot I_{tot;ver;sh;t} \tag{27}$$

where $I_{tot;ver;sh;t}$ [W/m_s^2] is the (by surface orientation) weighted hemispherical solar irradiance on vertical surfaces including shading effects, g_{bl} is the total solar energy transmittance of the window blinds (see Section 3.5), g_{gl} is the total solar energy transmittance of the glazing, and r_{gl} is the glazing-to-floor ratio. $I_{tot;ver;sh;t}$ is calculated as

$$I_{tot;ver;sh,t} = I_{dif;ver;t} + I_{dir;ver;t} \cdot F_{sh;ver;t} \tag{28}$$

where $I_{dif;ver;t}$ and $I_{dir;ver;t}$ are the total diffuse (inclusive ground reflectance and exclusive circum-solar) and the total direct (inclusive circum-solar) irradiance received on one square meter of weighted vertical surface area at time step t, and $F_{sh;ver;t}$ is a shading factor (see Section 3.4). Building elements are lumped according to their type, so solar irradiance variables also need to be allocated and weighted accordingly:

$$I_{dif;ver;t} = \sum_{k=1}^{n_\gamma} F_{\gamma_k} \cdot I_{dif;ver;\gamma_k;t} \qquad I_{dir;ver;t} = \sum_{k=1}^{n_\gamma} F_{\gamma_k} \cdot I_{dir;ver;\gamma_k;t} \tag{29}$$

where $I_{dir;ver;\gamma_k;t}$ is the total direct solar irradiance at a vertical surface with azimuth γ and time step t, F_{γ_k} is the surface area fraction of external walls at azimuth γ_k, and n_γ is the total number of surfaces of different azimuths to loop through. The same procedure is used for $I_{dif;ver;t}$. For example, solar irradiance for the cardinally oriented cube would be calculated using $F_\gamma = [0.25, 0.25, 0.25, 0.25]$ and $\gamma = [0°, 90°, -90°, 180°]$.

The solar irradiance absorbed by the roof needs its own variable. The calculation procedure is the same as that for the vertical surfaces. However, there is no need for area weighting as the roof is assumed to be representable with one horizontal surface:

$$I_{tot;hor;sh;t} = I_{dif;hor;t} + I_{dir;hor;t} \cdot F_{sh;hor;t}. \tag{30}$$

The solar irradiance at an arbitrarily tilted and oriented surface is calculated accordingly to the ISO 52010:2017 [25] standard. The calculation procedure was implemented in C++/R and validated against the accompanying test data of the the the standard. The code with validation/unit tests is published under a public domain license and is located at the first author's GitHub code repository (https://github.com/lukas-rokka/solarCalcISO52010).

3.4. Shading Reduction Factor

Distant obstacles (hills, trees, buildings etc) and obstacles on or nearby the building (balconies, rebates etc.) block parts of the solar radiation reaching a buildings surface. For best accuracy, these are calculated according to the actual surroundings of the building site. However, one of the goals with this work was to make reasonable estimates without actual information of the surroundings. Therefore, the shading factors are calculated based on assumptions about what the surroundings of a typical Swedish urban multifamily building typically look like. The resulting estimation can be expected to be more accurate on average than if shading were not accounted for at all. The shading reducing factor for the vertical envelope is calculated as

$$F_{sh;ver;t} = (\alpha_{sol;t} > 0) \cdot (F_{sh;obst;t} + F_{sh;ovh;t} - 1) \tag{31}$$

where $F_{sh;obst}$ is the shading from obstacles, and $F_{sh;ovh}$ is the shading from overhangs on the buildings. The shading factor for infinite length overhangs is calculated as described by Method 2 in Annex F of ISO 52016-1:2017 [10]:

$$F_{sh;ovh;t} = 1 - \max(0, \min(1, (D_{ovh} \cdot \tan(\alpha_{sol;t}) - L_{ovh}) / H_{gl})) \tag{32}$$

where D_{ovh} is the depth of the overhang, α_{sol} is the solar altitude, L_{ovh} is the vertical distance between the glazing and the overhang, and H_{gl} is the height of the window glazing. Many buildings have balconies; to account for this to at least some extent, the following values were used $D_{ovh} = 2\,\text{m}$, $L_{ovh} = 1\,\text{m}$, and $H_{gl} = 6\,\text{m}$. Setting $H_{gl} = 6\,\text{m}$ is approximately the same as having overhangs above every 4th window of a more typical window glazing height of 1.5 m.

The shading reduction factor calculation for distant obstacles is calculated as a mix of two semitransparent obstacles:

$$
\begin{aligned}
F_{sh;obst;t} = 1 &- \min(H_b, \max(0, H_{obst;1} - \lambda_{sol} \cdot L_{obst;1} \cdot \tan(\alpha_{sol;t}))) / H_b + \\
&\min(H_b, \max(0, H_{obst;2} - \lambda_{sol} \cdot L_{obst;2} \cdot \tan(\alpha_{sol;t}))) / H_b
\end{aligned}
\tag{33}
$$

where L_{obst} is the distance between the shading obstacle and the shaded object, H_{obst} is the height of the obstacle, H_b is the height of the shaded building, and λ_{sol} is a shading factor set based on the categorization of the surroundings of the building (see Table 2). Obstacle 1 models nearby objects such as trees, and its parameters are set as $L_{obst;1} = 15\,\text{m}$ and $H_{obst;1} = 20\,\text{m}$. Obstacle 2 models shading from other buildings, and its parameters are set as a function of the building height $L_{obst;2} = H_b$ and $H_{obst;2} = 1.3 \cdot H_b$. Shading at the roof level is calculated as

$$F_{sh;hor;t} = \min(3, H_b - H_{obst;1}) / 3 \cdot 0.5 + 0.5. \tag{34}$$

3.5. Window Blinds

Window blinds exist in most Swedish residential buildings. However, there is no comprehensive consensus about the way people operate blinds or the motivating factors that influence their decisions [33]. A Swedish survey by Sandberg and Engvall [34] showed a clear correlation between exposure to solar radiation and to both the occurrence and position of window blinds. The g-value of the window blinds at time interval t is calculated as

$$g_{bl;t} = 1 + (g_{bl;max} - 1) \cdot u_{bl;t} \tag{35}$$

where $g_{bl;max}$ is a parameter describing the g-value of the blinds when at maximum blocking capacity, and $u_{bl;t}$ is the control signal to the blinds. If the window blinds were automatically operated, the $u_{bl;t}$ could for example be modeled as an on/off signal depending on solar radiation and a threshold value. The following function attempts to deterministically estimate the control signal of occupant operated window blinds:

$$u_{bl;t} = \max(0.3, \min(0.7, I_{tot;wma;t} / 200)) \tag{36}$$

$$I_{tot;wma;t} = \sum_{k=0\,\text{h}}^{23\,\text{h}} \left(\frac{24 - k}{300} \cdot I_{tot;t-k} \right)$$

where the max and min functions limit the control signal between 30 and 70%, and $I_{tot;wma;t}$ is a weighted moving average of solar radiation exposure from the last 24 h. The motivation behind using weighted moving average was to model the position of the blinds as depending both on the current (active control) and recent (reactive control) solar radiation exposure. The use of a 24 h moving average will also result in a seasonal effect, as there are more hours with no solar radiation in winter than in the summer (modeling occupants as slightly more aware of the positions of the blinds in the summer season seems to be based on a reasonable assumption). The motivation behind limiting the control signal between the somewhat arbitrary 30 and 70% was that it can be expected that some blinds are also drawn in a situation of no solar exposure, that it is unlikely that all windows have blinds, and/or that these are in full position in a situation of high solar exposure.

4. Data Input

Sections 4.1 and 4.2 describe the environmental data sources proposed for the ISO14N modeling framework. For the comparison study, environmental data for Norrköping, Sweden (58.575° N, 16.15° E) and the year 2016 was used. Section 4.3 describes the example model building as implemented in IDA ICE and ISO14N.

4.1. Solar Irradiance

Direct normal and diffuse horizontal irradiance data were acquired from CAMS [35] (http://www.soda-pro.com/web-services/radiation/cams-radiation-service). This satellite-based solar irradiance data are available at a horizontal resolution of 5 km and 15 min time steps from 2004 until the present time, and covers Europe, Africa, and the Middle East.

4.2. Meteorological Reanalysis Data

Table 3 shows which variables were acquired from the reanalysis climate dataset ERA5 of the Copernicus Climate Change Service. ERA5 data are available at a global horizontal resolution of 31 km and hourly time steps [36] (https://cds.climate.copernicus.eu/cdsapp). Reanalysis uses numerical weather prediction schemes to assimilate historical observational data.

Table 3. Environmental variables derived from ERA5. The ISO14N column shows variable names as they are used in this paper, while the ERA5 column shows the variables sourced from ERA5 according to the notation convention of ERA5.

Description	ISO14N	ERA5	Transformation
External air temperature at 2 m	θ_e	2t	Kelvin to centigrade
Ground temperature at 1.0–2.89 m	θ_{gr}	stl4	Kelvin to centigrade
Wind speed at 10 m	U_{10m}	10u, 10v	$U_{10m} = \sqrt{10u^2 + 10v^2}$
Ground albedo	α_{gr}	ssr, ssrd	$\alpha_{gr} = 1 - ssr/ssrd$
Surface thermal radiation downwards	ϕ_{strd}	strd	Joule to Watt-hours

Apparent sky temperature was compared with results from IDA ICE but was not otherwise used in the ISO14N modeling framework. It is calculated as

$$\theta_{sky;t} = (\phi_{strd}/\sigma)^{0.25} - 273.15. \tag{37}$$

4.3. Example Building Model Data Input

This section describes the example building model used for comparison. Section 4.3.2 describes the example building model as constructed in IDA ICE. Section 4.3.1 describes the parameter input needed to construct the same building model in ISO14N format. The IDA ICE model was originally [37] based on an aerated concrete, four-floor high, multifamily building, typical for the early Swedish million homes programme era of the 1960s and 1970s. The material properties of walls, windows, and the roof were kept according to the original model, but the shape was simplified to a 20 × 10 m² rectangular, one-floor-high building. The simpler shape was chosen so that thermal zoning would not impact the results and to facilitate logging node temperatures in IDA ICE.

4.3.1. ISO14N

No shading was accounted for. Window blinds were assumed fully drawn. The supply temperature look-up table of Table 4 was used. Used parameter inputs are given in Table 5.

Table 4. Supply temperature look-up table, used for both IDA ICE and ISO14N models.

θ_e [°C]	−20	0	12	18
$\theta_{hyd;sup;set}$ [°C]	60	43	28	18

Table 5. Parameter input values used for the ISO14N model simulation.

Parameter	Description	Value(s)	Unit
N_{fl}	Number of floors	1	-
H_{fl}	Floor height	2.6	m
P	Building perimeter	60	m
A_{fl}	Floor area	2.6	m_{fl}^2
γ	Azimuth angles for the external walls	[0, 90, −90, 180]	°
F_γ	Surface area fraction for the external walls	[0.17, 0.33, 0.33, 0.17]	-
r_{gl}	Glazing to floor ratio	0.15	-
g_{gl}	Solar transmittance of the glazing	0.76	-
g_{bl}	Solar transmittance of the window blinds	0.53	-
U_{rf}	Thermal transmittance of the roof	0.20	$W/(°C \cdot m_s^2)$
U_{ew}	Thermal transmittance of the external walls	0.72	$W/(°C \cdot m_s^2)$
U_{gl}	Thermal transmittance of the glazing	2.9	$W/(°C \cdot m_s^2)$
U_{gf}	Thermal transmittance of the ground floor	0.23	$W/(°C \cdot m_s^2)$
$\kappa_{m;rf}$	Areal heat capacity of the external walls	56	$Wh/(°C \cdot m_s^2)$
$\kappa_{m;ew}$	Areal heat capacity of the external walls	15	$Wh/(°C \cdot m_s^2)$
$\kappa_{m;gf}$	Areal heat capacity of the external walls	56	$Wh/(°C \cdot m_s^2)$
Q_{ve}	Specific air flow rate of the ventilation system	0.35	$1/(s \cdot m_{fl}^2)$
η_{ve}	Efficiency of the ventilation heat recovery	0.0	-
C_{inf}	Normalized infiltration coefficient	0.0	$1/(s \cdot Pa^n \cdot m_{fl}^2)$
H_{hyd}	Radiator constant, normalized per floor area	0.66	$W/(°C \cdot m_{fl}^2)$
$\theta_{int;set}$	Internal air set-point temperature	21.0	°C
H_{tb}	Heat transfer due to thermal bridges	0.0	$W/(°C \cdot m_{fl}^2)$
ϕ_{int}	Internal heat gains	3.0	$W/(m_{fl}^2)$
Δt	Calculations interval	0.5	

4.3.2. IDA ICE

The model was constructed using IDA ICE version 4.8. The modeled building is a simple 20×10 m² rectangular, one-floor-high building, with the long sides oriented in an east–west direction and a window-to-floor ratio of 0.15. The following constructions were used (layers from outside to inside):

- Roof: 0.20 m insulation + 0.15 m concrete, U-value of 0.20 W/($m_s^2 \cdot$ K);
- External walls: 0.20 m aerated concrete, U-value of 0.72 W/($m_s^2 \cdot$ K);
- Glazing: 2-pane, U-value of 2.9 W/($m_s^2 \cdot$ K), g-value of 0.76;
- Ground floor: 0.1 m virtual ground layer + 0.5 m soil + 0.1 m insulation + 0.1 m concrete, U-value of 0.22 W/($m_s^2 \cdot$ K)(0.33 if not including the ground);
- Internal walls: 150 m² of 0.15 m thick aerated concrete;
- Furniture: 130 m² of 0.01 m thick default furniture material.

The heating system was modeled using the IDA ICE Water Radiator model (CeWatHet) with a nominal heat output of 50 W/m_{fl}^2 (at a supply temperature of 60 °C and a return at 40 °C), and the geometry of the radiator was set so that approximately half of the heat output was convective. The thermostatic control was modeled with a P-controller with a proportional band of 2 °C, a time constant of 20 min, and the set-point set to 21 °C. Table 4 shows the used look-up table for supply temperatures to the radiators. Ventilation was set to a constant 0.35 l/(s $\cdot$ m_{fl}^2) exhaust flow. Window blinds g-value was set to 0.53 and always fully drawn, internal heat gain was set to a constant of 3 W/m_{fl}^2, and thermal bridges were set to zero. Infiltration was nominally set to zero; however, for the case of studying the air infiltration (Section 5.4), the building envelope airtightness was set to 1 ACH at a 50 Pa pressure difference, and the surface pressure coefficient was semi-exposed.

5. Results

The developed ISO14N model and its procedures were validated by comparing it with simulations conducted with the detailed building energy simulation software IDA ICE.

5.1. Full-Year Simulation Comparison with IDA ICE

In this section, simulation results from the developed ISO14N model are compared with results from an IDA ICE simulation. As seen in Figure 3, the ISO14N is capable of reproducing results of the detailed IDA ICE model (see Appendix B for comparisons on monthly and daily scales). The first half of January 2016 had two successive cold spells. During this period, the models show larger deviations, suggesting that they differ in thermal capacity behavior in the lower frequencies. A perfectly sized heating system would result in internal air temperature close to the set-point during the heating season. The heating system was on purpose undersized to cause more deviations from the set-point temperature, which makes differences in dynamical behavior more apparent. Table 6 shows the root mean square deviation (RMSD) [38] and the coefficient of variation (CV) of the RMSD [38], calculated on hourly, daily, and monthly averages. As can be seen from the table, the ISO14N is able to replicate simulation results of IDA ICE on an hourly basis, which is much more sensitive to dynamics than daily or monthly averages.

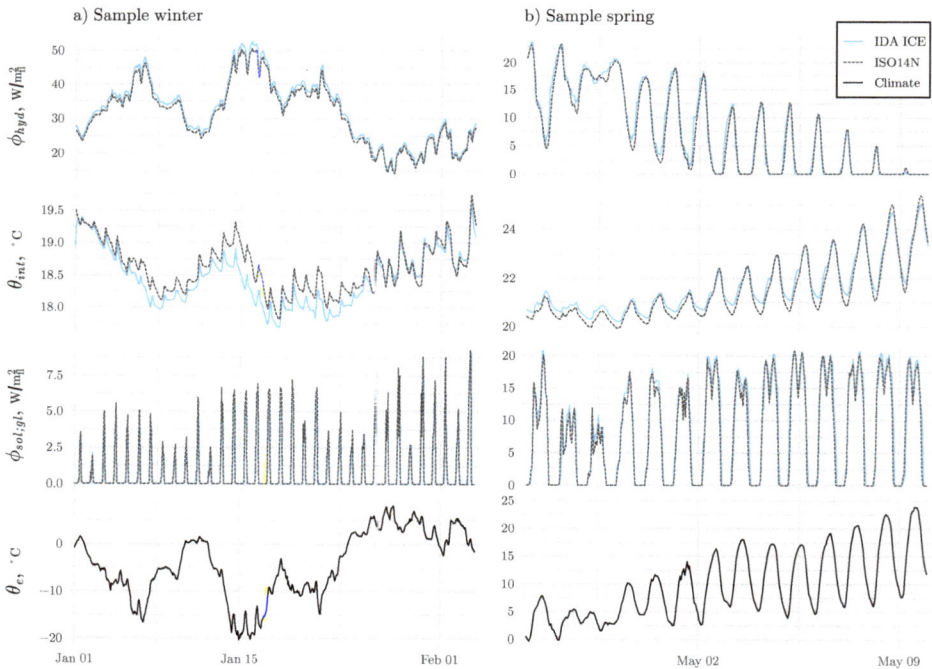

Figure 3. Hourly output variable comparison with IDA ICE. Diagram panels in column (**a**) shows a winter sample period, column (**b**) shows a spring sample period. Diagram panel rows compare following variables: energy use for space heating (ϕ_{hyd}), internal air temperature (θ_{int}), solar heat gains through glazing ($\phi_{sol;gl}$), and external air temperature (θ_e).

Table 6. Root mean square deviation (RMSD) for hourly, daily, and monthly averages on energy use (ϕ_{hyd}) and internal air (θ_{int}) and operative temperatures (θ_{op}) variables. For ϕ_{hyd}, CV(RMSD) is shown within parenthesis.

Variable	Hourly	Daily	Monthly
ϕ_{hyd}, W/m$^2_{fl}$	0.94 (2.6%)	0.74 (2.0%)	0.68 (1.8%)
θ_{int}, °C	0.30	0.26	0.17
θ_{op}, °C	0.34	0.28	0.19

5.2. Node Temperature Profiles of External Wall Elements

For the developed ISO14N model, the external walls are represented by one element with three temperature nodes, while the IDA ICE model has four external wall elements (one in each cardinal direction) with five temperatures nodes each. For better comparability, the external wall temperatures of the IDA ICE model is averaged by node position. As seen in Figure 4, the node temperature profiles of the two models behave quite differently, which is mostly due to the different number of nodes and different approaches to distribute thermal properties between the nodes. Still, the estimated internal air temperatures match well.

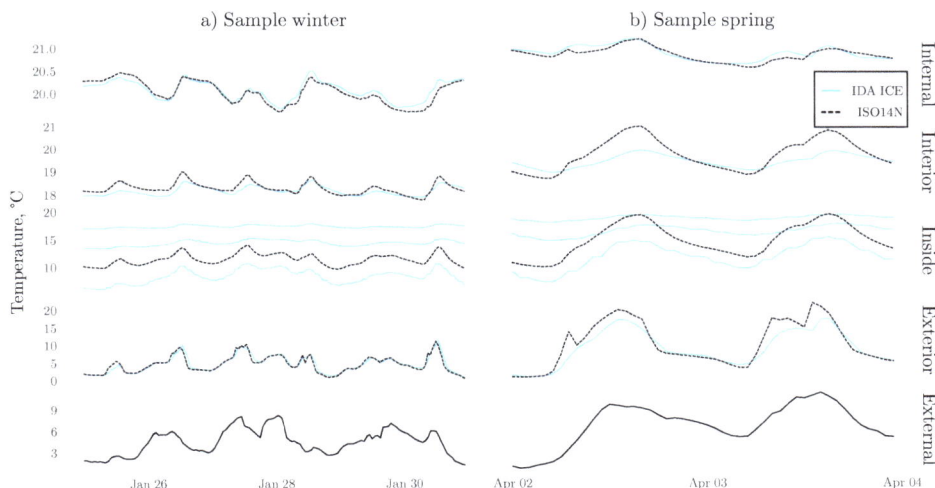

Figure 4. Comparison with IDA ICE on node temperatures for external walls. Diagram panels in column (**a**) shows a winter sample period, column; (**b**) shows a spring sample period. Diagram panel rows, from uppermost to lowermost: internal air nodes; interior surface nodes; inside nodes (one for ISO14N and three for IDA ICE); exterior surface nodes; and external air temperature.

5.3. Operative Temperature

Operative temperature was calculated in two different ways and compared with output from IDA ICE. "ISO14N alt. 1" in Figure 5 was calculated according to Equation (18), while "ISO14N alt. 2" was calculated using Equation (19) with the weighting/conversion factor $C_{hyd;r;mn}$ (expressing the impact of the hydronic heating system) set to $C_{hyd;r;mn} = 0.02$. As seen in Figure 5, the operative temperature of "ISO14N alt. 1" deviates from the operative temperature acquired from IDA ICE, especially during cold weather, which was expected as IDA ICE weights in the surface temperatures of the radiator panels in the calculation of operative temperatures. Using Equation (19) and setting $C_{hyd;r;mn} = 0.02$ results in a closer match. The $C_{hyd;r;mn}$ parameter could be derived from typical radiator panel configurations, but here it has been chosen by empirically matching it to the results of IDA ICE.

From Figure 5, it can be observed that IDA ICE appears to be more responsive to higher frequencies. This can partly be explained by the fact that IDA ICE splits the building elements into more nodes and uses surface nodes without mass. The other part of the explanation is that IDA ICE uses a fully dynamic simulation procedure (dynamic state transitions and variable heat transfer coefficients), while ISO14N linearizes the state transitions and uses constants for the interior surface heat transfer coefficients.

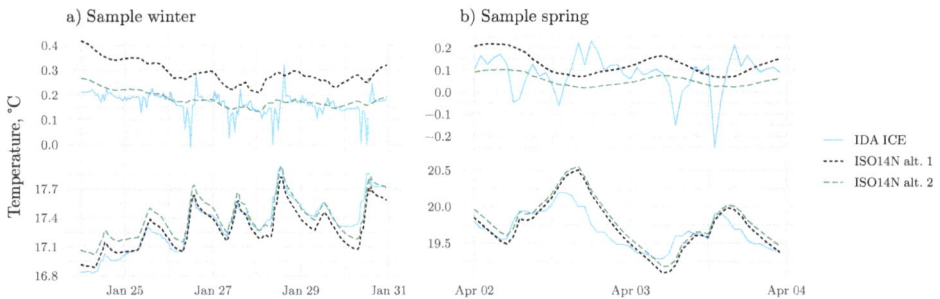

Figure 5. Comparison with IDA ICE on operative temperatures for two sample periods. The bottom row shows operative temperatures (θ_{op}) and the upper row shows the temperature difference between internal air temperature and operative temperatures ($\theta_{int} - \theta_{op}$).

5.4. Air Infiltration

The air infiltration implementation described in Section 2.4 (with shelter class set to 3 and the flow coefficient set to $C_{inf} = 0.08 \, l/(s \cdot Pa^n \cdot m_{fl}^2)$ is here compared with output from IDA ICE. As seen in Figure 6, the implemented infiltration model produces similar results as the single-zone, one-floor-high, IDA ICE building model.

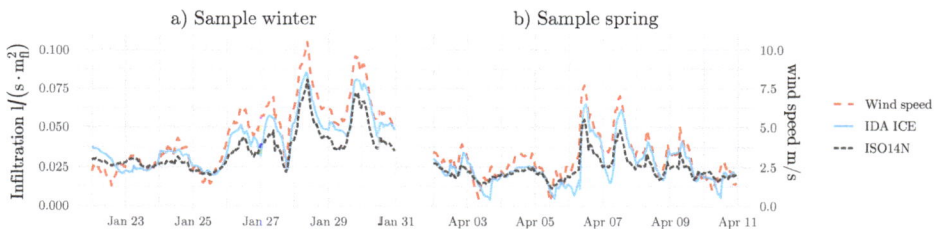

Figure 6. Comparison with IDA ICE on air infiltration. The solid blue line shows an infiltration level from IDA ICE, a gray-dotted line shows an infiltration level from the ISO14N model, and the dashed red line shows the meteorological wind speed (scale on right-hand y-axis).

5.5. Sky Temperature and Thermal Radiation to the Sky

Apparent sky temperature was calculated from the ERA5 variable "surface thermal radiation downwards" using Equation (37) and compared to output from IDA ICE. In IDA ICE, the apparent sky temperature was empirically derived from dry and dew point temperatures and cloud cover using the Walton–Clark–Allen model [39]. As seen in Figure 7, the proposed apparent sky temperature calculation differs substantially from that of IDA ICE, resulting in an hourly RMSD of 5.5 °C. Using the defaults suggested in the ISO 52016-1:2017 [10] standard, the extra thermal radiation to the sky variable ϕ_{sky} results in a constant of 45 W/m$_s^2$; deriving ϕ_{sky} from the "surface thermal radiation downwards" variable of ERA5 (as proposed in Section 2.5), this constant varies between -15 and 295 W/m$_s^2$, with an average of 61 W/m$_s^2$ for the used climate file.

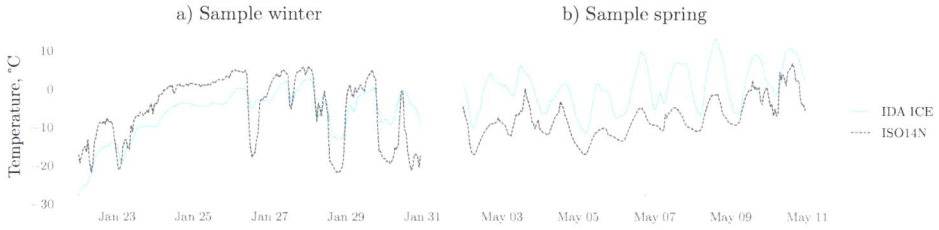

a) Sample winter   b) Sample spring

Figure 7. Comparison with IDA ICE on apparent sky temperatures for two sample periods. Solid blue line shows results from IDA ICE, gray dotted line shows from the ISO14N model.

5.6. Exterior Surface Convective Heat Transfer Coefficients

Exterior surface convective heat transfer coefficients calculated according to Section 3.1 are presented and compared with IDA ICE. IDA ICE refers to Clarke [40] for its calculation procedure, a method that accounts for wind direction but does not distinguish between building element type. As seen in Figure 8, our Liu et al. [32]-based method results in the roof element being more sensitive to wind speed than other elements. The lower values for the glazing element, compared to the external walls, is explained by its surface roughness multiplier being set to smooth. IDA ICE's method shows much less sensitivity to wind speed. For IDA ICE, the effect of wind direction is removed by showing smoothed lines of the average of the four external walls.

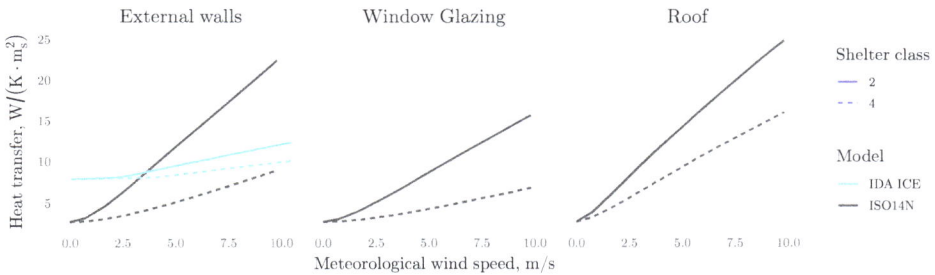

Figure 8. Exterior surface convective heat transfer coefficients calculated for two sheltering classes.

5.7. Shading Reduction Factor

Results of the proposed simplified shading calculation procedure (see Section 3.4) based on classification of the surrounding environment are presented. Table 7 shows the impact on annual solar irradiance received on vertical surfaces. The climate file described in Section 4 was used. The results can be compared to study conducted by Romero Rodríguez et al. [41], which showed an average of 35% reduction on solar irradiance for facades in the urban parts of Ludwigsburg. This part of the proposed modeling framework would benefit from further development, for example by employing aerial LiDAR data to model shading from the surrounding environment more accurately, see for example Lingfors et al. [17].

Table 7. Annual reduction of total solar irradiance on vertical surfaces, per sheltering class and building height.

Shading Class	1	1	2	2	3	3	4	4	5	5
Building height, m	18	6	18	6	18	6	18	6	18	6
Annual solar reduction, %	6	9	11	17	19	28	29	40	38	49

5.8. Window Blinds

Calculating the impact from window blinds according to Section 3.5, with the window blinds maximum g-value set to $g_{bl;max} = 0.4$, results in a 30% reduction of annual solar heat gains through window glazing. According to Van Den Wymelenberg [33], there is no comprehensive consensus about how to model occupant-operated window blinds. Nonetheless, the proposed procedure is anticipated to model the impact from window blinds more accurately than for example using a constant value. IDA ICE uses an on/off controller, which controls the window blinds position as a function of the outside solar radiation level (100 W/m^2 default set-point). This control scheme resulted in a 31% reduction of annual solar heat gains for the studied building and the climate file.

5.9. Computation Benchmark

This section show computation timings from conducted speed tests with the proposed ISO14N model, variants of it, and IDA ICE. The proposed ISO14N model pre-inversed its matrix **A** (the RC network) before the actual simulation run. For comparison, timing comparisons for model variants, where the matrix **A** of the RC network was inverted at each time-step, were also conducted. To compare speed in relation to a full ISO 52016-1:2017 implementation, an emulation was constructed and denoted "ISO41N" (due to its 41 temperature nodes). Table 8 shows the results. The largest speed gain was achieved from the matrix pre-inversion. Matrix inversion is computationally intensive and has a computation cost relationship of $O(n^2)$. The use of a pre-inverted matrix **A** breaks the $O(n^2)$ computational cost relationship and results in an approximately linear relationship between the RC network size and computation time. For the pre-inverted cases, it can be seen that the proposed simplification compared to a full ISO 52016-1:2017 implementation results in an approximately three-fold speed gain. The computation time difference between IDA ICE and the ISO14N is approximately 900 times (or an order of 1 magnitude).

Table 8. Computation benchmark, timing in milliseconds.

Model (Variant)	Computation Time, ms
ISO14N (matrix pre-inverted)	19
ISO41N (matrix pre-inverted)	56
ISO14N (matrix inversion each time-step)	83
ISO41N (matrix inversion each time-step)	755
IDA ICE	18,000

The benchmark was conducted on a standard laptop computer with an Intel Core i7-7500U processor using one thread. All models were simulated using one year of climate, plus 14 days for the adaptation period. IDA ICE simulation was performed within the user interface, using initial default settings and all variable logging turned off; timing was conducted manually with a stopwatch. Various types of the ISO14N model used a calculation interval of 0.5 h (18,384 calculation time-steps in total), and logged and returned energy use and internal air temperature.

6. Discussion

It has been shown that the proposed ISO14N model was able to reproduce the hourly energy use pattern obtained with the detailed building energy simulation software IDA ICE. The use of a hydronic radiator heating system limits the application to mainly multifamily buildings in Nordic climates. However, the proposed RC network itself is generic and could probably replace a full ISO 52016-1:2017 implementation in other applications, such as generic energy need calculations or as a basis for energy use calculation for cooling.

In the development of the proposed ISO14N model, effort was made to achieve fast simulation times without a significant loss of accuracy. The optimization for computational efficiency is motivated by that calibration approaches typically use algorithmic differentiation which requires the model to be simulated several times (10k or more for Monte Carlo simulation approaches). Speed gains were achieved by moving parts of the calculation load to the pre-processing, i.e., model nodal reduction, and by a simulation procedure that uses pre-inverted matrices for the RC network. The proposed simplification of the RC network consists of lumping building elements by type and decreasing the number of temperature nodes in opaque building elements (see Figure 2), resulting in a system of 14 temperature nodes. The largest speed gain was achieved from pre-inverting the matrix **A** of the RC network, breaking the $O(n^2)$ cost relationship and resulting in an approximately linear relationship between the RC network size and computation time. Pre-inversion requires the use of a constant matrix **A** element values, which results in a less flexible modeling structure. A constant **A** matrix, in turn, requires the use of node temperatures from the earlier time step for certain calculations (e.g., for the heating system control), which can result in time-shifts and numerical instabilities. However, our results show that the proposed ISO14N model achieves stable results without noticeable time-shifts.

During the development stage, a five-temperature node version (not included in this paper) was also tested and showed only limited performance gains compared to the proposed three-node version. Using five nodes adds flexibility and is likely to better model thermal admittance and more complex building structures such as sandwich walls (Akander [42]). However, for the intended use case of auto-generation of baseline models in large building stocks, it is anticipated that detailed information about construction will seldom be available in digital form. In the case of access to more detailed information, the proposed thermal network can be extended by adding more temperature nodes and/or by dividing the modeled building into more thermal zones.

Procedures to acquire and make use of satellite-based solar radiation data were presented, see Section 3.3. Solar radiation has a large impact on the thermal balance of buildings, and thermostatic radiators valves are often specifically used to avoid overheating due to solar heat gains. Thus, being able to model solar heat gains on an hourly basis (both directly through glazing and transmitted through opaque building elements) is anticipated to facilitate parameter estimation with hybrid/gray-box calibration methods. Reanalysis climate model data from ERA5 (see Section 4.2) was utilized. Conventional climate variables wind and air temperature were used as well as the more unconventional variables of ground temperature, ground albedo, and surface thermal radiation downward. Some of the benefits of using reanalysis data sources such as ERA5 are that the data are homogeneous and harmonized, covering the entire world, there is no need to deal with missing data, there is access to variables that are seldom measured locally, and the data are readily and publicly accessible. The temporal and spatial resolution of ERA5 is relatively rough, and in some cases it might be better to use local measurements (e.g., air temperature is readily measured locally at the building site).

7. Conclusions

A space heating model suitable for auto-generating baseline models of existing multifamily buildings has been proposed. As demonstrated in the result section, the proposed ISO14N model was able to reproduce the hourly energy use of space heating, indoor temperature, and operative temperature patterns obtained from the detailed building energy simulation software IDA ICE. Thus, the proposed model can be expected to model existing multifamily buildings, where space heating is provided by hydronic heating systems, to a similar degree of confidence as established simulation software.

The conducted computation timing showed that the proposed ISO14N was in the approximately 900 times faster than the detailed IDA ICE simulation. The largest speed gain was achieved from pre-inverting the matrix **A** of the RC network, resulting in an approximately linear cost relationship between the RC network size and computation time. Despite the use of constant pre-inverted **A** matrix, the proposed ISO14N model can achieve stable results without noticeable time-shifts. The achieved fast computation time enables using time-sensitive applications, such as Monte Carlo based calibration methods.

Solar heat gains have large impact on the thermal balance of buildings, and hydronic heating system often use thermostatic radiators valves to avoid overheating due to solar heat gains. Thus, it is necessary to model actual solar heat gains. We have presented procedures for converting sub-hourly satellite-based solar irradiation data to solar heat gains, while considering effects from building orientation, shading, and window blinds.

Future work will be conducted on calibrating the proposed model with actual energy meter readings, to ensure it represents the actual performance of the building and not only intended/designed performance. Hourly meter readings from district heating substations usually do not distinguish between domestic hot water and space heating. Thus, energy use for the occupant dependent domestic hot water needs to be modeled to be able to apply the proposed model on real buildings. Also, internal heat gains should be either measured (domestic electricity is always measured but not always readily available) or modeled to account for its occupant behavior dependent variations. The solar heat gain calculation could also be improved by employing aerial LiDAR data to model shading from the surrounding environment more accurately.

Supplementary Materials: The following are available online at https://www.mdpi.com/1996-1073/12/3/485/s1, Spreadsheet S1: Equation (11), empirical return temperature.

Author Contributions: For this article, the contributions of each author are listed as follows: L.L. contributed with conceptualization, coding, simulation, analysis, validation, and original draft preparation; J.A. contributed with knowledge, validation, and review and editing; J.Z. contributed with knowledge, validation, and review and editing.

Funding: The work was funded by Eskilstuna Kommunfastigheter, Eskilstuna Energy and Environment, and the Knowledge Foundation (KK-stiftelsen).

Acknowledgments: The work has been carried out under the auspices of the industrial post-graduate school Reesbe, and was funded by Eskilstuna Kommunfastigheter, Eskilstuna Energy and Environment, and the Knowledge Foundation (KK-stiftelsen).

Conflicts of Interest: The authors declare no conflict of interest.

Abbreviations

The following symbols are used in this manuscript:

α_{sol}	solar absorption [-] or solar altitude [°]
η	efficiency [-]
γ	azimuth angle [°]
Δt	time interval [h]
κ	areal heat capacity [Wh/(°C · m$_s^2$)]
$\kappa\rho_a$	heat capacity of air per volume [Ws/(l · °C)]
ϕ	normalized thermal power [W/m$_{fl}^2$]
ρ_a	density of air [kg/m^3]
σ	Stefan-Boltzmann constant, 5.67e-8 [W/(m^2 · K^4)]
θ	centigrade temperature [°C]
A, X, b	square matrix of system coefficients, vector of unknown node temperatures, vector of known terms
C	coefficient [-] or normalized heat capacity [Wh/(°C · m$_{fl}^2$)]
$\mathcal{D}$	Data
D, H, L	distance, height, length [m]
F, f	factor/fraction [-]
H	normalized heat transfer coefficient [W/(°C · m$_{fl}^2$)]
H^*	modified building's ceiling height [m]
I	solar or thermal radiation [W/m$_s^2$]
Q	specific air flow rate [l/(s · m$_{fl}^2$)]
Q^*	potential specific air flow rate [Pa^n/m$_{fl}^2$]
R	normalized thermal resistance [m$_{fl}^2$ · °C/W]
U	thermal transmittance [W/(°C · m$_s^2$)]
U_{loc}, U_{10m}	local wind speed [m/s], meteorological wind speed at 10 m height [m/s]
g	total solar energy transmittance [-]
h	surface coefficient of heat transfer [W/(°C · m$_s^2$)]
n	exponent

r	ratio [-]
u	control signal [-]

The following subscripts are used in this manuscript:

b	building
bl	(window) blinds
bou	boundary
c, ci	convective, convective interior surface
d	design (nominal)
dif, dir	diffuse, direct
e	external (as in outdoor)
el	(building) element
env	environment
ew	external walls
gf	ground floor
gl	glazing (windows, doors etc)
gr	ground
hor	horizontal
hyd	hydronic heating system
im	internal mass (internal walls, intermediate floors and adiabatic external walls)
inf	infiltration (uncontrolled air leakage)
int	internal (as in indoor)
$lmtd$	radiator logarithmic mean temperature difference
m	mass related conductance or capacitance
max	maximum
$obst, ovh$	obstacle, overhang
pb	proportional band
r, ri	radiative, radiative interior surface
ret	return
rf	roof
s	stack
se, si	surface exterior, surface interior
set	set-point
sh	shading or sheltering
sky	sky temperature or sky thermal radiation
sol	solar radiation/heat gain
$strd$	surface thermal radiation downwards
sup	supply
sys	system
t	time index
tb	thermal bridges
tot	total
trv	thermostatic radiator valve(s)
ve	ventilation
ver	vertical
vi	virtual ground layer
w	wind
wma	weighted moving average

Appendix A. Construct of the Linear Equations System

$$
\mathbf{A} =
\begin{bmatrix}
A_1 & -H_{1;rf} & 0 & 0 & 0 & 0 & 0 & 0 & 0 & 0 & 0 & 0 & 0 & 0 \\
-H_{1;rf} & A_2 & -H_{2;rf} & 0 & 0 & 0 & 0 & 0 & 0 & 0 & 0 & 0 & 0 & 0 \\
0 & -H_{2;rf} & A_3 & 0 & 0 & -f_{ew}\cdot H_{ri;rf} & 0 & -f_{gl}\cdot H_{ri;rf} & 0 & -f_{im}\cdot H_{ri;rf} & 0 & 0 & -f_{gf}\cdot H_{ri;rf} & -H_{ci;rf} \\
0 & 0 & 0 & A_4 & -H_{1;ew} & 0 & 0 & 0 & 0 & 0 & 0 & 0 & 0 & 0 \\
0 & 0 & 0 & -H_{1;ew} & A_5 & -H_{2;ew} & 0 & 0 & 0 & 0 & 0 & 0 & 0 & 0 \\
0 & 0 & -f_{rf}\cdot H_{ri;ew} & 0 & -H_{2;ew} & A_6 & 0 & -f_{gl}\cdot H_{ri;ew} & 0 & -f_{ew}\cdot H_{ri;ew} & 0 & 0 & -f_{gf}\cdot H_{ri;ew} & -H_{ci;ew} \\
0 & 0 & 0 & 0 & 0 & 0 & A_7 & -H_{1;gl} & 0 & 0 & 0 & 0 & 0 & 0 \\
0 & 0 & -f_{rf}\cdot H_{ri;gl} & 0 & 0 & -f_{ew}\cdot H_{ri;gl} & -H_{1;gl} & A_8 & 0 & -f_{im}\cdot H_{ri;gl} & 0 & 0 & -f_{gf}\cdot H_{ri;gl} & -H_{ci;gl} \\
0 & 0 & 0 & 0 & 0 & 0 & 0 & 0 & A_9 & -H_{1;im} & 0 & 0 & 0 & 0 \\
0 & 0 & -f_{rf}\cdot H_{ri;im} & 0 & 0 & -f_{ew}\cdot H_{ri;im} & 0 & -f_{gl}\cdot H_{ri;im} & -H_{1;im} & A_{10} & 0 & 0 & -f_{gf}\cdot H_{ri;im} & -H_{ci;im} \\
0 & 0 & 0 & 0 & 0 & 0 & 0 & 0 & 0 & 0 & A_{11} & -H_{1;gf} & 0 & 0 \\
0 & 0 & 0 & 0 & 0 & 0 & 0 & 0 & 0 & 0 & -H_{1;gf} & A_{12} & -H_{2;gf} & 0 \\
0 & 0 & -f_{rf}\cdot H_{ri;gf} & 0 & 0 & -f_{ew}\cdot H_{ri;gf} & 0 & -f_{gl}\cdot H_{ri;gf} & 0 & -f_{im}\cdot H_{ri;gf} & 0 & -H_{2;gf} & A_{13} & -H_{ci;gf} \\
0 & 0 & -H_{ci;rf} & 0 & 0 & -H_{ci;ew} & 0 & -H_{ci;gl} & 0 & -H_{ci;im} & 0 & 0 & -H_{ci;gf} & A_{14}
\end{bmatrix}
$$

$$
\mathbf{b}^\mathsf{T} = [b_1, \cdots, b_{14}], \qquad
\mathbf{X}^\mathsf{T} = \begin{bmatrix} \theta_{1;rf} & \theta_{2;rf} & \theta_{3;rf} & \theta_{1;ew} & \theta_{2;ew} & \theta_{3;ew} & \theta_{1;gl} & \theta_{2;gl} & \theta_{1;im} & \theta_{2;im} & \theta_{1;gf} & \theta_{2;gf} & \theta_{3;gf} & \theta_{int} \end{bmatrix}
$$

where the $A_1, \cdots, A_{14}$ elements of matrix **A** and the $b_1, \cdots, b_{14}$ elements of matrix **b** are detailed in Tables A1 and A2 respectively.

Table A1. Elements $A_1, \cdots, A_{14}$ of matrix **A**. *el* refers to the five building element types.

el	Exterior Surface Nodes	Inside Nodes	Interior Surface Nodes
rf	$A_1 = \frac{C_{1;rf}}{\Delta t} + H_{1;rf} + H_{se;rf;t}$	$A_2 = \frac{C_{2;rf}}{\Delta t} + H_{2;rf} + H_{1;rf}$	$A_3 = \frac{C_{3;rf}}{\Delta t} + H_{ci;rf} + (1-f_{rf})\cdot H_{ri;rf} + H_{2;rf}$
ew	$A_4 = \frac{C_{1;ew}}{\Delta t} + H_{1;ew} + H_{se;ew;t}$	$A_5 = \frac{C_{2;ew}}{\Delta t} + H_{2;ew} + H_{1;ew}$	$A_6 = \frac{C_{3;ew}}{\Delta t} + H_{ci;ew} + (1-f_{ew})\cdot H_{ri;ew} + H_{2;ew}$
gl	$A_7 = H_{1;gl} + H_{se;gl;t}$		$A_8 = H_{ci;gl} + (1-f_{gl})\cdot H_{ri;gl} + H_{1;gl}$
im		$A_9 = \frac{C_{1;im}}{\Delta t} + H_{1;im}$	$A_{10} = \frac{C_{2;im}}{\Delta t} + H_{ci;im} + (1-f_{im})\cdot H_{ri;im} + H_{1;im}$
gf	$A_{11} = \frac{C_{1;gf}}{\Delta t} + H_{se;gf} + H_{1;gf}$	$A_{12} = \frac{C_{2;gf}}{\Delta t} + H_{2;gf} + H_{1;gf}$	$A_{13} = \frac{C_{3;gf}}{\Delta t} + H_{ci;gf} + (1-f_{gf})\cdot H_{ri;gf} + H_{2;gf}$

Internal air node:
$$A_{14} = \frac{C_{int}}{\Delta t} + H_{ci;rf} + H_{ci;ew} + H_{ci;gl} + H_{ci;im} + H_{ci;gf} + H_{tb}$$

Table A2. Elements $b_1, \cdots, b_{14}$ of matrix **b**. *el* refers to the five building element types.

el	Exterior Surface Nodes	Inside Nodes	Interior Surface Nodes
rf	$b_1 = \frac{C_{1;rf}}{\Delta t}\cdot\theta_{1;rf;t-1} + H_{se;rf;t}\cdot\theta_{e;t} + \phi_{re;rf;t}$	$b_2 = \frac{C_{2;rf}}{\Delta t}\cdot\theta_{2;rf;t-1}$	$b_3 = \frac{C_{3;rf}}{\Delta t}\cdot\theta_{3;rf;t-1} + \phi_{ri;rf;t}$
ew	$b_4 = \frac{C_{1;ew}}{\Delta t}\cdot\theta_{1;ew;t-1} + H_{se;ew;t}\cdot\theta_{e;t} + \phi_{re;ew;t}$	$b_5 = \frac{C_{2;ew}}{\Delta t}\cdot\theta_{2;ew;t-1}$	$b_6 = \frac{C_{3;ew}}{\Delta t}\cdot\theta_{3;ew;t-1} + \phi_{ri;ew;t}$
gl	$b_7 = H_{se;gl;t}\cdot\theta_{e;t} + \phi_{re;gl;t}$		$b_8 = \phi_{ri;gl;t}$
im		$b_9 = \frac{C_{1;im}}{\Delta t}\cdot\theta_{1;im;t-1}$	$b_{10} = \frac{C_{2;im}}{\Delta t}\cdot\theta_{2;im;t-1} + \phi_{ri;im;t}$
gf	$b_{11} = \frac{C_{1;gf}}{\Delta t}\cdot\theta_{1;gf;t-1} + H_{se;gf}\cdot\theta_{gr;t}$	$b_{12} = \frac{C_{2;gf}}{\Delta t}\cdot\theta_{2;gf;t-1}$	$b_{13} = \frac{C_{3;gf}}{\Delta t}\cdot\theta_{3;gf;t-1} + \phi_{ri;gf;t}$

Internal air node:
$$b_{14} = \frac{C_{int}}{\Delta t}\cdot\theta_{int;t-1} + H_{tb}\cdot\theta_{e;t} - \phi_{ve;t} - \phi_{inf;t} + f_{c;int}\cdot\phi_{int;t} + f_{c;sol}\cdot\phi_{sol;t} + f_{c;hyd}\cdot\phi_{hyd;t}$$

Appendix B. Full-Year Comparison

Table A3. Energy use for space heating as monthly sums and deviations between the models.

Month	IDA ICE, kWh/m²$_{\text{fl}}$	ISO14N, kWh/m²$_{\text{fl}}$	Deviation, kWh/m²$_{\text{fl}}$	Deviation, %
1	24.8	23.9	0.9	3.6%
2	17.9	17.2	0.7	4.1%
3	15.4	14.7	0.7	4.5%
4	10.0	9.3	0.7	6.9%
5	3.0	2.9	0.1	4.3%
6	0.1	0.1	0.0	-
7	0.0	0.0	0.0	-
8	0.3	0.4	-0.1	-
9	1.5	1.5	0.0	0.8%
10	10.7	10.4	0.3	3.0%
11	16.7	16.2	0.5	3.1%
12	17.8	17.2	0.5	2.9%
Sum	118.3	114.0	4.4	3.7%

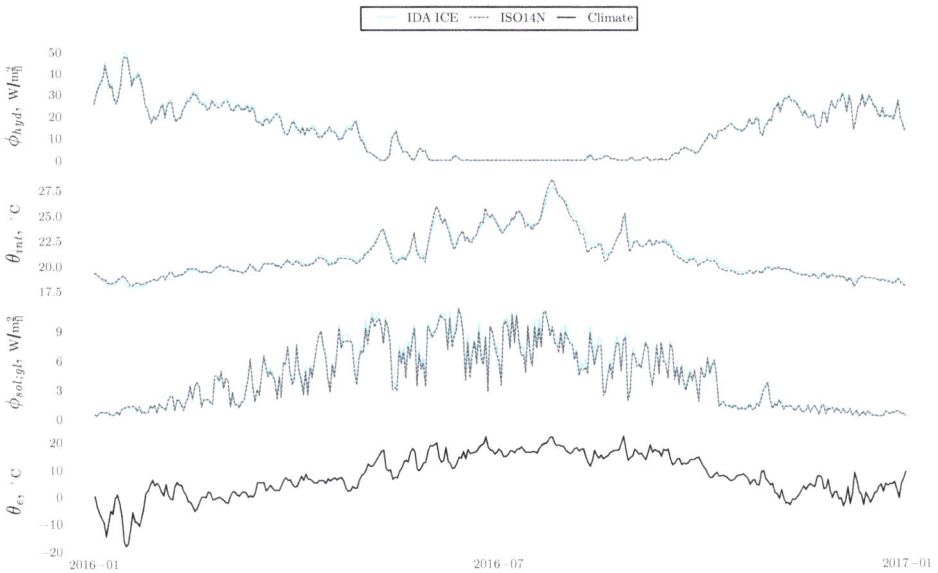

Figure A1. Full-year comparison with IDA ICE, using daily average values: energy use for space heating, (ϕ_{hyd}), internal air temperature (θ_{int}), solar heat gains through glazing ($\phi_{sol;gl}$) and external air temperature (θ_e).

References

1. Coakley, D.; Raftery, P.; Keane, M. A review of methods to match building energy simulation models to measured data. *Renew. Sustain. Energy Rev.* **2014**, *37*, 123–141. [CrossRef]
2. Ciancio, V.; Falasca, S.; Golasi, I.; Curci, G.; Coppi, M.; Salata, F. Influence of Input Climatic Data on Simulations of Annual Energy Needs of a Building: EnergyPlus and WRF Modeling for a Case Study in Rome (Italy). *Energies* **2018**, *11*, 2835. [CrossRef]
3. ASHRAE. *ASHRAE Handbook—Fundamentals (SI Edition)*; ASHRAE: Atlanta, GA, USA, 2017.
4. EnergyPlus. *EnergyPlus 8.9 Documentation—Engineering Reference*; Technical Report; U.S. Department of Energy: Washington, DC, USA, 2018.

5. Kang, Y.; Krarti, M. Bayesian-Emulator based parameter identification for calibrating energy models for existing buildings. *Build. Simul.* **2016**, *9*, 411–428. [CrossRef]

6. Tian, W.; Yang, S.; Li, Z.; Wei, S.; Pan, W.; Liu, Y. Identifying informative energy data in Bayesian calibration of building energy models. *Energy Build.* **2016**, *119*, 363–376. [CrossRef]

7. Amasyali, K.; El-Gohary, N.M. A review of data-driven building energy consumption prediction studies. *Renew. Sustain. Energy Rev.* **2018**, *81*, 1192–1205. [CrossRef]

8. Bacher, P.; Madsen, H. Identifying suitable models for the heat dynamics of buildings. *Energy Build.* **2011**, *43*, 1511–1522. [CrossRef]

9. Boodi, A.; Beddiar, K.; Benamour, M.; Amirat, Y.; Benbouzid, M. Intelligent Systems for Building Energy and Occupant Comfort Optimization: A State of the Art Review and Recommendations. *Energies* **2018**, *11*, 2604. [CrossRef]

10. ISO. *ISO 52016-1:2017—Energy Performance of Buildings—Energy Needs for Heating and Cooling, Internal Temperatures and Sensible and Latent Heat Loads—Part 1: Calculation Procedures (ISO 52016-1:2017)*; ISO: Geneva, Switzerland, 2017.

11. ISO. *ISO 13790:2008—Energy Performance of Buildings—Calculation of Energy Use For Space Heating and Cooling (ISO 13790:2008)*; ISO: Geneva, Switzerland, 2008.

12. Pavlak, G.S.; Florita, A.R.; Henze, G.P.; Rajagopalan, B. Comparison of Traditional and Bayesian Calibration Techniques for Gray-Box Modeling. *J. Archit. Eng.* **2014**, *20*, 04013011. [CrossRef]

13. Piotr, M. The simple hourly method of EN ISO 13790 standard in Matlab/Simulink: A comparative study for the climatic conditions of Poland. *Energy* **2014**, *75*, 568–578.

14. Fischer, D.; Wolf, T.; Scherer, J.; Wille-Haussmann, B. A stochastic bottom-up model for space heating and domestic hot water load profiles for German households. *Energy Build.* **2016**, *124*, 120–128. [CrossRef]

15. Vivian, J.; Zarrella, A.; Emmi, G.; De Carli, M. An evaluation of the suitability of lumped-capacitance models in calculating energy needs and thermal behaviour of buildings. *Energy Build.* **2017**, *150*, 447–465. [CrossRef]

16. Wang, R. 3D building modeling using images and LiDAR: A review. *Int. J. Image Data Fusion* **2013**, *4*, 273–292. [CrossRef]

17. Lingfors, D.; Bright, J.; Engerer, N.; Ahlberg, J.; Killinger, S.; Widén, J. Comparing the capability of low- and high-resolution LiDAR data with application to solar resource assessment, roof type classification and shading analysis. *Appl. Energy* **2017**, *205*, 1216–1230. [CrossRef]

18. Gadd, H.; Werner, S. Fault detection in district heating substations. *Applied Energy* **2015**. [CrossRef]

19. Werner, S. District heating and cooling in Sweden. *Energy* **2017**, *126*, 419–429. [CrossRef]

20. *Energy in the Buildings—Technical Properties and Calculations—Results from Project BETSI*; Technical Report; Swedish National Board of Housing, Building and Planning (Boverket): Stockholm, Sweden, 2010.

21. Olsson, D. Modellbaserad Styrning av Värmesystem Baserat på Prognostiserat väder Modellbaserad Styrning av Värmesystem Baserat på Prognostiserat Väder. Ph.D. Thesis, Chalmers University of Technology, Göteborg, Sweden, 2014.

22. Jangsten, M.; Kensby, J.; Dalenbäck, J.O.; Trüschel, A. Survey of radiator temperatures in buildings supplied by district heating. *Energy* **2017**, *137*, 292–301. [CrossRef]

23. Carpenter, B.; Gelman, A.; Hoffman, M.; Lee, D.; Goodrich, B.; Betancourt, M.; Brubaker, M.; Guo, J.; Li, P.; Riddell, A. Stan: A Probabilistic Programming Language. *J. Stat. Softw.* **2017**, *76*, 1–32. [CrossRef]

24. Wickham, H.; François, R.; Henry, L.; Müller, K. Dplyr: A Grammar of Data Manipulation; R Package Version 0.7.8. 2018. Available online: https://libraries.io/cran/dplyr (accessed on 2 February 2019).

25. ISO. *ISO 52010:2017—Energy Performance of Buildings—External Climatic Conditions—Part 1: Conversion of Climatic Data for Energy Calculations (EN ISO 52010:2017)*; ISO: Geneva, Switzerland, 2017.

26. ISO. *ISO 13370:2017—Thermal Performance of Buildings—Heat Transfer via the Ground—Calculation Methods (ISO 13370:2017)*; ISO: Geneva, Switzerland, 2017.

27. Walker, I.S.; Wilson, D.J. Field Validation of Algebraic Equations for Stack and Wind Driven Air Infiltration Calculations. *HVAC&R Res.* **1998**, *4*, 119–139. [CrossRef]

28. Wang, W.; Beausoleil-Morrison, I.; Reardon, J. Evaluation of the Alberta air infiltration model using measurements and inter-model comparisons. *Build. Environ.* **2009**, *44*, 309–318. [CrossRef]

29. Hayati, A.; Mattsson, M.; Sandberg, M. Evaluation of the LBL and AIM-2 air infiltration models on large single zones: Three historical churches. *Build. Environ.* **2014**, *81*, 365–379. [CrossRef]

30. Jokisalo, J.; Kurnitski, J.; Korpi, M.; Kalamees, T.; Vinha, J. Building leakage, infiltration, and energy performance analyses for Finnish detached houses. *Build. Environ.* **2009**, *44*, 377–387. [CrossRef]

31. Mirsadeghi, M.; Cóstola, D.; Blocken, B.; Hensen, J.L.M. Review of external convective heat transfer coefficient models in building energy simulation programs: Implementation and uncertainty. *Appl. Therm. Eng.* **2013**, *56*, 134–151. [CrossRef]

32. Liu, J.; Heidarinejad, M.; Gracik, S.; Srebric, J. The impact of exterior surface convective heat transfer coefficients on the building energy consumption in urban neighborhoods with different plan area densities. *Energy Build.* **2015**, *86*, 449–463. [CrossRef]

33. Van Den Wymelenberg, K. Patterns of occupant interaction with window blinds: A literature review. *Energy Build.* **2012**. [CrossRef]

34. Sandberg, E.; Engvall, K. *Delrapport 3 (MEBY): Beprövad Enkät-Hjälpmedel för Energiuppföljning*; Technical Report; MEBY: Stockholm, Sweden, 2002.

35. Hoyer-Klick, C.; Lefèvre, M.; Schroedter-Homscheidt, M.; Wald, L. MACC-III Deliverable D57.5: USER'S GUIDE to the MACC-RAD Services on Solar Energy Radiation Resources; Technical Report: Copernicus, European Union's Earth Observation Programme, 2015. Available online: https://www.researchgate.net/publication/281150019_USER'S_GUIDE_to_the_MACC-RAD_Services_on_solar_energy_radiation_resources_March_2015 (accessed on 2 February 2019).

36. Hersbach, H.; Dick, D. ERA5 reanalysis is in production. *ECMWF Newsl.* **2016**, *147*, 7.

37. Lundström, L.; Wallin, F. Heat demand profiles of energy conservation measures in buildings and their impact on a district heating system. *Appl. Energy* **2016**, *161*, 290–299. [CrossRef]

38. Ruiz, G.; Bandera, C. Validation of Calibrated Energy Models: Common Errors. *Energies* **2017**, *10*, 1587. [CrossRef]

39. Walton, G. *Thermal Analysis Research Program Reference Manual-NBSSIR 83-2655*; Technical Report; National Bureau of Standards: Washington, DC, USA, 1983.

40. Clarke, J.A. *Energy Simulation in Building Design*; Adam Hilger Lt: Bristol, UK, 1985.

41. Romero Rodríguez, L.; Nouvel, R.; Duminil, E.; Eicker, U. Setting intelligent city tiling strategies for urban shading simulations. *Sol. Energy* **2017**, *157*, 880–894. [CrossRef]

42. Akander, J. The ORC Method—Effective Modelling of Thermal Performance of Multilayer Building Components. Ph.D. Thesis, Royal Institute of Technology, Stockholm, Sweden, 2000.

energies

MDPI

Article

Experimental Investigation of Ventilation Performance of Different Air Distribution Systems in an Office Environment—Heating Mode

Arman Ameen *, Mathias Cehlin, Ulf Larsson and Taghi Karimipanah

Department of Building Engineering, Energy Systems and Sustainability Science, University of Gävle, 801 76 Gävle, Sweden; mathias.cehlin@hig.se (M.C.); ulf.larsson@hig.se (U.L.); taghi.karimipanah@hig.se (T.K.)
* Correspondence: arman.ameen@hig.se

Received: 30 April 2019; Accepted: 12 May 2019; Published: 15 May 2019

Abstract: A vital requirement for all-air ventilation systems are their functionality to operate both in cooling and heating mode. This article experimentally investigates two newly designed air distribution systems, corner impinging jet (CIJV) and hybrid displacement ventilation (HDV) in comparison against a mixing type air distribution system. These three different systems are examined and compared to one another to evaluate their performance based on local thermal comfort and ventilation effectiveness when operating in heating mode. The evaluated test room is an office environment with two workstations. One of the office walls, which has three windows, faces a cold climate chamber. The results show that CIJV and HDV perform similar to a mixing ventilation in terms of ventilation effectiveness close to the workstations. As for local thermal comfort evaluation, the results show a small advantage for CIJV in the occupied zone. Comparing C2-CIJV to C2-CMV the average draught rate (DR) in the occupied zone is 0.3% for C2-CIJV and 5.3% for C2-CMV with the highest difference reaching as high as 10% at the height of 1.7 m. The results indicate that these systems can perform as well as mixing ventilation when used in offices that require moderate heating. The results also show that downdraught from the windows greatly impacts on the overall airflow and temperature pattern in the room.

Keywords: corner impinging jet; corner mixing ventilation; hybrid displacement device; heating mode; thermal comfort; air exchange effectiveness; local air change effectiveness; draught rate,; downdraught

1. Introduction

Ventilation is one of the core systems that has a large impact on thermal comfort and indoor air quality (IAQ) in buildings. The design and implementation of air distribution systems require careful consideration, not only in terms of providing a good indoor environment, but also to be energy efficient. On top of these requirements the ventilation system needs to operate adequately both during cold and warm seasons.

One common air distribution system is called mixing ventilation (MV). This system is characterized by supplying air at high velocity into the room with the intended purpose of mixing the fresh supply air with the room air. This type of ventilation supply inlet is usually located high, close to the ceiling in the unoccupied area of the room. MV also creates a highly uniform vertical temperature field [1,2] which can result in slightly lower ventilation effectiveness when compared to other systems, e.g., impinging jet ventilation (IJV) and displacement ventilation (DV) [3–5].

DV and IJV are usually categorized as stratified ventilation systems when utilized in cooling mode [6–8]. The air from a DV supply device enters the room at a relatively low speed and at low height close to the floor when used for cooling. When entering the room, the fresh air will fall to the floor and continue flowing outward until it encounters a heat source. It will then start to heat up and start rising due to buoyancy effects, moving upwards to the upper parts of the room [9,10]. IJV works similarly in that respect but it uses impinging air jet with relatively high velocity and momentum which is discharged downwards close to a wall section at a distance from the floor area. IJV has been classified by several researchers as a hybrid system [10–12] in that it combines the positive effects of both MV and DV to overcome the shortcomings of the DV system, e.g., the limitation in covering the entire floor area due to the low momentum. Another downside of DV is the difficulty of utilizing the system during cold season when heating is required [6–8].

Most of the research around these ventilation systems has been conducted for cooling mode in a hot climate [13–15]. However, it is interesting to evaluate the IJV system when heating is required in a cold climate. Very few studies have been done to examine the ventilation performance of the IJV system when it operates in heating mode. Some researchers have stated that IJV can be used for heating due to its high momentum [9,16].

In a numerical study done by Ye et al. [17] they compared MV and IJV in order to evaluate the energy performance when used for heating in a large space with a high ceiling. Their results showed that IJV required less energy than MV for heating fresh air and re-circulating the return air. However, the fan power required more energy than MV. Adding these energy demands, the total heating energy usage for IJV was lower than that for MV. They concluded that the heating load index could be reduced by around 9–25 W/m^2 when the outdoor air temperature was in the range of −5 to 12 °C. Another study [18] also concluded that IJV is more energy efficient than MV in heating mode. This study was carried out in a climate chamber 3.0 (L) x 3.6 (W) x 2.6 (H) m which was placed in a laboratory space. It is worth mentioning that this study also included intermittent opening of a door that caused cold outside air to invade the heating space. One interesting observation in these two studies [17,18] was that MV created greater thermal stratification than IJV, which is the opposite of when these systems are used in cooling mode [11,19].

There have not been many studies conducted to evaluate typical ventilation systems for heating mode. Some of these studies have been focused on building optimization and control system which also included the control of the air handling unit [20–22]. Others have evaluated specific supply devices, such as stratum ventilation used for heating [23] or a low-temperature all-air heating system in an office cubicle that was equipped with an active supply device on the ceiling [24].

Due to the novelty of this research and to the authors' best knowledge there has not been any experimental research carried out to evaluate multiple IJV devices places in the corners of an office room for heating mode.

In a recent study Ameen et al. [19] evaluated and compared three different ventilation systems, corner impinging jet ventilation (CIJV), corner mixing ventilation (CMV) and DV. They evaluated heat removal effectiveness, local thermal comfort and indoor air quality in a mock-up medium-sized office room. The office contained two workstations, each with one mannequin and one piece of equipment. Nine different cases were examined with varying supply rates and heat sources. The results from this research showed that overall CIJV performed slightly better than the other two ventilation systems and there was a possibility of reducing the total energy usage. However, this research was conducted for summer cases, i.e., the systems were only evaluated for cooling mode.

The overall objective of this study is to continue the research done by Ameen et al. [19] and evaluate the same three types of air distribution systems for heating mode, i.e., winter conditions. The supply device for the DV evaluated in that study was a modified version that provided slightly higher supply velocity compared to traditional DV systems. This DV system is called hybrid displacement ventilation (HDV) in this study. These three different systems will be examined and compared to one another to evaluate their performance based on local thermal comfort and ventilation effectiveness in order to make an overall evaluation of their usability for both cooling and heating.

2. Theory and Mathematical Models

This section provides a brief overview and explanation of the key definitions of indoor climate indices which are used in this study. Since this study is a continuation of the experimental work done by Ameen et al. [19], a more in-depth explanation of these definitions can be found in that article.

According to ISO 7730 [25], draught rate (DR) describes the discomfort a person experiences due to unwanted cooling of the human body. This index is a function of air temperature, air velocity and turbulent intensity and predicts the percentage of dissatisfied due to draft. Another index, the percentage dissatisfied (PD), is related to the local discomfort due to high vertical air temperature between head and ankle. In this study the temperature difference, $\Delta T_{0.1-1.1}$ is used which is between ankle level (0.1 m) and neck level for a seated person (1.1 m).

Temperature effectiveness ($\varepsilon_{T'}$) [24] is an index that can be used to evaluate how effective space heating is in a space or location for heating mode. This is defined by

$$\varepsilon_{T'} = \frac{(T_i - T_o)}{\left(T_i - \overline{T}_{0.1,0.6,1.1}\right)} , \tag{1}$$

where T_i is the supply air temperature, $\overline{T}_{0.1,0.6,1.1}$ is the arithmetic mean air temperature of the heights 0.1, 0.6 and 1.1 m and T_o is the outlet air temperature. If $\varepsilon_{T'} > 1$, this indicates that the temperature in the occupied zone is higher than the outlet. If $\varepsilon_{T'} < 1$, this indicates that the temperature in the occupied zone is lower than the outlet which means lower utilization of the heat from the ventilation system to the occupied zone. For a perfect mixing ventilation system $\varepsilon_{T'} = 1$. This index is different from the one used in the cooling mode article [19] in that it can be used for heating mode.

The evaluation of ventilation effectiveness can be done in several ways. Two commonly used indexes related to IAQ are air exchange effectiveness (AEE) and air change effectiveness (ACE) [26–28]. The guidelines in ASHRAE Standard 129-1997 [26] require measuring ACE in 25% of the workstations or measuring a minimum of ten locations throughout the evaluated space. Another way to calculate AEE is to make measurements at the exhaust location. These indexes have been utilized by many researchers for evaluating indoor environments using different tracer gas techniques [29–31].

Inlet Archimedes number (Ar_i) [32,33] is a measure of the relative importance of buoyant and inertia forces. Ar_i is important in building airflows because it combines two important ventilation design parameters, supply air velocity and room temperature difference.

3. Experimental Set-Up

This study was conducted in a room 7.2 (L) × 4.1 (W) × 2.67 (H) m. The room resembled a medium-sized open-plan office space with three interior walls and one exterior wall. A climate chamber was built up in connection to the exterior wall of the test room as shown in Figure 1. For an in-depth description of the office wall materials, design setup, supply device dimensions, measuring equipment, etc. see [19].

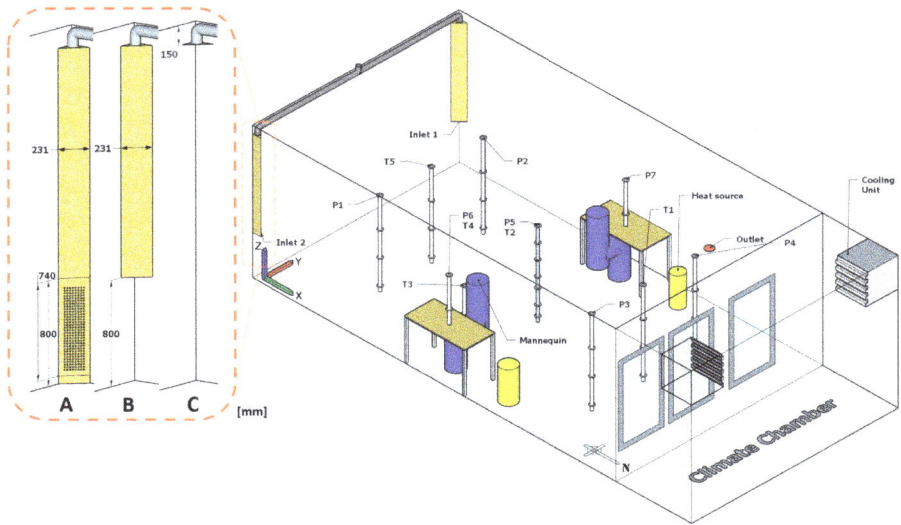

Figure 1. Layout of the office room and climate chamber. Three supply devices are illustrated for hybrid displacement ventilation (HDV, **A**), corner impinging jet ventilation (CIJV, **B**) and corner mixing ventilation (CMV, **C**).

Six cases were studied which are listed in Table 1. The primary supply air was maintained between 25.1–25.2 °C for C1 cases and 24.8 °C for C2 cases. It is important to mention that the comparisons were done in a non-dimensional form for all cases.

Table 1. Case conditions for different ventilation systems.

Case	Ventilation System	Supply Flow Rate [L/s]	Occupant [W]	Equipment [W]	Inlet Temp. [°C]	u_{in} [m/s]	$Ar_i \times 10^{-4}$
C1-HDV	HDV	2×15	2×100	2×75	25.2	0.50	−317
C1-CIJV	CIJV	2×15	2×100	2×75	25.2	1.13	−49
C1-CMV	CMV	2×15	2×100	2×75	25.1	2.98	−5
C2-HDV	HDV	2×20	2×100	2×75	24.8	0.67	−140
C2-CIJV	CIJV	2×20	2×100	2×75	24.8	1.51	−19
C2-CMV	CMV	2×20	2×100	2×75	24.8	3.98	−2

The wall facing the climate chamber contained three windows. The size of each window was 1.61 m × 0.91 m with a frame to glass ratio of 31.7%. Single pane windows were used with a total U-value of 4.6 W/m^2 °C. The surface temperatures of the windows were maintained at 10.1 ± 0.3 °C during the measurement periods. The location of the room was inside a large laboratory hall with a steady temperature condition of 22.3 ± 0.3 °C during the measurement periods. The heat transfer between the cold climate chamber and the office room was measured by several heat flux sensors of the type HFP01 made by Hukseflux. A total of three heat flux sensors were used. One was placed on the external wall 130.5 cm above the floor level and 47.2 cm from the west wall. Another sensor was placed in the dead center of the window located closest to the west wall. The last sensor was placed on the westside lower frame corner of the same window. The uncertainty of the heat flux sensor is ±3%. The total heat transfer from the office to the climate chamber and the surrounding surface is presented in Table 2. The climate chamber was maintained at −6.2 ± 0.3 °C during measurement periods. Two cooling units were used, one on each side of the climate chamber to provide an even cooling of the air inside the chamber. The heat loss through the rest of the office surfaces, excluding the external wall, amounted to -0.2 ± 0.1 W/m^2. The measurement positions and a top view layout of the experimental set-up are shown in Figure 2.

Table 2. Energy balance overview of the office room.

Case	Internal H. Generation [1] [W]	Ventilation [2] [W]	External Wall [3] [W]
C1-HDV	389	49.9	−417.2
C1-CIJV	389	52.7	−411.3
C1-CMV	389	54.2	−426.7
C2- HDV	389	48.5	−419.2
C2-CIJV	389	48.6	−424.5
C2-CMV	389	58.6	−411.3

[1] The internal heat was generated from the mannequins (2 × 100 W), two pieces of equipment (2 × 75 W) and from measuring equipment (39 W). [2] The ventilation effect was calculated from the flow rate and the temperature difference between the inlet and outlet. [3] This also includes the windows.

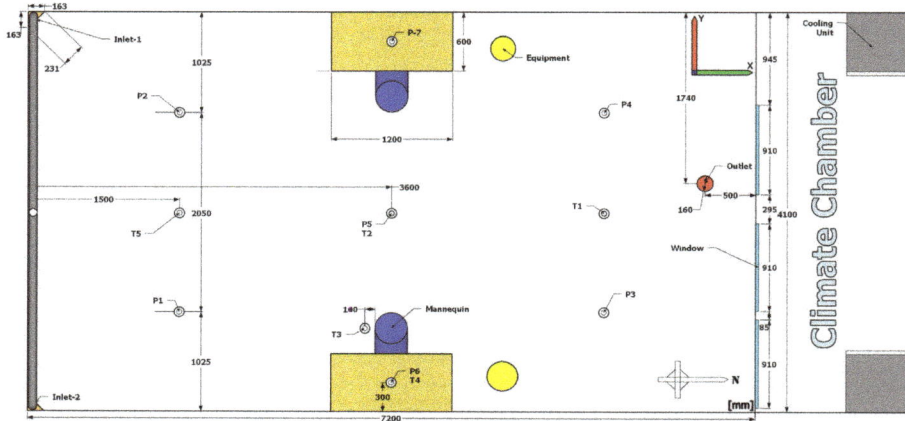

Figure 2. Measurement positions and schematic top view layout of office room and the climate chamber.

4. Results and Discussion

4.1. Flow Pattern and Thermal Conditions

The results of the air temperature for all the cases are shown in Figure 3. In position P1 (Figure 3a) and P2 (Figure 3b) which are close to the inlets, CMV together with C2-CIJV shows the lowest vertical temperature gradient. One possible reason for this is that the center velocities of the HDV system bypass the measurement probes in those positions. HDV shows the highest temperature gradient compared to CMV or CIJV. One important difference between these three systems is that HDV is designed to deliver the airstream perpendicular to the supply device surface, see Figures 1 and 2, compared to the other two systems where the flow is spread out in all directions when reaching the floor surface.

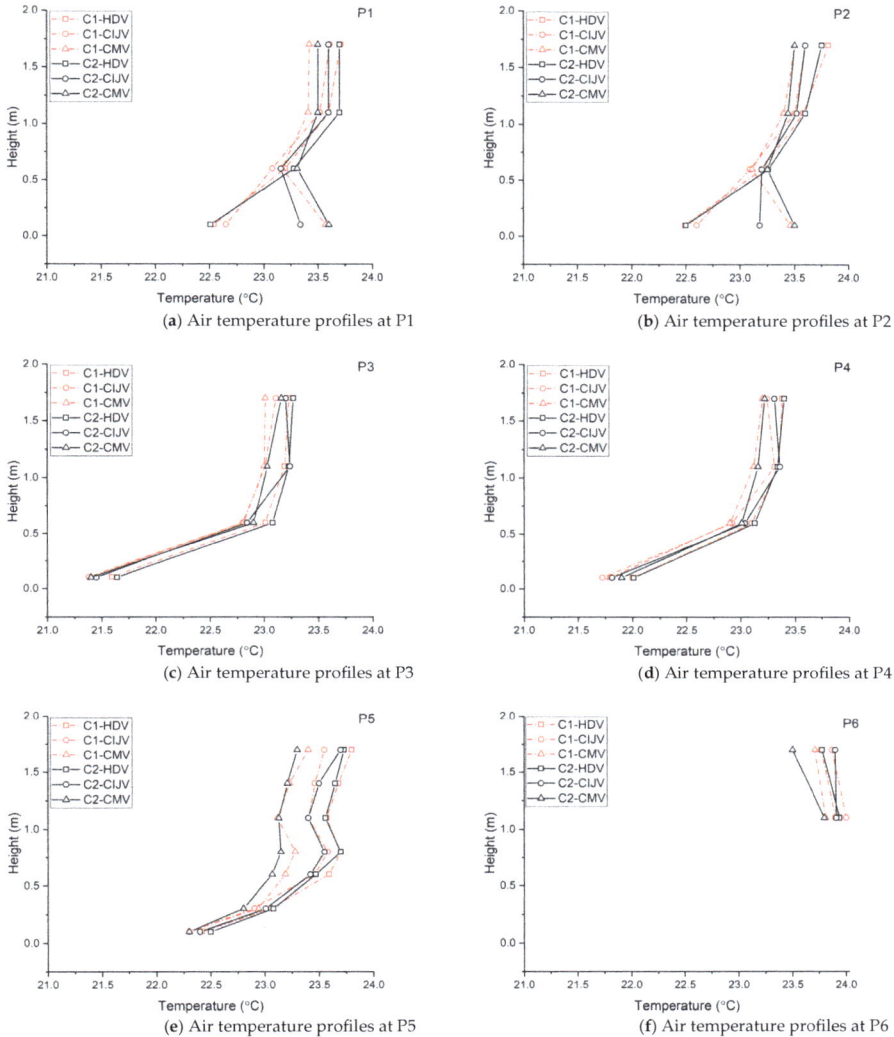

(**a**) Air temperature profiles at P1

(**b**) Air temperature profiles at P2

(**c**) Air temperature profiles at P3

(**d**) Air temperature profiles at P4

(**e**) Air temperature profiles at P5

(**f**) Air temperature profiles at P6

Figure 3. Air temperature profiles at position P1 (**a**), P2 (**b**), P3 (**c**), P4 (**d**), P5 (**e**) and P6 (**f**) for all cases.

In position P5 (Figure 3e) which is in the center of the office, the graph shows that the HDV cases have slightly higher temperatures and higher temperature gradient compared to the other systems. Another observation that can be seen is that the CMV cases have a lower temperature and temperature gradient at this position compared to the other ventilation systems. As suggested previously, the possible reason for this is the high level of entrainment created by this type of air distribution system.

Position P3 (Figure 3c) and P4 (Figure 3d) show a high temperature stratification. These locations are heavily affected by the external wall and the cold windows. The main driving force behind the airflows in this region is probably the downdraught flow from the cold windows, which is also shown in other studies [34–36]. It is also worth noting that P3 has a slightly lower temperature in the lower part of the room compared to P4. The reason for this is that P3 is adjacent to two windows compared to position P4 which is only close to one window, as can be seen in Figure 2.

The velocity profiles at P1 (Figure 4a) and P2 (Figure 4b) showed that the highest velocities were measured at 0.1 m above the floor level for C2-CMV, C1-CMV and C2-CIJV. In contrast, HDV has very low velocity compared to the other two ventilation systems at that height. A probable explanation is that the centerline of the HDV airstream bypasses the P1 and P2 measuring poles. The CMV cases have the highest velocities of all, reaching as high as 0.5 m/s for C2-CMV. This is explained by the special configuration of the CMV inlets. By being placed high up in the corner of the room and having a high supply velocity, the inlet air jets create a high level of entrainment. This results in an airstream with higher momentum and higher boundary layer thickness compared to the other systems, which was also evident in the cooling mode study [19].

In position P5 (Figure 4e), CIJV shows slightly lower velocity in the upper part of the room compared to the other systems. This also results in lower ACE_p value for position P5 (T2) compared to the other systems as seen in Table 4.

The velocities in P3 (Figure 4c) and P4 (Figure 4d) show a stagnation of the air movement. One probable explanation for this is that the cold air movement is below 0.1 m in the direction towards the occupied zone and warm air above 1.7 m in the opposite direction towards the ventilation outlet and the wall facing the climate chamber [37].

The draught levels at P1 (Figure 5a) and P2 (Figure 5b) show a strong connection to the velocity profiles as expected. Due to high velocities at P1 and P2 the draught levels are higher in this part of the room for CIJV and CMV.

P5 (Figure 5e) shows acceptable DR levels for all cases, with CIJV showing excellent levels for both C1 and C2. When comparing C2-CIJV to C2-CMV the results show that the average DR rate at P5 is 0.3% for C2-CIJV and 5.3% for C2-CMV with the highest difference reaching as high as 10% at the height of 1.7 m.

In position P3 (Figure 5c) and P4 (Figure 5d) the DR is at acceptable levels for all cases. However, there is a high possibility of cold air coming from the cold side of the office as suggested previously.

Another way to illustrate the correlation between high velocities and high draught rates can be seen in Figure 6. Figure 6a shows the maximum draught rate (DR_{max}) based on all the points in each location at P1–P2, P5 and P3–P4. Figure 6b shows the maximum velocity (U_{max}) at the same locations. The strong connection between the high velocities and the high DR is shown in the graph. Another interesting observation is that in P3–P4 the differences between the different ventilation systems in terms of (ΔT_{max}) are almost nonexistent. One probable reason for this is that in this part of the room the cold wall and windows are having major impact on the flow and temperature pattern. Since the setting of the outside cold temperature is the same for all cases, the impact should be equal for cases with the same flow rate and inlet temperature, i.e., C1 and C2.

Table 3 shows that PD in all cases are within category A classification.

Table 3. Local discomfort (PD) due to high vertical air temperature between head and ankle.

Case	Position	C1-HDV	C1-CIJV	C1-CMV	C2-HDV	C2-CIJV	C2-CMV
	P1	0.8%	0.7%	0.4%	0.9%	0.4%	0.3%
	P2	0.8%	0.7%	0.3%	0.8%	0.4%	0.3%
PD	P3	1.2%	1.3%	1.2%	1.2%	1.4%	1.3%
	P4	1.0%	1.2%	1.0%	1.0%	1.2%	0.9%
	P5	0.8%	0.8%	0.6%	0.8%	0.7%	0.6%

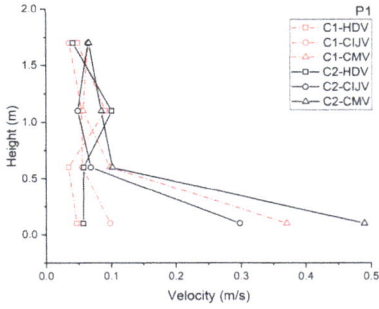

(**a**) Velocity profiles at P1

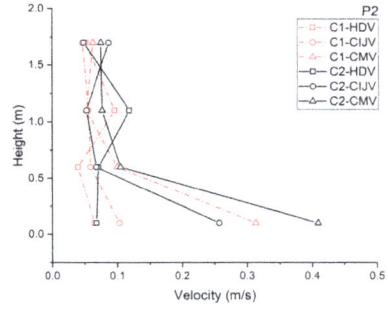

(**b**) Velocity profiles at P2

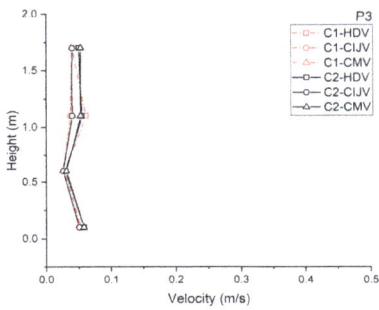

(**c**) Velocity profiles at P3

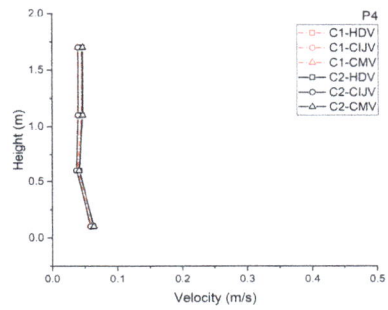

(**d**) Air temperature profiles at P4

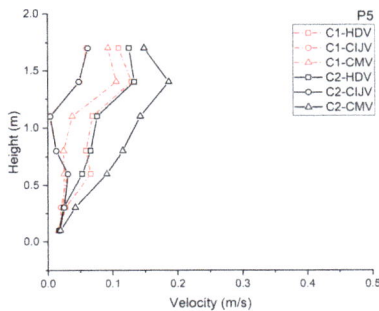

(**e**) Air temperature profiles at P5

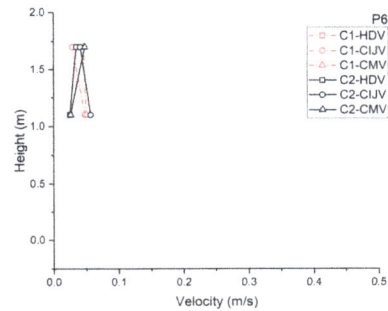

(**f**) Air temperature profiles at P6

Figure 4. Velocity profiles at position P1 (**a**), P2 (**b**), P3 (**c**), P4 (**d**), P5 (**e**) and P6 (**f**) for all cases.

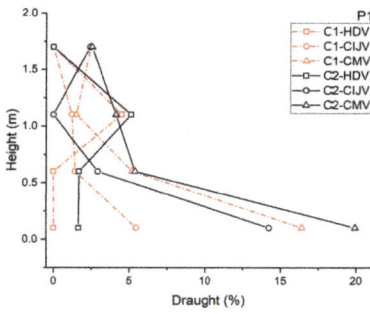

(a) Draught levels at P1

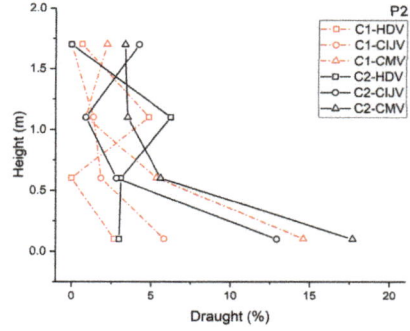

(b) Draught levels at P2

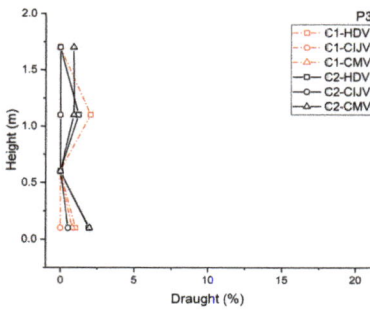

(c) Draught levels at P3

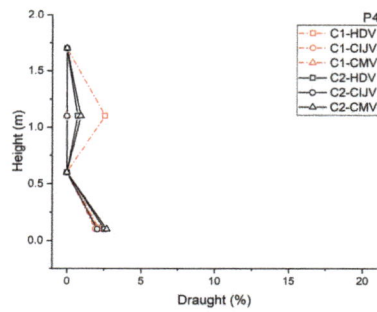

(d) Draught levels at P4

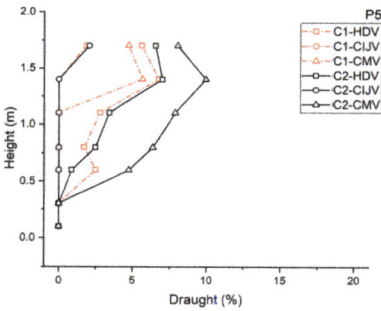

(e) Draught levels at P5

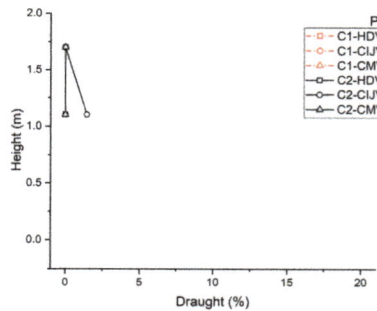

(f) Draught levels at P6

Figure 5. Draught levels at position P1 (**a**), P2 (**b**), P3 (**c**), P4 (**d**), P5 (**e**) and P6 (**f**) for all cases.

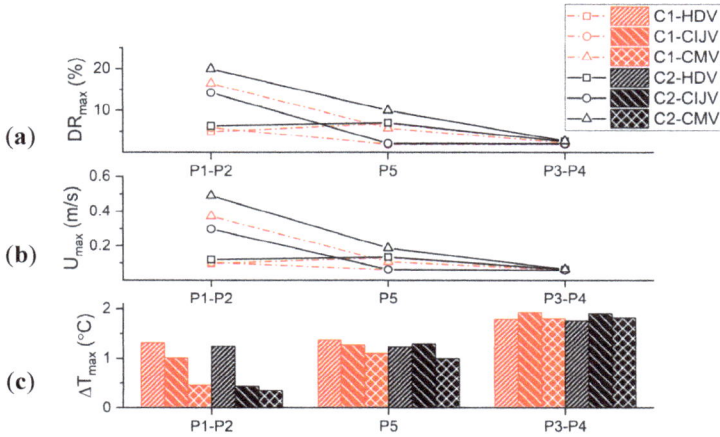

Figure 6. Maximum draught rate (**a**), maximum velocity (**b**) and maximum temperature difference (**c**) in positions P1–P2, P5 and P3–P4.

4.2. Ventilation Effectiveness

The ACE_p values presented in Table 4 show that C2-CMV has the most uniform ACE_p when compared to the other systems due to the high entrainment it creates when entering the room. However, all the cases show similar ACE_p. This indicates that the air is equally "fresh" at breathing level in all the measuring locations, corresponding to an IAQ that equals to a fully mixed condition.

Table 4. Air change effectiveness (ACE), local and average, and air exchange effectiveness (AEE) for all cases.

Case	Position	C1-HDV	C1-CIJV	C1-CMV	C2-HDV	C2-CIJV	C2-CMV
	T1	1.05	1.02	1.03	1.04	1.01	1.03
	T2	1.09	1.05	1.07	1.09	1.06	1.07
ACE_p	T3 [1]	1.05	1.06	1.01	1.02	0.98	1.02
	T4	1.01	1.06	1.06	1.03	1.00	1.00
	T5	1.15	1.14	1.13	1.19	1.12	1.08
ACE_{avg} [2]		1.07	1.07	1.06	1.07	1.03	1.04
AEE		0.51	0.51	0.52	0.51	0.49	0.51

[1] The location was in the occupied zone close to the mannequin, as also shown in Figure 2. [2] ACE_{avg} is the average ACE value for the measuring points T1–T5.

At position T3 located close to the mannequin the best performance is achieved by C1-CIJV and C1-HDV, although by a very small margin. Not surprisingly at T5, which is close to the inlets, the highest values of ACE_p are obtained. Overall, the ventilation systems all perform very similar, at breathing level, to a MV with values close to 1. This is different when compared to the previous study in which HDV and CIJV operated in cooling mode [19].

The AEE values are close to each other in all cases. The reason for this is that AEE takes into account the mean age of air for the entire room. This means that a case can have high ACE_p values in some zones, but lower values in others.

Table 5 shows the average $\varepsilon_{T'}$ of the heights 0.1, 0.6 and 1.1m for all studied cases. These results suggest that the studied ventilation systems produce similar results in heating mode when evaluating the ventilation effectiveness in the occupied zone.

Table 5. Average ε_T in locations P1–P7 for all cases.

Case	Position	C1-HDV	C1-CIJV	C1-CMV	C2-HDV	C2-CIJV	C2-CMV
ε_T [1]	P1	0.66	0.70	0.87	0.61	0.70	0.93
	P2	0.66	0.69	0.84	0.59	0.67	0.88
	P3	0.53	0.53	0.56	0.46	0.44	0.52
	P4	0.58	0.58	0.60	0.50	0.49	0.58
	P5	0.69	0.69	0.67	0.61	0.58	0.63
ε_T [2]	P6	1.07	1.24	1.15	1.13	1.11	1.24
	P7	0.93	1.01	1.00	0.95	0.95	1.04

[1] calculated by using the arithmetic mean air temperature of the heights 0.1, 0.6, and 1.1 m. [2] calculated by using the arithmetic mean air temperature of the height 1.1 m only.

To summarize the results, HDV and CIJV provide similar ventilation effectiveness to CMV. As for local thermal comfort evaluation, the results show a small advantage for CIJV in the occupied zone. It is worth mentioning that the flow pattern of the downdraught from the windows has a major impact on the overall airflow pattern in the room. Hence, further studies are recommended in order to fully evaluate the effects of downdraught on the performance of these air distribution systems operating in heating mode.

5. Conclusions

These are the most significant conclusion:

- CIJV and HDV perform similar to a mixing ventilation in terms of ventilation effectiveness close to the workstations.
- CIJV performs slightly better that the other systems regarding local thermal comfort close to the workstations

This indicates that these systems can perform as good as MV when used in offices that requires moderate heating. This also provides the possibility of using CIJV and HDV both for heating and cooling. It is important to note that this study was based only on one-room geometry and one configuration of workstation placement in the room. Further studies have to be conducted in order to evaluate corner impinging jet ventilation and hybrid displacement ventilation in more details and also to evaluate different location for the workstations, different heating demands, different locations for the supply inlets, different supply temperatures, etc.

Author Contributions: Main contribution was made by A.A.; The other three authors contributed equally.

Funding: Internal research funding from the University of Gävle.

Acknowledgments: The authors are grateful for the valuable help from University of Gävle's laboratory staff and the possibility of using the university's laboratory facilities.

Conflicts of Interest: The authors declare no conflicts of interest.

Nomenclature

ACE	air change effectiveness [-]
ACE$_{avg}$	average spatial air change effectiveness in a region [-]
ACE$_p$	local air change effectiveness [-]
AEE	air exchange effectiveness [-]
Ar$_i$	inlet Archimedes number [-]
CIJV	corner impinging jet ventilation
CMV	corner mixing ventilation
DR	draught rate [%]

DR_{max}	maximum draught rate between 0.1 m and 1.7 m above floor level [%]
DV	displacement ventilation
HDV	hybrid displacement ventilation
IAQ	indoor air quality
IJV	impinging jet ventilation
MV	mixing ventilation
PD	percentage dissatisfied due to vertical air temperature difference [%]
$\overline{T}_{0.1,0.6,1.1}$	arithmetic mean air temperature based on the values at height 0.1, 0.6 and 1.1 m [°C]
T_i	mean supply air temperature [°C]
T_o	mean outlet air temperature [°C]
$\Delta T_{0.1-1.1}$	vertical air temperature gradient between 0.1 m and 1.1 m above floor level [°C]
ΔT_{max}	maximum air temperature gradient between 0.1 m and 1.7 m above floor level [°C]
u_{in}	nominal inlet air velocity [m/s]
U_{max}	maximum air velocity between 0.1 m and 1.7 m above floor level [m/s]
W	power [kg·m^2·s^{-3}]
$\varepsilon_{T'}$	temperature effectiveness [–]

References

1. Etheridge, D.W.; Sandberg, M. *Building Ventilation: Theory and Measurement*; John Wiley & Sons: Chichester, UK, 1996; ISBN 978-0-471-96087-4.
2. Awbi, H.B. *Ventilation of Buildings*; Routledge: New York, NY, USA, 2002; ISBN 1135817413.
3. Larsson, U.; Moshfegh, B. Comparison of ventilation performance of three different air supply devices: A measurement study. *Int. J. Vent.* **2017**, *16*, 244–254. [CrossRef]
4. Cho, Y.; Awbi, H.; Karimipanah, T. A comparison between four different ventilation systems. In Proceedings of the 8th International Conference on Air Distribution in Rooms (Roomvent 2002), Copenhagen, Denmark, 8–11 September 2002; pp. 181–184.
5. Amai, H.; Novoselac, A. Experimental study on air change effectiveness in mixing ventilation. *Build. Environ.* **2016**, *109*, 101–111. [CrossRef]
6. Serra, N.; Semiao, V. Comparing displacement ventilation and mixing ventilation as HVAC strategies through CFD. *Eng. Comput.* **2009**, *26*, 950–971. [CrossRef]
7. He, G.; Yang, X.; Srebric, J. Removal of contaminants released from room surfaces by displacement and mixing ventilation: Modeling and validation. *Indoor Air* **2005**, *15*, 367–380. [CrossRef] [PubMed]
8. Lin, Z.; Chow, T.; Fong, K.; Tsang, C.; Wang, Q. Comparison of performances of displacement and mixing ventilations. Part II: Indoor air quality. *Int. J. Refrig.* **2005**, *28*, 288–305. [CrossRef]
9. Awbi, H.B. *Ventilation Systems: Design and Performance*; Routledge: New York, NY, USA, 2007; ISBN 1135815313.
10. Cao, G.; Awbi, H.; Yao, R.; Fan, Y.; Sirén, K.; Kosonen, R.; Zhang, J.J. A review of the performance of different ventilation and airflow distribution systems in buildings. *Build. Environ.* **2014**, *73*, 171–186. [CrossRef]
11. Chen, H.; Janbakhsh, S.; Larsson, U.; Moshfegh, B. Numerical investigation of ventilation performance of different air supply devices in an office environment. *Build. Environ.* **2015**, *90*, 37–50. [CrossRef]
12. Awbi, H.B. Ventilation for good indoor air quality and energy efficiency. *Energy Procedia* **2017**, *112*, 277–286. [CrossRef]
13. Chen, H.; Moshfegh, B.; Cehlin, M. Numerical investigation of the flow behavior of an isothermal impinging jet in a room. *Build. Environ.* **2012**, *49*, 154–166. [CrossRef]
14. Chen, H.; Moshfegh, B.; Cehlin, M. Investigation on the flow and thermal behavior of impinging jet ventilation systems in an office with different heat loads. *Build. Environ.* **2013**, *59*, 127–144. [CrossRef]
15. Kobayashi, T.; Sugita, K.; Umemiya, N.; Kishimoto, T.; Sandberg, M. Numerical investigation and accuracy verification of indoor environment for an impinging jet ventilated room using computational fluid dynamics. *Build. Environ.* **2017**, *115*, 251–268. [CrossRef]
16. Karimipanah, T.; Awbi, H. Theoretical and experimental investigation of impinging jet ventilation and comparison with wall displacement ventilation. *Build. Environ.* **2002**, *37*, 1329–1342. [CrossRef]
17. Ye, X.; Zhu, H.; Kang, Y.; Zhong, K. Heating energy consumption of impinging jet ventilation and mixing ventilation in large-height spaces: A comparison study. *Energy Build.* **2016**, *130*, 697–708. [CrossRef]

18. Ye, X.; Kang, Y.; Yang, X.; Zhong, K. Temperature distribution and energy consumption in impinging jet and mixing ventilation heating rooms with intermittent cold outside air invasion. *Energy Build.* **2018**, *158*, 1510–1522. [CrossRef]

19. Ameen, A.; Cehlin, M.; Larsson, U.; Karimipanah, T. Experimental investigation of ventilation performance of different air distribution systems in an office environment—Cooling mode. *Energies* **2019**, *12*, 1354. [CrossRef]

20. Mandin, C.; Trantallidi, M.; Cattaneo, A.; Canha, N.; Mihucz, V.G.; Szigeti, T.; Mabilia, R.; Perreca, E.; Spinazzè, A.; Fossati, S. Assessment of indoor air quality in office buildings across Europe–The OFFICAIR study. *Sci. Total Environ.* **2017**, *579*, 169–178. [CrossRef]

21. Baldi, S.; Michailidis, I.; Ravanis, C.; Kosmatopoulos, E.B. Model-based and model-free "plug-and-play" building energy efficient control. *Appl. Energy* **2015**, *154*, 829–841. [CrossRef]

22. Michailidis, I.T.; Baldi, S.; Pichler, M.F.; Kosmatopoulos, E.B.; Santiago, J.R. Proactive control for solar energy exploitation: A german high-inertia building case study. *Appl. Energy* **2015**, *155*, 409–420. [CrossRef]

23. Kong, X.; Xi, C.; Li, H.; Lin, Z. A comparative experimental study on the performance of mixing ventilation and stratum ventilation for space heating. *Build. Environ.* **2019**. [CrossRef]

24. Rabani, M.; Madessa, H.B.; Nord, N.; Schild, P.; Mysen, M. Performance assessment of all-air heating in an office cubicle equipped with an active supply diffuser in a cold climate. *Build. Environ.* **2019**. [CrossRef]

25. *ISO 7730—Ergonomics of the Thermal Environment-Analytical Determination and Interpretation of Thermal Comfort Using Calculation of the PMV and PPD Indices and Local Thermal Comfort Criteria*; ISO: Geneva, Switzerland, 2005.

26. *ASHRAE Standard 129-1997 (RA 2002)—Measuring Air Change Effectiveness*; ASHRAE: Atlanta, GA, USA, 2002.

27. Mundt, M.; Mathisen, H.M.; Moser, M.; Nielsen, P.V. *Ventilation Effectiveness*; REHVA: Brussels, Belgium, 2004; ISBN 978-29-6004-680-9.

28. Breum, N. Ventilation efficiency in an occupied office with displacement ventilation—A laboratory study. *Environ. Int.* **1992**, *18*, 353–361. [CrossRef]

29. Cehlin, M.; Karimipanah, T.; Larsson, U.; Ameen, A. Comparing thermal comfort and air quality performance of two active chilled beam systems in an open-plan office. *J. Build. Eng.* **2019**, *22*, 56–65. [CrossRef]

30. Sandberg, M.; Blomqvist, C. A quantitative estimate of the accuracy of tracer gas methods for the determination of the ventilation flow rate in buildings. *Build. Environ.* **1985**, *20*, 139–150. [CrossRef]

31. Krajčík, M.; Simone, A.; Olesen, B.W. Air distribution and ventilation effectiveness in an occupied room heated by warm air. *Energy Build.* **2012**, *55*, 94–101. [CrossRef]

32. Both, B.; Szántó, Z. Objective investigation of discomfort due to draught in a tangential air distribution system: Influence of air diffuser's offset ratio. *Indoor Built Environ.* **2018**, *27*, 1105–1118. [CrossRef]

33. Cehlin, M. *Visualization of Air Flow, Temperature and Concentration Indoors: Whole-Field Measuring Methods and CFD*; KTH: Stockholm, Sweden, 2006.

34. Ge, H.; Fazio, P. Experimental investigation of cold draft induced by two different types of glazing panels in metal curtain walls. *Build. Environ.* **2004**, *39*, 115–125. [CrossRef]

35. Manz, H.; Frank, T. Analysis of thermal comfort near cold vertical surfaces by means of computational fluid dynamics. *Indoor Built Environ.* **2004**, *13*, 233–242. [CrossRef]

36. Jurelionis, A.; Isevičius, E. CFD predictions of indoor air movement induced by cold window surfaces. *J. Civ. Eng. Manag.* **2008**, *14*, 29–38. [CrossRef]

37. Heiselberg, P. Draught risk from cold vertical surfaces. *Build. Environ.* **1994**, *29*, 297–301. [CrossRef]

Article

Experimental Investigation of the Ventilation Performance of Different Air Distribution Systems in an Office Environment—Cooling Mode

Arman Ameen *, Mathias Cehlin, Ulf Larsson and Taghi Karimipanah

Department of Building Engineering, Energy Systems and Sustainability Science, University of Gävle,
801 76 Gävle, Sweden; mathias.cehlin@hig.se (M.C.); ulf.larsson@hig.se (U.L.); taghi.karimipanah@hig.se (T.K.)
* Correspondence: arman.ameen@hig.se

Received: 17 March 2019; Accepted: 4 April 2019; Published: 9 April 2019

Abstract: The performance of a newly designed corner impinging jet air distribution method with an equilateral triangle cross section was evaluated experimentally and compared to that of two more traditional methods (mixing and displacement ventilation). At nine evenly chosen positions with four standard vertical points, air velocity, turbulence intensity, temperature, and tracer gas decay measurements were conducted for all systems. The results show that the new method behaves as a displacement ventilation system, with high air change effectiveness and stratified flow pattern and temperature field. Both local air change effectiveness and air exchange effectiveness of the corner impinging jet showed high quality and promising results, which is a good indicator of ventilation effectiveness. The results also indicate that there is a possibility to slightly lower the airflow rates for the new air distribution system, while still meeting the requirements for thermal comfort and indoor air quality, thereby reducing fan energy usage. The draught rate was also lower for corner impinging jet compared to the other tested air distribution methods. The findings of this research show that the corner impinging jet method can be used for office ventilation.

Keywords: corner impinging jet; mixing ventilation; displacement ventilation; tracer gas; air exchange effectiveness; local air change effectiveness; draught rate

1. Introduction

As humans spend more time indoors than outdoors, the importance of having a good indoor environment becomes crucial when designing or renovating buildings. Whether we are at home, at the office, or at school, a good indoor climate is important for several reasons, such as having fresh air, maintaining a high work performance [1] and a good cognitive function [2], and complying with state and local regulations. One of the most important systems for controlling the indoor climate is the ventilation system. This is also one of the components in a building that requires a lot of energy, which is also an important issue to address, since buildings accounted for roughly 22% of total world energy use in 2016 [3]. There are several types of air distribution systems that work in different ways. Some of these have been researched extensively, while others are still in the evaluation stage.

One of these systems is called displacement ventilation (DV). Ventilation based on this type of device enters the room at a relatively low speed and at low height, close to the floor. When used for cooling the space, the inlet air will fall to the floor level and continue flowing out in the space until it encounters heat sources. It will then start to heat up and rise as a result of buoyancy effects, moving to the upper part of the space, where it will usually exit from an extraction point located close to the ceiling. There has been a lot of research on this type of ventilation. In the early 2000s, a research group examined whether the indoor air quality (IAQ) in the breathing zone was better than the average IAQ in the occupied zone, and the results showed that there was a 35–50% improvement in the breathing zone due to the buoyancy effect around the mannequin body [4]. In a more recent publication, four different ventilation systems were compared in terms of ventilation efficiency, thermal comfort and energy-saving potential by using numerical simulations. The four systems compared were DV, mixing ventilation (MV), wall confluent jets ventilation (WCJ), and impinging jet ventilation (IJV). The comparison was done for an office environment. The results showed that DV was better than the other air delivery systems in all areas except when evaluating the vertical temperature gradient between ankle and neck levels for a standing person [5]. Other research groups have also concluded that DV is more suitable for cooling by exhibiting higher ventilation and energy effectiveness. It also creates a temperature stratification which facilitates the concentration of pollutants in the lower strata of the occupied space [6–8].

Another type of air distribution system is IJV, which has been the subject of much research [9–13]. However, very few studies have examined multiple inlet devices based on IJV. In an early study, Karimipanah and Awbi [14] compared IJV to wall displacement ventilation in a laboratory classroom. They tested several key parameters such as ventilation efficiency, local mean age of air, and other characteristic parameters both experimentally and by numerical simulations. One of the conclusions of their research was that the IJV system showed a slight improvement in mean age of air and velocity distributions due to a better balance between buoyancy and momentum forces. Similarly, Koufi et al. [15] also reached the conclusion that IJV has higher ventilation effectiveness. They conducted a numerical simulation by comparing two types of MV, DV and IJV, investigated under isothermal conditions.

In another study, numerical simulations were carried out to evaluate IJV, MV, and DV. One of the results of that study showed that IJV has an advantage over DV in that it can also be used for heating in winter time; also, IJV was found to distribute air more efficiently to the occupied zone when compared to a top–top-configured MV [16].

IJV has been classified by some researchers as a hybrid system [5,17,18] in that it combines the positive effects of both MV and DV to overcome the shortcomings of the DV system, for example the limitation in covering the entire floor area due to low velocity.

An interesting placement of an IJV device is in the corners of a room. In this study, this configuration is called corner impinging jet ventilation (CIJV). One aspect of CIJV is the possibility of having a non-intrusive supply device, i.e., the device can almost be hidden in the corner of the room. This property increases the value of this device compared to others that are installed in the middle of a wall section. To the authors' best of knowledge, there has not been any experimental research carried out to evaluate multiple IJV devices placed in the corners of an office room.

It is also important to consider some shortcomings of this type of system. The region in front of the jet impact area must be cleared from any furniture or objects for proper operation, and the possibility of stratification discomfort and draft might occur near the supply device [19].

Finally, the most common ventilation system used in buildings today is MV. This system can be installed in various configurations. This type of ventilation has been researched extensively. In one study [20], an experimental investigation was conducted comparing MV to confluent jet ventilation and a floor-mounted underfloor air distribution system. This investigation was conducted in a medium-sized open-plan office which included six workstations. The study focused on the thermal comfort performance of these systems, and the conclusion reached was that MV had some shortcomings in terms of higher draught in the occupied zone and lower heat removal capacity compared to the other systems. Similar investigative comparisons between MV and other types of ventilation system to evaluate thermal comfort and/or ventilation effectiveness have been made by many research groups [6,21–28].

This article focuses on the experimental evaluation of the potential benefits of a newly designed impinging jet ventilation system located in the corners of an office-type environment. The CIJV system was compared to two other separate systems based on DV and MV, to evaluate heat removal effectiveness, local thermal comfort, and indoor air quality in the occupants' breathing zone. The DV supply device used in this study was designed for relatively high momentum compared to traditional DV devices. The supply devices for MV were located high in the room corners, providing a type of ventilation hereby called corner mixing ventilation (CMV).

2. Theory and Mathematical Models

This section provides an overview and explanation of the key definitions of indoor climate indices which are used in this study. According to ISO 7730 [29], the draught rate (DR) describes the discomfort a person experiences due to unwanted cooling of the human body. This index is a function of air temperature, air velocity, and turbulent intensity, and predicts the percentage of dissatisfaction due to draft. This is estimated by

$$DR = (3.14 + 0.37 \cdot u_a \cdot I_p)(34 - T_a)(u_a - 0.05)^{0.62}$$
$$\text{For } u_a < 0.05 \text{ m/s use } u_a = 0.05 \text{ m/s} \tag{1}$$
$$\text{For } DR > 100\% \text{ use } DR = 100\%,$$

where u_a is the mean air velocity, I_p is the local turbulence intensity, and T_a is the local temperature.

Another index, the percentage dissatisfied (PD), is related to the local discomfort due to a high vertical air temperature difference between head and ankle. In this study, the temperature difference, $\Delta T_{0.1-1.1}$ between the ankle level (0.1 m) and the neck level for a seated person (1.1 m) was used. According to ISO 7730, PD is estimated by

$$PD = \frac{100(\%)}{1 + exp(5.76 - 0.856 \cdot \Delta T_{0.1-1.1})} . \tag{2}$$

In ISO 7730, the local thermal comfort is categorized into three levels (A, B, and C) for office environments. Table 1 shows the criteria for each category.

Table 1. Local thermal comfort category based on ISO 7730. DR: draught rate PD: percentage dissatisfied.

Category	DR	Maximum Mean Air Speed (m/s)		PD
		Summer	Winter	
A	<10%	0.12	0.10	<3%
B	<20%	0.19	0.16	<5%
C	<30%	0.24	0.21	<10%

Temperature effectiveness (ε_T) [30–33] is a parameter that can be used to evaluate the effectiveness of heat removal and is defined by

$$\varepsilon_T = \frac{(T_o - T_i)}{(\overline{T}_{0.1, \, 0.6, 1.1} - T_i)} , \tag{3}$$

where $\overline{T}_{0.1,0.6,1.1}$ is the arithmetic mean air temperature of the heights 0.1, 0.6, and 1.1 m, T_o is the outlet air temperature, and T_i is the supply air temperature.

The evaluation of ventilation effectiveness can be done in multiple ways. Two commonly used indices related to IAQ are air exchange effectiveness (AEE) and air change effectiveness (ACE) [34–36]. The guidelines in ASHRAE Standard 129-1997 [34] require measuring ACE in 25% of the workstations or measuring a minimum of 10 locations throughout the evaluated space. Another way to calculate AEE is to make measurements at the exhaust location. These indices have been utilized by many researchers for evaluating indoor environments using different tracer gas techniques [21,22,37–44].

The definition of local mean age of air is the average time it takes for fresh air to travel from an inlet to any place in the room [40,45,46]. The air at the examined place is a mixture of components of the air present in the room for different lengths of time. The supply air should be distributed in such a way that the occupants are "flushed" with fresh air at the breathing level. The local mean age of air,

$\overline{\tau}_p$, at a specific point is calculated from sampled tracer gas concentration histories. By utilizing the step-down tracer method, the local mean age of air is obtained by

$$\overline{\tau}_p = \frac{1}{C_p(0)} \int_0^\infty C_p(t)dt \, , \tag{4}$$

where the start of the decay corresponds to a time of zero with initial concentration $C_p(0)$, and $C_p(t)$ is the tracer gas concentration at time t.

The mean age of air for the entire space, $\langle \overline{\tau} \rangle$, is calculated by

$$\langle \overline{\tau} \rangle = \frac{\int_0^\infty t C_0(t)dt}{\int_0^\infty C_0(t)dt} \, . \tag{5}$$

AEE is calculated by

$$\text{AEE} = \frac{\tau_n}{2\langle \overline{\tau} \rangle} \, , \tag{6}$$

where τ_n is the nominal time constant defined as the reciprocal of the nominal air exchange rate. $\langle \overline{\tau} \rangle$ represents the mean age of air at the ventilation outlet obtained from tracer gas measurements [45]. τ_n is calculated by

$$\tau_n = \frac{V}{q_v} \, , \tag{7}$$

where V is the volume of the room, and q_v is the ventilation flow rate.

ACE relates to local air change effectiveness (ACE_p) or average spatial air change effectiveness in a region (ACE_{avg}) within the breathing level against the nominal time constant of the ventilation system. They are defined by

$$\text{ACE}_p = \frac{\tau_n}{\tau_p} \tag{8}$$

and

$$\text{ACE}_{avg} = \frac{\tau_n}{\overline{\tau}_{avg}} \, . \tag{9}$$

Here, τ_p is the local age of air in a considered point, and $\overline{\tau}_{avg}$ is the arithmetic average age of air for several points. A reference case (ACE = 1.0) is used, which is correspond to perfect air mixing.

The inlet Archimedes number (Ar_i) [47,48] is a measure of the relative importance of buoyant and inertia forces. Ar_i is important in building airflows because it combines two important ventilation design parameters, i.e., supply air velocity and room temperature difference. Ar_i is defined by

$$\text{Ar}_i = g \cdot \frac{(T_r - T_i)}{T_r} \cdot \frac{\sqrt{A_e}}{(u_{in})^2} \, , \tag{10}$$

where g is the gravitational acceleration, T_r [K] is the mean air temperature in the center of the room at 1.7 m above floor level, T_i [K] is the mean supply air temperature, A_e is the inlet supply opening area, and u_{in} is the nominal inlet air velocity.

3. Experimental Set-Up

The study was carried out in a medium-sized mock-up of an open-plan office room, 7.2 (L) × 4.1 (W) × 2.67 (H) m. The room resembled a medium-sized open office space with three interior walls and one exterior wall. The room was divided into two sections that were occupied by two workstations, each containing a desk, a chair, and a seated thermal mannequin, as can be seen in Figure 1. The composition of the side walls from the inside to the outside were as follows: 15 mm wood sheet, 35 mm air gap, 15 mm wood sheet, 190 mm insulation, and 5 mm wood sheet. The floor and the main ceiling were insulated by a 150 mm-thick layer of mineral wool and covered by a layer of plastic sheet to reduce air infiltration. The suspended ceiling consisted of 60 cm × 60 cm fiberglass tiles and was located 31 cm below the main ceiling. The location of the room was inside a large laboratory hall with a steady temperature of 24.6 °C ± 0.9 °C during the measurement periods. The supply inlets were installed in the corners of the south wall. The main air inlet was in the middle of the wall section, and air was delivered to each device through well-insulated (20 mm mineral wool) ventilation tubes. The first supply devices (A) were installed in the lower section of the corners. The height location of these devices was similar to those used by many other researchers [4,8,14,17,49,50]. These devices represent DV. The second set of supply devices (B) represents CIJV. The shape of the outlet was an equilateral triangle, and the air entered the room at the height of 80 cm above the floor level. The last supply devices (C) were suspended 15 cm from the ceiling. These devices represent CMV, and the air entered the room from an inlet which had a circular shape and a diameter of 80 mm. In addition to the two mannequins, two heat sources were also placed on the side of the tables. There was only one outlet, which was located on the ceiling close to the north wall. A climate chamber was built up in connection to the north wall of the test room where three windows were located. For velocity and temperature measurements, 27 low-velocity omnidirectional thermistor anemometers were used. The thermistor and the logger system, CTA88, were designed and calibrated for velocities between 0.0 and 1.0 m/s. The turbulence intensity was calculated by the following equation

$$I_p = \frac{u_{rms}}{\overline{U}} \cdot 100(\%) \,, \tag{11}$$

where u_{rms} is the root mean square of the turbulent velocity fluctuations, and $\overline{U}$ is the mean velocity. Measurements were performed at seven locations in the room. For position P1 through P4, the heights used were 0.1, 0.6, 1.1, and 1.7 m. For P5, 0.1, 0.3, 0.6, 0.8, 1.1, 1.4, and 1.7 m were used, and for P-6 and P-7, only 1.1 and 1.7 m were used. The sampling interval for all measurements was set to 600 s. The velocity was measured with an accuracy of ±0.05 m/s excluding the directional error with the response time of 0.2 s to 90% of a step change. The uncertainty of temperature measurements was ±0.2 °C with the response time of 12 s to 90% of value in still air. The temperatures of the supply inlets, exhaust, and surrounding laboratory were measured by using T-type (copper–constantan) thermocouples connected to an Agilent 34970A data logger and a computer. In order to confirm that the accuracy of the thermocouples and logger were in the expected range, all measuring devices were calibrated before and after the measurements.

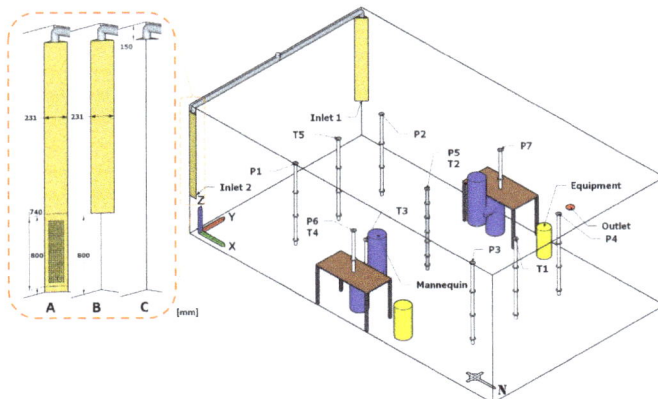

Figure 1. Layout of the test chamber. Three supply devices are illustrated for displacement ventilation (DV, A), corner impinging jet ventilation (CIJV, B), and corner mixing ventilation (CMV, C).

Nine cases were studied which are listed in Table 2. The primary supply air temperature was maintained around 17.6 °C except for case C1-CMV which was 0.3 °C lower. It is important to mention that the comparisons were done in a non-dimensional form for all cases. The mannequins used in the experiments had the same surface area as a human, and each produced 100 W of heat in a sitting position. They were made of galvanized tube 0.32 m in diameter and covered with fabric to emit the same level of radiation as a normal person. Two black painted metal cylinders containing a halogen lamp generated 75 W of heat each when used in the experiments. There was also heat generation from the measuring equipment that amounted to 39 W. The nominal inlet air velocity and inlet Archimedes number were calculated on the basis of A_e for each supply device and case. The A_e for DV, CIJV, and CMV were 299, 133, and 50 cm^2, respectively.

Table 2. Case conditions for different ventilation systems.

Case	Ventilation System	Supply Flow Rate [L/s]	Occupant [W]	Equipment [W]	Inlet Temp. [°C]	u_{in} [m/s]	$Ar_i \times 10^{-4}$
C1-DV	DV	2×20	2×100	-	17.6	0.67	649
C1-CIJV	CIJV	2×20	2×100	-	17.6	1.51	91
C1-CMV	CMV	2×20	2×100	-	17.3	3.98	8
C2-DV	DV	2×20	2×100	2×75	17.6	0.67	787
C2-CIJV	CIJV	2×20	2×100	2×75	17.6	1.51	104
C2-CMV	CMV	2×20	2×100	2×75	17.6	3.98	9
C3-DV	DV	2×30	2×100	2×75	17.6	1.00	317
C3-CIJV	CIJV	2×30	2×100	2×75	17.6	2.26	39
C3-CMV	CMV	2×30	2×100	2×75	17.6	5.97	3

The measurement positions and other components of the experimental set-up are shown in Figure 2. Sulfur hexafluoride (SF$_6$) was used as the tracer gas in this study. The measurements were performed at six locations in the room, at a height of 1.1 m. These were labeled T1 up to T5. The sixth location was at the outlet. During the experiments, the test room was exposed to about 350 ppm of SF$_6$. Gas chromatography (GC) was used to measure the concentration of the gas in air samples. In each tracer gas test, air samples were collected via a pump connected to the GC and analyzed from five fixed locations in the room and one in the outlet. The measurements were repeated three times to ensure the validity of the results. The average deviation between the three measurements ranged between 1 to 3%. The uncertainty of measurements of mean age of air was ±2.5%, but, when including airflow variation, pressure balancing, air leakage, etc., this value might increase. The uncertainty of ACE in this study was estimated to be in compliance with Appendix E of ASHRAE Standard 129 [34].

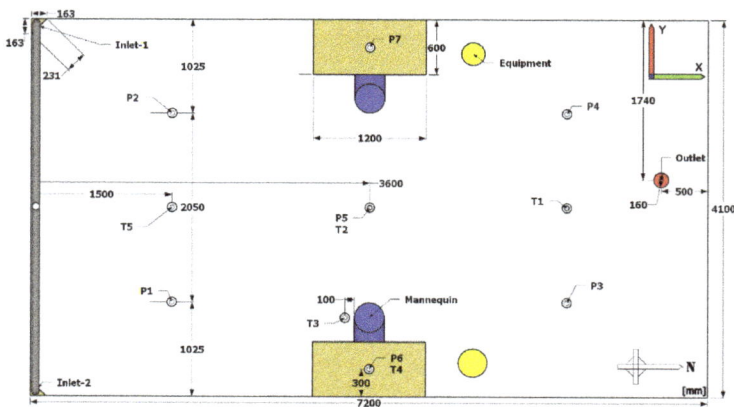

Figure 2. Measurement positions and schematic top-view layout of the office room.

The final estimated uncertainty of measured values of ACE was around 7% which was based on the negligible air leakage and the measuring accuracy of the equipment. This result is close to those reported by some other laboratory studies [42–45].

4. Results and Discussion

4.1. Flow Pattern and Thermal Conditions

The results of the dimensionless air temperature (DAT) for all the cases are shown in Figure 3. DAT is defined as a dimensionless value to compare the vertical air temperature profile between different cases and is defined as

$$DAT = \frac{T_a - T_i}{T_o - T_i} . \tag{12}$$

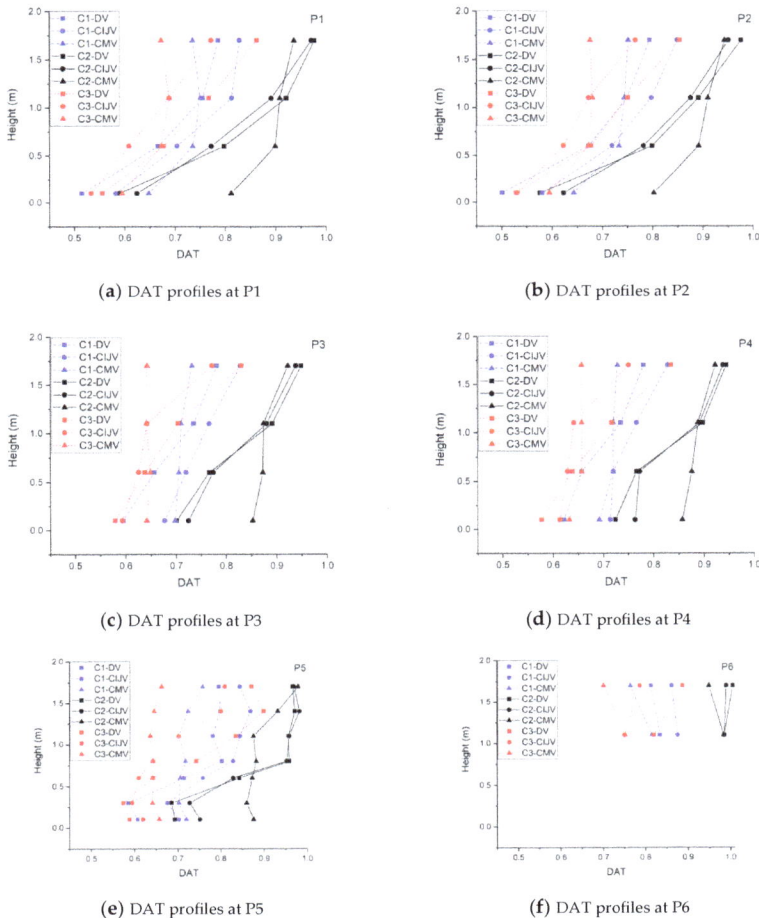

(**a**) DAT profiles at P1

(**b**) DAT profiles at P2

(**c**) DAT profiles at P3

(**d**) DAT profiles at P4

(**e**) DAT profiles at P5

(**f**) DAT profiles at P6

Figure 3. Dimensionless vertical air temperature (DAT) profiles at positions P1 (**a**), P2 (**b**), P3 (**c**), P4 (**d**), P5 (**e**), and P6 (**f**) for all cases.

In position P1 (Figure 3a) and P2 (Figure 3b), which were close to the inlets, C2-DV showed the largest vertical temperate gradient, followed by C2-CIJV. The cases that showed the lowest gradient were C1-CMV and C3-CMV. The air in C2-CIJV entered the room closely attached to the wall before hitting the floor area in the corner of the room. It then moved and spread out as a layer over the floor area.

The air entrainment for C2-DV and C2-CIJV streams was lower compared to that for C2-CMV. Another observation is that the temperature profiles of C2-DV and C2-CIJV were very similar. When comparing C1 to C2 for DV and CIJV, the extra added heat in C2 generated more stratification. The results of P1 and P2 also showed a good level of symmetry.

In position P5 (Figure 3e), which was in the center of the office, the DAT values decreased compared to P1 and P2 because this area was located further away from the inlets, and the mannequins were in close proximity to this location. Although the stratification levels decreased for DV and CIJV, they were still larger than for CMV.

Positions P3 (Figure 3c) and P4 (Figure 3d) showed a similar pattern as P1 and P2 but with slightly lower gradients. The reason for this is that the temperature of the airstream along the floor increased with the distance from the supply devices.

The thermal stratification enhancement created by DV and CIJV are in agreement with other research results [5,10,51–54]. P6 (Figure 3f) location was above the table and was similar to P7.

The velocity profiles at P1 (Figure 4a) and P2 (Figure 4b) showed that the highest velocities were measured at 0.1 m above the floor level in all cases. The CMV cases showed the highest velocities, reaching 0.7 m/s for C3-CMV. This is explained by the special configuration of the CMV inlets. By being placed high up in the corner of the room and having a high supply velocity, the inlet air jets had a high level of entrainment. This created an airstream with higher momentum and higher boundary layer thickness compared to the other systems. In comparison, the airstreams for DV and CIJV had a thinner boundary layer when reaching P1 and P2. One can assume that the locations of the maximum air velocities were below 0.1 m for these systems.

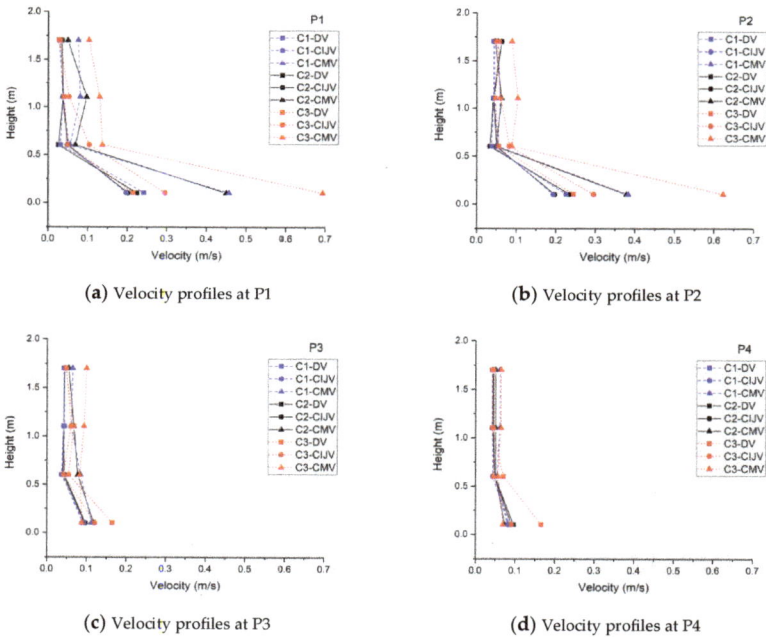

(**a**) Velocity profiles at P1   (**b**) Velocity profiles at P2

(**c**) Velocity profiles at P3   (**d**) Velocity profiles at P4

Figure 4. *Cont.*

(**e**) Velocity profiles at P5

(**f**) Velocity profiles at P6

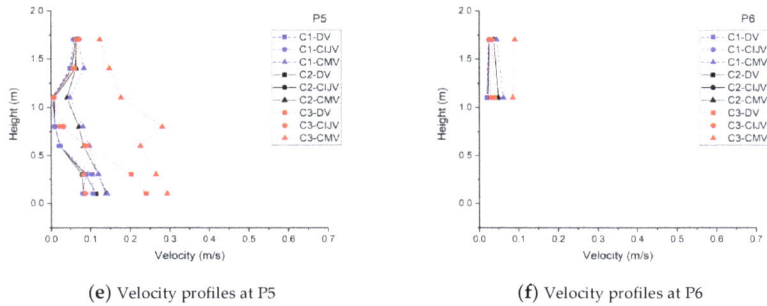

Figure 4. Velocity profiles at positions P1 (**a**), P2 (**b**), P3 (**c**), P4 (**d**), P5 (**e**), and P6 (**f**) for all cases.

In position P5 (Figure 4e), the velocities decreased in all cases compared to P1 and P2, except for C3-DV. One possible explanation for this is that the two airstreams merged at some point before reaching P5, with a combined momentum from both streams. It is also worth mentioning that P1 and P2 were not in the direct centerline of the airstreams and were not recording the center velocities of these streams.

The draught levels at P1 (Figure 5a) and P2 (Figure 5b) showed a strong connection to the velocity profiles. Because of the high velocities at P1 and P2, the draught levels were higher than normal in this part of the room. According to the ISO 7730 classifications, none of the cases was able to obtain category A. In one case, C3-CMV, it did not even pass the lowest category C. Continuing to P5 (Figure 5e), the DR decreased considerably, except for C3-DV and C3-CMV. The DR for all other cases was below 10% (category A). High levels of DR for MV compared to IJV have been shown in a previous study as well [5].

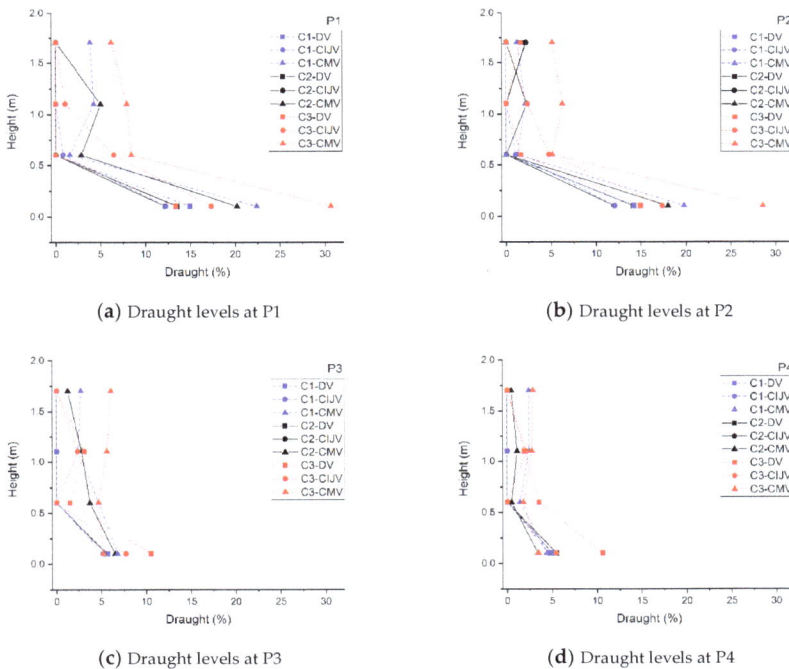

(**a**) Draught levels at P1

(**b**) Draught levels at P2

(**c**) Draught levels at P3

(**d**) Draught levels at P4

Figure 5. *Cont.*

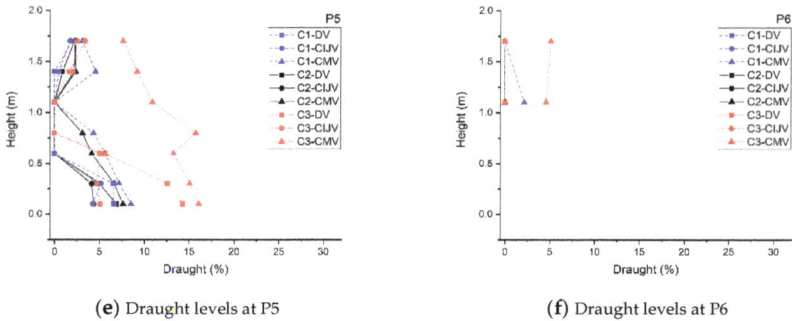

(**e**) Draught levels at P5   (**f**) Draught levels at P6

Figure 5. Draught levels at position P1 (**a**), P2 (**b**), P3 (**c**), P4 (**d**), P5 (**e**), and P6 (**f**) for all cases.

In position P3 (Figure 5c) and P4 (Figure 5d), the DR were at acceptable levels, except for C3-DV, for which it was slightly above 10% at 0.1 m.

Another way to illustrate the correlation between high velocities and high draught rates can be seen in Figure 6. Figure 6a shows the maximum draught rate (DR_{max}) based on all the points in each location at P1–P2, P5, and P3–P4. Figure 6b shows the maximum velocity (U_{max}) at the same locations, and Figure 6c shows the maximum temperature difference (ΔT_{max}). In the case of C3-CMV, it can be observed that ΔT_{max} did not change between P5 and P3–P4 locations, but DR_{max} decreased considerably. This indicates a strong dependency of draught on velocity rates. Table 3 shows that PD was within category A classification for all cases.

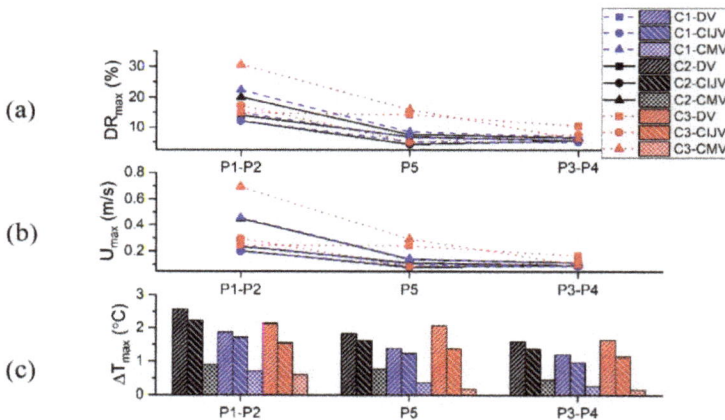

Figure 6. (**a**) Maximum draught rate, (**b**) maximum velocity, and (**c**) maximum temperature difference in positions P1–P2, P5, and P3–P4.

Table 3. Local discomfort (PD) due to high vertical air temperature difference between head and ankle.

Case	Position	C1-DV	C1-CIJV	C1-CMV	C2-DV	C2-CIJV	C2-CMV	C3-DV	C3-CIJV	C3-CMV
PD	P1	1.2%	1.1%	0.6%	1.9%	1.3%	0.5%	1.0%	0.7%	0.5%
	P2	1.2%	1.0%	0.5%	1.7%	1.2%	0.6%	1.1%	0.7%	0.5%
	P3	0.7%	0.5%	0.3%	0.9%	0.7%	0.4%	0.6%	0.4%	0.3%
	P4	0.6%	0.4%	0.4%	0.8%	0.6%	0.4%	0.7%	0.4%	0.4%
	P5	0.8%	0.7%	0.3%	1.3%	1.0%	0.3%	1.2%	0.5%	0.3%

4.2. Ventilation Effectiveness

The ACE_p values presented in Table 4 show that CMV had the most uniform ACE_p compared to the other systems. This indicated that the air was equally "fresh" in all the measuring locations and was well mixed. At position T3 located close to the mannequin, the best performance was achieved by DV, followed closely by CIJV. In DV and CIJV, the air was distributed as a layer over the floor area. The thickness and velocity of the layer was dependent on the shape and configuration of the supply device, and, as mentioned previously, the supply air in DV had a narrow and straightforward trajectory when entering the room, while, in CIJV, it spread out evenly across the floor when the air jet from the inlet hit the floor surface, as shown in previous studies [5,11].

Table 4. Air change effectiveness (ACE), local and average, and air exchange effectiveness (AEE) for all cases.

Case	Position	C1-DV	C1-CIJV	C1-CMV	C2-DV	C2-CIJV	C2-CMV	C3-DV	C3-CIJV	C3-CMV
	T1	1.64	1.47	1.05	1.15	1.39	1.13	1.34	1.13	1.03
	T2	1.41	1.46	1.08	1.09	1.25	1.15	1.25	1.21	1.06
ACE_p	T3 [1]	1.61	1.55	1.07	1.50	1.37	1.14	1.30	1.17	1.04
	T4	1.47	1.54	1.03	1.32	1.38	1.11	1.32	1.14	1.01
	T5	1.33	1.38	1.06	1.04	1.17	1.13	1.21	1.26	1.03
ACE_{avg} [2]		1.48	1.47	1.06	1.20	1.31	1.13	1.28	1.18	1.03
AEE		0.58	0.61	0.52	0.56	0.54	0.55	0.55	0.53	0.50

[1] The location was in the occupied zone close to the mannequin, as also shown in Figure 2. [2] ACE_{avg} is the average ACE value for the measuring points T1–T5.

The AEE values were close to each other in all cases. The reason for this is that AEE takes into account the mean age of air for the entire room. This means that it is possible to have high ACE_p values in some zones, but lower values in others. C1-DV and C1-CIJV had the highest ACE_{avg}, which were the average ACE_p values for T1–T5. Higher ACE_{avg} also led to slightly higher AEE values, as seen in Table 4.

Table 5 shows that ε_T was lower for the CMV cases compared to the other systems at P1–P5. DV and CIJV performed best in the occupied zone (P5), excluding P6–P7. This result followed the same pattern as ACE_p results, when comparing the three systems. Close to the heat sources, for example, the mannequins, DV and CIJV outperformed CMV in terms of both ACE and ε_T.

Table 5. Average ε_T in locations P1–P7 for all cases.

Case	Position	C1-DV	C1-CIJV	C1-CMV	C2-DV	C2-CIJV	C2-CMV	C3-DV	C3-CIJV	C3-CMV
	P1	1.25	1.24	1.08	1.30	1.30	1.14	1.31	1.31	1.00
	P2	1.26	1.24	1.08	1.32	1.31	1.15	1.34	1.32	1.01
ε_T [1]	P3	1.22	1.2	1.09	1.27	1.25	1.15	1.37	1.29	1.02
	P4	1.21	1.18	1.08	1.26	1.23	1.14	1.36	1.28	1.01
	P5	1.15	1.13	1.08	1.20	1.17	1.14	1.27	1.24	1.01
ε_T [2]	P6	0.97	0.99	0.95	1.01	1.01	1.02	1.07	1.07	0.87
	P7	1.00	1.04	1.01	1.05	1.06	1.08	1.12	1.14	0.92

[1] calculated by using the arithmetic mean air temperature of the heights 0.1, 0.6, and 1.1 m. [2] calculated by using the arithmetic mean air temperature of the height 1.1 m only.

To summarize the results, we found that DV and CIJV provided better heat removal efficiency and fresher air in the occupied areas of the room compared to CMV. However, DV had some shortcomings when evaluating the draught rates which were highly dependent on the velocity rates.

CMV behaved very similar to a regular mixing system, with low temperature stratification and even levels of ACE and ε_T in the evaluated locations, compared to the other systems.

5. Conclusions

The findings show that the corner impinging jet air distribution system behaves very similar to the DV system and performs slightly better considering the draught rate. This study also shows that CIJV is a viable option when choosing an air distribution system for a medium-sized office room. The draught rates for this system were within the required levels in ISO 7730 for the occupied zone. Since CIJV obtained high values for both ACE and ε_T, there is a possibility to slightly lower the supply rates in order to reduce fan energy usage, while still meeting the requirements for thermal comfort and IAQ. Although DV performed similar to CIJV, the special design of the supply inlets for this system resulted in an increase of DR in the occupied zone. It is important to note that this study was based only on one-room geometry and one configuration of workstation placement in the room. Typical DV systems are not designed for heating mode, which is believed to be a shortcoming of the system because of the dominance of thermal forces to momentum forces. In contrast, the impinging jet system, with its higher momentum, overcomes this shortcoming. Further studies have to be conducted in order to evaluate the corner impinging jet when the outside temperature is lower than the inside one (heating mode), with different room geometries, and with different set-ups of the workstations. Finally, the promising results and the simplicity of installation and design make corner impinging jet ventilation an interesting research topic for the scientific community.

Author Contributions: Main contribution was made by A.A.; The other three authors contributed equally.

Funding: Internal research funding from the University of Gävle.

Acknowledgments: The authors are grateful for the valuable help from University of Gävle's laboratory staff and the possibility of using the university's laboratory facilities.

Conflicts of Interest: The authors declare no conflicts of interest.

Nomenclature

ACE	air change effectiveness [-]
ACE_{avg}	average spatial air change effectiveness in a region [-]
ACE_p	local air change effectiveness [-]
AEE	air exchange effectiveness [-]
Ar_i	inlet Archimedes number [-]
CIJV	corner impinging jet ventilation
CMV	corner mixing ventilation
DAT	dimensionless vertical air temperature [-]
DR	draught rate [%]
DR_{max}	maximum draught rate between 0.1 m and 1.7 m above floor level [%]
DV	displacement ventilation
IAQ	indoor air quality
IJV	impinging jet ventilation
MV	mixing ventilation
PD	percentage dissatisfied due to vertical air temperature difference [%]
WCJ	wall confluent jets ventilation
A_e	inlet supply opening area [m^2]
C_p	local tracer gas concentration [ppm]
C_o	tracer gas concentration at outlet [ppm]
I_p	local turbulent intensity [-]
g	gravitational acceleration [m/s^2]
q_v	ventilation flow rate [m^3/s]
u_{rms}	the root mean square of turbulent velocity fluctuations [m/s]
t	time [s]

$\overline{T}_{0.1,0.6,1.1}$	arithmetic mean air temperature based on the values at the heights of 0.1, 0.6, and 1.1 m [°C]
T_a	local air temperature [°C]
T_i	mean supply air temperature [°C], [K]
T_o	mean outlet air temperature [°C]
T_r	mean air temperature in the center of the room at 1.7 m above floor level [°C], [K]
$\Delta T_{0.1-1.1}$	vertical air temperature gradient between 0.1 m and 1.1 m above floor level [°C]
ΔT_{max}	maximum air temperature gradient between 0.1 m and 1.7 m above floor level [°C]
u_a	local air velocity [m/s]
u_{in}	nominal inlet air velocity [m/s]
$\overline{U}$	mean air velocity [m/s]
U_{max}	maximum air velocity between 0.1 m and 1.7 m above floor level [m/s]
V	volume [m^3]
ε_T	temperature effectiveness [-]
$\langle \overline{\tau} \rangle$	room mean age of air [s]
τ_n	nominal time constant [s]
$\overline{\tau}_{avg}$	arithmetic average age of air [s]
$\overline{\tau}_p$	local mean age of air [s]

References

1. Vimalanathan, K.; Babu, T.R. The effect of indoor office environment on the work performance, health and well-being of office workers. *J. Environ. Health Sci. Eng.* **2014**, *12*, 113. [CrossRef] [PubMed]

2. Allen, J.G.; MacNaughton, P.; Satish, U.; Santanam, S.; Vallarino, J.; Spengler, J.D. Associations of cognitive function scores with carbon dioxide, ventilation, and volatile organic compound exposures in office workers: A controlled exposure study of green and conventional office environments. *Environ. Health Perspect.* **2015**, *124*, 805–812. [CrossRef]

3. IEA World Energy Balances 2018: Overview. 2018, p. 24. Available online: https://webstore.iea.org/world-energy-balances-2018-overview (accessed on 12 October 2018).

4. Xing, H.; Hatton, A.; Awbi, H. A study of the air quality in the breathing zone in a room with displacement ventilation. *Build. Environ.* **2001**, *36*, 809–820. [CrossRef]

5. Chen, H.; Janbakhsh, S.; Larsson, U.; Moshfegh, B. Numerical investigation of ventilation performance of different air supply devices in an office environment. *Build. Environ.* **2015**, *90*, 37–50. [CrossRef]

6. Serra, N.; Semiao, V. Comparing displacement ventilation and mixing ventilation as HVAC strategies through CFD. *Eng. Comput.* **2009**, *26*, 950–971. [CrossRef]

7. He, G.; Yang, X.; Srebric, J. Removal of contaminants released from room surfaces by displacement and mixing ventilation: Modeling and validation. *Indoor Air* **2005**, *15*, 367–380. [CrossRef] [PubMed]

8. Lin, Z.; Chow, T.; Fong, K.; Tsang, C.; Wang, Q. Comparison of performances of displacement and mixing ventilations. Part II: Indoor air quality. *Int. J. Refrig.* **2005**, *28*, 288–305. [CrossRef]

9. Chen, H.; Moshfegh, B.; Cehlin, M. Numerical investigation of the flow behavior of an isothermal impinging jet in a room. *Build. Environ.* **2012**, *49*, 154–166. [CrossRef]

10. Chen, H.; Moshfegh, B.; Cehlin, M. Investigation on the flow and thermal behavior of impinging jet ventilation systems in an office with different heat loads. *Build. Environ.* **2013**, *59*, 127–144. [CrossRef]

11. Chen, H.; Moshfegh, B.; Cehlin, M. Computational investigation on the factors influencing thermal comfort for impinging jet ventilation. *Build. Environ.* **2013**, *66*, 29–41. [CrossRef]

12. Haghshenaskashani, S.; Sajadi, B.; Cehlin, M. Multi-objective optimization of impinging jet ventilation systems: Taguchi-based CFD method. *Build. Simul.* **2018**, *11*, 1207–1214. [CrossRef]

13. Cehlin, M.; Larsson, U.; Chen, H. Numerical investigation of Air Change Effectiveness in an Office Room with Impinging Jet Ventilation. In Proceedings of the 4th international Conference on Building Energy & Environment (COBEE2018), Melbourne, Australia, 5–9 February 2018; pp. 641–646.

14. Karimipanah, T.; Awbi, H. Theoretical and experimental investigation of impinging jet ventilation and comparison with wall displacement ventilation. *Build. Environ.* **2002**, *37*, 1329–1342. [CrossRef]

15. Koufi, L.; Younsi, Z.; Cherif, Y.; Naji, H.; El Ganaoui, M. A numerical study of indoor air quality in a ventilated room using different strategies of ventilation. *Mech. Ind.* **2017**, *18*, 221. [CrossRef]

16. Ye, X.; Zhu, H.; Kang, Y.; Zhong, K. Heating energy consumption of impinging jet ventilation and mixing ventilation in large-height spaces: A comparison study. *Energy Build.* **2016**, *130*, 697–708. [CrossRef]
17. Cao, G.; Awbi, H.; Yao, R.; Fan, Y.; Sirén, K.; Kosonen, R.; Zhang, J.J. A review of the performance of different ventilation and airflow distribution systems in buildings. *Build. Environ.* **2014**, *73*, 171–186. [CrossRef]
18. Awbi, H.B. Ventilation for good indoor air quality and energy efficiency. *Energy Procedia* **2017**, *112*, 277–286. [CrossRef]
19. Larsson, U. *On the Performance of Stratified Ventilation*; Linköping University Electronic Press: Linköping, Sweden, 2018.
20. Arghand, T.; Karimipanah, T.; Awbi, H.B.; Cehlin, M.; Larsson, U.; Linden, E. An experimental investigation of the flow and comfort parameters for under-floor, confluent jets and mixing ventilation systems in an open-plan office. *Build. Environ.* **2015**, *92*, 48–60. [CrossRef]
21. Wu, X.; Fang, L.; Olesen, B.W.; Zhao, J. Air distribution and ventilation effectiveness in a room with floor/ceiling heating and mixing/displacement ventilation. In Proceedings of the 8th International Symposium on Heating, Ventilation and Air Conditioning (ISHVAC2013), Xi'an, China, 19–21 October 2013; pp. 59–67.
22. Amai, H.; Novoselac, A. Experimental study on air change effectiveness in mixing ventilation. *Build. Environ.* **2016**, *109*, 101–111. [CrossRef]
23. Conceição, E.Z.; Lúcio, M.M.J.; Awbi, H.B. Comfort and airflow evaluation in spaces equipped with mixing ventilation and cold radiant floor. *Build. Simul.* **2013**, *6*, 51–67. [CrossRef]
24. Shan, X.; Zhou, J.; Chang, V.W.-C.; Yang, E.-H. Comparing mixing and displacement ventilation in tutorial rooms: Students' thermal comfort, sick building syndromes, and short-term performance. *Build. Environ.* **2016**, *102*, 128–137. [CrossRef]
25. Ren, S.; Tian, S.; Meng, X. Comparison of Displacement Ventilation, Mixing Ventilation and Underfloor Air Distribution System. In Proceedings of the 2015 International Conference on Architectural, Civil and Hydraulics Engineering (ICACHE 2015), Guangzhou, China, 28–29 November 2015; pp. 79–82.
26. Zhou, J.; Chau, H.; Kang, Y.; Hes, D.; Noguchi, M.; Aye, L. Comparing mixing ventilation and displacement ventilation in university classrooms. In Proceedings of the Zero Energy Mass Custom Home (ZEMCH2018), Melbourne, Australia, 29 January–1 February 2018; pp. 263–274.
27. Cheng, Y.; Lin, Z. Experimental study of airflow characteristics of stratum ventilation in a multi-occupant room with comparison to mixing ventilation and displacement ventilation. *Indoor Air* **2015**, *25*, 662–671. [CrossRef]
28. Vachaparambil, K.; Cehlin, M.; Karimipanah, T. Comparative Numerical Study of the Indoor Climate for Mixing and Confluent Jet Ventilation Systems in an Open-plan Office. In Proceedings of the 4th International Conference On Building Energy & Environment (COBEE2018), Melbourne, Australia, 5–9 February 2018; pp. 73–78.
29. ISO 7730. *Ergonomics of the Thermal Environment-Analytical Determination and Interpretation of Thermal Comfort Using Calculation of the PMV and PPD Indices and Local Thermal Comfort Criteria*; ISO: Geneva, Switzerland, 2005.
30. Nielsen, P.V. *Displacement Ventilation: Theory and Design*; Aalborg University: Aalborg, Denmark, 1993; p. 42. Available online: https://www.forskningsdatabasen.dk/en/catalog/2389386598 (accessed on 5 October 2018).
31. Qingyan, C.; Van Der Kooi, J.; Meyers, A. Measurements and computations of ventilation efficiency and temperature efficiency in a ventilated room. *Energy Build.* **1988**, *12*, 85–99. [CrossRef]
32. Artmann, N.; Jensen, R.L.; Manz, H.; Heiselberg, P. Experimental investigation of heat transfer during night-time ventilation. *Energy Build.* **2010**, *42*, 366–374. [CrossRef]
33. Olesen, B.W.; Simone, A.; Krajcfík, M.; Causone, F.; Carli, M.D. Experimental study of air distribution and ventilation effectiveness in a room with a combination of different mechanical ventilation and heating/cooling systems. *Int. J. Vent.* **2011**, *9*, 371–383. [CrossRef]
34. ASHRAE Standard 129-1997 (RA 2002). *Measuring Air Change Effectiveness*; ASHRAE: Atlanta, GA, USA, 2002.
35. Mundt, M.; Mathisen, H.M.; Moser, M.; Nielsen, P.V. *Ventilation Effectiveness*; REHVA: Brussels, Belgium, 2004; ISBN 978-29-6004-680-9.

36. Breum, N. Ventilation efficiency in an occupied office with displacement ventilation—A laboratory study. *Environ. Int.* **1992**, *18*, 353–361. [CrossRef]

37. Krajčík, M.; Simone, A.; Olesen, B.W. Air distribution and ventilation effectiveness in an occupied room heated by warm air. *Energy Build.* **2012**, *55*, 94–101. [CrossRef]

38. Fisk, W.; Faulkner, D.; Pih, D.; McNeel, P.; Bauman, F.; Arens, E. Indoor air flow and pollutant removal in a room with task ventilation. *Indoor Air* **1991**, *1*, 247–262. [CrossRef]

39. Larsson, U.; Moshfegh, B. Comparison of ventilation performance of three different air supply devices: A measurement study. *Int. J. Vent.* **2017**, *16*, 244–254. [CrossRef]

40. Sandberg, M.; Blomqvist, C. A quantitative estimate of the accuracy of tracer gas methods for the determination of the ventilation flow rate in buildings. *Build. Environ.* **1985**, *20*, 139–150. [CrossRef]

41. Almesri, I.; Awbi, H.; Foda, E.; Sirén, K. An air distribution index for assessing the thermal comfort and air quality in uniform and nonuniform thermal environments. *Indoor Built Environ.* **2013**, *22*, 618–639. [CrossRef]

42. Faulkner, D.; Fisk, W.; Sullivan, D. Indoor airflow and pollutant removal in a room with floor-based task ventilation: Results of additional experiments. *Build. Environ.* **1995**, *30*, 323–332. [CrossRef]

43. Buratti, C.; Mariani, R.; Moretti, E. Mean age of air in a naturally ventilated office: Experimental data and simulations. *Energy Build.* **2011**, *43*, 2021–2027. [CrossRef]

44. Cehlin, M.; Karimipanah, T.; Larsson, U.; Ameen, A. Comparing thermal comfort and air quality performance of two active chilled beam systems in an open-plan office. *J. Build. Eng.* **2019**, *22*, 56–65. [CrossRef]

45. Sandberg, M.; Sjöberg, M. The use of moments for assessing air quality in ventilated rooms. *Build. Environ.* **1983**, *18*, 181–197. [CrossRef]

46. Sandberg, M. What is ventilation efficiency? *Build. Environ.* **1981**, *16*, 123–135. [CrossRef]

47. Both, B.; Szánthó, Z. Objective investigation of discomfort due to draught in a tangential air distribution system: Influence of air diffuser's offset ratio. *Indoor Built Environ.* **2018**, *27*, 1105–1118. [CrossRef]

48. Cehlin, M. Visualization of Air Flow, Temperature and Concentration Indoors: Whole-Field Measuring Methods and CFD. Ph.D. Thesis, KTH, Stockholm, Sweden, 2006.

49. Novoselac, A.; Srebric, J. A critical review on the performance and design of combined cooled ceiling and displacement ventilation systems. *Energy Build.* **2002**, *34*, 497–509. [CrossRef]

50. Loveday, D.; Parsons, K.; Taki, A.; Hodder, S. Displacement ventilation environments with chilled ceilings: Thermal comfort design within the context of the BS EN ISO7730 versus adaptive debate. *Energy Build.* **2002**, *34*, 573–579. [CrossRef]

51. Schiavon, S.; Bauman, F.; Tully, B.; Rimmer, J. Room air stratification in combined chilled ceiling and displacement ventilation systems. *HVAC&R Res.* **2012**, *18*, 147–159.

52. Li, Y.; Sandberg, M.; Fuchs, L. Vertical temperature profiles in rooms ventilated by displacement: Full-scale measurement and nodal modelling. *Indoor Air* **1992**, *2*, 225–243. [CrossRef]

53. Cermak, R.; Melikov, A.K. Air quality and thermal comfort in an office with underfloor, mixing and displacement ventilation. *Int. J. Vent.* **2006**, *5*, 323–352. [CrossRef]

54. Yin, Y.; Xu, W.; Gupta, J.K.; Guity, A.; Marmion, P.; Manning, A.; Gulick, B.; Zhang, X.; Chen, Q. Experimental study on displacement and mixing ventilation systems for a patient ward. *HVAC&R Res.* **2009**, *15*, 1175–1191. [CrossRef]

Article

Simulation of Temperature Distribution on the Face Skin in Case of Advanced Personalized Ventilation System

Ferenc Szodrai * and Ferenc Kalmár

Department of Building Services and Building Engineering, University of Debrecen; Otemeto str. 2-4,
4028 Debrecen, Hungary; fkalmar@eng.unideb.hu
* Correspondence: szodrai@eng.unideb.hu

Received: 4 March 2019; Accepted: 21 March 2019; Published: 27 March 2019

Abstract: Energy saving is one of the most important research directions in the building sector. Personalized ventilation systems are energy conscious solutions providing fresh air for the occupants. As a side effect, cooling energy can be saved due to higher convective heat removal. Using the data gathered from previous experiments performed with the developed personalized ventilation system, a ± 1.408 °C accurate simulation model was created in ANSYS 19.2 Academic version in order to determine the temperature distribution on the face. In this paper, the method and the first results are presented. It was clearly demonstrated by measurements and simulations that the personalized ventilation equipment used has a considerable effect on the skin temperature of the face. The developed model can be used to analyze the skin temperature on the faces of people using the novel, personalized ventilation equipment. This way the time spent on examination can be reduced considerably.

Keywords: air jet; personalized ventilation; skin temperature; CFD; thermal analysis

1. Introduction

Buildings are responsible for about 40% of the total energy consumption in the European Union. The situation is similar in the United States and other countries. This is the reason why energy saving in the building sector has been getting more and more important in the last decades. The share of heating and cooling energy demand depends on the local climate conditions. However, it can be stated that in summer periods, the number of heat waves and the temperature amplitude is increasing [1]. In European countries the energy performance directive encourages member states to only build nearly zero-energy buildings in the future [2,3]. In countries where heating represents 60–70% of the total energy use of a building, severe requirements were adopted regarding thermal performance of the building envelope in order to reduce heat losses. New insulation materials are tested in order to meet the requirements with lower thicknesses [4–6]. However, in such buildings small heat loads can lead to high indoor temperatures. To optimize the facade solutions, including window properties, external wall insulation, window-to-wall ratio, and external shading, simulations and cost optimization calculations were performed even in cold climates [7]. In case of free-running office or educational buildings with large glazed areas, extreme high indoor temperatures may appear [8,9]. In case of new buildings, the improved air tightness of the envelope may lead to the increase of carbon dioxide concentration and humidity of the indoor air [10]. Complex studies have to be performed in order to choose the appropriate ventilation strategy [11]. Advanced personalized ventilation (PV) systems may be an energy conscious solution to assure proper air quality and thermal comfort in buildings. According to Melikov, the focus must be shifted from total volume air distribution to advanced air distribution based on the following principles [12]:

- remove/reduce the air pollution and generated heat (when not needed) locally;
- provide clean air, also heating and cooling, where, when, and as much as needed;
- make active control of the air distribution possible;
- involve each occupant in creating his/her own preferred microenvironment.

Schiavon et al. found that the energy consumption of PV is 51% lower compared to mixing ventilation [13]. Having lower air flow will lead to lower energy consumption for cooling. Pan et al. proved that energy savings up to 45% can be obtained by comparing a partition-type fan coil unit with a central air conditioning system [14].

PV has the advantage that each occupant is authorized to optimize and control the temperature, flow rate (local air velocity), and direction of the locally supplied air flow [15]. Zhang et al. demonstrated that the local discomfort caused by stratification of the air temperature can be reduced by PV's and the stratification can be higher [16]. It was shown that by applying 0.8 m s^{-1} air velocity around the head, the acceptable stratification goes up to 6 °C if the air temperature around the head is 26.8 °C. In the face skin, the number of thermo-receptors is high in comparison to other body segments. According to Lynette Jones, there are more cold spots than warm spots, the density of spots varies across the body, and the time to respond to a cold stimulus is significantly shorter than to a warm stimulus [17]. This is the reason why PV systems may improve thermal sensation of occupants in warm indoor environments. Another advantage of these systems is that PV can help to improve work performance. Maula et al. performed a study in order to analyze the effect of a temperature of 29 °C on performance in tasks involving different cognitive demands. They aimed to assess the effect on perceived performance, subjective workload, thermal comfort, perceived working conditions, cognitive fatigue, and somatic symptoms, in a laboratory with a realistic office environment [18]. They made a comparison to a temperature of 23 °C. It was shown that performance was negatively affected by slightly warmer temperatures in the N-back working memory task. The effect of a cooling jet on performance and comfort in a warm office environment (29.5 °C air temperature) was analyzed by Maula et al. and it was demonstrated that the jet improved the speed of response in a working memory task with increasing exposure time [19].

Because of elevated air velocities around the head and chest of the occupants, draft may appear and can lead to discomfort. Griefahn et al. studied the significance of air velocity and turbulence intensity on responses to horizontal drafts in a constant air temperature of 23 °C [20]. They found that draft-induced general annoyance and draft-induced local annoyance, as stated for the neck and for the forearm, increased with air velocity and/or with turbulence intensity. The decrease in skin temperature, however, was only related to air velocity but not to turbulence intensity. However, draft sensation is related to general thermal sensation [21]. Moreover, with special air terminal devices, better thermal comfort sensation is obtained and draft might be avoided [22,23].

Most PV systems have one air terminal device [24–28]. Conceicao et al. presented the results of their study on comfort level in desks equipped with two personalized ventilation systems [29]. In the experimental tests the mean air velocity and the turbulence intensity in the upper air terminal device were 3.5 m s^{-1} and 9.7%, while in the lower air terminal device they were 2.6 m s^{-1} and 15.2%. The mean air temperature in the air terminal devices was around 28 °C, while the mean radiant temperature in the occupation area, the mean air temperature far from the occupation area, and the internal mean air relative humidity were, 28 °C, 28 °C and 50% respectively. They found that The Predicted Percentage of Dissatisfied people reduce from 27.77% (without personalized ventilation) to 16.1% (with personalized ventilation).

The aim of our research was to develop a simulation model in order to analyze the skin temperature distribution on the face of a sitting person at the desk, where the air is introduced around the head alternatively from different directions. Based on previous measurements carried out, the model was created in ANSYS environment.

2. ALTAIR PV System

At the University of Debrecen, Department of Building Services and Building Engineering an advanced personalized ventilation system (ALTAIR) was developed [30,31]. The novelty of ALTAIR PV system is providing the air flow jet around the head of the occupants alternatively from different directions (left-front-right). ALTAIR operates like a hand-held fan, the time steps of changing the air flow direction and the air flow velocity can be chosen by the user (Figure 1).

Figure 1. ALTAIR PV system.

In the Indoor Environment Quality laboratory of the University of Debrecen, numerous measurements were carried out testing the ALTAIR equipment under elevated operative temperatures and asymmetric radiations [32–34].

It was clearly demonstrated that the reduction of the skin temperature on the face was minimum 0.5 K or higher depending on the indoor temperature, which varied continuously during the operation of ALTAIR to avoid adaptation. In Figure 2, the temperature distribution on the face was presented when the air jet was blown on the right side of the face, respectively when the air jet was not blown on the right side of the face. The air temperature and the mean radiant temperature in the room were 28 °C.

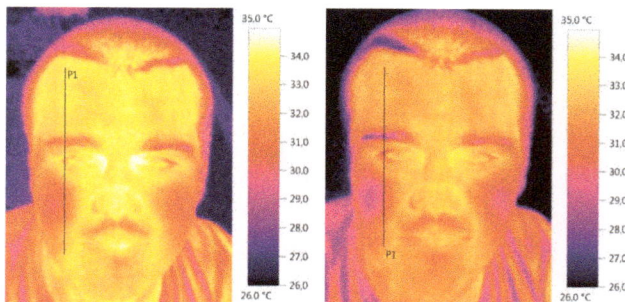

Figure 2. Skin temperature on the right side of the face (**left**) 32.7 °C average temperature along the line without air jet; (**right**) 31.4 °C average temperature along the line with air jet).

The variation of the skin temperature on the face over 30 min can be observed in Figure 3 in a closed space with 28 °C operative temperature (data were gathered every 10 s). The ALTAIR PV system was in operation with 20 m^3 h^{-1} air jet and the direction was varied (in this case) every 20 s (left-front-right-front-left-front-right-...). The air jet temperature was equal to the temperature of indoor air.

One of the biggest challenges of the PV systems is to avoid draft. Even though draft might be favorable from a thermal comfort point of view in warm environments, in most cases occupants claim discomfort perceiving the draft. Having a database of skin temperatures in different environmental conditions created the basis for a numerical model in order to determine the distribution of the face skin temperatures in different environments. In the following chapters the methodology and first results are presented.

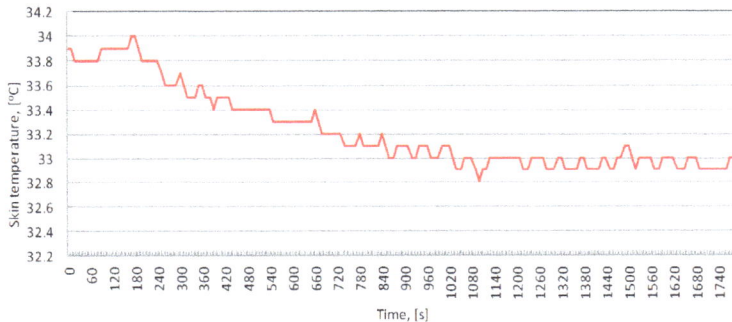

Figure 3. Face skin temperature variation.

3. Methods

3.1. Geometry and Mesh

The numerical model was made in ANSYS 19.2 Academic version. Our first step was to define the flow domain where the analysis was carried out. The geometry was a digital copy of the Indoor Environment Quality Laboratory of the University of Debrecen, where the measurements were taken with the ALTAIR [34]. In this 2.5 m wide, 3.65 m long, 2.55 m high room (Figure 4), the ALTAIR can be found. The PV system was situated in front of a 900 mm × 630 mm large glazed surface.

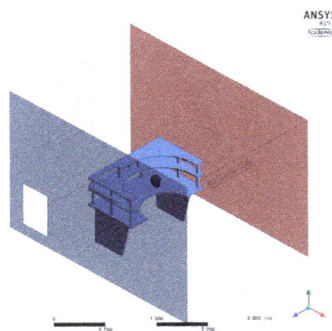

Figure 4. Room geometry.

At the ALTAIR a hypothetical person was seated whose head was defined, although the rest of the body was not modelled. This head was a royalty free 3D model of a male head (Figure 5). It can be seen from Figures 4 and 5 that a simplified geometry was created to reduce the complexity of the geometric model. The tolerance was less then ±2 mm of the 3D model.

The mesh was generated in such a density that the distance of the nodes did not affect the resulting values. For this reason, several mesh independency examinations were made when the cases converged. In the final mesh, element sizes of 20 mm and 2 mm were applied around the desk and around the head respectively.

125

Figure 5. Face geometry.

3.2. Physics Model

The physics model was created in ANSYS Fluent 19.2, where two independent airflows, thermodynamic changes and diffuse solar radiation were modelled. Due to the complex ventilation systems and fine geometries, robust and turbulent flows were produced. It gave us a reason to use k-ε model. With this model it was possible to take the kinetic energy change and the turbulence dissipation into account.

The examined face was placed in the room (Figure 4) where the wall was cooled from the left (blue surface) and heated (red surface) from the right. By the cooling, an average cold surface temperature (T_{SC}) and by the heating, an average warm surface temperature (T_{SW}) was achieved. In front of the face there was a window from which diffuse solar radiation $I_{solar} = 20$ W m^{-2} was emitted in to the room. The surface of the face had a 70 W m^{-2} heat load, while the rest of the surfaces in the room had 0 W m^{-2} heat losses. 0.98 [35] was chosen to be the absorption coefficient of the skin and 0.9 was chosen for the rest of the surfaces according to the MSZ EN ISO 6946:2017 Hungarian Standard [36].

Two ventilation systems were placed in the examined room. One supplied $V_{AF} = 50$ m^3 h^{-1} fresh air with a temperature of T_{AF} and a personalized ventilation system (ALTAIR) that circulated V_{AV} [m^3 h^{-1}] air in the room with a temperature of T_{AV}.

The minimum and the maximum values of the previously mentioned parameters are presented in Table 1.

Table 1. Input parameters.

Parameters	T_{SC}	T_{SW}	T_{AF}	T_{AV}	V_{AV}
Units	[°C]	[°C]	[°C]	[°C]	[m^3 h^{-1}]
Validation case	17.5	31.1	24	24	20
Minimum value	16	30	22	22	0
Maximum value	20	36	26	26	40

We wanted to create an accurate model that can be considered to be valid. To validate the model, we have chosen an important parameter that was not given as an input parameter. In the validation process we examined how closely we could approach this value. It was known from a previous research paper [34] that the average face temperature (AFT) was 32 °C for male participants and this parameter was chosen as the validation parameter. For validation 1.6 °C (5% of 32 °C) accuracy was chosen because the complexity of the model could cause such a degree of deviation. When the mesh and the physical model were considered to be validated, combinations of the minimum and maximum parameter values were made to examine the sensitivity of the AFT.

3.3. Radiation Model

During the simulation, radiation had to be modelled due to the diffuse solar radiation and heat radiation from the warm surfaces. For the solar radiation simulation three parameters had to be defined. Direction, direct and diffuse irradiation, and the transmissivity factor of the semi-opaque surface from where the radiation was excepted [37]. In the examined case the window had a transmissivity factor of 1 and there was no direct solar radiation. For radiation modelling, the Monte Carlo and the surface to surface method was examined.

The Monte Carlo is a probability-based model. The basis of this model is that it calculates the discrete particle (photons) energy that is randomly emitted from the surface. To achieve good convergence with the measured data, the number of cases should be increased, however the large number of photons requires large computational power [37,38]. To reduce the processing power this method was not preferred. When the model was examined, several hotspots were generated on the face due to the amount of the simulated photons being low. By increasing the number of the photons the peak temperatures were lowered but the temperature distribution was not considered to be adequate. This model was not used for the final presented cases.

The surface to surface (S2S) model ignores the movement of the photons and it calculates radiation between two surfaces. These surfaces are Lambert-surfaces that are defined by the absorption factor. With this factor absorption, emission, and scattering of the radiation can be modelled. When the absorption coefficients were defined, the view factors were also calculated for all surfaces. This method showed smooth distribution without considerable peaks. Because of the fact that only diffuse radiation and smooth temperature distribution occurred, the S2S radiation model was used for the validated and also for the following cases.

4. Results

The ALTAIR changed the direction of the flow in every 20 s during the operation. Consequently, the validation had to be done for the average values of three cases, when the air flow came from the left, the right, and the front of the face. The AFT in Celsius and in Kelvin and the calculated errors can be seen in Table 2. The average error from the three directions was 4.4%, this value was mostly increased by the frontal flow. In the further cases similar phenomena occurred during the analysis. This 4.4% error means that with the presented simulation model we could predict the AFT with ±1.408 °C accuracy under the circumstances that the freedom of the model allows. It has to be mentioned that in all three cases the calculated AFT values were lower than the validation value.

Table 2. Average face temperature values for validation.

Parameters	[K]	Error	[°C]	Error
left	303.92	0.40%	30.78	3.81%
front	302.69	0.80%	29.55	7.66%
right	304.59	0.18%	31.45	1.72%
validation [35]	305.14	-	32	-
average	303.73	0.46%	30.59	4.40%

The AFT distribution is shown in Figure 6. This figure was made to depicting that when the face was blown from the left side, it caused a considerable cooling for this side of the face, while large hotspots appeared on the right side that was protected from the flow.

The flow around the head created by ALTAIR can be seen on Figure 7. The streamlines are not appearing on the other side of the head. Due to this phenomenon, the thermal surface resistance of that side increases for a short period of time which can lead to temperature rise. Since the three directions are constantly alternating, the shown temperature raise easily dissipates.

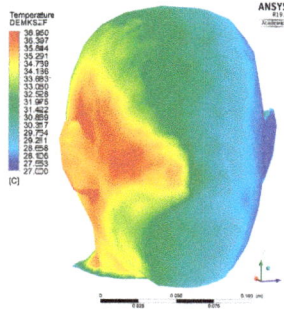

Figure 6. Face surface temperature distribution, blown from left.

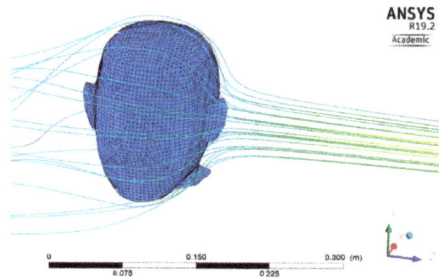

Figure 7. Streamlines around the head.

The histogram of the face temperatures (Figure 8) shows that the lowest temperatures appeared when the flow was initiated from the front, while the highest temperatures were achieved when the flow was formed from the cold (left) side of the room. The phenomenon that increased the surface resistance helped the already warm (right) side of the face to be heated by the warm wall. The lowest temperature occurred when the flow of ALTAIR touched the face from the warm side (right).

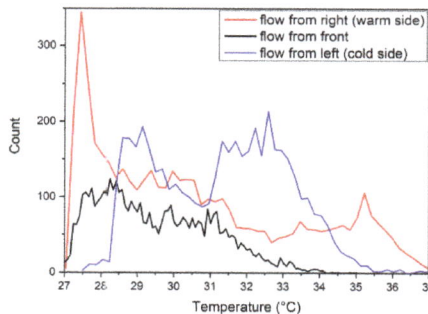

Figure 8. Histogram of the surface temperatures from the face.

Results from the Cases

56 cases were made using the validated model by changing the temperature values and the direction of the flow of ALTAIR, and the AFT of all cases is presented in Table 3. The cases are sorted into four categories when the ALTAIR was turned off (series 1), when the ALTAIR blew from the left position (series 2), the front position (series 3) and the right position (series 4) as shown in Figure 9.

Table 3. Average face temperature values.

T_{SC}	T_{SW}	T_{AV}	T_{AF}	Series 1	Series 2	Series 3	Series 4	Average
[°C]	[°C]	[°C]	[°C]	[°C]	[°C]	[°C]	[°C]	[°C]
16	30	22	22	33.69	27.59	25.99	27.45	27.01
20	30	22	22	35.05	28.45	26.83	28.34	27.88
16	36	22	22	36.21	29.63	27.31	28.78	28.57
20	36	22	22	37.55	30.49	28.14	29.67	29.43
16	30	22	26		28.00	26.49	28.19	27.56
20	30	22	26		28.86	27.33	29.08	28.42
16	36	22	26		30.04	27.80	29.52	29.12
20	36	22	26		30.89	28.64	30.40	29.98
16	30	26	22	-	28.95	27.91	28.89	28.58
20	30	26	22		29.80	28.67	29.77	29.41
16	36	26	22		30.96	29.11	30.21	30.09
20	36	26	22		31.81	29.91	31.08	30.93
16	30	26	26	34.39	29.34	28.35	29.61	29.10
20	30	26	26	35.75	30.20	29.11	30.49	29.93
16	36	26	26	36.90	31.36	29.57	30.93	30.62
20	36	26	26	38.24	32.21	30.39	31.80	31.47

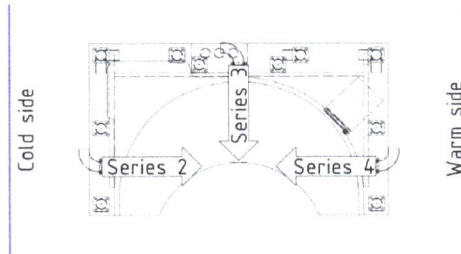

Figure 9. Sketch of the measurement setup.

The box chart (Figure 10) reveals that without the produced ventilation from the ALTAIR the mean temperature would raise up by an average of 6.72 °C. In all the examined series the maximum AFT differences were 4.48 ± 0.09 °C while the average standard error was ±1.04 °C. This means that by changing the wall and air temperatures the value deviations were roughly the same, although the mean temperatures were different. Series 2 had the highest mean temperatures, while series 3 had the lowest. It can also be observed that when the flow was attacking from the side, the temperature values had a slight deviation compared to each other, although data from series 2 displays warmer temperatures. Considerable (in average 1.5 °C) AFT changes occurred when the warm surface temperature and the ventilation temperature were alternated, in the rest of the cases the deviation was less than ±1 °C.

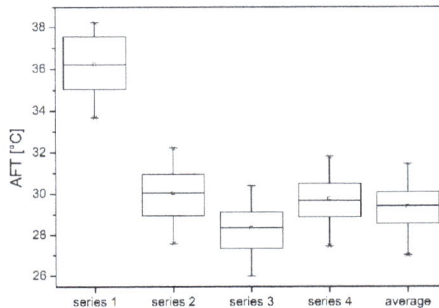

Figure 10. AFT values.

5. Conclusions

Personalized ventilation systems may improve the thermal comfort and the indoor air quality in closed spaces with elevated indoor temperatures. However, draft has to be avoided otherwise users will experience discomfort. The skin temperature on the face is an important parameter which gives useful information about thermal comfort sensation. In this paper, skin temperatures were calculated with a ±1.408 °C accuracy in ANSYS 19.2 Academic version. In the simulation two independent airflows, surface to surface radiation and various surface temperatures, were modelled with ANSYS Fluent in 56 different cases. It was shown that flow from the personal ventilation system has a considerable effect on the average face temperature. The average flow temperature of the personalized ventilation has a strong connection to the average face temperature. We predicted that the lowest average face temperature occurs when the flow originates from the front of the face.

Author Contributions: Conceptualization, Funding Acquisition, Methodology, Project Administration, Writing, Supervision, Review F.K.; Editing, Investigation, Software, Validation, Writing-Original Draft Preparation, F.S.; Data curation, Formal Analysis, Methodology F.K. and F.S.

Funding: This research was funded by Ministry of Human Capacities grant number 20428-3/2018/FEKUTSTRAT and the APC was funded by Ministry of Human Capacities and University of Debrecen.

Acknowledgments: The research was financed by the Higher Education Institutional Excellence Programme of the Ministry of Human Capacities in Hungary, within the framework of the Energetics Thematic Programme of the University of Debrecen.

Conflicts of Interest: The authors declare no conflict of interest.

Nomenclature

AFT	average face skin temperature [°C]
ε	emission coefficient [-]
I_{solar}	solar radiation [W m^{-2}]
T_{AF}	fresh air temperature [°C]
T_{AV}	ventilated air temperature [°C]
T_{SC}	average cold surface temperature [°C]
T_{SW}	average warm surface temperature [°C]
V_{AV}	ventilated air volume flow [m^3 h^{-1}]
V_{AF}	fresh air volume flow [m^3 h^{-1}]

References

1. Schär, C.; Vidale, P.L.; Lüthi, D.; Frei, C.; Häberli, C.; Liniger, M.A.; Appenzeller, C. The role of increasing temperature variability in European summer heatwaves. *Nature* **2004**, *427*, 332–336. [CrossRef]
2. Kurnitski, J.; Saari, A.; Kalamees, T.; Vuolle, M.; Niemelä, J.; Tarke, T. Cost optimal and nearly zero (nZEB) energy performance calculations for residential buildings with REHVA definition for nZEB national implementation. *Energy Build.* **2011**, *43*, 3279–3288. [CrossRef]
3. Kurnitski, J.; Allard, F.; Braham, D.; Goeders, G.; Heiselberg, P.; Jagemar, L.; Kosonen, R.; Lebrun, J.; Mazzarella, L.; Railio, J.; Seppänen, O.; Schmidt, M.; Virta, M. How to define nearly net zero energy buildings nZEB– REHVA proposal for uniformed national implementation of EPBD recast. *Rehva J.* **2011**, *48*, 6–11.
4. Riffat, S.B.; Qiu, G. A review of state-of-the-art aerogel applications in buildings. *Int. J. Low-Carbon Technol.* **2013**, *8*, 1–6. [CrossRef]
5. Umberto, B.; Lakatos, Á. Thermal Bridges of Metal Fasteners for Aerogel-enhanced Blankets. *Energy Build.* **2019**, *185*, 307–315.
6. Lakatos, Á. Stability investigations of the thermal insulating performance of aerogel blanket. *Energy Build.* **2019**, *185*, 103–111. [CrossRef]
7. Thalfeldt, M.; Pikas, E.; Kurnitski, J.; Voll, H. Facade design principles for nearly zero energy buildings in a cold climate. *Energy Build.* **2013**, *67*, 309–321. [CrossRef]
8. Kalmár, F. Interrelation between glazing and summer operative temperatures in buildings. *Int. Rev. Appl. Sci. Eng.* **2016**, *7*, 51–60. [CrossRef]

9. Kalmár, F. Summer operative temperatures in free running existing buildings with high glazed ratio of the facades. *J. Build. Eng.* **2016**, *6*, 236–242. [CrossRef]
10. Zemitis, J.; Borodinecs, A.; Frolova, M. Measurements of moisture production caused by various sources. *Energy Build.* **2016**, *127*, 884–891. [CrossRef]
11. Baranova, D.; Sovetnikov, D.; Semashkina, D.; Borodinecs, A. Correlation of energy efficiency and thermal comfort depending on the ventilation strategy. *Procedia Eng.* **2017**, *205*, 503–510. [CrossRef]
12. Melikov, A.K. Advanced air distribution. *ASHRAE J.* **2011**, *53*, 73–78.
13. Schiavon, S.; Melikov, A.K.; Sekhar, C. Energy analysis of the personalized ventilation system in hot and humid climates. *Energy Build.* **2010**, *42*, 699–707. [CrossRef]
14. Pan, C.S.; Chiang, H.C.; Yen, M.C.; Wang, C.C. Thermal comfort and energy saving of a personalized PFCU air conditioning system. *Energy Build.* **2005**, *37*, 443–449. [CrossRef]
15. Melikov, A.K. Personalized ventilation. *Indoor Air* **2004**, *14*, 157–167. [CrossRef] [PubMed]
16. Zhang, H.; Huizenga, C.; Arens, E.; Yu, T. Modeling thermal comfort in stratified environments. In Proceedings of the Indoor Air, Beijing, China, 4–9 September 2005; pp. 133–137.
17. Jones, L. Thermal touch. *Scholarpedia* **2009**, *4*, 7955. [CrossRef]
18. Maula, H.; Hongisto, V.; Ostman, L.; Haapakangas, A.; Koskela, H.; Hyöna, J. The effect of slightly warm temperature on work performance and comfort in open-plan offices—A laboratory study. *Indoor Air* **2016**, *26*, 286–297. [CrossRef]
19. Maula, H.; Hongisto, V.; Koskela, H.; Haapakangas, A. The effect of cooling jet on work performance and comfort in warm office environment. *Build. Environ.* **2016**, *104*, 13–20. [CrossRef]
20. Griefahn, B.; Künemund, C.; Gehring, U. The significance of air velocity and turbulence intensity for responses to horizontal drafts in a constant air temperature of 23 °C. *Int. J. Ind. Ergon.* **2000**, *26*, 639–649. [CrossRef]
21. Toftum, J.; Nielsen, R. Draught sensitivity is influenced by general thermal sensation. *Int. J. Ind. Ergon.* **1996**, *18*, 295–305. [CrossRef]
22. Melikov, A.K.; Cermak, R.; Majer, M. Personalized ventilation: evaluation of different air terminal devices. *Energy Build.* **2002**, *34*, 829–836. [CrossRef]
23. Sun, W.; Tham, K.W.; Zhou, W.; Gong, N. Thermal performance of a personalized ventilation air terminal device at two different turbulence intensities. *Build. Environ.* **2007**, *42*, 3974–3983. [CrossRef]
24. Cermak, R.; Holsoe, J.; Mayer, E.; Melikov, A.K. PIV measurements at the breathing zone with personalized ventilation. In Proceedings of the Eight International Conference Air Distribution in Rooms, Roomvent, Copenhagen, Denmark, 8–11 September 2002; pp. 349–352.
25. Gao, N.; Niu, J. CFD study on micro-environment around human body and personalized ventilation. *Build. Environ.* **2004**, *39*, 795–805. [CrossRef]
26. Gao, N.; Niu, J. Modeling the performance of personalized ventilation under different conditions of room air and personalized air. *Int. J. Heat. Vent. Air-Cond. Refrig. Res.* **2005**, *11*, 587–602. [CrossRef]
27. Kaczmarczyk, J.; Melikov, A.K.; Fanger, P.O. Human response to personalized ventilation and mixing ventilation. *Indoor Air* **2004**, *14*, 1–13. [CrossRef]
28. Sekhar, S.C.; Gong, N.; Tham, K.W.; Cheong, K.W.; Melikov, A.K.; Wyon, D.P.; Fanger, P.O. Findings of personalized ventilation studies in a hot and humid climate. *Int. J. Heat. Vent. Air-Cond. Refrig. Res.* **2005**, *114*, 603–620. [CrossRef]
29. Conceiçao, E.Z.E.; Manuela, M.; Lúcio, J.R.; Rosa, S.P.; Custódio, A.L.V.; Andrade, R.L.; Meira, M.J.P.A. Evaluation of comfort level in desks equipped with two personalized ventilation systems in slightly warm environments. *Build. Environ.* **2010**, *45*, 601–609. [CrossRef]
30. Kalmár, F.; Kalmár, T. Alternative personalized ventilation. *Energy Build.* **2013**, *65*, 37–44. [CrossRef]
31. Kalmár, F. Innovative method and equipment for personalized ventilation. *Indoor Air* **2015**, *25*, 297–306. [CrossRef] [PubMed]
32. Kalmár, F. An indoor environment evaluation by gender and age using an advanced personalized ventilation system. *Build. Serv. Eng. Res. Technol.* **2017**, *38*, 505–521. [CrossRef]
33. Kalmár, F. Impact of elevated air velocity on subjective thermal comfort sensation under asymmetric radiation and variable airflow direction. *J. Build. Phys.* **2018**, *42*, 173–193. [CrossRef]
34. Kamár, F.; Kalmár, T. Study of human response in conditions of surface heating; asymmetric radiation and variable air jet direction. *Energy Build.* **2018**, *179*, 133–143. [CrossRef]

35. Buzug, M.B.T.; Holz, D.; Kohl-Bareis, M.; Schmitz, G. Frontiers in Medical Imaging. In Proceedings of the VDE Kongress 2004, Berlin, Germany, 18–20 October 2004; pp. 37–42.

36. MSZ EN ISO 6946:2017. *Building Components and Building Elements. Thermal Resistance and Thermal Transmittance. Calculation Method (ISO 6946:2017)*; ISO: Geneva, Switzerland, 2017.

37. Baehr, H.D.; Stephan, K. *Heat and Mass Transfer*; Springer: Berlin/Heidelberg, Germany, 2006.

38. Frank, A.; Heidemann, W.; Spindler, K. Modeling of the surface-to-surface radiation exchange using a Monte Carlo method. *J. Phys. Conf. Ser.* **2016**, *745*, 032143. [CrossRef]

energies

MDPI

Article

Correlation of Ventilative Cooling Potentials and Building Energy Savings in Various Climatic Zones

Haolia Rahman [1] and Hwataik Han [2,*]

[1] Mechanical Engineering, Politeknik Negeri Jakarta, Depok 16425, Jawa Barat, Indonesia;
   haolia.rahman@mesin.pnj.ac.id
[2] Mechanical Engineering, Kookmin University, 77 Jeungneung-ro, Seongbuk-gu, Seoul 02707, Korea
*   Correspondence: hhan@kookmin.ac.kr

Received: 12 February 2019; Accepted: 10 March 2019; Published: 13 March 2019

Abstract: The introduction of cool outdoor air can help in reducing the energy consumption for cooling during summer. Ventilative cooling potentials (VCPs) have been defined in various ways in the literature to represent potential cooling hours in specified outdoor temperature ranges. However, the energy-saving potential of ventilative cooling can differ between buildings in the same climatic zone depending on the buildings' thermal characteristics and system operations. In this study, new VCPs are introduced with an index of temperature shift based on adaptive thermal comfort. This index can be determined based on the balance temperature difference of the buildings, which is defined as the heat gain in the building divided by the thermal transmission and air exchange characteristics of the building envelope under quasi-steady state conditions. The proposed method was also compared with those reported in the literature, including a computer-based VCP tool. It is the objective of the present study to investigate the correlation between VCPs and actual energy savings via ventilative cooling. Simulations were conducted in an office building for a four-month period during summer to calculate the energy saved via ventilative cooling in comparison with that achieved with a mechanical cooling system. Eight cities representing four different climatic conditions were considered: tropical, dry, temperate, and continental. Our results revealed a strong correlation between the energy savings and the proposed VCPs in the case of a proper temperature shift estimation in all climatic zones. The computerized VCP tool also exhibited good correlation with the calculated energy savings and with the VCPs proposed herein.

Keywords: energy; building; ventilation; cooling; outdoor air

1. Introduction

Research on exploiting the climatic cooling potential is progressively increasing toward achieving buildings with low energy consumption. In most modern buildings, the highest amount of energy is consumed for cooling purposes. Ventilative cooling in which natural or mechanical ventilation is used when adequate outdoor air is available is a way to reduce the operation of mechanical cooling [1]. Thus, it is necessary to quantify the potential of ventilative cooling in a specific climate with a standard index.

Several indices for quantifying the climatic cooling potential have been introduced in accordance with the energy saving [2,3]. Yao [4] assessed an index of the natural ventilation cooling potential (NVCP) for an office building—the ratio of the number of hours within the comfort zone to the total occupied hours. Building characteristics, ventilation type, and internal heat load must be defined in advance to match the natural ventilation with the expected occupancy thermal comfort. Without including the building model, Causone [5] proposed an index of the climatic potential for natural ventilation (CPNV). The index is based on the number of hours that natural ventilation agreed with the temperature and humidity constraints. The defined acceptable supply air conditions were within the lower and upper temperature limits of 10 °C and 3.5 °C higher than the adaptive thermal comfort temperature, respectively, and the humidity level was within 30% and 70% relative humidity (RH). However, the wide acceptable temperature range in CPNV may create overcapacity in the design of ventilation systems or cause occupant dissatisfaction. A climatic cooling potential (CCP) has been introduced by Campanico [6,7] as a climatic index in the unit of kWh. The index represents the climatic condition for passive cooling systems depending on the airflow rate, the comfort set point, and various building characteristics. In the Annex 62 project, experts from 13 countries developed a VCP tool to assess the VCPs considering the building characteristics, loads, and ventilation systems [8].

However, research on assessing the actual energy savings based on VCPs is yet to be conducted. This study investigates the correlation between various VCP models and actual energy savings. Energy simulations were conducted to calculate the energy savings of a model office building in four climates during the daytime of summer.

2. Methodology

2.1. Balance Temperature Difference

The balance temperature difference (BTD) is defined as the indoor–outdoor equilibrium temperature difference when the total heat gain of an indoor space equals the heat losses through the building as formulated in Equation (1). The heat losses comprise heat transmission through envelopes and heat infiltration due to ventilation air exchange.

$$\widehat{UA}_{bldg}\Delta T_{bal} + \rho C_p Q \Delta T_{bal} = W_{IHG}, \tag{1}$$

where $\widehat{UA}_{bldg}$ is the overall heat transmission factor through the building envelope; Q is the ventilation rate; W_{IHG} is the indoor heat generation rate; and ρ and C_p are the density and specific heat of air, respectively. The BTD expresses the overall thermal characteristics of a building with a single parameter and can be calculated by rearranging and simplifying Equation (1) into Equation (2):

$$\Delta T_{bal} = \frac{W_{IHG}}{\widehat{UA}_{bldg} + \rho C_p Q} = \frac{1}{\frac{\widehat{UA}_{bldg}}{W_{IHG}} + \frac{\rho C_p Q}{W_{IHG}}} = \frac{1}{\frac{1}{\Delta T_{bal,U}} + \frac{1}{\Delta T_{bal,Q}}}. \tag{2}$$

The overall BTD is half of the harmonic mean of the BTD caused by wall transmission and that caused by air exchange. The BTD caused by transmission (Equation (3)) indicates the temperature difference when there is no infiltration and/or ventilation, whereas the BTD by air exchange (Equation (4)) is the temperature difference when there is no heat transmission through the building envelopes.

$$\Delta T_{bal,U} = \frac{W_{IHG}}{\widehat{UA}_{bldg}}, \ Q \approx 0 \tag{3}$$

$$\Delta T_{bal,Q} = \frac{W_{IHG}}{\rho C_p Q}, \ U \approx 0 \tag{4}$$

For a given internal heat generation, well-insulated but leaky buildings have high $\Delta T_{bal,U}$ but low $\Delta T_{bal,Q}$, whereas poorly insulated but air-tight buildings have low $\Delta T_{bal,U}$ but high $\Delta T_{bal,Q}$. The BTD distribution in a parametric zone of $\Delta T_{bal,U}$ and $\Delta T_{bal,Q}$ is shown in Figure 1.

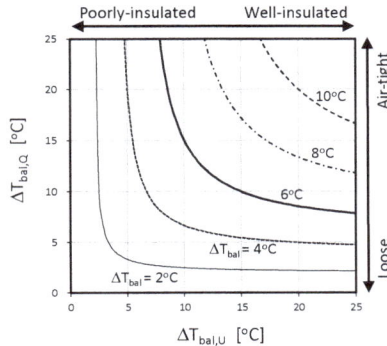

Figure 1. Balance temperature difference dependence on a building's thermal characteristics.

2.2. Definition of VCPs

The VCP term used to evaluate climatic data is defined as the number of potential ventilative cooling hours divided by the total hours, although the potential cooling hours has varied in the literature depending on the application. The VCP can be formulated as follows:

$$VCP = \frac{1}{H} \sum_{d=d_i}^{d_f} \sum_{h=h_i}^{h_f} h_{vc}, \tag{5}$$

where H is the total observed hours; h_{vc} is the number of hours when ventilative cooling is possible; d and h are the standard time parameters for day and hour, respectively; and the subscripts i and f denote the initial and final time variables for day and hour, respectively. Although this definition is widely accepted in the community, the resulting value widely varies depending on the varying temperature ranges and durations of the VCP model used.

To standardize and better represent the energy-saving potential, new VCPs are proposed based on the thermal comfort zone shifted by the amount of BTD representing the thermal characteristics of a building. Two VCPs, that is, VCP_1 and VCP_2, were calculated and compared with the CPNV proposed by Causone [5] and the VCP tool developed by IEA Annex 62 [9]. VCP_1 was defined as the number of hours in the comfort zone shifted to a lower temperature by ΔT_{bal}; both the lower and upper limits were shifted. To calculate VCP_2, the lower limit of the comfort zone (T_{lc}) was shifted to a lower temperature by ΔT_{bal} and the upper limit (T_{uc}) was shifted by half of ΔT_{bal}, as shown in Figure 2. The figure shows an example of hourly climate data for one year in Seoul on a psychrometric chart. The number of data points in the zone represents the VCP_2 during the period of interest.

The thermal comfort zones for VCP_1 and VCP_2 were determined according to the adaptive thermal comfort model of ASHRAE standard 55 [10] for a naturally ventilated building, as originally proposed by de Dear and Brager [11]. Occupant acceptability was set to 80% with a $\pm 3.5\,^\circ$C band gap, as shown in Equation (6), where $T_{a,out}$ represents the mean outdoor air temperature. Occupants were assumed to adapt their clothing to the thermal conditions and be sedentary, with a metabolic rate of 1.0~1.3.

$$T_{comf} = 0.31 T_{a,out} + 17.8. \tag{6}$$

Two other climate cooling approaches were investigated for comparative purposes. The first approach is the CPNV defined as the region above a lower limit of 10 °C and below an upper limit of T_{uc}, which counts the number of hours of thermal comfort in the region for a naturally ventilated building. The temperature comfort range is the same as given in Equation (6), but the humidity constraints by Causone [5] are not considered.

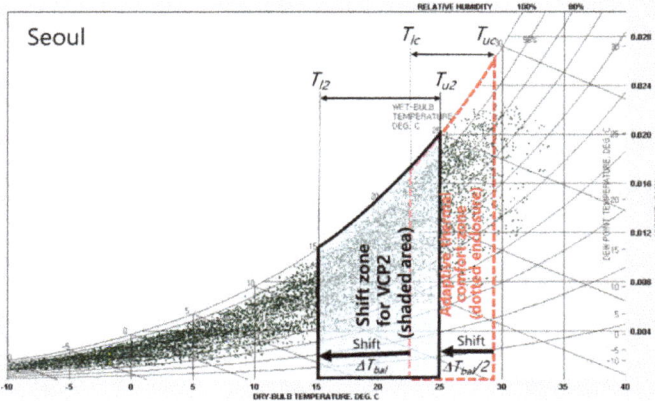

Figure 2. Shift zone for ventilative cooling potential (VCP$_2$) and adaptive thermal comfort zone along with hourly climatic data shown on a psychrometric chart.

The second approach is the VCP tool of which the evaluation criteria is based on user inputs, including building thermal model and climatic data on an hourly basis. Summer hours are categorized into four modes, which is related to indoor, outdoor, and set temperature, as well as cooling rate by ventilation. Only "mode 2", in which the outdoor temperature can meet the indoor comfort with increased ventilation rate, was evaluated in the VCP tool herein. Thus, out of the output datasets provided by the tool, only some of datasets were selected in the range compatible with the other approaches. The VCP tool refers to the adaptive thermal comfort model in the EN 1521:2007 standard [11] with a ±3 °C band gap, which is expressed as follows:

$$T_{comf} = 0.33T_{rm} + 18.8, \tag{7}$$

where T_{rm} is the outdoor running mean temperature. Evaluation was conducted within office hours (08:00~16:00). An illustration of the comfort zone and four climate evaluation approaches are shown in Figure 3.

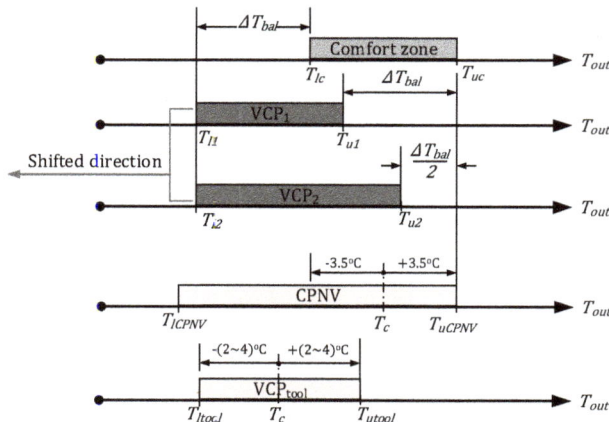

Figure 3. Ventilative cooling potential (VCP) evaluation methods based on temperature ranges. CPNV—climatic potential for natural ventilation.

2.3. Climates

The Köppen climate classification [12] specifies five main climate groups: tropical, dry, temperate, continental, and polar. The fifth group, polar, is not considered herein because ventilative cooling is not quite necessary. Thus, eight cities were analyzed to include two of each of the four main climate groups studied. A list of the cities considered herein and their climates are summarized in Table 1. The table also lists the average summertime outdoor temperature and wind speed.

Table 1. Climate data for the eight cities analyzed (data source: Energyplus [13]).

Climate Zone	City	Location	Average Outdoor Temperature (°C)				Average Wind Speed (m/s)			
			Jun	Jul	Aug	Sep	Jun	Jul	Aug	Sep
Tropical (Megathermal)	Jakarta (Indonesia)	6.13 S, 106.75 E	29.0	29.0	29.4	29.6	4.51	4.76	5.11	4.89
	Mumbai (India)	18.9 N 72.82 E	29.0	27.8	27.2	27.6	2.74	3.26	3.23	2.08
Dry (Arid)	Madrid (Spain)	40.45 N, 3.55 W	23.2	27.0	20.6	25.5	2.73	3.26	3.61	3.46
	Alice (Australia)	23.8 S 133.88 E	11.5	12.0	13.1	20.4	2.62	1.39	1.66	3.52
Temperate (Mesothermal)	Los Angles (USA)	33.93 N, 118.4 W	24.7	20.1	21.9	21.6	4.54	5.00	5.10	4.49
	Yunnan (China)	22.78 N 100.97 E	22.5	22.0	21.8	21.2	1.04	0.76	0.76	0.66
Continental (Microthermal)	Seoul, (Korea)	37.57 N, 126.97 E	23.2	26.2	27.0	22.3	2.46	2.60	2.25	2.17
	Prague (Czech)	50.1 N 14.28 E	15.6	17.3	17.6	13.3	4.03	3.07	3.40	3.80

2.4. Building and Energy Simulation Model

A medium-sized, three-story, 4982 m² office building [14] with a 5 m long central atrium and 3 m ceilings was used for simulation, as shown in Figure 4. The total load produced by occupants, lights, and equipment was 31.24 W/m² and lasted from 08.00 to 16.00. The glazed and open areas represented 33% and 11%, respectively, of the wall per floor area ratio. The building was located in a rural area with low buildings and faced 90° to the north.

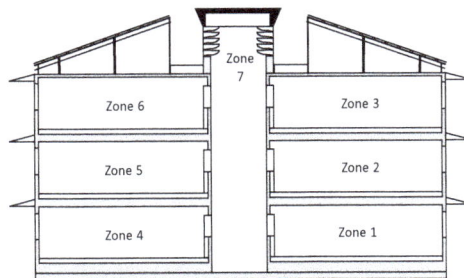

Figure 4. Building model for validation.

Two cooling schemes were separately run to perform the energy-saving analysis. The first operated as the control in which an air conditioner with a COP of 3.0 was used to meet all cooling requirements. In the second, a mechanical fan cooled the indoor space using outdoor air at a constant flow rate of 14 m³/s when the internal temperature was between 22 °C and T_{uc}. If the indoor temperature rose above T_{uc}, the fan was replaced with air conditioning. The energy consumption and indoor conditions were calculated using the CoolVent software package [15] from June to September on an hourly basis. ΔT_{bal} was manually calculated by considering the energy balance between the internal heat gain and the building heat transfer. The UA value of the building was estimated to be 3600 W/K ($\Delta T_{bal,U}$ = 43 °C), and the $\rho C_p Q$ value was 16,800 W/K ($\Delta T_{bal,Q}$ = 9 °C). In practice, users can use the maximum fan capacity to define the air exchange rate for balance temperature difference calculation. The natural infiltration has not been taken into account because the rates are not controlled and can be neglected compared with the mechanical ventilation rates for cooling purposes. Solar radiation was not included in the calculation for simplification, but it can be included as a part of the indoor heat generation rate in Equation (1). It is a matter of how to model complicated solar heat gains varying considerably depending on various parameters into a single parameter. In this paper, ΔT_{bal} was thus calculated to be approximately 7 °C without considering solar gains. A ventilation rate of 2.81 L/s·m^2 was employed in the VCP tool.

3. Results

The ventilative cooling potentials, VCP$_1$ and VCP$_2$, for the four representative climates in various temperature shifts are plotted in Figure 5, where ΔT is used as an index of temperature shift. The lower and upper limits were both shifted by ΔT in the calculation of VCP$_1$. For VCP$_2$, the upper limit was only shifted by half of ΔT.

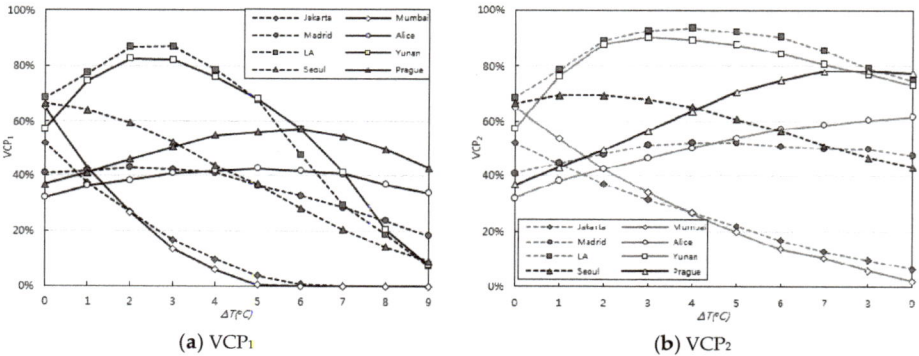

(a) VCP$_1$ (b) VCP$_2$

Figure 5. Variations of VCP$_1$ and VCP$_2$ with respect to ΔT in various cities.

Both VCPs exhibited similar patterns according to the temperature shift, as shown in Figure 5a,b. In the tropical climate, both VCPs began at a moderate level and decreased rapidly with increasing ΔT; the greater the BTD, the lesser the cooling benefits. Thus, buildings with large ΔT in tropical climates cannot extract the advantage of outdoor cooling. However, in the dry climate, both VCPs began at a low value and slowly increased with increasing ΔT. In the temperate climate group, the VCP values remained high over a wide range of ΔT, reaching over 80% when ΔT was between 1 °C and 7 °C. The two cities of the continental climate group presented opposite trends. Seoul began with a moderate VCP and decreased slightly with increasing ΔT, whereas Prague began with a low VCP and increased with increasing ΔT. This was likely caused by their heat levels (third classification scheme in Köppen climate); Seoul is classified as having a "hot summer", and Prague is classified as having a "cold summer". Because each city shows its own characteristics of VCP variations according to ΔT, a VCP lookup table can be generated with an index of the temperature shift in all cities so that users can easily determine the cooling potential according to their ΔT_{bal} building design plan [16].

Seoul has an extreme temperature variation between summer and winter. Furthermore, the summer period has a wide temperature distribution, as shown in Figure 6. In the early and late summer months (June and September), VCP$_2$ experienced only slight changes with increasing ΔT and remained above 60%; this pattern is similar to that observed in the dry climate. Meanwhile, in

July and August, the months experiencing high summer, both proposed VCPs sharply decreased with increasing ΔT, indicating that similar to tropical climates, outdoor cooling cannot be reliably used for buildings with large balance temperature differences.

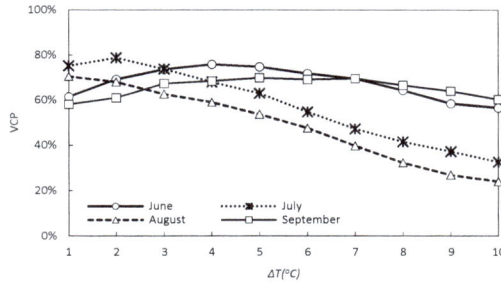

Figure 6. Monthly breakdown of the VCP_2 profile in Seoul.

The weekly values of VCP_1, VCP_2, and energy savings in Seoul at $\Delta T = 7$ °C are plotted in Figure 7a from June to September. The VCP patterns agreed with the calculated energy savings and were high in the early and late summer months and relatively low during the high summer months, as is typical for hot-summer continental climates. In comparison with VCP_1, the magnitude of VCP_2 was much closer to the energy savings. Additionally, VCP_2 had a stronger correlation with the energy savings, as shown in Figure 7b.

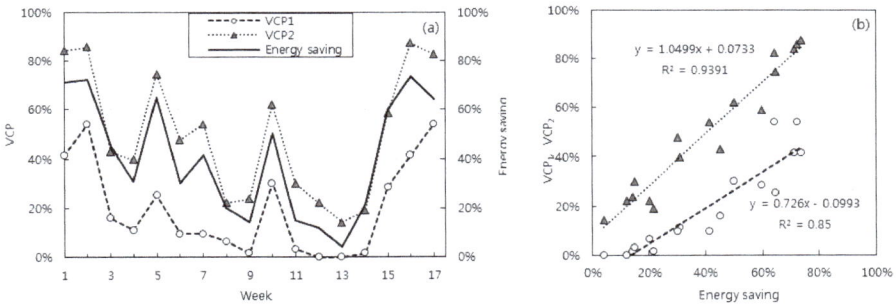

Figure 7. Summary of VCPs and energy savings in Seoul: (**a**) weekly VCPs and energy savings and (**b**) correlation between VCPs and energy savings.

Correlations between the calculated energy savings for the eight cities during the 17 weeks of summer with each climate evaluation method are presented in Figure 8. VCP_2 was found to have the highest correlation with energy savings for the given building model. In VCP_2, outdoor air can be partially used for ventilative cooling as the temperature difference is not sufficiently large to completely cover the cooling load. Assuming that the outdoor temperature is equally distributed in the selected region statistically, as much as half of ΔT_{bal} can be added to the shifted zone for the VCP calculation. No significant correlation was found between energy savings and VCP_1. Unlike VCP_2, the narrower selection of outdoor temperatures may have omitted the outdoor cooling potential.

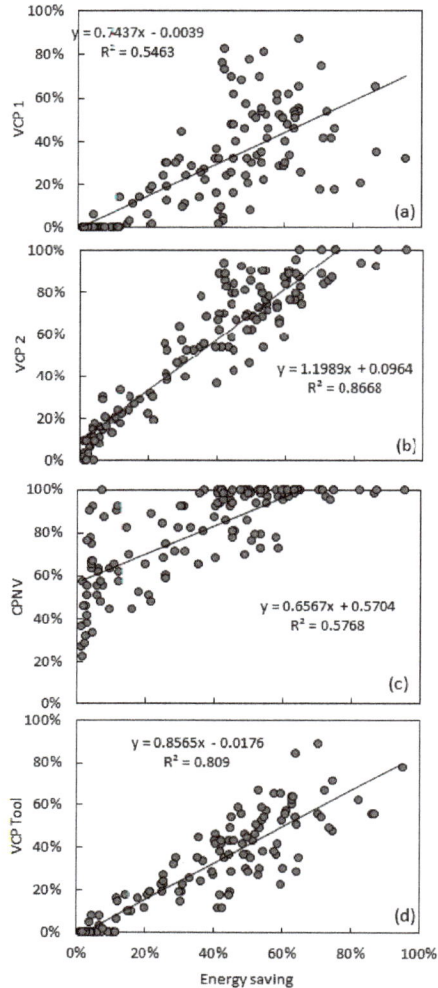

Figure 8. Correlations between VCPs and energy savings: (a) VCP_1, (b) VCP_2, (c) climatic potential for natural ventilation (CPNV), and (d) VCP tool.

Similarly, a weak correlation was found between CPNV and energy saving, possibly because of the wide boundary conditions of the CPNV evaluation design. CPNV does not account for the building characteristics or ventilation systems and relies only on weather, unlike the VCP tool, which includes these characteristics. Thus, a moderate correlation was found between energy savings and the VCP tool, indicating that the number of hours for which the ventilation rate should be increased correlated with the energy saved because of ventilation.

A summary of the correlation between the proposed VCPs and the energy savings at various temperature shifts ΔT is presented in Figure 9. Although the building simulation was performed with a known ΔT value of the building, it will be useful to how the ventilation strategy (i.e., energy reduced as a result of ventilation) behaves at various ΔT. The correlation between VCP_2 and ΔT gradually increased with increasing ΔT until ΔT reached 6 °C; a slight change in the correlation was observed afterward. It is important to accurately evaluate the BTD. Even though the BTD is not evaluated precisely plus minus a few degrees, the correlation between VCP_2 and energy savings remains nearly constant, with an R-squared value of over 85%.

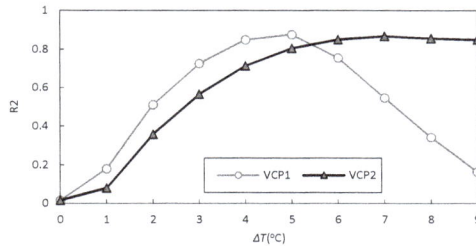

Figure 9. Correlation of VCP_1 and VCP_2 with energy savings at various ΔT.

4. Conclusions

The two methods proposed herein for evaluating the ventilative cooling potential were assessed in eight cities across four climate zones: tropical, dry, temperate, and continental. Both methods classified the outdoor temperature on an hourly basis based on the temperature shift of the comfort zone. The temperature shift was determined based on a single parameter—the balance temperature difference between the indoor and outdoor temperature—which varies in terms of the envelope design and ventilation operation of the building analyzed. The two representative cities of each climate group exhibited a similar pattern of VCPs according to the temperature shifts.

The VCP distributions with respect to the temperature shift show unique patterns depending on climatic groups. VCPs stay nearly constant over a wide range of temperature shifts in a Mediterranean climate. In a tropical climate, small temperature shifts are preferred for taking full advantage of ventilative cooling. A similar conclusion can be drawn for a continental climate, but annual energy consumption should be addressed, including heat loss in winter. Ventilative cooling is best applicable to a dry semi-arid climate where daily temperature fluctuations are large.

The proposed VCPs were validated by performing an energy simulation on a building to determine the potential for mechanical cooling reduction. The amount of energy saved using outdoor ventilation was found to be in good correlation with the proposed VCPs, particularly VCP_2. The proposed VCPs can be used in early building design stages to predict the amount of energy savings by ventilative cooling without the use of a computer-based tool. A look-up table can be provided for various cities with an index of temperature shifts, so that design engineers can optimize the balance temperature difference of the building they design.

Author Contributions: Conceptualization, H.H.; Methodology, H.H. and H.R.; Software, H.R.; Validation, H.H. and H.R.; Formal Analysis, H.H.; Investigation, H.H. and H.R.; Resources, H.R.; Data Curation, H.R.; Writing—Original Draft Preparation, H.R.; Writing—Review & Editing, H.R. and H.H.; Supervision, H.H.; Project Administration, H.H.

Funding: This work was supported by the Basic Science Research Program through the National Research Foundation of Korea (NRF), funded by the Ministry of Education (2016R1D1A1B01009625) and by Global scholarship of Kookmin University.

Conflicts of Interest: The authors declare no conflict of interest.

References

1. Kolokotroni, M.; Heiselberg, P. *Ventilative Cooling State of the Art Review*; Annex 62 Ventilative Cooling: Aalborg, Denmark, 2015.
2. Artmann, N.; Manz, H.; Heiselberg, P. Climatic potential for passive cooling of buildings by night-time ventilation in Europe. *Appl. Energy* **2007**, *84*, 187–201. [CrossRef]
3. Haase, M.; Amato, A. An investigation of the potential for natural ventilation and building orientation to achieve thermal comfort in warm and humid climates. *Sol. Energy* **2009**, *83*, 389–399. [CrossRef]

4. Yao, R.; Li, B.; Steemers, K.; Short, A. Assessing the natural ventilation cooling potential of office buildings in different climate zones in China. *Renew. Energy* **2009**, *34*, 2697–2705. [CrossRef]

5. Causone, F. Climatic potential for natural ventilation. *Arch. Sci. Rev.* **2016**, *59*, 212–228. [CrossRef]

6. Campaniço, H.; Hollmuller, P.; Soares, P.M.M. Assessing energy savings in cooling demand of buildings using passive cooling systems based on ventilation. *Appl. Energy* **2014**, *134*, 426–438. [CrossRef]

7. Campaniço, H.; Soares, P.M.M.M.; Hollmuller, P.; Cardoso, R.M. Climatic cooling potential and building cooling demand savings: High resolution spatiotemporal analysis of direct ventilation and evaporative cooling for the Iberian Peninsula. *Renew. Energy* **2016**, *85*, 766–776. [CrossRef]

8. Belleri, A.; Avantaggiato, M.; Psomas, T.; Heiselberg, P. Evaluation tool of climate potential for ventilative cooling. *Int. J. Vent.* **2017**, 2044–4044. [CrossRef]

9. British Standards Institution. PrEN 16798-1. (2016-02-07 under Approval). In *Energy Performance of Buildings—Part 1: Indoor Environmental Input Parameters for Design and Assessment of Energy Performance of Buildings Addressing Indoor Air Quality, Thermal Environment, Lighting and Acoustics*; British Standards Institution: London, UK, 2016.

10. ASHRAE. Thermal Environmental Conditions for Human Occupancy. Available online: http://www.aicarr.org/Documents/Editoria_Libri/ASHRAE_PDF/STD55-2004.pdf (accessed on 12 February 2019).

11. De Dear, R.J.; Brager, G.S.; Artmann, N. Thermal comfort in naturally ventilated buildings: Revisions to ASHRAE Standard 55. *Energy Build.* **2002**, *34*, 549–561. [CrossRef]

12. Peel, M.C.; Finlayson, B.L.; Mcmahon, T.A. Updated world map of the Köppen-Geiger climate classification. *Hydrol. Earth Syst. Sci. Discuss.* **2007**, *4*, 439–473. [CrossRef]

13. Energyplus, Weather Data. 2016. Available online: https://energyplus.net/weather (accessed on 12 February 2019).

14. Deru, M.; Field, K.; Studer, D.; Benne, K.; Griffith, B.; Torcellini, P.; Liu, B.; Halverson, M.; Winiarski, D.; Rosenberg, M.; et al. *U.S. Department of Energy Commercial Reference Building Models of the National Building Stock*; NREL/TP-5500-46861; NREL: Lakewood, CO, USA, 2011; p. 19.

15. Maria-Alenjandra, B.M.; Glicksman, L. Coolvent: A multizone airflow and thermal analysis simulator for natural ventilation in building. In Proceedings of the 3rd National Conference of IBPSA-USA, Berkeley, CA, USA, 30 July–1 August 2008.

16. Rahman, H.; Han, H. Ventilative cooling potential based on climatic condition and building thermal characteristics. In Proceedings of the AIVC Conference, Nottingham, UK, 13–14 September 2017.

Article

Field Measurements and Numerical Simulation for the Definition of the Thermal Stratification and Ventilation Performance in a Mechanically Ventilated Sports Hall

Lina Seduikyte [1,*], Laura Stasiulienė [1], Tadas Prasauskas [2], Dainius Martuzevičius [2], Jurgita Černeckienė [1], Tadas Ždankus [1], Mantas Dobravalskis [1] and Paris Fokaides [1,3,*]

[1] Faculty of Civil Engineering and Architecture, Kaunas University of Technology, Studentu str. 48, LT-51367 Kaunas, Lithuania; laura.stasiuliene@ktu.lt (L.S.); jurgita.cerneckiene@ktu.lt (J.Č.); tadas.zdankus@ktu.lt (T.Ž.); mantasdob@inbox.lt (M.D.)

[2] Department of Environmental Technology, Kaunas University of Technology, Radvilenu str. 19, LT-50254 Kaunas, Lithuania; Tadas.Prasauskas@ktu.lt (T.P.); Dainius.Martuzevicius@ktu.lt (D.M.)

[3] School of Engineering, Frederick University, Nicosia 1036, Cyprus

* Correspondence: lina.seduikyte@ktu.lt (L.S.); eng.fp@frederick.ac.cy (P.F.)

Received: 26 April 2019; Accepted: 8 June 2019; Published: 12 June 2019

Abstract: Sports halls must meet strict requirements for energy and indoor air quality (IAQ); therefore, there is a great challenge in the design of the heating, ventilation, and air conditioning (HVAC) systems of such buildings. IAQ in sports halls may be affected by thermal stratification, pollutants from different sources, the maintenance of building, and the HVAC system of the building, as well as by the activities performed inside the building. The aim of this study is to investigate thermal stratification conditions in accordance with the performance of the HVAC systems in the basketball training hall of Žalgirio Arena, Kaunas in Lithuania. Field measurements including temperature, relative humidity, and CO_2 concentration were implemented between January and February in 2017. The temperature and relative humidity were measured at different heights (0.1, 1.7, 2.5, 3.9, 5.4, and 6.9 m) and at five different locations in the arena. Experimental results show that mixing the ventilation application together with air heating results in higher temperatures in the occupied zone than in the case of air heating without ventilation. Computational fluid dynamics (CFD) simulations revealed that using the same heating output as for warm air heating and underfloor heating, combined with mechanical mixing or displacement ventilation, ensures higher temperatures in the occupied zone, creating a potential for energy saving. An increase of air temperature was noticed from 3.9 m upwards. Since CO_2 concentration near the ceiling was permissible, the study concluded that it is possible to recycle the air from the mentioned zone and use it again by mixing with the air of lower layers, thus saving energy for air heating.

Keywords: indoor air quality; stratification; basketball hall; CFD; field measurement

1. Introduction

Large, open indoor spaces are found in shopping malls, arenas, sports halls, theatres, factory workshops, railway stations, airports, etc. This type of building is usually mechanically ventilated to ensure appropriate indoor air quality (IAQ) conditions. IAQ is influenced by several factors, including thermal stratification, pollutants from different sources, the building's maintenance, the ventilation type, and the activities performed inside the building. Large spaces must meet requirements for energy and IAQ; therefore, the design of heating, ventilation, and air conditioning (HVAC) systems in such buildings constitutes a major challenge. Poorly designed mechanical ventilation in large spaces may result to insufficient IAQ and energy losses [1]. Mainstream research about the thermal conditions of large indoor spaces focuses on IAQ and health problems, as well as on HVAC systems and energy efficiency.

The parameters that affect the indoor thermal conditions in sports halls are well documented in the literature [2]. Andrade et al. [3] present a review of 1281 scientific studies in 18 countries related to IAQ of environments used for physical activities. The analysis of scientific papers considered studies published from 1975 [4] onwards. Most of the articles discuss health, respiratory problems, and pollutants in environments used for physical exercise and sport activities. Physical activities implemented in polluted indoor spaces set people at health risk, as the air is inhaled through the mouth, and not through the nose, remaining in this manner unfiltered [4]. Branis et al. [5] deals with the indoor–outdoor relationship and the potentially negative effect on the health of schoolchildren, aged 11–15 years, when exercising in naturally ventilated sports halls of elementary schools. The concentration of indoor particulate matter of aerodynamic diameter less than 2.5 µm (PM 2.5) was an indicator of potentially negative health risks. In that study, it was revealed that the amount of particulate matter in the respiratory tract during inhalation might be five times higher during intensive exercise compared with the steady conditions. The results of measured PM 2.5 concentration showed that concentrations exceeded World Health Organization (WHO) recommendations on 50% of measure days indoors and 48.6% of measured days outdoors. The concentration of indoor particles was found to depend on the season, the location of the school, and the amount of people in the sports hall.

Several studies have been performed to evaluate the IAQ and ventilation effectiveness when different ventilation types are installed in large indoor spaces. Although mixing ventilation dominates in large indoor spaces, in the literature studies considering alternative ventilation solutions are also found, including natural and mechanical ventilation [1,6–8], displacement ventilation [8,9], and changing of the position of extract fans [10,11]. The use of mechanical ventilation with the contributions of different heat-gain sources [12] are also reported in the literature.

Some studies were implemented in large space buildings with glass facades, where high intensity solar radiation must also be considered [13]. Studies employing new methods for assessing ventilation in large spaces, aiming to measure the local age of the air by tracer gas step-down or the decay method inside a control volume, were also implemented [1]. In most cases, field measurements were implemented. Computational fluid dynamics (CFD) methods are usually employed to further expand the investigated cases; however, in some cases studies are based merely on computational analysis [14,15]. Indoor thermal conditions indicators, such as the predicted mean vote (PMV) and the predicted percentage of dissatisfaction (PPD), introduced by Fanger in 1970 [16] and adopted by ISO Standard 7730:1994 [17], were also extracted in some cases [8,14].

Stathopoulou et al. [6] conducted their research in two large sports halls. The investigated halls had different ventilation types (natural and mechanical). Also, one hall was surrounded by heavy traffic. The temperature levels in the sports hall with the mechanical ventilation system were not stratified, and stratification was not intense. Although pollution stratification was observed in both halls, this phenomenon was more intense in the naturally ventilated hall. Air stratification results from the influence of buoyancy and the stack effect [18]. Rajagopalan and Elkadi [11] present a study, based on CFD simulation, which aims to test the thermal and ventilation performance of a sports hall within an aquatic center, located in Australia. Seven different scenarios were investigated, which were based on the different positioning of extract fans and the incorporation of natural ventilation. The results showed that the lower the position of the exhaust fans, the better the comfort at the occupants' level. Also, the scenario with the exhaust fans located on the roof was the most undesirable. Similar conclusions were also determined for natural ventilation. Rajagopalan and Luther [7] present the results of a study performed in a naturally ventilated (sometimes assisted with exhaust fans) sports hall, using field measurements and CFD simulations. The results showed that for those periods that the hall was occupied, the CO_2 concentration in the hall was below 800 ppm, which is an indication of good ventilation and IAQ. According to Kamar et al. [8], installing exhaust fans with a 1 m diameter, at the height of 6 m from the floor, has the potential to reduce the PMV index by 75–95% and the PPD index by 87–91%. Another important aspect that significantly affects the indoor thermal conditions of sport halls concerns air recirculation. The analysis of energy efficiency measures in the study of Nord et al. [19] shows that air recirculation has the greatest effect on total energy use in sports halls, and that air recirculation could give an energy savings of 27% when 50% of the indoor air is recirculated. Even though the use of the PMV and the PPD index is regularly found in studies concerning the thermal comfort of sports halls, the outcome of the study of Revel and Arnesano [20] demonstrates that these indicators cannot be applied in sports halls. The same team investigated the problem of indoor temperature distributions in large sport spaces and its impact on energy efficiency and thermal comfort [21].

Another recent trend in the scientific literature of sports halls energy assessment concerns the indoor air conditions of passive buildings. In the study of Kisilewicz and Dudzińska [22], under high ambient air temperature and switched-off ventilation, acute overheating was observed. Due to the widespread implementation of the nearly-zero-energy building concept, often-observed overheating is expected to become an important issue concerning the indoor thermal conditions of sports halls. In that study, it was also determined that indoor thermal condition measurements are affected by window location and solar radiation geometry. In the field of passive buildings, innovative air-supported membrane structures have also been recently employed in low-energy sports halls to investigate their physical and thermal behavior. In the study of Suo et al. [23], it was proven that such double-layered envelopes allow savings of 11–18% of the heating energy compared to single-layer. Moreover, in that study further energy-saving strategies are proposed and quantified, considering low-emissivity coatings, reduction of cracks' areas, and modifying indoor air set points.

The literature review reveals that although there is a significant number of previous studies concerning the assessment of the IAQ conditions in sports halls, a comprehensive assessment concerning the optimal HVAC configurations for achieving ideal IAQ is missing. The definition and choice of HVAC systems in sports halls in the scientific literature is not based on a specific rationale and is rather arbitrary. This aspect is particularly important, in view of the European challenge of 2021 to achieve nearly-zero-energy buildings, including commercial buildings and sports halls. The necessity of studies that will focus on a comparative assessment between different HVAC technologies and their performance for sports halls is obvious. To this end, the aim of this study is to investigate IAQ conditions in accordance with thermal stratification and the performance of the HVAC systems in the basketball training hall of Žalgirio Arena, Kaunas in Lithuania, with the use of field measurements (temperature and CO_2), and to test different ventilation scenarios with the application of CFD simulations.

2. Methodology

For this investigation, the following methods were used:

- Field measurements for the measurement of the temperature, relative humidity, and the CO_2 concentration;
- CFD simulations for the calculation of the temperature distribution in the training hall.

In this section, a short description of the adopted methods is presented.

2.1. Žalgirio Arena Basketball Training Hall

Žalgirio Arena is one of the biggest multifunctional arenas in the Baltic region, located on Nemunas Island in the centre of the city of Kaunas in Lithuania. Different events, including basketball games, volleyball games, handball games, and ice-skating competitions, as well as theatrical events and expos are organized on a regular basis in Žalgirio Arena. The building also hosts offices, sport clubs, restaurants, an amphitheater, and a training hall. The total area of the building is 39,684.2 m^2, of which the arena's area is 2841.72 m^2 and the heated area is 28,297.75 m^2. There are up to 20,000 seats for concerts, and up to 15,708 seats for basketball games.

The basketball training hall (Figure 1) has dimensions of 20.8 × 33.9 × 8.4 m. The floor area of the training hall is 705 m², and the volume of the hall is 5923 m³. The training hall has a 1.5 m width balcony along the longer side of the hall. The distance between the balcony and the floor is 4.5 m. The walls and the ceiling of the hall are made of concrete. The training hall is illuminated by florescent, hanging light fixtures. The temperature of the lighting fixtures during the time of measurements was around 72 °C. The training hall has two exposed and two internal walls. During the measurement campaign, the temperature of the exposed walls and floors was 18 °C, and the temperature of the interior walls and the ceiling was 21 °C. The arena construction started in 2008. At that period, in Lithuania, according to the valid regulations, the heat transfer coefficients for walls, floors in contact with soil, roofs, and windows were 0.25, 0.30, 0.20, and 1.60 W/(m²·K) respectively.

The training hall is ventilated by a mechanical ventilation system. Air is supplied by eight swirl diffusers of 0.4 m diameter located at 6.6 m height. Diffusers ensure mixing ventilation airflow patterns. For air extraction, six duct grills are used, with dimensions of 0.225 m × 1.025 m and located at 6.2 m height. The air change rate in the training hall was 1.1 h⁻¹. The flow rate for both the air supply and the extracted air was 6600 m³/h. The training hall is equipped with two air heaters that have axial flow fans and water-based heating coils. Each heater has a heating capacity of 10.25 kW.

Figure 1. Basketball training hall view.

2.2. Field Measurements

The field measurements were performed between January and February 2017. The parameters measured were the air temperature, relative humidity (RH), and CO_2 concentration. The layout of the heating and ventilation equipment, as well as the measurement equipment, are presented in Figure 2. The temperature and the relative humidity were measured on five stands (S1–S5) at six heights (0.1, 1.7, 2.5, 3.9, 5.4, and 6.9 m) above the floor. The measurements were conducted for the conditions described in Table 1.

Table 1. Conditions under which the field measurements of indoor temperature were conducted.

Symbol	Condition
B_T1	Before basketball training, without operating ventilation systems or people
B_T2	Before basketball training, with operating ventilation systems, without people
T	During basketball training, with operating ventilation systems and people (at heights above 5.4 m)
A_T1	After basketball training, with operating ventilation systems, without people
A_T2	After basketball training, without operating ventilation systems or people

For temperature and relative humidity measurement, HOBO MX1101 data loggers (±0.2 °C, ±2% RH accuracy) were used. CO_2 concentration was measured at four locations: in the occupied zone at a height of 1.7 m (S6 and S8), in the ceiling zone at a height of 7.1 m (S7), in the main extract air duct (E), and in the main supply air duct (S). For measurements of CO_2 concentrations, IAQ-CALC 7545 IAQ meters (TSI, United States), with an accuracy of ±3% were used. The temperature of the surfaces was measured with a thermal camera (Type Fluke Tir1). The air velocity was measured in the occupied zone at a height of 1.7 m. The temperature and the relative humidity were recorded at 5 min intervals, and the CO_2 concentration at 1 min intervals. During the period of field measurements, according to the local weather station, the average outside temperature in Kaunas city varied between −7 and +4 °C. The lowest average temperature varied between −11 and −17 °C.

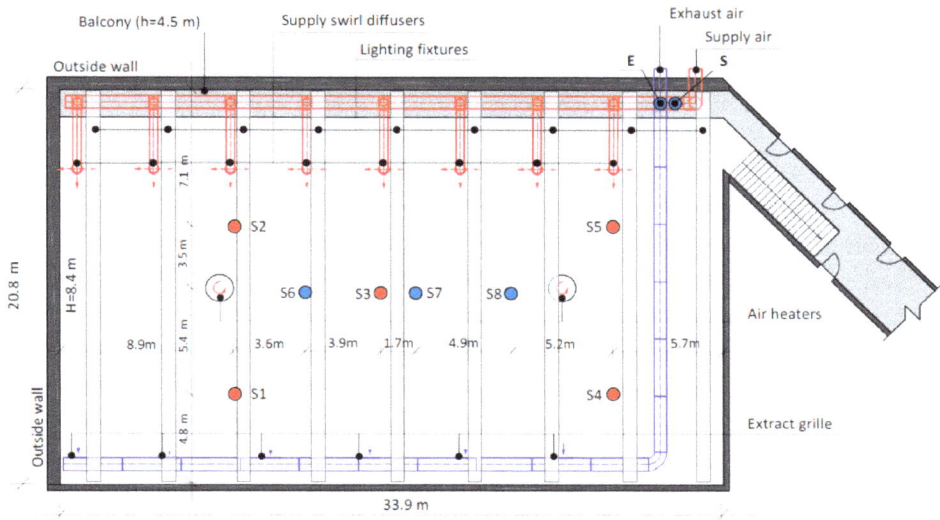

Figure 2. Heating and ventilation equipment, as well as measurement equipment locations in the training hall (S1–S8: locations of stands; E, S: locations of CO_2 m).

2.3. Numerical Method

The commercial CFD tool FloVENT (Mentor Graphics, United States) was chosen to visualise and calculate the temperature distribution in the training hall. FloVENT is a tool used to investigate the airflow, contamination, and heat performance of both individual thermal zones and whole buildings. In this study, the k-ε turbulence model (LVEL k-ε) was employed. The k-ε turbulence model shows good performance in predicting the behavior of indoor airflows, temperatures, and contaminant distribution in buildings [24]. To achieve steady-state simulation results, the double-precision solver of FloVENT was applied. Flovent's function of "kE model stratification" was employed to include buoyancy generation terms. Contaminants were not modelled in CFD calculations, as the idea of CFD simulations were to visualise and calculate the temperature distribution in the training hall. CFD simulations were performed for the conditions before basketball training, with operating heating and ventilation systems, and without people.

The basketball training hall of the Žalgirio Arena was replicated in a three-dimensional CFD model (Figure 3). All boundary and initial conditions replicate parameters measured during field measurements. The temperature for all the exposed walls and the floor of the training hall was set to 18 °C. For the internal walls and the ceiling, the temperature was set to 21 °C. Swirl diffusers and fixed flow openings were used to create the supply and extract air terminals (diffusers and grilles), respectively. Supply airflow was set to 825 m^3/h for each swirl diffuser, for the case of mixing ventilation. Displacement ventilation diffusers were modelled as fixed-flow air supply openings with a 0.7 free area ratio, and supply airflow was set to 2200 m^3/h. Supply air temperature, in cases with mechanical mixing ventilation, was set at 18 °C, and at 19 °C in cases with mechanical displacement ventilation. In all cases, the air exhaust grilles were designed with the 0.5 free area ratio, and exhaust airflow was set at 1100 m^3/h for each grille. These airflows ensured an air change rate of 1.1 h^{-1}. A solid cuboid with a thermal attribute of 30 W/m^2 was used to simulate underfloor heating. Solid cuboids were also used to simulate the lighting fixtures, with a thermal attribute of 72 °C.

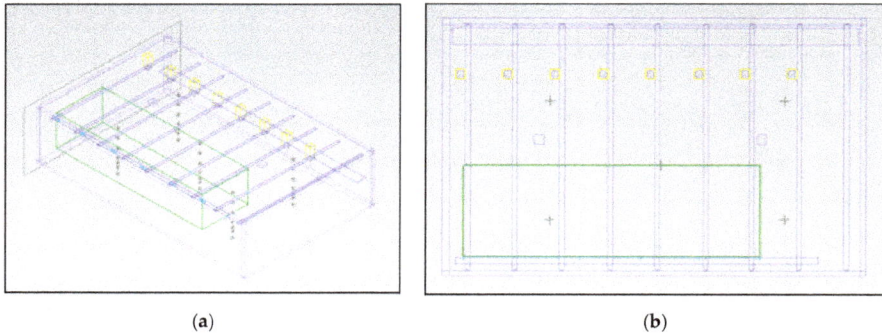

| (a) | (b) |

Figure 3. Example of training hall geometry in a computational fluid dynamics (CFD) model (isometric view (**a**) and plan (**b**) of the case with warm air heating and mechanical mixing ventilation). The "+" indicates the location of the monitor points.

A variable Cartesian grid was used to divide the geometry into regions. The density of the grid was increased close to the air supply terminals, the heated floor, and the exposed walls. As grid quality in CFD simulations is an important factor that affects the results, a grid sensitivity analysis was also performed, concluding that a 100,000-cell grid would be appropriate for this study. Refining the grid further gives negligible changes in results and complicates the calculation.

Three cases with different heating and ventilation systems were studied:

- The AH+MMV case, in which the simulation was in accordance with the field measurements. Mechanical mixing ventilation combined with warm air heating was presumed.
- The UFH+MMV case, which considered mechanical mixing ventilation combined with underfloor heating.
- The UNF+MDV case, which represented mechanical displacement ventilation combined with underfloor heating.

SPSS 23 (IBM Corp., Armonk, NY, USA) and Excel 365 (Microsoft Corp., Albuquerque, NM, USA) packages were used for post-processing of the numerical results. Data was tested for distribution using the Shapiro–Wilk test. Nonparametric Kruskal-Wallis (for multiple samples) and Mann–Whitney (for two unpaired groups) tests were also used to test the hypothesis of the difference in means, setting a 5% level of significance.

3. Results and Discussion

3.1. Field Results

3.1.1. Indoor Temperature

In Figure 4, the indoor temperature measured at different conditions for the same location is presented. Figure 5 presents the indoor temperature measurements at the same height for different conditions. The indoor temperature measurements before the beginning of the basketball training, with operating ventilation systems and without people in the hall (scenario B_T2), was statistically analyzed. A significant difference in the spatial distribution of the indoor temperature in the basketball training hall, considering all measurement stands at the same height at once, was observed ($p < 0.05$). This data is presented in Table 2. The paired data from the stands also revealed statistically significant differences in most, except for stands 2 and 3 at a height of 1.7 m and 5.4 m, respectively, as well as for stands 3 and 4 at heights of 0.1, 1.7, and 5.4 m, respectively. Similarly, a significant difference in the overall stratification distribution of the indoor temperature ($p < 0.05$) as measured in each stand was revealed. The p-value represents the maximal probability that, when the null hypothesis is true, the statistical summary would be greater than or equal to the actual observed results. In Table 3, the Kruskal-Wallis test statistics for stratification distribution are presented. Pairwise, in each stand the temperature differed significantly between heights of 0.1 and 1.7 m (all stands), 1.7 and 2.5 m (except stands 2 and 5), 2.5 and 3.9 m (except stands 1, 2 and 5), 3.9 and 5.4 m (all stands), and 5.4 and 6.9 m (all stands).

Table 2. Kruskal-Wallis test statistics for spatial distribution.

	Measurement Points				
	S1	**S2**	**S3**	**S4**	**S5**
Kruskal-Wallis H	256.616	193.943	130.158	211.743	239.548
df	5	5	5	5	5
Asymp. Sig. (p-values)	$p < 0.01$	$p < 0.01$	$p < 0.01$	$p < 0.01$	$p < 0.01$

Table 3. Kruskal-Wallis test statistics for stratification distribution.

	Measurement Height					
	0.1	**1.7**	**2.5**	**3.9**	**5.4**	**6.9**
Kruskal-Wallis H	104.044	86.204	54.542	82.484	133.129	204.204
df	4	4	4	4	4	4
Asymp. Sig. (p-values)	$p < 0.01$	$p < 0.01$	$p < 0.01$	$p < 0.01$	$p < 0.01$	$p < 0.01$

3.1.2. Relative Humidity and CO_2 Concentration

There was no significant difference in the measured relative humidity (RH). The average RH values ranged between a minimum value of 23.4% and a maximum value of 25.5%. The range of the measured CO_2 concentrations in the extract air was found between 372–868 ppm, in the supply air from 365–457 ppm, in the occupied zone from 386–951 ppm, and in the ceiling zone from 417–629 ppm. The measured CO_2 values are lower than 1000 ppm. Therefore, the indoor air quality is assigned to medium air quality class (IDA2) in accordance with the EN 13779 standard [25].

3.2. Numerical Results

The vertical indoor temperature gradient was also numerically calculated with a CFD simulation, and the results were compared with the measured values. Figure 6 presents the vertical temperature gradient for the following cases:

(a) air heating combined with mechanical mixing ventilation from field measurements;
(b) air heating combined with mechanical mixing ventilation from CFD predictions;
(c) underfloor heating combined with mechanical mixing ventilation from CFD predictions;
(d) underfloor heating combined with mechanical displacement ventilation from CFD predictions.

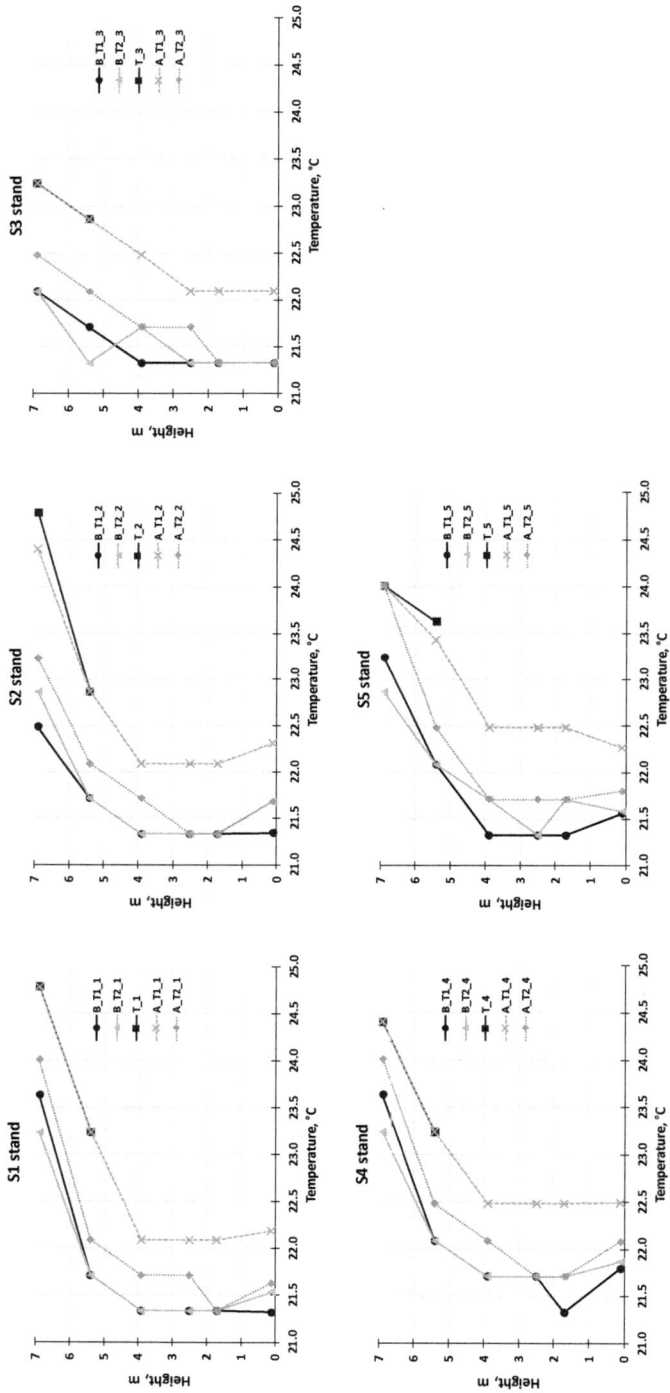

Figure 4. Indoor temperature measured at different conditions for the same location.

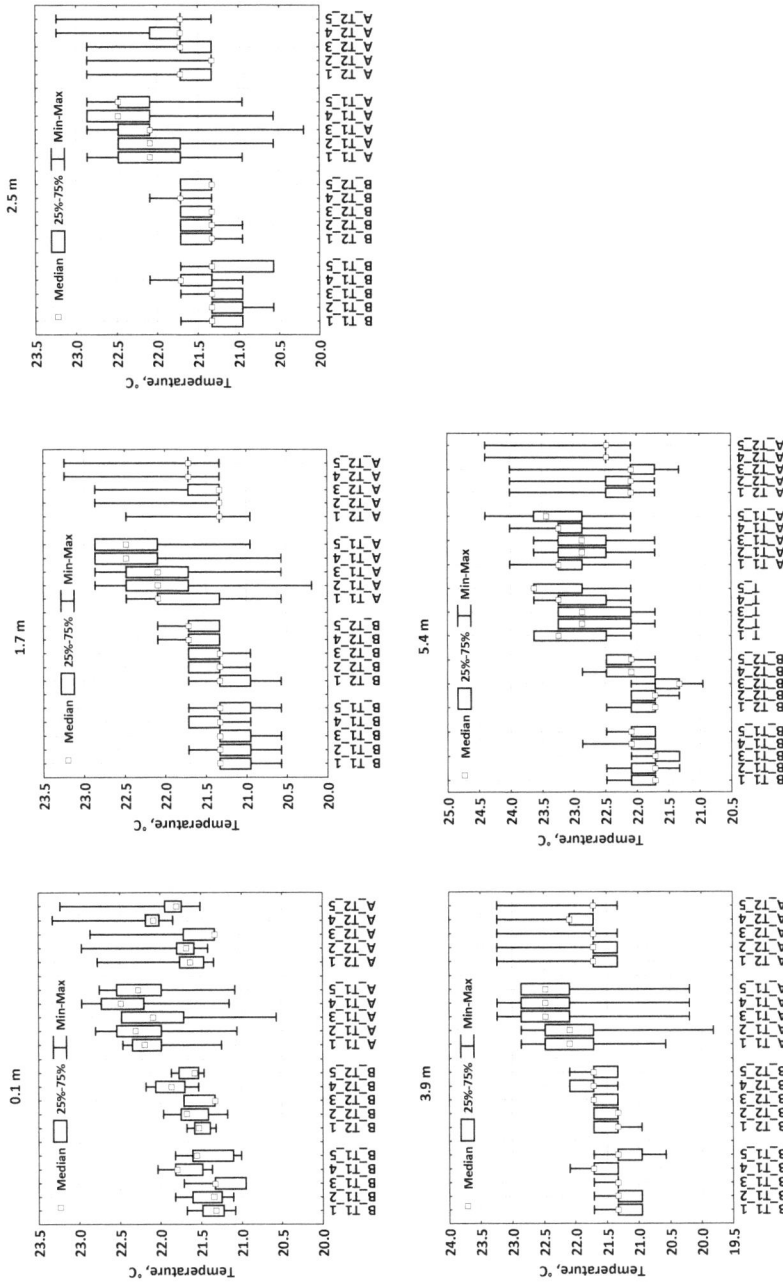

Figure 5. Indoor temperature measured at different conditions for the same height.

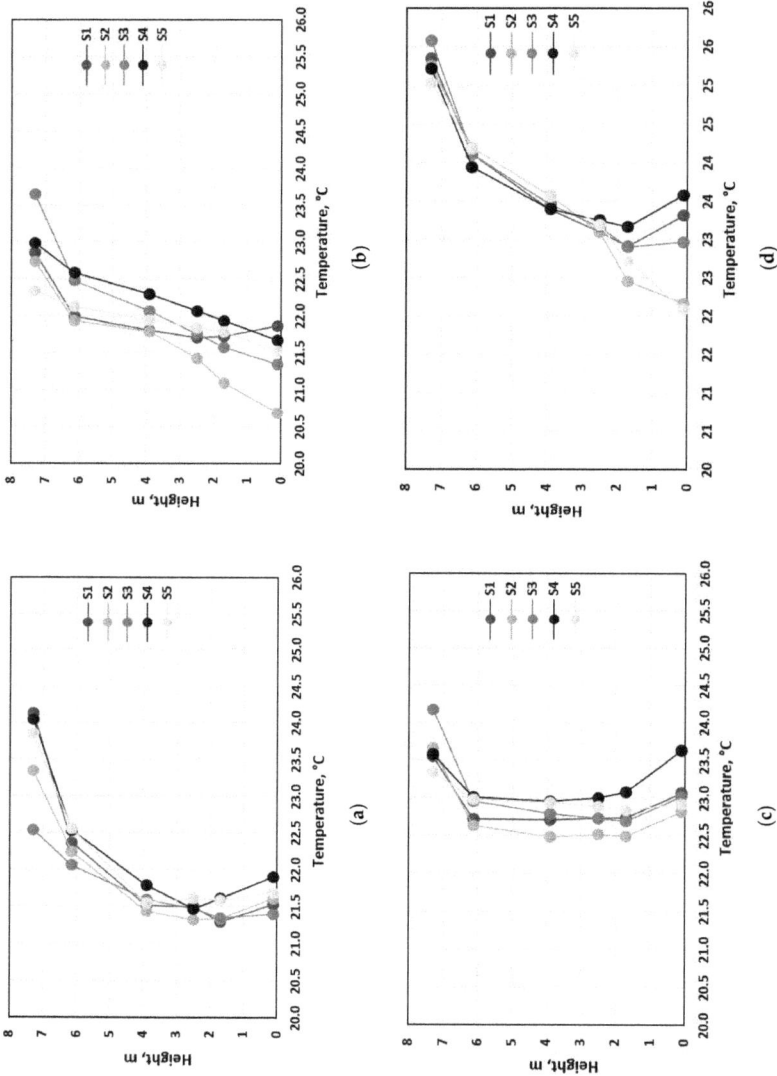

Figure 6. Numerical calculation of the vertical temperature gradient for the case with warm air heating combined with mechanical mixing ventilation from field measurements (**a**), warm air heating combined with mechanical mixing ventilation from CFD predictions (**b**), underfloor heating combined with mechanical mixing ventilation from CFD predictions (**c**), and underfloor heating combined with mechanical displacement ventilation from CFD predictions (**d**).

According to the field measurements, the vertical gradient of the case with air heating combined with mechanical mixing ventilation was measured to be 0.26 °C/m. The vertical temperature gradient, according to the CFD simulation, was calculated to be 0.16 °C/m, being in a good agreement with the measured value of 0.27 °C. In Figure 6, the vertical temperature gradients calculated for other HVAC configurations are presented. The vertical temperature gradient for the case with underfloor heating and mechanical mixing ventilation was calculated at 0.11 °C/m, and at 0.12 °C/m for the case with underfloor heating combined with mechanical displacement ventilation. As can be seen from the graphs presented in Figure 6, the temperature at the height of 1.7 m was calculated to be 22 °C with air heating combined with mechanical mixing ventilation. In the case of underfloor heating and mechanical mixing or displacement ventilation, the temperature at the same height was 23 °C. Lower temperatures in occupied zones is common in rooms heated by warm air heating systems. This difference shows that a lower heating load could be needed in cases with underfloor floor heating.

Figure 7 presents the vertical temperature distribution in the training hall for all analysed cases. CFD predictions show that more intense temperature stratification occurs in cases when air heating or displacement ventilation is used. Underfloor heating and the mixing ventilation pattern ensure an even distribution of the temperature in the hall. In the case of displacement ventilation, higher temperatures occur in the higher parts of the hall. The analysis of the results of the temperature measurement at the basketball training hall showed a slight change of the indoor temperature in the layer from the floor to a height of 3.9 m. An increase of the indoor temperature was observed in the zone from 3.9 m to the ceiling. In this layer, measurements were provided at heights equal to 5.4 m and 6.9 m. These indoor thermal conditions are in line with the Lithuanian norm requirement to keep indoor temperature between 15 °C and 25 °C.

Since CO_2 concentration near the ceiling is permissible, it is possible to recycle the air from this zone and reuse it, thus saving energy for air heating. Another important criterion of the optimal heating and ventilation performance is the exploitation energy cost. Indoor temperature measurements at a height of 6.9 m show that temperature stratification indicates energy saving potential for high premises that could be combined with the option of periodic heating application, as the sport halls have quite uneven workload.

The installation of additional engineering equipment, such as destratification fans, can be analyzed for sports halls in order to achieve the most energy-effective solutions for such a type of premises. More intensive temperature stratification in the analyzed cases between 5.4 and 6.9 m height might be caused by the artificial lighting, since the analyzed hall has no natural lightning and all the experiments were performed with the lights switched on.

Despite high expectations for indoor air quality for professional athletes, soft floor surfaces are important, to avoid physical trauma. Soft floor surfaces can be achieved by keeping the surface warm, which avoids moisture condensation and ensures faster sweat evaporation. Fast sweat evaporation is especially relevant during the intensive workouts of basketball games, cases when the area employment is relatively dense. Additional analysis of relative humidity distribution and air movement speed at nearly ground level (0.1 m) would allow us to predict and evaluate the best heating and ventilation systems case for the sports hall.

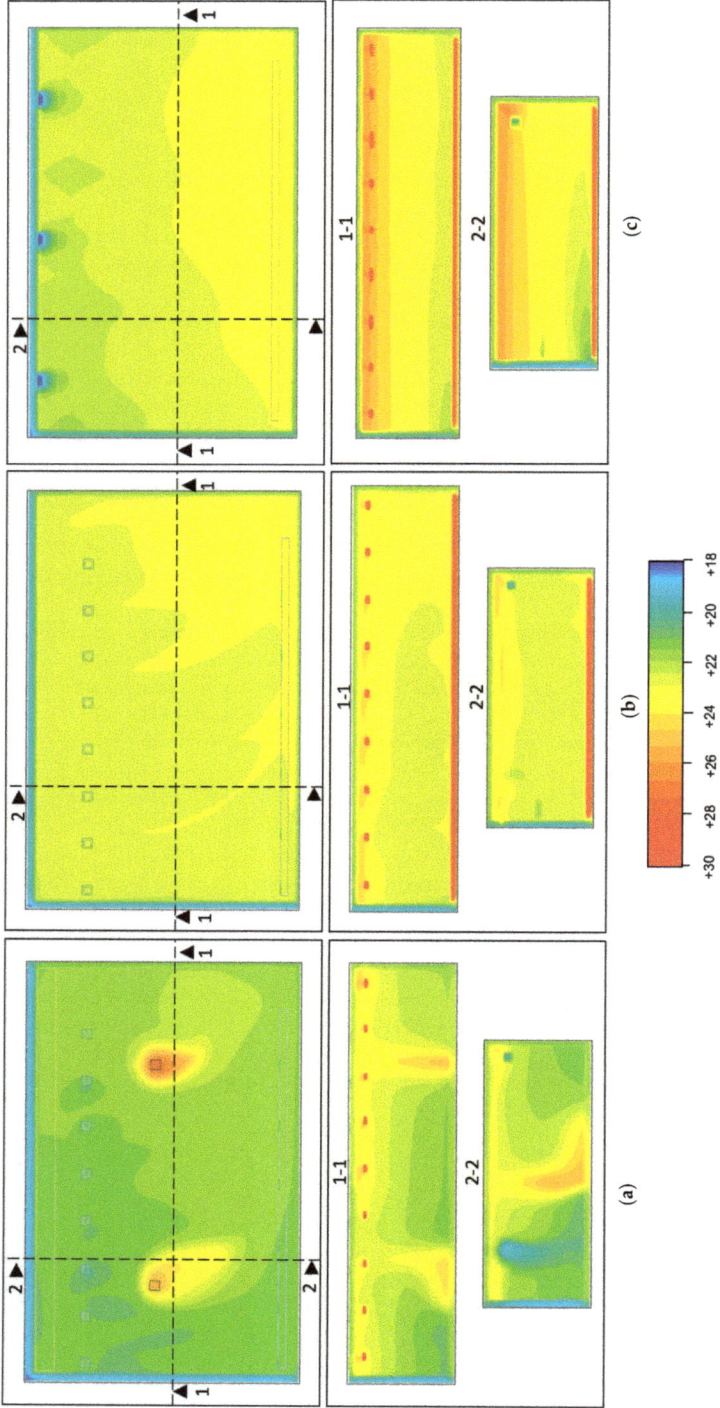

Figure 7. Temperature distribution for the cases with warm air heating combined with mechanical mixing ventilation (**a**), underfloor heating combined with mechanical mixing ventilation (**b**), and underfloor heating combined with mechanical displacement ventilation (**c**).

4. Conclusions

In this study, the indoor thermal conditions in terms of thermal stratification in the basketball training hall of Žalgirio Arena, Kaunas in Lithuania, was assessed with the use of field measurements and numerical simulation. The main conclusions of this study can be summarized as follows:

- Lithuanian normal requirements to keep indoor temperatures between 15 °C and 25 °C are met in the analyzed sports hall space for heights up to 5.4 m, using an air heating and mechanical mixing ventilation combination.
- CO_2 concentration was measured at a maximum value of 951 ppm in the occupied zone, indicating that the combination of the air heating and the mechanical mixing ventilation ensures high indoor air quality, with an air change rate of 1.1 h^{-1}.
- The experimental results show that the mixing ventilation application, together with air heating, allows higher temperatures in the occupied zone than in the case of air heating without ventilation.
- CFD simulations revealed that using the same heating output for air heating, underfloor heating combined with mechanical mixing, or displacement ventilation ensures higher temperatures in the occupied zone, creating the potential for energy savings.
- The numerical analysis also showed that the highest temperature in the 0.1 m level to avoid a slippery floor surface could be reached with an underfloor heating and mechanical mixing ventilation combination case.

The findings of this study support the necessity for the integration of numerical assessment methods into the design stage of large spaces and public buildings. This practice would assist in the optimal choice of HVAC equipment for large halls, as well as other design aspects, such as the optimal position or operational conditions of the equipment. In view of the pan-European target for nearly-zero-energy buildings, the findings of this study emphasize the gaps and limitations of the energy performance of sports halls. Future work on this subject should include the investigation of additional HVAC systems, as well as the measurement and simulation of more properties of sports halls, including air pollutants, radiant temperature, and air velocities. The hypothesis that the comfort indicators of the Fanger system may not apply in sports halls should also be further investigated in a future work.

Author Contributions: Conceptualization, L.S. (Lina Seduikyte); L.S. (Laura Stasiulienė); methodology, L.S. (Lina Seduikyte); L.S. (Laura Stasiulienė); validation, T.P., D.M., M.D.; formal analysis, L.S. (Lina Seduikyte); L.S. (Laura Stasiulienė); investigation, J.Č., T.Ž., M.D.; resources, D.M., T.P., M.D.; data curation, T.P., D.M., J.Č., T.Ž., M.D.; writing—original draft preparation, L.S. (Lina Seduikyte); L.S. (Laura Stasiulienė); writing—review and editing, P.F.; visualization, L.S. (Laura Stasiulienė); supervision, L.S. (Lina Seduikyte), P.F.; project administration, L.S. (Lina Seduikyte), P.F.

Funding: This research received no external funding.

Conflicts of Interest: The authors declare no conflict of interest.

References

1. Chow, W.K.; Fung, W.Y.; Wong, L.T. Preliminary studies on a new method for assessing ventilation in large spaces. *Build. Environ.* **2002**, *37*, 145–152. [CrossRef]
2. Trianti-Stourna, E.; Spyropoulou, K.; Theofylaktos, C.; Droutsa, K.; Balaras, C.A.; Santamouris, M.; Papanikolaou, N. Energy conservation strategies for sports centers: Part A. Sports halls. *Energy Build.* **1998**, *27*, 109–122. [CrossRef]
3. Andrade, A.; Dominski, F.H.; Coimbra, D.R. Scientific production on indoor air quality of environments used for physical exercise and sports practice: Bibliometric analysis. *J. Environ. Manag.* **2017**, *196*, 188–200. [CrossRef] [PubMed]
4. Johnson, C.J.; Moran, J.C.; Paine, S.C.; Anderson, H.W.; Breysse, P.A. Abatement of toxic levels of carbon monoxide in Seattle ice-skating rinks. *Am. J. Public Health* **1975**, *65*, 1087–1090. [CrossRef] [PubMed]

5. Braniš, M.; Šafránek, J.; Hytychová, A. Exposure of children to airborne particulate matter of different size fractions during indoor physical education at school. *Build. Environ.* **2009**, *44*, 1246–1252. [CrossRef]
6. Stathopoulou, O.I.; Assimakopoulos, V.D.; Flocas, H.A.; Helmis, C.G. An experimental study of air quality inside large athletic halls. *Build. Environ.* **2008**, *43*, 834–848. [CrossRef]
7. Rajagopalan, P.; Luther, M.B. Thermal and ventilation performance of a naturally ventilated sports hall within an aquatic centre. *Energy Build.* **2013**, *58*, 111–122. [CrossRef]
8. Kamar, H.M.; Kamsah, N.B.; Ghaleb, F.A.; Alhamid, M.I. Enhancement of thermal comfort in a large space building. *Alex. Eng. J.* **2019**, *58*, 49–65. [CrossRef]
9. Mateus, N.M.; Da Graça, G.C. Simulated and measured performance of displacement ventilation systems in large rooms. *Build. Environ.* **2017**, *114*, 470–482. [CrossRef]
10. Gil-Lopez, T.; Galvez-Huerta, M.A.; O'Donohoe, P.G.; Castejon-Navas, J.; Dieguez-Elizondo, P.M. Analysis of the influence of the return position in the vertical temperature gradient in displacement ventilation systems for large halls. *Energy Build.* **2017**, *140*, 371–379. [CrossRef]
11. Rajagopalan, P.; Elkadi, H. Thermal and Ventilation Performance of a Multi-functional Sport Hall within an Aquatic Centre. In Proceedings of the Building Simulation 2011: 12th Conference of International Building Performance Simulation Association, Sydney, Australia, 14–16 November 2011; pp. 2815–2821.
12. El-Kadi, E.W.M.A.; Fanny, M.A. Architectural designs and thermal performances of school sports-halls. *Appl. Energy* **2003**, *76*, 289–303. [CrossRef]
13. Zhao, K.; Liu, X.H.; Jiang, Y. Application of radiant floor cooling in large space buildings—A review. *Renew. Sustain. Energy Rev.* **2016**, *55*, 1083–1096. [CrossRef]
14. Stamou, A.I.; Katsiris, I.; Schaelin, A. Evaluation of thermal comfort in Galatsi Arena of the Olympics Athens 2004 using a CFD model. *Appl. Therm. Eng.* **2008**, *28*, 1206–1215. [CrossRef]
15. Tsoka, S. Optimizing indoor climate conditions in a sports building located in Continental Europe. *Energy Procedia* **2015**, *78*, 2802–2807. [CrossRef]
16. Fanger, P.O. *Thermal Comfort*; Danish Technical Press: Copenhagen, Denmark, 1970; p. 242.
17. ISO. *Moderate Thermal Environments—Determination of the PMV and PPD Indices and Specifications of the Conditions for Thermal Comfort*; International Organization for Standardization: Geneva, Switzerland, 1994; Standard 7730:1994.
18. Fokaides, P.A.; Jurelionis, A.; Gagyte, L.; Kalogirou, S.A. Mock target IR thermography for indoor air temperature measurement. *Appl. Energy* **2016**, *164*, 676–685. [CrossRef]
19. Nord, N.; Mathisen, H.M.; Cao, G. Energy cost models for air supported sports hall in cold climates considering energy efficiency. *Renew. Energy* **2015**, *84*, 56–64. [CrossRef]
20. Revel, G.M.; Arnesano, M. Perception of the thermal environment in sports facilities through subjective approach. *Build Environ.* **2014**, *77*, 12–19. [CrossRef]
21. Arnesano, M.; Revel, G.M.; Seri, F. A tool for the optimal sensor placement to optimize temperature monitoring in large sports spaces. *Autom. Constr.* **2016**, *68*, 223–234. [CrossRef]
22. Kisilewicz, T.; Dudzińska, A. Summer overheating of a passive sports hall building. *Arch. Civ. Mech. Eng.* **2015**, *15*, 1193–1201. [CrossRef]
23. Suo, H.; Angelotti, A.; Zanelli, A. Thermal-physical behavior and energy performance of air-supported membranes for sports halls: A comparison among traditional and advanced building envelopes. *Energy Build.* **2015**, *109*, 35–46. [CrossRef]
24. Nielsen, P.V. Fifty years of CFD for room air distribution. *Build. Environ.* **2015**, *91*, 78–90. [CrossRef]
25. EN 13779:2007. *Ventilation for Non-residential Buildings—Performance Requirements for Ventilation and Room-Conditioning Systems*; European Committee for Standardization: Brussels, Belgium, 2007.

energies

MDPI

Article

Numerical Simulations and Empirical Data for the Evaluation of Daylight Factors in Existing Buildings in Sweden

Sara Eriksson [1]**, Lovisa Waldenström** [1]**, Max Tillberg** [1,*]**, Magnus Österbring** [2] **and Angela Sasic Kalagasidis** [3]

[1] Bengt Dahlgren AB, Krokslätts Fabriker 52, 431 37 Gothenburg, Sweden; sara.eriksson@bengtdahlgren.se (S.E.); lovisa.waldenstrom@bengtdahlgren.se (L.W.)

[2] NCC Sweden, Gullbergs Strandgata 2, 411 04 Gothenburg, Sweden; magnus.osterbring@ncc.se

[3] Chalmers University of Technology, Department of Architecture and Civil Engineering, Sven Hultins gata 6, SE-41296 Gothenburg, Sweden; angela.sasic@chalmers.se

* Correspondence: max.tillberg@bengtdahlgren.se

Received: 26 April 2019; Accepted: 5 June 2019; Published: 10 June 2019

Abstract: Point Daylight Factor (DF_P) has been used for daylighting design in Sweden for more than 40 years. Progressive densification of urban environments, in combination with stricter regulations on energy performance and indoor environmental quality of buildings, creates complex daylight design challenges that cannot be adequately solved with DF_P. To support a development of the current and future daylight indicators in the Swedish context, the authors have developed a comprehensive methodology for the evaluation of daylight levels in existing buildings. The methodology comprises sample buildings of various use and their digital replicas in 3D, detailed numerical simulations and correlations of diverse DF metrics in existing buildings, a field investigation on residents' satisfaction with available daylight levels in their homes, and a comparison between the numerical and experimental data. The study was deliberately limited to the evaluation of DF metrics for their intuitive understanding and easy evaluation in real design projects. The sample buildings represent typical architectural styles and building technologies between 1887 and 2013 in Gothenburg and include eight residential buildings, two office buildings, two schools, two student apartment buildings, and two hospitals. Although the simulated DF_P is 1.4% on average, i.e., above the required 1%, large variations have been found between the studied 1200 rooms. The empirical data generally support the findings from the numerical simulations, but also bring unique insights in the residences' preferences for rooms with good daylight. The most remarkable result is related to kitchens, typically the spaces with the lowest DF values, based on simulations, while the residents wish them to be the spaces with the most daylight. Finally, the work introduces a new DF metric, denoted DF_W, which allows daylighting design in early stages when only limited data on the building shape and windows' arrangement are available.

Keywords: daylight; existing buildings; daylight factor; daylight simulations; daylight survey

1. Introduction

Point Daylight Factor (DF_P) is an official criterion for daylighting design of buildings in Sweden. It was introduced after the 1970's energy crisis to secure a minimum daylight quantity in indoor environments and, thereby, to counterbalance stricter regulations on limited energy use for space heating [1]. The minimum requirement for DF_P has been set to 1% at a specified point in an indoor space where people stay for a longer time and where the nature of their activities requires access to daylight. Despite its known limitations, such as the inability to describe temporal variations and perceived quality of daylight in a space, where the latter according to [2–4] is composed of both physical metrics and psychological criteria, its simple definition and resource-efficient evaluation during the building design have determined its decades long presence in the Swedish building regulations [5].

DF_P may be pragmatic, but it is also a rather rigid design criterion, which creates various indecisive situations during a building's design. As pointed by national experts in daylighting design [6], the 1% target is challenging to achieve in buildings located in densified urban areas, in situations where architectural appearance and functional aspects of a building are differently valued in the building's design, or in indoor spaces of irregular shapes. Sweden experiences a positive internal migration of population to larger cities [7], which, in combination with high prices of municipal land and a need for better techno-economic utilization of the existing infrastructure, leads to an increased densification of urban areas [8]. Placement of new buildings among the existing ones, rather than in the outskirts of a city, decreases the amount of available daylight in these areas over time. Similar outcomes are noticeable after buildings' refurbishment projects, when the addition of new floors on tops of the existing buildings is needed to balance the costs of the refurbishment. Furthermore, trends in architecture that favor vertical window shapes, deep volumes, and less reflective interior finishes are more likely to create larger contrasts and uneven daylight distribution indoors [9], which are difficult to describe by DF_P. Moreover, DF_P is subjective to different interpretations in rooms with irregular, e.g., polygon-type layouts [10]. Finally, the current DF_P requirement, or any other daylight metrics, is literally not achievable in northern parts of the country during the several-months-long periods of total darkness at high geographical latitudes. These everyday design challenges have encouraged national experts in daylighting design to review the current and future development needs for daylight indicators in the Swedish context [6,11].

Since the late 1990s, climate-based modelling software has provided fairly accurate predictions of daylight distribution in interiors by taking into account realistic sun and sky conditions [12]. This has allowed various spatial and temporal averaging of daylight availability in a space and definitions of new daylight criteria such as daylight coefficient (DC), useful daylight illuminance, etc. [13–17]. Prior to choosing one or another refined daylight indicator, at least the following two practical aspects need to be carefully considered: Resources needed for the implementation of new criteria in real projects and satisfaction of end-users (residents) with the available daylight levels in existing buildings. The former is of a large importance in the early design stages, when the building geometry and fenestration are reconfigured many times before a satisfactory architectural solution is found. If energy and daylighting design of the building cannot follow the other design activities due to, e.g., time-consuming simulations, the targeted energy and daylight performances will be coarsely estimated, and any mistakes made thereby will be difficult to correct during the detailed building design. As for the daylight levels in existing buildings, a common sense would say the more the better. At the same time, the most densified urban areas in Sweden, like in many other parts of the world, are the most attractive for habitation and work despite a generally low availability of daylight. This indicates that residents adapt to a range of daylight metrics instead of just a single value, as it is commonly defined in building regulations and certification systems. This paper aims to clarify these practical aspects based on the results of a comprehensive study conducted by [10].

1.1. Daylighting Indicators and Calculation Techniques in Use

Along with introducing DF_P in the Swedish building regulations, daylight protractors were recommended for the quantification of DF_P. The manual reading of daylight components and correction factors for various design situations was shown to be time-consuming, i.e., expensive in respect to the projects' costs and, thus, not used by practitioners in a desired extent [6]. To simplify the daylighting design, the so called 'AF method' was introduced in 1988, with 'AF' referring to the window-to-floor ratio [18].

$$A_{glazing} \geq f \cdot A_{floor}, \qquad (1)$$

where $A_{glazing}$ and A_{floor} are the window and the floor area in [m^2], respectively, and f is a constant to be chosen with respect to obstructions in front of the window. To reach a DF$_P$ of about 1% in a room, the values for f should be chosen between 0.075 to 1.5, for low to high obstruction angles.

Since the AF method was introduced, only small adjustments in the daylight design were made. Both DF$_P$ and AF are still in use, with the difference being that DF$_P$ is considered a more precise criterion whose value should be quantified by means of validated software such, as Daylight visualizer [19] and Radiance [20], while AF is adequate for simpler design situations [18]. For example, the Swedish environmental certification method Miljöbyggnad (Green building) provides target values for both DF$_P$ and AF for the basic grade, 'bronze', which indicates a compliance with ruling building regulations, as well as for the medium grade, 'silver', while only DF$_P$ is accepted for the highest grade, 'gold' [21]. The Nordic Swan Ecolabel aims at stringent environmental and functional qualities and thus approves only DF$_P$ [22].

Sweden has systematically worked on the regulatory policies on energy use in buildings over the past decades, which has resulted in low energy use in buildings despite the cold climate [23]. Similar positive outcomes on indoor environmental quality can be expected from the stricter national regulations and standards related to daylighting design. However, new demands will be challenged by budget constraints in a design process. With the prevailing performance-based design, i.e., the lack of formative standards on how the buildings' energy design should be conducted, one can expect that stricter regulations on daylighting design will, sooner or later, trigger the invention of simplified design indicators and calculation methods.

1.2. Daylight Availability in Existing Buildings

Despite affordable daylight simulation tools and a satisfactory accuracy between the simulated and measured daylight levels in buildings [24], comprehensive studies on actual daylight levels in existing buildings are rare. Most of the available studies are performed on representative models of rooms or buildings, e.g., [25–29]. In that sense, the study done by [30], who considered DF$_P$ in several apartments of an existing residential building in Hong Kong, was rather unique. To the best of the authors' knowledge, comparable results for Sweden were produced for the first time by [10], who calculated different daylight metrics in 16 sample buildings of various purposes in the city of Gothenburg. By using different techniques, including the AF method and detailed numerical simulations in Radiance, both point and area averaged DF (mean and median) were quantified in about 1200 rooms in the sample buildings. Furthermore, a qualitative assessment of the daylight availability was obtained from interviews with tenants in selected sample buildings. Moreover, a new DF metric, DF$_W$, has been proposed for daylighting design in the early stages. The numerical approach of [10] was adopted by [31], who quantified the same daylight metrics, i.e., the point and area averaged DF, in the 54 sample buildings of the city of Stockholm, expanding the pool of numerical results for Sweden. For supporting future studies on daylight availability in existing buildings by using DF metrics, this paper introduces the methodology of [10] in a full extent, including both simplified and detailed numerical calculations, questionnaires and results from the field survey, and suggestions on how DF$_W$ can be used in early stages. To the best of the authors' knowledge, the only example of DF metrics like DF$_W$ can be found in the work by [32] which is one of the follow-up studies of [10].

1.3. The Layout of the Paper

This paper presents two types of results, as follows: The calculated, i.e., theoretically possible indoor illuminance levels in Swedish buildings, and a qualitative assessment of the resident's satisfaction with the available indoor illuminance in selected sample buildings. Section 2 presents the overall research methodology and starts with a rationale behind the sample buildings, including a brief description of their morphology and technical characteristics. Thereafter, the procedure of constructing CAD replicas of the sample buildings is introduced together with the calculation settings. A definition of the used DF metrics, reference calculation points with examples of situations where intentional adjustments of the reference points were made is presented in the continuation. Finally, an in-house questionnaire that was used in the field studies is presented at the very end of Section 2. Results of numerical simulations of DF_p in the sample buildings are enclosed in Section 3. Correlations between various point and area averaged DF metrics are presented and discussed in Section 4. At the very end of Section 4, a correlation between the calculated DF_p levels and those from the field study is shown. Conclusions from the entire work are shown in Section 5.

2. Research Methodology

To get good insight into the available amounts of daylight in buildings, it was decided the study would be performed on existing buildings and by using DF metrics that were known to daylight designers in Sweden. Since all authors of this work are from Gothenburg, the sample buildings were selected in the same city for purely practical reasons, namely a good insight in daylight challenges in the city and an easy access to the buildings' documentation from the city administration and private archives.

A substantial part of the work was spent on creating computer aided drawings (CAD) of the sample buildings and their surroundings, in three-dimensions (3D). Generally, floor plans were created in AutoCAD (2014), while 3D drawings were done in Rhinoceros 5. The latter was chosen strategically for its compatibility with the daylight simulation software Daysim, which was used for outdoor daylight simulations, and Radiance, for indoor daylight distributions. Daysim and Radiance are accessed through Diva and used in conjunction with the Grasshopper plugin [33]. The buildings surrounding the sample buildings were reconstructed as simpler 3D objects by using a digital 3D map of the city [34].

Since the aim of this study was to get an insight in theoretically possible indoor illuminance in the sample buildings, rather than to quantify the actual indoor illuminance, no specific validation of the calculated results was performed. The daylight simulations used in this work are well-known and widely used for daylight design investigations, and for which a variety of validation results exist. Based on the results from the reported validation studies, such as [13,25,33,35–37] it was concluded that the accuracy of Diva is sufficient for the complexity of the daylighting design investigations enclosed in the study.

Definitions and a practical assessment of the used DF metrics for various daylight situations in the studied buildings is presented by means of illustrative examples. The survey on residents' satisfaction with the daylight availability in their homes was limited to three residential buildings due to a short time frame (about three months). The average response rate was 67%, which was found satisfactory for making comparisons with the results of numerical studies.

2.1. Sample Buildings and Their CAD Replicas

Given that this numerical study was the first of its kind in Sweden, the sample buildings were selected in a close cooperation with local daylight experts to assure a good representation of daylighting design challenges. The available buildings' documentation and the limited time for the study, six months in total, also influenced the choice of buildings. In total, sixteen existing multistore buildings were included in the study, wherein eight were residential, while the remaining types included student apartments, offices, schools, and hospitals, each represented by two samples.

When selecting the buildings, their primary use and age were considered together, as a compound criterion reflecting the buildings' architectural and technical standards, as well as their location in the city. It should be noted that Gothenburg is a medium size city (the second largest in the country) with about 600,000 inhabitants and a rather fragmented urban fabric. Although all studied buildings are in the city area (see Figure 1), those grouped at the south bank of the river (ID 1, 3, 2, 10, 12, and 16) are in the oldest, somewhat more densified area than the other buildings. Further details about the sample buildings are summarized in Table 1 and complementary street photos can be found in Section 4.

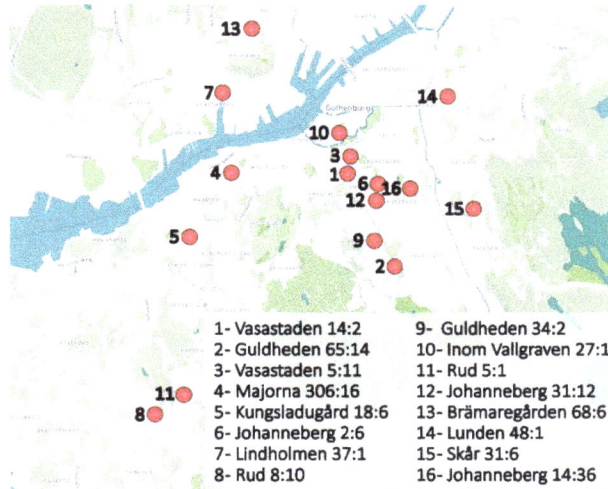

1- Vasastaden 14:2   9- Guldheden 34:2
2- Guldheden 65:14   10- Inom Vallgraven 27:1
3- Vasastaden 5:11   11- Rud 5:1
4- Majorna 306:16   12- Johanneberg 31:12
5- Kungsladugård 18:6   13- Brämaregården 68:6
6- Johanneberg 2:6   14- Lunden 48:1
7- Lindholmen 37:1   15- Skår 31:6
8- Rud 8:10   16- Johanneberg 14:36

Figure 1. Location of the studied buildings in Gothenburg. The cadastral numbers are shown next to the ID.

Table 1. Main characteristics of the studied buildings. Year = year of construction. Use: R-residential, S-school, O-office, H-hospital, SA-student apartment (micro apartment). Only the floors with the same layout were used, in accordance with the main use of the building.

ID	Characteristics	Year	Use	Total Number of Floors	Number of Evaluated Floors	Number of Evaluated Rooms
1	Elongated apartment block	1972	R	6	4	36
2	Tower block	1960	R	12	8	200
3	Townhouse with a courtyard	1887	R	4	4	42
4	Governor's house with a courtyard	1897	R	4	3	140
5	Governor's house with a courtyard	1923	R	3	3	37
6	L-shaped apartment block	1928	R	6	5	70
7	Compact tower block	2013	R	11	4	68
8	Compact tower block	1960	R	10	4	100
9	Compact block	2004	SA	5	3	97
10	Townhouse with two street sides	1863	O	5	4	47
11	Low-rise, elongated	1962	S	2	2	14
12	L-shaped compact block	2006	SA	13	5	230
13	Low-rise compact block	1966	H	4	4	38
14	Low-rise, indented top floor	2006	O	3	3	36
15	U-shaped, low-rise with a school yard	2001	S	2	2	30
16	Block with a tower on one side	1927	H	4	3	52
	Total					1237

Typical features of the buildings built around the turn of the 20th century (IDs 3, 4, 5, 6, 10 and, 16 in Table 1.) are brick masonry, deep floor plans, low window-to-floor ratios, and sides facing courtyards. The buildings built between the years 1960 and 1972 (IDs 1, 2, 8, 11, and 13) are typically high-rise or elongated tower blocks of pre-fabricated concrete elements, with shallow floor plans and rather large windows, often arraigned in serials. Finally, the new buildings built after the year 2000 (IDs 7, 12, 14, and 15) are of higher energy standards compared to the other sample buildings, which is reflected in deeper window niches due to thick thermal insulation in walls. Other features of the modern buildings include energy windows, with U-values around 1 W/m^{2}K, and deep and open floor plans.

To optimize the drawing process, only floors with the same layout were considered, following the main purpose of a building. Floors excluded from the study were those with storage or rentable spaces, i.e., of another purpose than the main use of a building in question. In total, more than 1200 spaces were evaluated.

As mentioned earlier, the quality of available blueprints was also an influential criterion, when selecting the sample buildings, in order to enable the creation of precise CAD drawings of the buildings' interiors and exteriors for the daylight simulations. The blueprints were mostly obtained as scanned floor plans and sections from the City Planning Authority of Gothenburg. If several blueprints from different periods were available for a building, the newest ones were used for its CAD replica. Older blueprints were consulted in cases where some information was missing in the newer blueprints. The process of creating CAD drawings of the studied buildings is depicted in Figure 2. Two-dimensional floorplans were drawn in AutoCAD 2014, using the available blueprints as templates. These were then imported to Rhinoceros 5 to create three dimensional drawings by adding sections and facades.

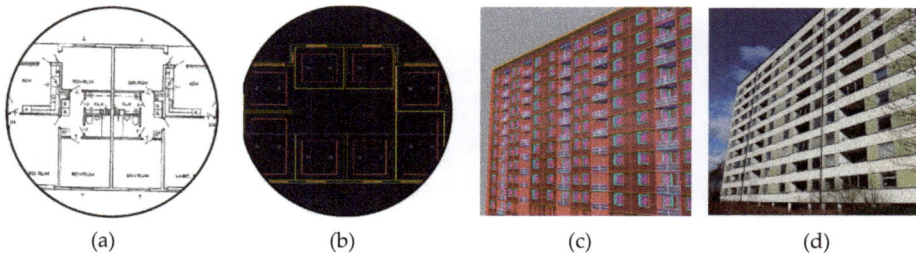

Figure 2. Creating CAD replicas of a building: (**a**) the original (scanned) blueprint of a floor plan; (**b**) the same floor plan redrawn in AutoCAD; (**c**) a 3D model of the building in Rhino, reconstructed from the floor plans; and (**d**) a photo of the actual building (ID 2).

2.2. Optical Properties of Surfaces

Since the aim of the study was to show the access to daylight in the existing buildings, rather than the exact levels of daylight, two major simplifications were made in the simulations. The first relates to the optical properties of interior and exterior surfaces. The light transmission of glass is assumed to be 0.70 since the window glass currently used in Swedish housing is normally between 0.65 and 0.75 [11]. Reflectance of specific interior and exterior surfaces was decided in consultations with the daylight designers [6], as presented in Table 2, and used as set of constants throughout the daylight simulations. This simplification was found necessary for both practical and methodological reasons. At the time of the study, exact data about the optical properties of the surfaces were not available. By assigning different yet assumed optical properties, the calculated DF metrics would be systematically biased from the presented results, but the relative difference between the studied rooms would remain the same. Another advantage of using the same optical properties was in reducing the number of possible causes for differences between the studied rooms. An exception was made for balcony railings, since their optical properties vary greatly with their opacity. For completely opaque balcony railings, a reflectance of 0.3 was assumed, i.e., the same as for external facades. For semi-transparent and transparent balcony railings, the reflectance and transmittance were assumed in the range 0.3 to 0.7. The second simplification refers to the interior geometrical objects, such as furniture. These were neglected in the simulations for the sake of simplicity.

Table 2. Reflectance of different objects as used in this study. Reproduced with permission from [6].

Object	Reflectance	Object	Reflectance
Outside ground	0.2	Window frame	0.8
External façades	0.3	Side of window	0.5
Surrounding buildings and objects	0.2	Balcony	0.3
Floor	0.3	Balcony bottom	0.7
Walls	0.7	Water	0.5
Ceiling	0.8	Roof	0.3

2.3. Calculation of Indoor and Outdoor Illuminance

When quantifying DF metrics, the illuminance from the overcast sky reaching a surface or a point on a building was calculated by the sky view factor (SVF) and sky exposure factor (SEF), respectively. For all sample buildings, SEF and SVF were calculated in Grasshopper by considering the buildings' surroundings (neighboring buildings, streets, ground, and sky), and by using the same calculation points and grids as used for the definition of area-averaged DFs. As mentioned earlier, the shape of surrounding buildings was based on a 3D digital map of the city [34] and introduced in Grasshopper as three-dimensional CAD-drawings (see 'Surroundings' in Figure 3). Based on the analysis in Section 2.6, SVF and SEF produced similar results.

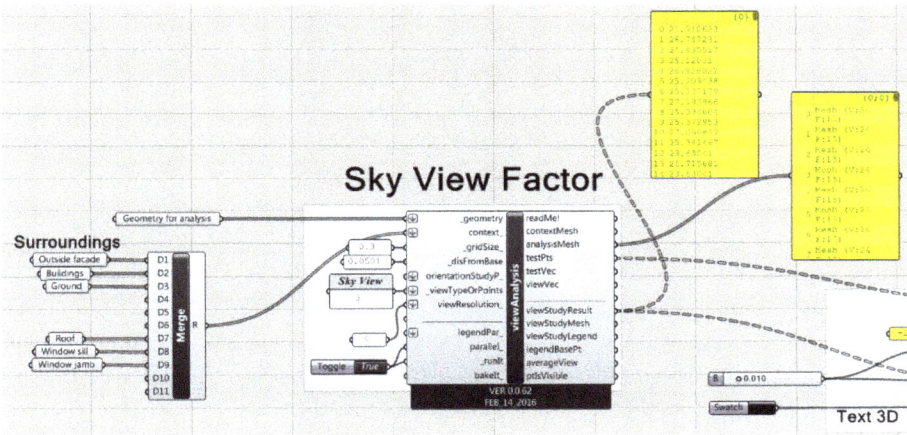

Figure 3. Calculation of the sky view factor in Grasshopper.

For the grid-based simulations of indoor illuminance in Radiance, the following five settings were used: The number of ambient bounces (-ab 7), the ambient divisions (-ad 2048), the ambient super-samples (-as 512), the ambient resolution (-ar 256), and the accuracy (-aa 0.1). These settings were found in an iterative manner, by optimizing results' convergence in relation to the simulation time. It is worth noting that all calculation points and grids were defined in two-dimensional models in AutoCAD to simplify the creation of computational grids in the three-dimensional models in Radiance.

2.4. Control Points for DF$_P$ in Complex Rooms

Based on the instructions in standard [38], the control point for DF$_P$ calculations should be placed at half of the room depth, one meter from the darkest inner wall and 0.85 m above the floor. While this rule is easily applied on rectangular, single-purpose rooms, it requires certain interpretations for rooms with complex layouts or in multi-purpose rooms. During this work, these challenges were encountered numerous times and examples on how these were solved are provided below.

A rectangular multi-purpose room, composed of a kitchen and a dining room, is shown in Figure 4. By following the instructions factually, the control point for this room should be placed between these two partitions where it would not have any practical value. Instead, a fictitious wall was added between the kitchen and the dining room in the daylight simulations. The two new control points were defined in accordance with the standard, each better describing the areas where people would stay for an extended time.

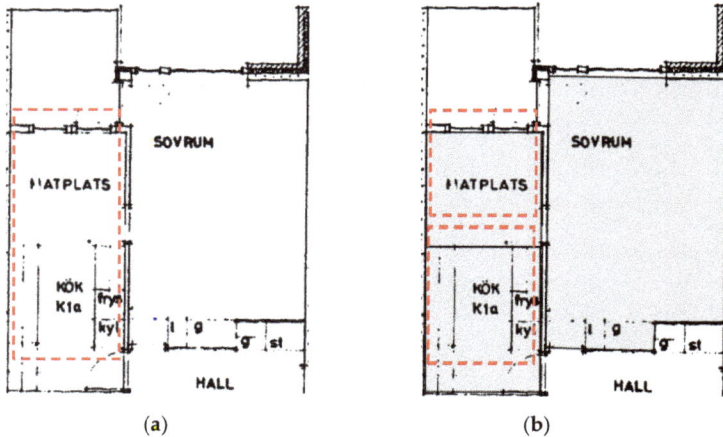

(a)   (b)

Figure 4. Separation of an open floor plan into two entities: (**a**) The original plan with a kitchen and a dining room (within the dashed line) and (**b**) a wall added between these two rooms for DF$_P$ calculations.

Single-purpose rooms of polygonal shape typically have an unclear definition of room depth, as shown in Figure 5a, or a clear direction of the room depth but variable depths, as in Figure 5b. In these and similar cases, DF$_P$ was evaluated at an average room depth, calculated from the viable options.

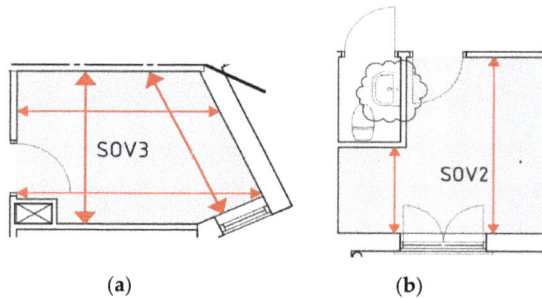

(a)   (b)

Figure 5. Rooms with varying room depths in: (**a**) Different directions and (**b**) one direction.

In multi-purpose rooms with a polygonal layout, typically found in new buildings where a kitchen and a living room are combined into one open space, as shown in Figure 6, DF$_P$ was calculated at several control points (marked in red circles), all fulfilling the requirement presented in the standard. The point with the lowest DF$_P$ value was then chosen as the final control point (red dots).

Figure 6. Examples of combined kitchens and living rooms that result in complex floor layouts. Acceptable positions for the evaluation of single-point DF are indicated with circles. Positions with the lowest calculated single-point DF are shown by dots.

2.5. Control Surfaces and Grids for Area-Averaged DF

Due to the uncertainties in finding a control point for the evaluation of DF_P in complex rooms, described above, complementary calculations of area-averaged DF-values were conducted for all the studied rooms. The two following types of control surfaces were used, as shown in Figure 7: A whole horizontal section at 0.85 m from the floor and a horizontal area at the same height, but retracted 0.5 m from each wall. The rationale behind the retracted control surface was to focus on the area wherein people would spend longer times.

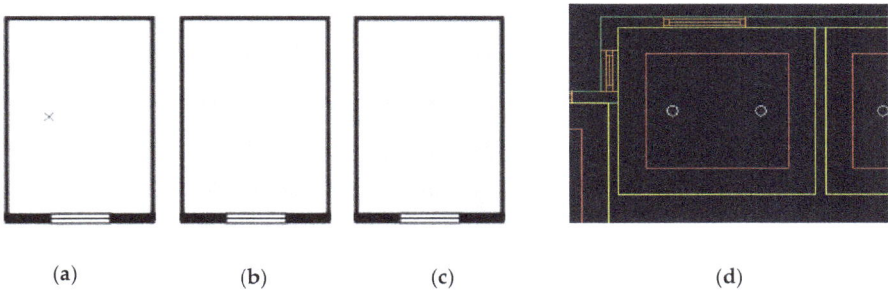

(a) (b) (c) (d)

Figure 7. Control points and control surfaces for DF calculations in a room: (**a**) The control point for DF_P; (**b**), and the retracted and (**c**) whole control surface for DF_A and DF_M. (**d**) A drawing of a floor plan in AutoCAD with the control points (white circles) and the retracted and the whole control surfaces, enclosed by the red and yellow rectangles, respectively.

When calculating the indoor illuminance, calculation points in each control surface were placed in a uniform rectangular grid, at 0.3 m distance. From the grid-point values, the mean DF_A and median DF_M were calculated. For each room, four area-averaged DF values, namely DF_A and DF_M for both the full and the retracted control surface, and one DF_P were found.

2.6. New Daylight Metrics: DF_W

Traditionally, DF metrics are calculated for buildings' interiors. However, in early stage designs, only a rough building shape and window arrangements are known, but not the floor layouts. To avoid conflicting situations between the energy and daylight performance of a building, which are typically revealed during the detailed design stage, i.e., after the architectural design is finalized, a new daylight design criterion was introduced, reading as follows:

$$DF_W = \frac{\text{Area averaged outdoor illuminance in front of a window}}{\text{Outdoor illuminance (unobstructed)}} \times 100 \ [\%]. \qquad (2)$$

This definition is based on a hypothesis that it is possible to establish a unique and predictive correlation between DF_W and DF_P for a room. To calculate the area-averaged illuminance in front of a window, a control surface was defined at 0.05 m distance from the vertical plane on the exterior side of the window. As in the case of area-averaged DF inside a room, calculation points were placed in a rectangular grid at 0.3 m distance. An example of the calculation grid is shown in Figure 8. The correlation of DF_W and DF_P is evaluated in Section 3, by using the residential buildings as case studies. From the obtained correlations, the usefulness of DF_W is evaluated.

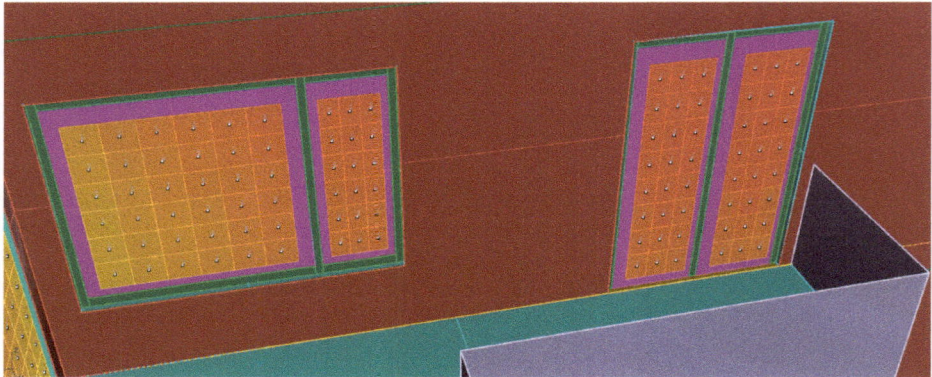

Figure 8. Calculation grids outside the windows in Rhinoceros.

When calculating DF_W, it is of large importance that the outdoor illuminance is correctly modelled. For that purpose, the illuminance from the overcast sky reaching a surface or a point on a building was calculated by the sky view factor (SVF) and the sky exposure factor (SEF), and compared to DF_W from Equation (2). The result of this comparison is presented in Figure 9 for buildings with ID 1, 2, and 3. The values are organized in ascending order, after DF_W. As it can be seen, these three methods give nearly the same values. The somewhat larger difference between DF_W and the other two metrics for the lower percentages on the y-axis can be explained by external reflections from the surroundings, which are only considered when calculating DF_W. In the higher range, all three indicators almost coincide since there are nearly no external reflections.

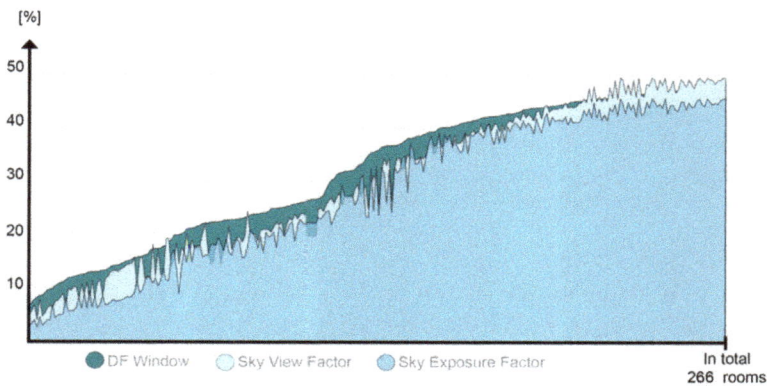

Figure 9. Daylight factor, sky view factor, and sky exposure factor measured outside the windows. The values are organized after the size of the daylight factor.

2.7. Field Surveys

To gain a better understanding of the calculation results, a survey for the residents of three residential buildings of substantially different age and design was distributed, with IDs 2 (tower block from 1960), 6 (L-shaped apartment block from 1928), and 7 (compact tower from 2013). More details about the buildings can be found in Table 1 and Figure 1.

The survey was composed of nine questions focusing on the residents' satisfaction with the amount of available daylight in specific rooms. Answers were collected as grades and then compiled and compared to the simulated results, as shown in Section 4.4. Answers were received from 45 out of 67 apartments, corresponding to an average response rate of 67% (see Table 3).

Table 3. Number of distributed and collected surveys for the selected residential buildings.

ID	Number of Distributed Surveys	Number of Collected Surveys	Response Rate
2	36	24	67%
6	15	8	53%
7	16	13	81%
All	67	45	67%

The questions were asked in Swedish and their provisional translations to English read as follows:

Q1. How do you perceive the access to daylight in the following rooms?
Q2. Would you prefer more or less daylight in these following rooms?
Q3. Which room type do you think is the most important room to have the greatest access to daylight?
Q4. Do you normally use electrical lighting daytime in the following rooms?
Q5. Do you have access to direct sunlight in the following rooms?
Q6. Do you often use curtains, blinds or other sun shadings in the following rooms?
Q7. In case of using a sun shading, why is it used?
Q8. Do you consider the view out to be interesting in the following rooms? (Quantity)
Q9. Do you consider the view out to be enough in the following rooms? (Quality)

Except for questions Q3 and Q7, all other questions were to be answered with a scaled answer and for a specific room, to allow comparisons with the simulated daylight factors in the room.

3. Results: Simulated DF$_P$ in the Sample Buildings

A selection of the most indicative results and data analysis is presented hereafter. As DF is commonly calculated at a single point in Sweden, the calculated single-point DF values are firstly presented and grouped per the building use. Comparisons between the single-point and area-averaged DF values is shown in the next section.

3.1. Residential Buildings (IDs 1–8)

The distribution of the calculated single-point DF$_P$ in all considered rooms in the residential buildings with the IDs 1–8 is shown in Figure 10. The current threshold value on 1% is shown by the vertical dashed line. Only the habitable rooms, i.e., bedrooms, kitchens, living rooms, and dining rooms in the residential buildings, were simulated. The results indicate a significant difference between the buildings, but also within the same building. The most striking are the results for building number 3, without a single room fulfilling the current requirement on DF$_P$. This is a direct consequence of large room depths and ceiling heights in comparison to the position and size of the windows. Narrow buildings, such as IDs 2, 4, and 8, have a significant percentage of rooms fulfilling the current daylight requirement. However, none of the buildings meet the requirement in all the studied rooms, not even in the newest building from 2013 (ID 7).

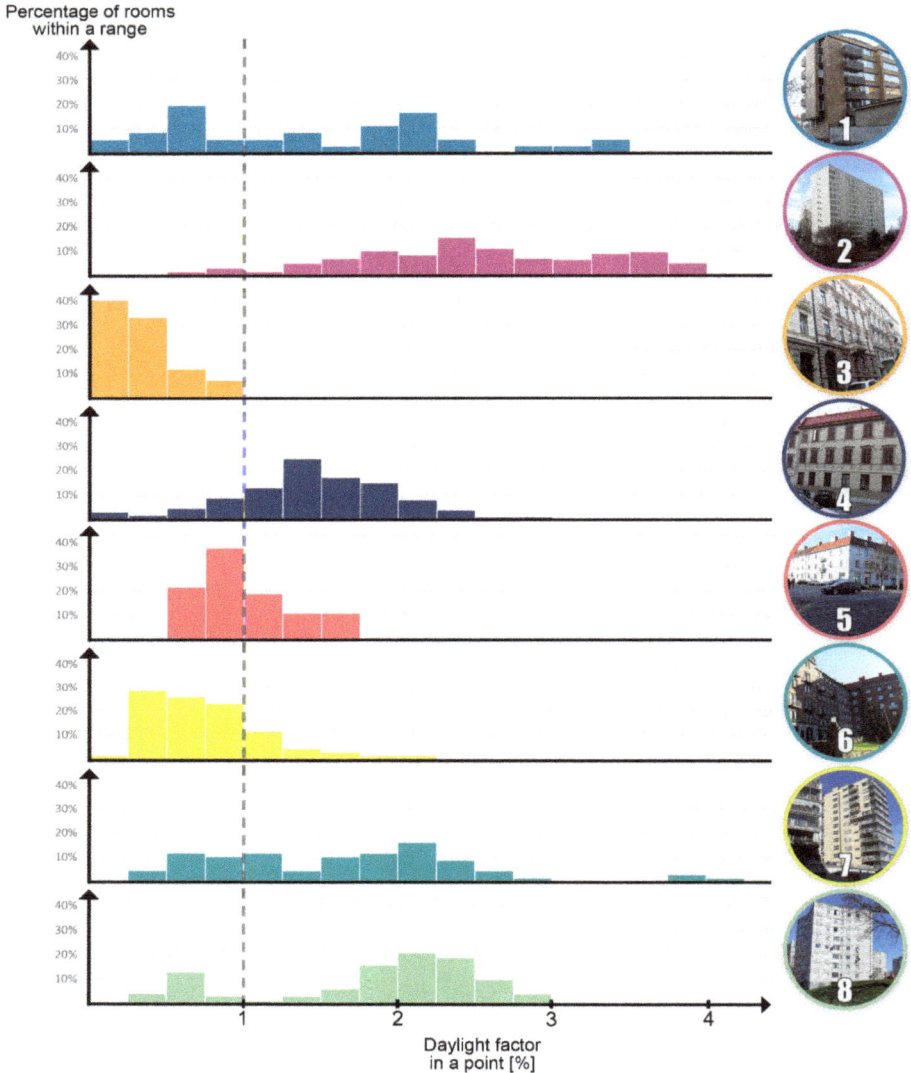

Figure 10. Distribution of DF$_P$ in all the simulated rooms in the residential buildings.

The percentage of the studied rooms with a DF$_P$ greater than 1% and the average DF$_P$ for each studied building is shown in Table 4. The total averages are also included, showing that the requirement is fulfilled in 71% of the studied rooms at the average DF$_P$ of 1.67%. However, given that the number of rooms varies largely between the buildings (see Table 1), these simple averages can be biased in favor of the buildings with larger number of rooms. This is confirmed by the weighted averages in Figure 11, which show that 56% of rooms fulfil the requirement of 1% and that weighted average DF$_P$ is about 1.4%.

Table 4. The percentage of rooms with the single-point DF greater than 1% in the residential buildings.

Building ID	Percentage of All Rooms with a DF > 1%	Average DF for a Building
1	61%	1.47%
2	96%	2.50%
3	0%	0.31%
4	83%	1.45%
5	41%	1.01%
6	21%	0.74%
7	74%	1.65%
8	80%	1.85%
Average for all rooms/buildings	71%	1.67%

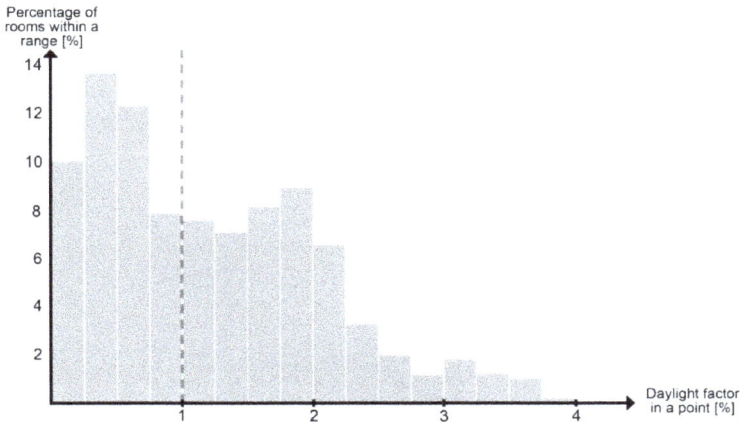

Figure 11. Distribution of the weighted single-points DF in the residential buildings.

The DF_P results have been further split per room type and are shown in Figure 12. Since the studied rooms have different names in the original drawings, they have been grouped into four main types, namely bedroom, living room, kitchen, and dining room, as presented in Table 5. Based on the calculations, kitchen is the room type that has the least amount of daylight in general. A reasonable explanation for this can be found in the kitchen placement, which is often further inside the building, sometimes behind other rooms. There are few results for the dining rooms, mainly because there were very few distinct rooms of this type in the analysis.

Table 5. Shows all the different room types studied, divided into the four main categories.

Bedroom	Living Room	Kitchen	Dining Room
Bedroom	Living room	Kitchen	Dining room
Small room	Family room	Divided kitchen	Divided dining
	Living room/Bedroom	Divided kitchenette	
	Living room/Kitchen	Living room/Kitchen	

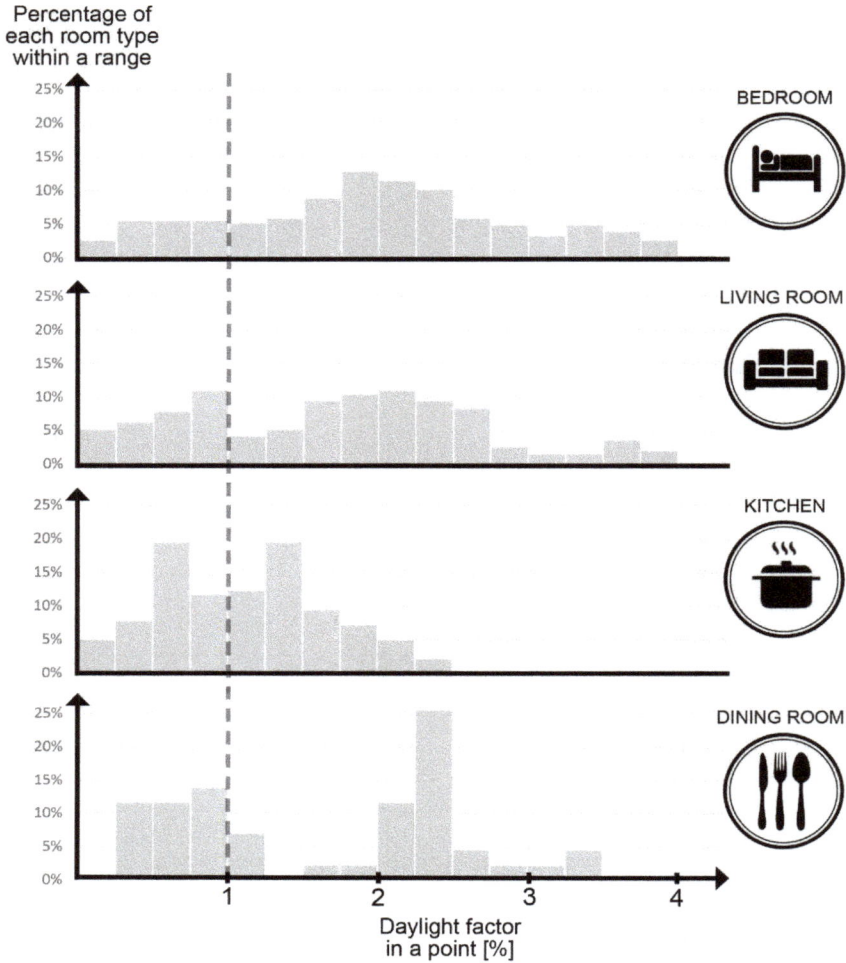

Figure 12. Distribution of DF$_p$ per room type.

3.2. Non-Residential Buildings (ID 9 to ID 16)

The distribution of DF$_p$ in the non-residential buildings is shown in Figure 13. It is worth noting that the requirement for DF$_p$ in student apartment buildings (ID 9 and ID12) is basically the same as in residential buildings, except in rooms intended for cooking, where it is enough to have access to indirect daylight. Regardless, the kitchenettes were also included in this study for comparisons. The dark grey bars represent kitchenettes and the light grey ones are for oher rooms, such as bedrooms, living rooms, and combined rooms (bedroom and living room). Based on the results, DF$_p$ in the bedrooms, living rooms, and combined rooms is mostly above the requirement. All kitchenettes, on the other hand, have DF$_p$ below 0.75% because they are typically placed furher into the building, behind other rooms.

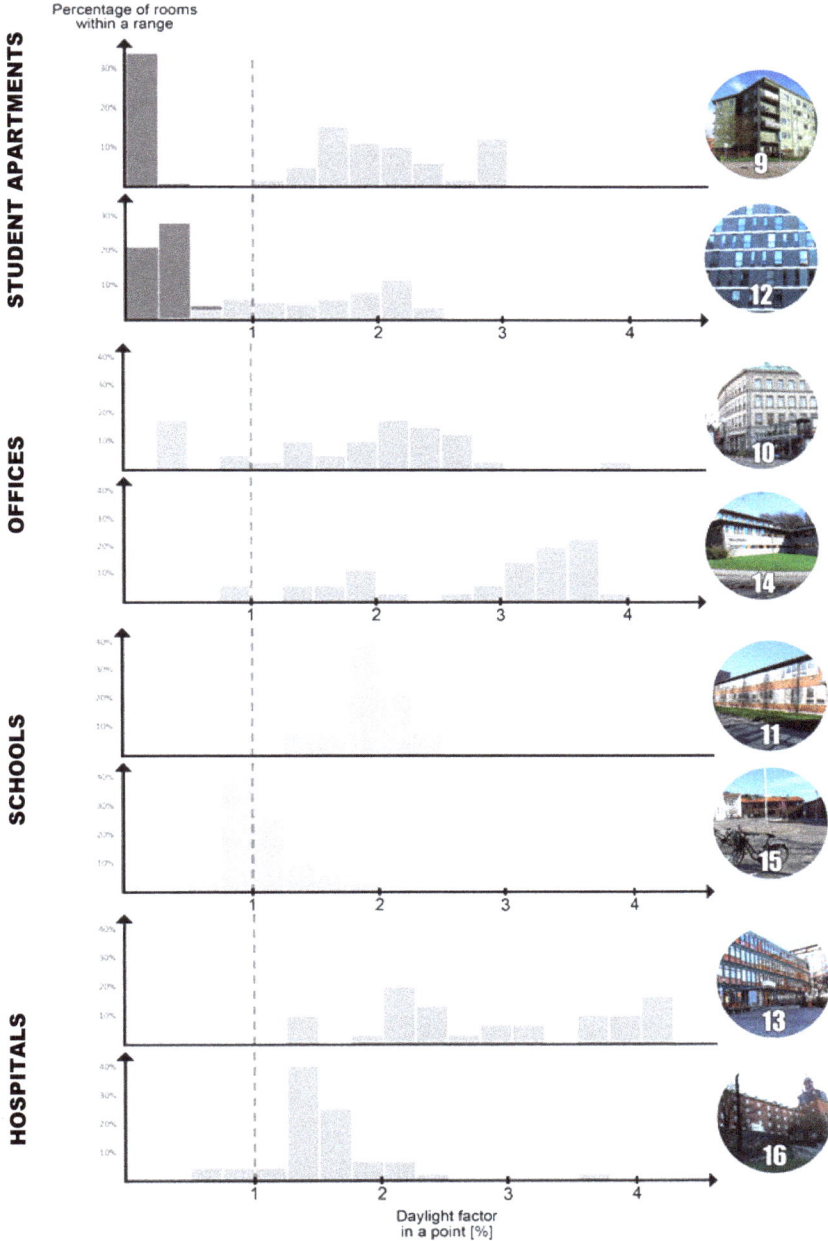

Figure 13. Distribution of DF_p in the student non-residential buildings. For the student apartment buildings, the kitchenettes are marked in dark grey and all other rooms in light grey.

Moving on to the office buildings with IDs 10 and 14, the results indicate that the most of the rooms have a rather high DF_p. An explanation for this can be found in the small size of the cell-type offices and, particularly, their small depth in comparison to the size of the windows. In both schools

(ID 11 and ID 15), merely the classrooms have been studied. In building 11, the daylight factor is around 2% while, in building 15, the daylight factor is a bit lower in general, around 1%. Finally, the majority of studied rooms in the hospitals (ID 13 and ID 16), i.e., examination rooms, patient rooms, and offices, have a daylight factor within a range. In contrast to the student apartment buildings, offices, and hospitals, none of the rooms in the schools have a daylight factor above 2.5%. A reasonable explanation for this is that the depths of the classrooms are generally greater, in comparison to other room types, and, therefore, the point where the daylight factor is measured is placed further into the buildings.

4. Results – Correlation Between Different DF Metrics

4.1. Correlation Between DF_P, DF_A and DF_M

The area-averaged DFs from Section 2.4, i.e., DF_A and DF_M are alternative metrics for describing the daylight availability in rooms. Therefore, the correlations between the single-point DFs for the residential buildings, from Section 3, and their area-averaged equivalents are shown in Figure 14. The latter comprise the following four values: An average or a median value, over a whole or a retracted horizontal area. Only the rooms (more than 95%, approximately) where the position of a control-point for DF_P could be clearly defined were included in the comparisons.

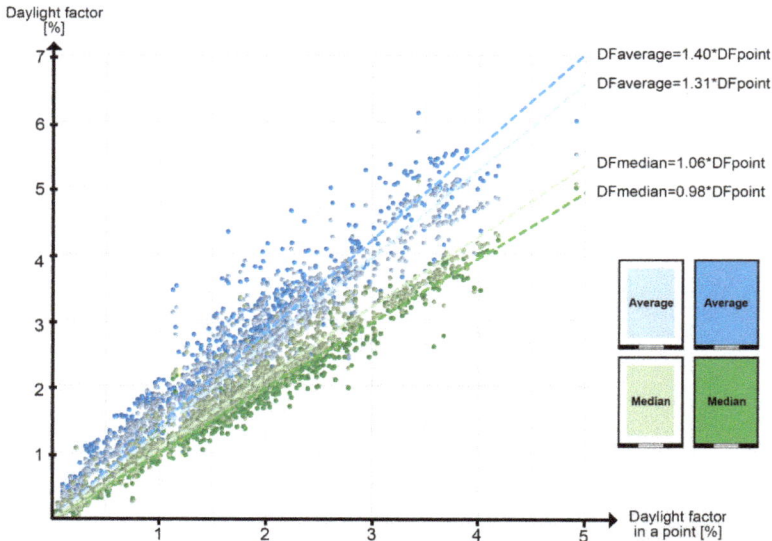

Figure 14. Comparison between the point-value DF_P (horizontal axis) and area-averaged DFs (DF_A = DFaverage, DF_M = DFmedian, vertical axis), for both full and retracted control surfaces.

As it can be seen, there is almost a 1:1 correlation between the DF_P and DF_M for both control surfaces, while DF_A is about 30–40% larger than DF_P. This is because the smaller (retracted) control surface affects DF_A and DF_M differently; DF_M generally increases while DF_A decreases. This trend can be explained by the example in Figure 15, which shows a DF distribution in a simple rectangular room with a window. When retracting the control surface by 0.5 m from each wall, a greater part of the values below the median value is removed. Consequently, the median value for the retracted control surface increases.

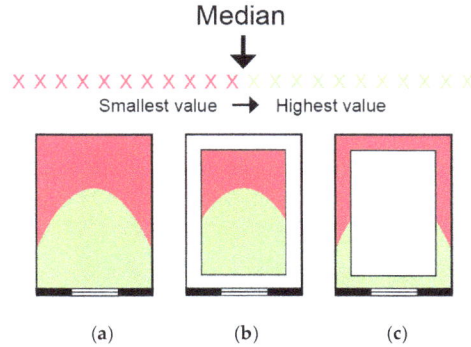

Figure 15. Impact of the size of a control area on the results of ordinary mean and median DF: (**a**) the control area covers the whole horizontal section; (**b**) the control area is retracted 0.5 m from the inner walls; (**c**) the difference between (**a**) and (**b**). The green zone represents the area in the room were the daylight factor is above the median value, while the red zone represents the area in the room were the daylight factor is below the median value.

One needs to look at DF values at each point in the grid to understand the effects of the control surface retraction on the DF-averages. This has been done for four different room types, as shown in Figure 16. Each graph in the figure shows the calculated grid-points DFs in ascending order. The points with the highest DFs, placed closest to the windows, are basically truncated when the control surface is reduced. This applies also to the points with the lowest DFs, which are furthest from the windows. In the rooms with non-linear distributions of DFs, the truncation of the highest DF-values has a greater impact on the average DF (i.e., it decreases) then in the rooms with a linear distribution of DFs.

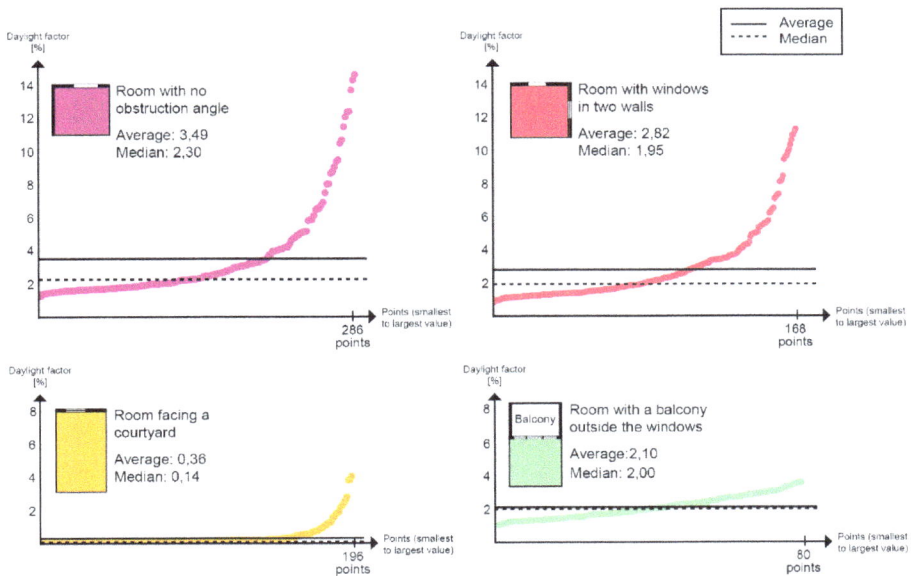

Figure 16. Grid-point values of DF arraigned in ascending order, for four different room types. The corresponding average and median values for the same control surfaces are also included.

4.2. Correlation Between DF_p and AF

As presented in the introductory section, the AF method is a simpler and, thus, widely used method for the daylight design of buildings. To evaluate its reliability, the AF method was applied to the studied buildings and the results were compared to the single-point DF calculations presented in Section 3. It is worth noting that the AF method only indicates a probability that the DF in a room will be greater than 1%, rather than the actual amount of daylight indoors.

Based on the instructions in [38], the AF method can be applied to simpler room geometries and shading situations, which, in this study, stands for about 61% of the studied rooms. Among those, in 78% of cases there was an agreement between the calculated single-point DF and the assessment by the AF-method; 70% with good daylight (both DF ≥1% and the requirement in equation 2 fulfilled) and 8% with poor daylight. In the remaining 22%, there was a disagreement between these two methods as presented in Table 6.

Table 6. Comparison between the results from the AF-method and single-point DF calculations.

Results for AF	Results for DF$_p$	Percentage of the Studied Rooms	Agreement between the Methods
$A_{glazing} \geq f \cdot A_{floor}$	$DF_p \geq 1\%$	70%	Yes
$A_{glazing} < f \cdot A_{floor}$	$DF_p < 1\%$	8%	Yes
$A_{glazing} \geq f \cdot A_{floor}$	$DF_p < 1\%$	12%	No
$A_{glazing} < f \cdot A_{floor}$	$DF_p \geq 1\%$	10%	No

The main reason for this disagreement can be found in the limitations of the AF method. The AF method works only in two dimensions, i.e., in a vertical section, without considering lateral objects and situations, as shown in Figure 17. In addition, rooms with windows with different directions and heights, or rooms placed behind other rooms, are basically not possible to evaluate using this method.

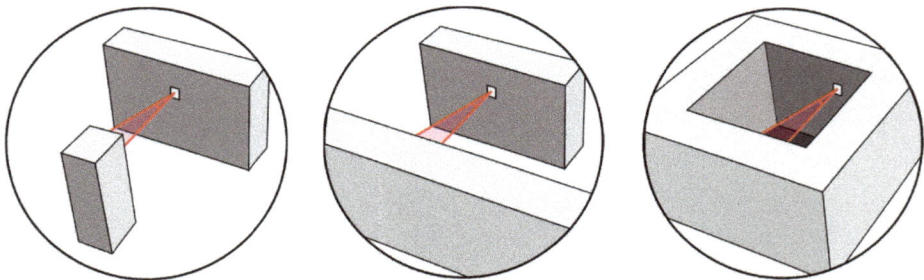

Figure 17. Three situations for which the AF-method gives the same results (by anticipating the same obstruction angle in front of the window), while the DF$_p$ method gives the different results.

4.3. Correlation Between Daylight Factors Indoors and DF_W

The available daylight in front of windows, described by Equation (3), DF$_W$, was calculated for the residential buildings with IDs 2, 3, and 6, with the largest, medium, and lowest DF$_p$ metrics indoors, respectively (see Figure 10). Examples of DF$_W$ for selected windows are shown in Figure 18. As expected, DF$_W$ reaches much higher values than DF inside the rooms. The range of calculated values was between 7% and 46%.

Figure 18. Examples of different daylight factors outside selected windows in buildings 2, 3, and 6.

Amounts of daylight inside a room and in front of windows are strongly correlated to the floor and window size. When multiplying DF_W and the area of windows, the resulting value (in $\%\cdot m^2$) can be interpreted as an indication of the available daylight entering a room. A similar result can be obtained when DF_P and the area-averaged DFs are multiplied with the full floor area in the room. The corresponding products are compared in Figure 19. Based on the results of the regression analyses, the following relations can be established:

$$DF_P \cdot A_{floor} \approx 0.22 \cdot DF_W \cdot A_{window},$$
$$DF_M \cdot A_{floor} \approx 0.24 \cdot DF_W \cdot A_{window}, \quad\quad (3)$$
$$DF_A \cdot A_{floor} \approx 0.35 \cdot DF_W \cdot A_{window}.$$

Although the established relations are rough, with the coefficient of determination about 0.8, they are also indicative and could be of use as guiding values in the early design stages, in the Swedish design context.

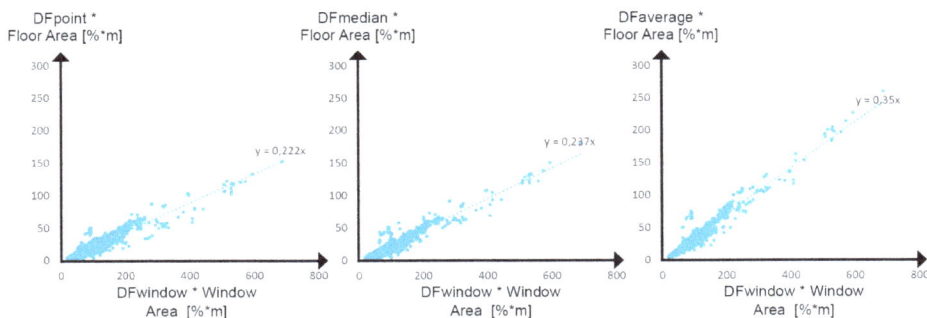

Figure 19. Correlation between the simulated amount of daylight outside the windows (DF_W multiplied by the window area) and indoors. The latter is given, for three DF indicators, each multiplied by the floor area.

4.4. Correlation Between DF_P and Results of the Field Survey

As mentioned in Section 2.7, answers on all questions but Q3 and Q7 were in form of grades, to allow comparisons with the simulated daylight factors in the room. An example of such a comparison can be found in Figure 20, which shows the DF_P (in ascending order) for all 124 evaluated rooms, together with the scaled answers from question Q1. These were further correlated by a polynomial trend line, included in the figure as a dashed line to indicate a general trend of the answers.

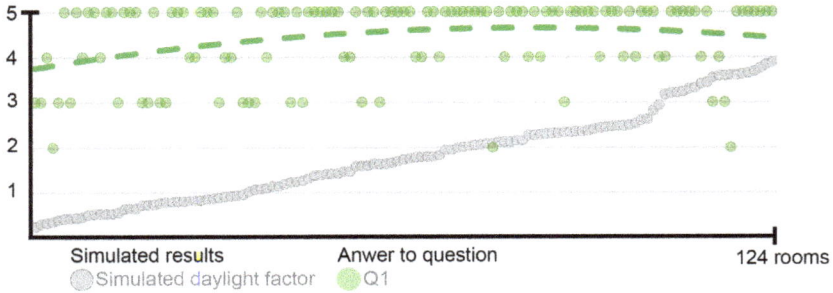

Figure 20. Simulated single-point DF and the scaled answers to question Q1 from the residents. The dashed line is a fitted polynomial to the answers from residents.

The same procedure was used to process and compare the results on questions Q2, Q4–Q6, and Q8–Q9 with the calculated DF, as shown in Figure 21. As can be seen in the figure, there is a general agreement between the perceived access to daylight (Q1) and the calculated DF; people give higher grades to the rooms with higher DF. A similar but stronger correlation can be found between the perceived access to direct sunlight (Q5) and the DF, which can be explained in two ways, as follows: People may find the access to daylight better in sunlit rooms because the rooms get brighter at these moments, or because the sunlit rooms are truly exposed to more daylight due to fewer shading objects in the surroundings.

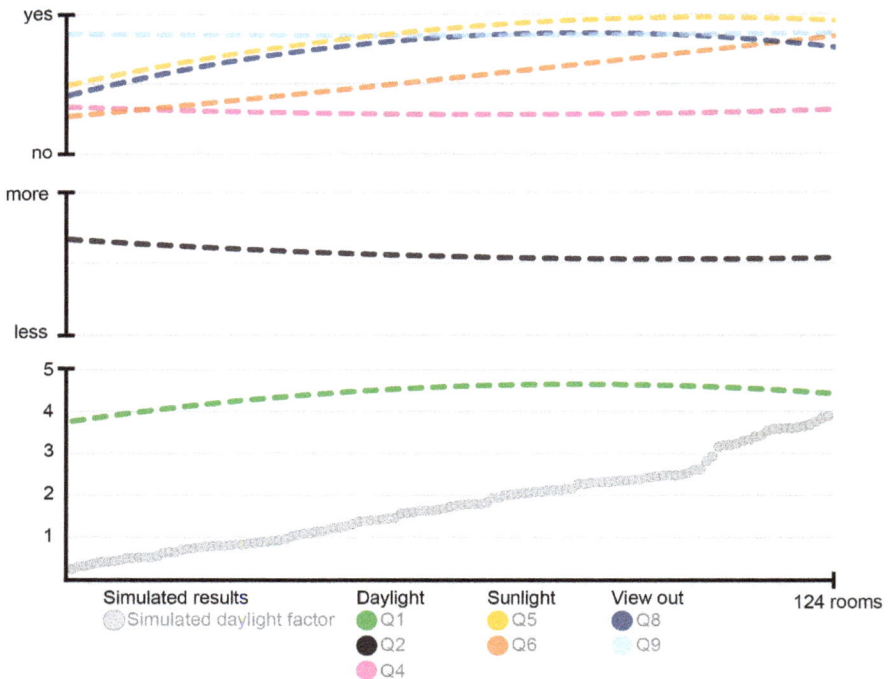

Figure 21. Trendlines (dashed lines) based on the scaled answers from the survey, compared to the calculated single-point DF for all 124 rooms.

Answers to question Q2 reveal how content the residents had been when assessing the access of daylight in their apartments. Most of respondents, i.e., 79%, were pleased with the current daylight levels, while 17% and 4% would like to have more and less daylight, respectively. These results are also in agreement with the answers to Q1.

Question Q3 showed in which rooms the residents would like to have the most daylight. Four different room types were considered (kitchen, living room, bedroom, and dining room) and the residents ranked them 1–4, where 1 was the most important room. The average grades for each room type, in Figure 22, show that the kitchen was ranked as the most important room to have access to a lot of daylight, while the bedroom as the least important one. These results are very interesting because they are in a direct contradiction with the findings from Figure 12, i.e., the kitchens normally have the least access to daylight (the lowest single-point DF), much lower than the bedrooms (the majority with a DF larger than 1%). The results from the surveys alone are not enough to make reliable conclusions about how the residents perceive the daylight in their homes. They can, however, indicate what people desire, in general.

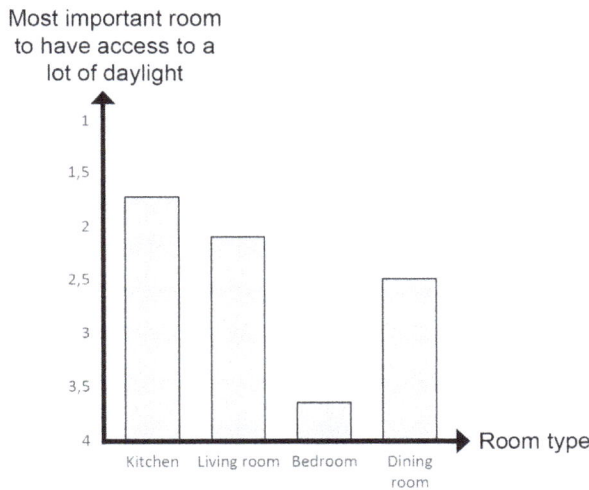

Figure 22. The residents' priorities regarding the most important room to have much access to daylight.

5. Conclusions

To evaluate daylight levels in existing buildings in the Swedish context and, thereby, to support an ongoing review of the current and future daylight indicators in Sweden, comprehensive numerical simulations of various DF metrics in more than 1200 rooms of 16 sample buildings were conducted. The study was deliberately limited to the evaluation of DF metrics for their intuitive understanding and easy evaluation in real projects. The sample buildings represent typical architectural styles and building technologies, from between 1887 and 2013, in Gothenburg and include eight residential buildings, two office buildings, two schools, two student apartment buildings, and two hospitals. Although the simulated point daylight factor DF_P is found to be 1.4% on average, which is above the required 1%, large variations were found between the studied rooms.

For overcoming various indecisive situations when evaluating DF_P, alternative DF metrics were introduced, i.e., the mean DF_A and median DF_M, both averaged over the same horizontal surface in a room, by considering either the full surface or the retracted area (by 0.5 m from all walls). Based on regression analyses, almost 1:1 correlation was found between DF_P and DF_M, while DF_A gave typically 30–40% larger amounts of daylight in the rooms, compared to DF_P. It was shown that this difference is a consequence of a non-linear distribution of the daylight over the control surface, for which DF_M is more suitable. In addition, results for DF_P in selected sample buildings were compared to the ones obtained by the AF method, which is another broadly used daylight design method in Sweden. This comparison basically confirmed the known limitations of the AF method, i.e., that the AF method is suitable only for simpler daylight design tasks.

The field investigation, aimed at revealing the residents' satisfaction with available daylight levels in selected sample buildings by means of an in-house questionnaire, brought some further and unique insights in daylight design challenges. It was found that the empirical data generally supported the findings from the numerical simulations of DF_P. This is particularly valid for kitchens, the spaces with the lowest DF_P values, based on simulations. While the empirical data confirm that kitchens are the spaces with the lowest amounts of daylights, they also indicate that the residents would like kitchens to be the spaces with the most daylight.

A new DF metric, denoted DF_W, allowing for daylighting design in early stages, when only limited data on the building shape and windows arrangement are available, is introduced and evaluated. The latter was done through a regression analysis with the results based on the calculated DF_P, DF_A, and DF_M values. Rough, but rather indicative, correlations were found between DF_W and other DF metrics, indicating that the former could be of use in the early design stages in the Swedish design context.

It is worth noticing that the findings may look different for different sample buildings and urban constellations, as was shown in [31], who used the methodology developed by the authors of this work. Yet, the developed methodology is general and can be applied for further studies. The combination of numerical and field studies is particularly important for spaces where lower DF levels are identified by simulations. In this regard, the future work should focus on field studies of perceived daylight levels in schools, since these were shown to be generally lower than in the other non-residential buildings studied.

Author Contributions: Conceptualization and methodology, all authors; formal analysis, investigation and visualization, L.W. and S.E.; validation, L.W., S.E., and M.T.; writing—original draft, L.W., S.E., A.S.K., and M.Ö.; writing—review and editing, A.S.K. and M.Ö.; supervision, M.T., M.Ö., and A.S.K.

Funding: This research received no external funding.

Acknowledgments: Anna Larsson and Mats-Inge Olsson from Bengt Dahlgren AB in Gothenburg are acknowledged for providing supervision and necessary drawings.

Conflicts of Interest: The authors declare no conflict of interest.

References

1. Boverket. *SBN 1975 Svensk Byggnorm*. 1975. Available online: https://www.boverket.se/contentassets/c4c3f9ae57294ae889bfaf710b08b125/sbn-1975-utg-3.pdf (accessed on 30 May 2019).
2. Xue, P.; Mak, C.M.; Cheung, H.D. The effects of daylighting and human behavior on luminous comfort in residential buildings: A questionnaire survey. *Build. Environ.* **2014**, *81*, 51–59. [CrossRef]
3. Cheung, H.D.; Chung, T.M. A study on subjective preference to daylit residential indoor environment using conjoint analysis. *Build. Environ.* **2008**, *43*, 2101–2111. [CrossRef]
4. Cantin, F.; Dubois, M.-C. Daylighting metrics based on illuminance, distribution, glare and directivity. *Light. Res. Technol.* **2011**, *43*, 291–307. [CrossRef]
5. Boverket. BBR Boverket's building regulations. BFS 2011:6 with amendments up to 2018:4 (in Swedish). 2018. Available online: https://www.boverket.se/globalassets/publikationer/dokument/2018/bbr-2018_konsoliderad-version.pdf (accessed on 30 May 2019).
6. Rogers, P.; Tillberg, M.; Bialecka-Colin, E.; Österbring, M.; Mars, P. En Genomgång av Svenska Dagsljuskrav 2015. Available online: http://www.acc-glas.se/wp-content/uploads/2013/12/SBUF-12996-Slutrapport-Förstudie-Dagsljusstandard.pdf (accessed on 30 May 2019).
7. Boverket. Housing, Internal Migration and Economic Growth in Sweden. (In Swedish). 2016. Available online: https://www.boverket.se/globalassets/publikationer/dokument/2016/housing-internal-migration-and-economic-growth-in-sweden.pdf (accessed on 30 May 2019).

8. Strømann-Andersen, J.; Sattrup, P. The urban canyon and building energy use: Urban density versus daylight and passive solar gains. *Energy Build.* **2011**, *43*, 2011–2020. [CrossRef]

9. Robinson, A.; Selkowitz, S. Tips for Daylighting with Windows. Available online: https://buildings.lbl.gov/sites/default/files/ellen_thomas_lbnl-6902e.pdf (accessed on 14 April 2019).

10. Eriksson, S.; Waldenström, L. Daylight in Existing Buildings A Comparative Study of Calculated Indicators for Daylight. Master's Thesis, Chalmers University of Technology, Gothenburg, Sweden, 2016.

11. Rogers, P.; Dubois, M.-C.; Tillberg, M.; Östbring, M. Moderniserad Dagsljusstandard. no. SBUF ID: 13209. 2018. Available online: http://www.acc-glas.se/wp-content/uploads/2013/12/SBUF-12996-Slutrapport-Förstudie-Dagsljusstandard.pdf (accessed on 30 May 2019).

12. Mardaljevic, J.; Christoffersen, J. A Roadmap for Upgrading National/EU Standards for Daylight. In Proceedings of the CIE Centenary Conference "Towards a New Century of Light", Paris, France, 15–16 April 2013; pp. 1–10.

13. Reinhart, C.; Breton, P.-F. Experimental validation of 3DS MAX ® DESIGN 2009 and DAYSIM 3.0 1 2. In Proceedings of the 11th International IBPSA Conference, Glasgow, Scotland, 27–30 July 2009.

14. Mardaljevic, J.; Andersen, M.; Roy, N.; Christoffersen, J. Daylighting metrics for residential buildings. In Proceedings of the 27th Session CIE, Sun City, South Africa, 11–15 July 2011; p. 18.

15. Hellinga, H.; Hordijk, T. The D&V analysis method: A method for the analysis of daylight access and view quality. *Build. Environ.* **2014**, *79*, 101–114.

16. Yu, X.; Su, Y. Daylight availability assessment and its potential energy saving estimation -A literature review. *Renew. Sustain. Energy Rev.* **2015**, *52*, 494–503. [CrossRef]

17. Tregenza, P. Uncertainty in daylight calculations. *Light. Res. Technol.* **2017**, *49*, 829–844. [CrossRef]

18. SIS. *SS 914201 Building design - Daylighting - Simplified method for checking required window glass area;* SIS Swedish Standards Institute: Stockholm, Sweden, 1988.

19. Velux. *Daylight Visualizer.* 2019. Available online: https://www.velux.com/article/2016/daylight-visualizer (accessed on 30 May 2019).

20. Radsite. *Radiance.* 2019. Available online: https://www.radiance-online.org// (accessed on 20 April 2016).

21. SGBC. *Miljöbyggnad 3.0;* Sweden Green Building Council: Stockholm, Sweden, 2017.

22. Ecolabelling, N. Small Houses, Apartment Buildings and Buildings for Schools and Pre-Schools. 2016. Available online: https://www.ecolabel.dk/kriteriedokumenter/089e_2_11.pdf (accessed on 20 March 2016).

23. Mata, É.; Kalagasidis, A.S.; Johnsson, F. Energy usage and technical potential for energy saving measures in the Swedish residential building stock. *Energy Policy* **2013**, *55*, 404–414. [CrossRef]

24. Yu, X.; Su, Y.; Chen, X. Application of RELUX simulation to investigate energy saving potential from daylighting in a new educational building in UK. *Energy Build.* **2014**, *74*, 191–202. [CrossRef]

25. Mardaljevic, J. Simulation of annual daylighting profiles for internal illuminance. *Light. Res. Technol.* **2000**, *32*, 111–118. [CrossRef]

26. Li, D.; Cheung, G. Average daylight factor for the 15 CIE standard skies. *Light. Res. Technol.* **2006**, *38*, 137–152. [CrossRef]

27. Li, D.H.W.; Wong, S.L. Daylighting and energy implications due to shading effects from nearby buildings. *Appl. Energy* **2007**, *84*, 1199–1209. [CrossRef]

28. Dubois, M.-C.; Flodberg, K. Daylight utilisation in perimeter office rooms at high latitudes: Investigation by computer simulation. *Light. Res. Technol.* **2013**, *45*, 52–75. [CrossRef]

29. Thomsen, K.E.; Rose, J.; Mørck, O.; Jensen, S.Ø.; Østergaard, I.; Knudsen, H.N.; Bergsøe, N.C. Energy consumption and indoor climate in a residential building before and after comprehensive energy retrofitting. *Energy Build.* **2016**, *123*, 8–16. [CrossRef]

30. Li, D.; Wong, S.; Tsang, C.; Cheung, G.H. A study of the daylighting performance and energy use in heavily obstructed residential buildings via computer simulation techniques. *Energy Build.* **2006**, *38*, 1343–1348. [CrossRef]

31. Bournas, I.; Dubois, M.-C. Daylight regulation compliance of existing multi-family apartment blocks in Sweden. *Build. Environ.* **2019**, *150*, 254–265. [CrossRef]

32. Jacobsson, E.; Eriksson, F. Evaluation of Sun-and Daylight Availability in Early Stages of Building Development A Method Based on Correlations of Interior and Exterior Metrics. Master's Thesis, Chalmers University of Technology, Gothenburg, Sweden, 2017.

33. Reinhart, C.F.; Lagios, K.; Niemasz, J.; Jakubiec, A. DIVA for Rhino Version 2.0. 2011. Available online: http://www.diva-for-rhino.com/ (accessed on 30 April 2016).
34. Chalmers Geodata Portal. Available online: https://geodata.chalmers.se/ (accessed on 22 April 2019).
35. Bellia, L.; Pedace, A.; Fragliasso, F. The impact of the software's choice on dynamic daylight simulations' results: A comparison between Daysim and 3ds Max Design®. *Sol. Energy* **2015**, *122*, 249–263. [CrossRef]
36. Jones, N.L. Validated Interactive Daylighting Analysis for Architectural Design. Ph.D. Thesis, Massachusetts Institute of Technology, Cambridge, MA, USA, 2017.
37. Nocera, F.; Faro, A.L.; Costanzo, V.; Raciti, C. Daylight Performance of Classrooms in a Mediterranean School Heritage Building. *Sustainability* **2018**, *10*, 3705. [CrossRef]
38. SIS. *"SS-EN 17037-2018 Daylight in buildings."*; SIS Swedish Standards Institute: Stockholm, Sweden, 2018.

Article

An Efficient Power Scheduling in Smart Homes Using Jaya Based Optimization with Time-of-Use and Critical Peak Pricing Schemes

Omaji Samuel [1], Sakeena Javaid [1], Nadeem Javaid [1], Syed Hassan Ahmed [2], Muhammad Khalil Afzal [3] and Farruh Ishmanov [4,*]

[1] Department of Computer Science, COMSATS University Islamabad, Islamabad 44000, Pakistan; omajiman1@gmail.com (O.S.); sakeenajavaid@gmail.com (S.J.); nadeemjavaid@comsats.edu.pk (N.J.)
[2] Department of Computer Science, Georgia Southern University, Statesboro, GA 30460, USA; sh.ahmed@ieee.org
[3] Department of Computer Science, COMSATS University Islamabad, Wah Campus, Wah Cantonment 47040, Pakistan; khalilafzal@ciitwah.edu.pk
[4] Department of Electronics and Communication Engineering, Kwangwoon University, Seoul 01897, Korea
* Correspondence: farruh.uzb@gmail.com

Received: 30 September 2018; Accepted: 5 November 2018; Published: 14 November 2018

Abstract: Presently, the advancements in the electric system, smart meters, and implementation of renewable energy sources (RES) have yielded extensive changes to the current power grid. This technological innovation in the power grid enhances the generation of electricity to meet the demands of industrial, commercial and residential sectors. However, the industrial sectors are the focus of power grid and its demand-side management (DSM) activities. Neglecting other sectors in the DSM activities can deteriorate the total performance of the power grid. Hence, the notion of DSM and demand response by way of the residential sector makes the smart grid preferable to the current power grid. In this circumstance, this paper proposes a home energy management system (HEMS) that considered the residential sector in DSM activities and the integration of RES and energy storage system (ESS). The proposed HEMS reduces the electricity cost through scheduling of household appliances and ESS in response to the time-of-use (ToU) and critical peak price (CPP) of the electricity market. The proposed HEMS is implemented using the Earliglow based algorithm. For comparative analysis, the simulation results of the proposed method are compared with other methods: Jaya algorithm, enhanced differential evolution and strawberry algorithm. The simulation results of Earliglow based optimization method show that the integration of RES and ESS can provide electricity cost savings up to 62.80% and 20.89% for CPP and ToU. In addition, electricity cost reduction up to 43.25% and 13.83% under the CPP and ToU market prices, respectively.

Keywords: Jaya algorithm; smart grid; optimal energy management; demand response; demand side management

---

1. Introduction

Presently, the aging power grid infrastructure is now gradually improving due to the emerging technological innovations. This improvement witnessed by the recent innovations does not only revive the interest of researchers and socio-economic development. However, it has a great benefit to the society at large. Despite this improvement, the majority of power generation relied on the conventional electric power distribution network. This network is very complicated and cannot meet the demands of the 21st century [1]. More so, this intricate network is influenced by the following factors. Firstly, the growing population, global climatic change, equipment malfunction and exponential demands for electricity. Lastly, the electric storage problems, strength of the generating plants, one-way communication, resilience problem and the shortage of fossil fuel. To address the above limitations, the concept of smart grid (SG) is adopted to include the renewable energy sources (RES) [2]. The advent of SG is not only beneficial to all those involved in the electric power industry. However, it has numerous stakeholders.

To actualize the capabilities of SG, the demand side management (DSM) is included in the existing SG, since it provides alteration in the consumer demand for electricity. This can be achieved through several methods like financial incentives and behavioral change via education. Generally, the aim of DSM is to motivate consumers to reduce their electricity consumption during peak hours as well as in critical periods. DSM uses demand response (DR) which plays a significant role in the operation of the power grid by shifting the electricity usage during the peak period in response to the market prices. This also involves other forms of financial incentives as well as balancing supply and demand.

To efficiently implement the DSM strategies, home energy management system (HEMS) is used to minimize the generation cost of electricity by shifting certain household loads to off-peak hours. This can be done by optimally adjusting the energy obtained from the power grid. HEMS involves scheduling of certain household loads that have high priority as they are important for meeting the consumer's satisfaction. Thus, the request of such loads has to be met instantly and the energy minimization potency of such loads is not easily rescheduled. However, such loads can be combined with RES and energy storage system (ESS) for which peak hour consumption is minimized. Nevertheless, some loads can be rescheduled to the period of low consumption. Scheduling periods are determined by the consumers' choice settings. In this paper, we describe such household loads as shiftable appliances. On the contrary, some household loads have controllable patterns. Hence, the electricity cost can be minimized based on the electricity consumption profile of such loads.

Reliable operations of HEMS require coordination among the ESS and loads which can be achieved through different optimization techniques. These techniques are of two types. The traditional techniques require the exact rules for changing one solution into another. It is fundamentally deterministic in nature and has shown tremendous achievement in the field of science and engineering. Examples of these optimization techniques are dynamic programming, generalized reduced gradient method, nonlinear and linear programming, and quadratic and geometric programming [3–6]. Other optimization techniques are meta-heuristic with an inherent probabilistic transformation rules [7–10].

The commonly used heuristic optimization algorithms can be evolutionary or iterative based methods. However, as the number of optimization variables increases, the computational complexities increase as well, especially with iterative methods. On the contrary, it is not with the case of evolutionary methods. However, evolutionary methods are faced with local optimum problem. Examples of heuristic algorithms are either swarm intelligence-based, biological-inspired, nature-inspired, population-based or musical-inspired phenomena. These heuristic optimization algorithms are stochastic with common controlling parameters like generation size, elite size, and population size while others have algorithmic specific control parameters [11]. The best possible adjustment of these algorithmic specific control parameters is the essential factor that can influence algorithm's performance. This adjustment avoids high computational complexity and prevents yielding a local optimum solution.

Several pricing schemes are now introduced by utility to calculate the electricity tariffs of different periods of the day. This is done to reshape the consumer consumption patterns. Some electricity tariffs are computed on an hourly or day-ahead basis, whereas a day is divided into periods such as off-peak, mid peaks and on-peak hours, and the electricity tariff is calculated for such periods. During critical events or hours, consumers are charged with peak rate that is higher than any other rate.

This paper is the extension of work in [12] which includes shared RES and ESS to supply energy to multiple residential households. For the load scheduling, an Earliglow based optimization algorithm is proposed to shift loads within the time-of-use (ToU) and critical peak pricing (CPP) scheduling horizon for 24 h. The Earliglow algorithm has both flavors of the existing techniques, Jaya and strawberry algorithm (SBA). The proposed algorithm reduces the number of iterations created by SBA since SBA duplicates every computational individual at each iteration as well as local and global search problem. Furthermore, the proposed algorithm ensures that the solution moves towards the ideal solution by circumventing the worse solution. The proposed algorithm becomes efficient, if the best optimal solution is obtained. In addition, a battery is considered for energy storage and it gets charged only if the energy produced by the RES is greater than the load demands. Otherwise, the battery will be discharging. The state of charge (SOC) of the battery depends on the history of battery and the differences between the power generated by RES and the load demands.

This paper is organized as follows: Section 2 presents the related work. Afterwards, Section 3 includes the problem statement and the system model description. In Section 4, the proposed model is given in detail. Next, simulation results and discussions, and the feasible region (FR) is discussed in Section 5. Finally, Section 6 concludes the work with future work.

2. Related Work

The implementation of heuristic optimization algorithm has been proposed in order to derive solutions within a reasonable amount of execution time. Because of its ability to handle complex problems of nonlinear nature, extra decision variables can be included without increasing the execution time. However, considering the stated pros, there is the need for dynamic and deterministic algorithms to provide an accurate optimal solution to home energy management (HEM) scheduling problems.

Maytham et al. [3] propose a real optimal controller for HEMS using new binary backtracking search algorithm (BBSA). The proposed controller reduces the energy consumption and electricity cost, and ensures energy savings at peak hours for weekdays and weekends. The proposed algorithm is compared with binary particle swarm optimization (BPSO) to ascertain the accuracy of their proposed controller in the HEMS. Within household surroundings, smart meters coordinate with both real time and appliances scheduling via HEMS. Abdul et al. [4] propose an autonomous energy management system, which is formulated by the mixed integer linear programming (MILP) and the Dijkstra algorithm for reducing the electricity cost of the real time load. An efficient solution is obtained with lower complexity via the proposed HEMS without disturbing the operation of non shiftable appliances.

Energy monitoring plays a vital role in energy management. As such, it is required to monitor the energy consumption of household devices before commencing on the technical measures to reduce energy consumption. Abubakar et al. [5] demonstrate energy management via intrusive load monitoring (ILM) and non-intrusive load monitoring (NILM). The proposed techniques reduce electricity cost, provide cost savings and minimize the greenhouse emission. Addressing the challenges involving high dimensional optimization problems for iterative HEM, Hepeng et al. [6] propose an approximate dynamic programming (ADP). It is used as the mechanism for energy management of vehicle-to home (V2home) and vehicle-to grid (V2grid). Due to the large number of DR capable devices, the proposed method is useful for minimizing the energy cost and providing peak load shifting for the iterative HEM.

Currently, DSM schemes have been proposed for residential, commercial and industrial sectors. These schemes are useful in reducing the energy profile of consumers in the grid area network. In this respect, work in [9] implements a hybrid genetic wind-driven optimization (GWD) technique for scheduling the residential household loads. The GWD technique is able to shift the load in a real-time pricing (RTP) environment between on-peak and off-peak hours. In addition, it reduces electricity cost and the consumer's comfort is protected. In order to show a higher search efficiency and dynamic capability to achieve an optimal solution, Hafiz et al. [13] present an efficient HEMS (EHEMS) based on a genetic harmony search algorithm (GHSA). This proposed method reduces the electricity cost, peak-to-average ratio (PAR) and maximizes the consumer comfort via the real time electricity pricing (RTEP) and CPP.

In [14], HEMS is proposed to allow individual smart household device interaction with data collecting module in the form of Internet of Things (IoT). This motivates consumers to locally monitor and control devices, and online costing generation through a mobile web mobile application. This module can be extended to work for a multi-home with distributed energy resources. The central module is decomposed into a two-level optimization problem that corresponds to the local HEM at the first level and a global HEM at the second level. As such, a distributed two-level HEMS algorithm is proposed in [15] to reduce the aggregate electrical cost of a few households. Moreover, it coordinates the operations of ESS, and electric power of the neighboring household while the consumer's choice comfort level remains sustained.

HEM enables consumers to engage in a DR program (DRP) in an active manner. Conversely, some methods are faced with the challenge of uncertainties with respect to consumer behavior as well the RES. As a consequence, a stochastic model for HEMS while considering the limited size of renewable energy generation is proposed in [16]. This model optimizes the consumer's electricity cost of several DRPs and occupant satisfaction is acquired via a response fatigue index. DRP was achieved via HEM, which helps consumers achieve electricity price savings and also minimizes peak load demand for the power grid. To further achieve these benefits, an evolutionary algorithm-based optimization models like the genetic algorithm (GA), Cuckoo, and the BPSO search are proposed for the intelligent management of load scheduling for residential users [17]. These models decrease the electricity cost with intense peaks.

Nikolaos et al. [18] propose HEMS for optimal day-ahead controllable appliance scheduling, with a distributed generation and ESS is integrated into a dynamic pricing environment. The proposed HEMS minimizes electricity cost that is required to meet the load demands of the consumers. However, the consumer load demand continues to increase due to the growing population, buildings and industries. As such, the utility cannot withstand the consumer's consumption requirement. Therefore, the utility must resolve load balancing and threshold problem. As a consequence, the authors have proposed a multi-objective evolutionary algorithm [19] to address the load balancing and threshold problem. The proposed solution minimizes the cost of energy usage as well as the waiting time for appliances' execution. Obviously, consumers are concerned about the safety of appliance operations; thus, safety risk of appliances is influenced by the increased in continuous operation, since consumers have no control over it. In this way, consumers are charged based upon appliance continuous operations. To reduce the electricity cost, a Pareto-optimal front is proposed to provide scheduling decisions based on the relationship between two multi-objectives: the electricity cost and operational delay. The proposed approach is compared to the weight and constraint approaches, and the simulation results show that electricity cost is reduced and operational delay is enhanced [20].

Javaid et al. [21] propose four heuristic optimization methods for HEMS: bacterial foraging optimization algorithm (BFOA), GA, BPSO, wind driven optimization (WDO) and a hybrid (genetic BPSO). The proposed algorithms reduce the electricity cost and PAR under the RTP market prices. The consumer's real-time demand and energy consumption are unpredictable, and manually operated appliances (MOA) are also difficult to schedule day-ahead shiftable appliances.

The MOA is the classes of appliances that are manually controlled by real-time demands of consumers. Yuefang et al. [22] propose an optimization approach formulated as a MOA scheduling problem under the RTP and inclining block rate (IBR) market prices. The proposed approach reduces electricity costs as compared to MOA approach without certainty. Li et al. [23] propose a fuzzy logic controller for dynamic adjustment of the quality of experience threshold to optimize users' comfort. In addition, peak load and electricity bill are minimized. A multi-objective DR optimization model is proposed in [24] to minimize customers' convenience level as well as the electricity cost using the non-dominated sorted genetic algorithm (NSGA-II). Oprea et al. [25] propose informatics solution that optimizes daily operational appliances, minimizes consumption peak and reduces stress on the main grid using the artificial neural network (ANN). Table 1 presents the summary of the related work with respect to techniques, achievements, pricing schemes and limitations.

Table 1. Summary of related work.

Technique(s)	Achievement(s)	Pricing Schemes	Limitation(s)
BBSA [3]	Reduces the energy consumption, electricity cost and achieves energy savings for week days and weekend	RTP	Consumer comfort and RES are not considered
MILP, Dijkstra [4]	Minimizes the total cost, trades the computational complexity for lower performance	ToU	Inconsideration of consumer comfort, PAR and RES
ILM, NILM [5]	Reduces the electricity cost, provides cost saving and minimizes the greenhouse gas emission	-	Ignored consumer comfort and RES
ADP [6]	Considered V2home and V2grid for energy management, provides cost reduction, peak load shifting and ESS is also considered using electric vehicle battery	RTP	Inconsideration of consumer comfort
GWD [9]	Minimizes electricity cost and protects user comfort	-	Energy consumption and PAR are ignored
GHSA [13]	Minimizes electricity cost, PAR and maximizes the consumer comfort	RTEP and CPP	Inconsideration of RES
IoT [14]	Motivates users to locally monitor and control devices	-	Ignored load scheduling, consumer comfort and electricity cost reduction
MILP [15]	Minimizes the total electricity cost of multiple households with distributed energy resources while maintaining the consumer's thermal comfort level	ToU	Inconsideration of PAR
Stochastic model [16]	Considers limited size of RES to minimize electricity cost	ToU and CPP	Computational complexity
GA, Cuckoo and BPSO [17]	Reduces electricity cost, PAR and provides energy saving using RES	ToU	Inconsideration of power loss
MILP [18]	Incorporates a distributed generation, ESS and minimizes the electricity cost	Day-ahead RTP	Computational complexity
Multi-objective evolutionary algorithm [19]	Addresses load balancing and threshold problem; reduces electricity cost as well the waiting time	ToU	Inconsideration of RES
Pareto-optimal front [20]	Electricity cost is reduced and operational delay is enhanced	Day-ahead RTP	Ignored user comfort and RES
BFOA, GA, BPSO, WDO and hybrid (genetic BPSO) [21]	Reduces electricity cost and PAR	RTP	Ignored consumer comfort and RES
MOA [22]	Reduces electricity cost	RTP-IBR	Inconsideration of RES, ESS and consumer comfort
Fuzzy controller [23]	Optimal users' comfort, peak load and electricity bill minimization	Dynamic price	Required high computational time
NSGA-II [24]	Ensures customers' convenience and electricity cost minimization	RTP	Inconsideration of RES
Informatics solution and ANN [25]	Performs consumption forecast, reduces peak consumption, optimizes daily appliances' operation and lessen main grid burden	ToU	Inconsideration of RES

This paper provides the following contributions:

1. Similar schemes used in [26] focus on the supplier side, which minimize the daily fuel cost, production cost and maximizes the sales revenue for grid-connected micro-grid; meanwhile, this paper expands the work carried out in [12] by incorporating the RES using ToU and CPP to schedule household appliances.

2. A model is proposed to provide scheduling of appliances within the smallest execution time via the Earliglow optimization algorithm. In addition, it provides a platform that enables a shared RES and ESS.
3. Including RES as well as ESS encourages the generation of on-site power which further alleviate the electricity cost and PAR with a minimal user waiting time, simultaneously.
4. The proposed model elaborates the individual appliances' energy consumption behavior, which provides hourly appliances scheduling and operations.

3. Problem Statement

In order to move ahead, we should be desperate for a new power grid, which is built from a bottom up approach to address the increase of computerized and digital household appliances with technological dependence. This can automatically monitor and manage the electricity demands of the 21st century. Different works in literature have proposed methods for DSM: work in [3] proposes a scheme that reduces electricity cost and consumption. However, consumer's comfort is not considered while there is so much reliance on power grid for electricity, which they fail to integrate RES to augment power supply. Work in [9] proposes a technique that schedules residential household appliances to minimize electricity cost and protect consumer's comfort. However, they ignore PAR, electricity consumption and also do not take into account the need of RES. Work in [16] presents a strategy that considers the limited size of RES for electricity cost optimization. However, it is insufficient to address the computational complexity of the proposed stochastic model.

The limitations of existing techniques in the related work are considered and a profound solution to overcome these limitations are proposed. The scheduling problem of household appliance operations in a given time horizon of 24 hours is expressed as a multi-objective optimization problem consisting of (1) the electricity consumption minimization, (2) electricity cost reduction, (3) user comfort maximization, and (4) load balancing. The proposed solution to the problem is based on the multi-objective scheduling problem. We additionally assess the performance of HEMS and optimize the execution of various kinds of household appliances associated within. The household appliance's power rating is taken to generate the time of operations fulfilling all the time requirements given by the consumers. To further reduce the reliance on electricity from the power grid, we integrate ESS and RES for better energy distribution. This incorporation of RES and ESS will provide load handling, since the overall power grid load is not stable and can fluctuate over time, thus creating a decentralized grid system that encourages the generation of on-site power.

Table 2 and Sections 3.1–3.5 provide a detailed description of the household appliance specifications and the formulation of multi-objective scheduling problem.

Table 2. Appliances specification details [12]. h: (hours); LOT: Length of operational time.

Appliance Class	Appliance Name	Power Rating (kW)	Starting Time (h)	Ending Time (h)	LOT (h)
Shiftable	Cloth dryer	1.5	06	14	04
	Vacuum cleaner	1	06	15	30 min
	Refrigerator	0.125	06	15	24
	Air conditioner	1	12	24	10
	Dish washer	1	08	22	30 min
	Pool pump	2	12	21	08
	Electric vehicle	2.5	16	24	2.5
	Television	0.25	01	16	6 h 45 min
	Iron	1	06	16	30 min
	Hair dryer	1	06	13	1 h 30 min
	Water heater	1.5	06	23	03
	Other	1.5	06	24	24
Nonshiftable	Light	0.5	16	24	6 h 15 min
	Electric stove	1.5	06	14	05
	Personal computer	0.25	08	24	04
	Heater	1.5	03	15	03

3.1. Appliance Specification

The household appliances are classified on their energy consumption patterns and operational behavior. The description of each classification is given below.

3.1.1. Shiftable Appliances

Shiftable household appliances consist of the interruptible and uninterruptible loads. The uninterruptible loads have a flexible finishing time with certain consumption period and a specified consumption rate. Examples are the dishwasher, washing machine, etc. The interruptible loads have a fixed consumption rate, and the execution periods depend upon consumer choice setting. Examples are the refrigerator, water heater, etc. Let $N_{\text{shiftable}}$ be presented as the number of shiftable appliances which belong to the overall household appliances. From Equation (1), $S_{\text{shiftable}}$ denotes the set of shiftable household appliances, where $\wp$ denotes the power rating of the individual appliance and $X^{\text{app}}(t)$ denotes the status of an appliance at any time slots $t \in T$. Equation (2) shows the cases when the appliance status is OFF and ON:

$$P_{\text{shiftable}}^{\text{consumption}} = \sum_{t=1}^{T} \sum_{S_{\text{shiftable}}=1}^{N_{\text{shiftable}}} \wp \times X^{\text{app}}(t), \tag{1}$$

$$X^{\text{app}}(t) = \begin{cases} 1 \text{, if appliance is turn ON,} \\ 0 \text{, if otherwise.} \end{cases} \tag{2}$$

3.1.2. Non Shiftable Appliances

Non shiftable appliances consist of unmanageable loads and weather-based loads. It relies on weather and energy consumption. It is also known as the fixed household appliances. Televisions, air conditioners, etc. are listed as non shiftable appliances. Let $N_{\text{nonshiftable}}$ denote the number of non shiftable appliance which belong to overall household appliances. From the Equation (3), $S_{\text{nonshiftable}}$ denotes the set of non shiftable appliances. The power rating of each appliance is denoted as $\wp$, and $X^{\text{app}}(t)$ is the status of appliance at any time slot $t \in T$:

$$P_{\text{nonshiftable}}^{\text{consumption}} = \sum_{t=1}^{T} \sum_{S_{\text{nonshiftable}}=1}^{N_{\text{nonshiftable}}} \wp \times X^{\text{app}}(t). \tag{3}$$

3.2. Electricity Cost

The electricity cost reduction is defined as the minimum charges on consumed loads issued to the consumers by the utility. For the electricity cost minimization problem, the shiftable and non shiftable loads are considered, and it is derived using Equation (4):

$$\text{Minimize} \sum_{a=1}^{N} \sum_{t=1}^{T} (X_{a,t}^{\text{app}}(t) \times \wp \times E_{a,t}^{\text{Price}}), \tag{4}$$

where $X_{a,t}^{\text{app}}(t)$ denotes the state of appliances as OFF or ON (0 = OFF and 1 = ON) and $E_{a,t}^{\text{Price}}$ denotes the price at any time interval t for the consumed electric energy. t is the index of time that has the upper limit of $T(T = 24)$ h of a day and a is the index of the total number of household appliances.

3.3. Energy Consumption

The proposed HEMS is designed to shift loads from on-peak to off-peak hours in a stable manner. This shifting depends upon the variation of demand over specific hours and is inversely proportional to the electricity market price. Mathematically, it is computed in Equation (5):

$$P^{consumption} = \sum_{t=1}^{T} \sum_{S=1}^{N} \wp \times X^{app}(t), \tag{5}$$

where $P^{consumption}$ denotes

energy consumption for the shiftable and non shiftable loads. N denotes the number of Sth household appliances and T denotes the tth time slots. For the optimization model, the load is classified according to the operation of household appliances and the behavior of consumers. Table 2 provides details of load categorization.

3.4. Load Balancing

The grid stability is important to ensure sustainability and reliability of the grid management and operations. Reduction in the PAR helps utility to retain the stability and ultimately leads to the reduction in electricity cost. It is mathematically calculated using Equation (6):

$$PAR = \frac{max(P^{consumption})2}{avg(P^{consumption})2}, \tag{6}$$

where $P^{consumption}$ denotes the list of hourly load calculated using Equation (5).

3.5. Objective Function

The overall objective function is expressed as a multi-objective optimization function to minimize electricity cost with reasonable energy consumption from the power grid. This also minimizes the frustration at consumer end. In addition, incorporating the RES is useful to reduce the greenhouse gas emission. The objective function is modeled as minimization of Equations (4) and (5), as well the waiting time.

3.6. Electricity Price Models

Presently, most of the smart households have the advanced metering infrastructure (AMI) installed which allows bidirectional communication with the utility. The utility uses information regarding consumption from AMI for efficient management of energy resources in order to maintain demand and supply. The utility provides strategies that regulate energy consumption of consumers through different pricing schemes which are essential for DR implementation.

Several electricity pricing schemes have been proposed by the utility. However, ToU and CPP are the focus of this paper. The CPP pricing scheme is commonly used by commercial and industrial sector to reduce peak loads, especially in an event-based situation [27]. In this pricing scheme, consumers are charged with higher electricity price during peak hours especially for winter and summer seasons, and for the power system emergency conditions. On the other hand, consumers are charged with lower electricity prices during other periods of the year.

In the ToU pricing scheme, the pricing rate is divided into scheduling time horizons such as on-peak, mid-peak and off-peak time slots [28]. The on-peak time slots receive the highest electricity price as compared to the off-peak time slots, whereas the mid-peak time slots receive an electricity price that falls within the on-peak and off-peak time slots.

Today, moving load to off-peak from on-peak time slots is most effective for the ToU pricing scheme as compared to the flat rate pricing scheme. In addition, this scheme is simple to implement and it is not controlled by different cost conditions. In addition, the scheme encourages the use of RES and ESS especially when electricity prices are high during on-peak and low during off-peak time slots. Furthermore, ToU provides extreme peak reduction through the following mechanism, finds average electricity price during intense peak time slots or average electricity price when the number of higher prices is minimal [26].

3.7. RES

If a system already had a large share of PV energy and more PV is added to the system, then the additional increment in the renewable energy penetration will have impacts on the system from hour to hours. This additional PV and other RES may make the system complex and may require grid stabilization.

The renewable energy assumes a vital part in reducing the greenhouse effect. At the point, when RES is utilized, the request for fossilized energy is reduced. Not like the non-biomass and fossil fuels, the renewable sources of energy (solar, wind, geothermal, and hydro-power) do not immediately generate greenhouse gases. Frequently used RES are the PV, hydroelectric, and wind turbine. Over time, the RES has experienced large acceptance being compatible with other energy sources like the coal and lignite, however, behind the natural gas. In previous years, the world renewable energy share is calculated to indicate high percentage usage in hydroelectric and PV [29]. However, wind turbine and PV are the most encouraging energy generation and they are still very recent. However, the two encouraging RES have variate requirements. On the other hand, PV is widely used in most residential and industrial areas for making it the most promising energy generation.

Generally, RES has inherent variability and uncertainties that will affect the total energy production planning which is common among renewable energy and distributed energy [30]. Nevertheless, the deployment of RES has brought extensive changes to the present electrical power grid system [31], for making it a reliable and consumer-oriented system. This deployment of RES exhibits uncertainty that depend upon the share of input. For example, the PV incorporated into a system with only a small share of the PV energy. The system will only respond in the same manner during all hours of a day as well as all hours throughout the year. The technical impact of this incorporation is easy to understand in terms of cost saving on a monthly and annual basis [32]. Moreover, the renewable energy input does not impact challenges to the entire operation and balancing of the power grid.

This paper considers battery capacity of 200 kW per 4 h for the RES. Obviously, it will provide efficiency to the HEMS. ESS is an example of interruptible load shifted to any time slots of a day (i.e., SOC). The maximum charging must be greater than or equal to the SOC given in Equation (7):

$$ESS_t^{charging} \leq ESS^{max}, \tag{7}$$

$$ESS_t^{charging} < ESS^{limit}, \tag{8}$$

$$ESS_t^{discharging} \geq ESS^{min}. \tag{9}$$

The maximum and minimum electricity storage level is 90% and 10%, respectively [10]. The ESS is charged based on the electricity price. Equation (8) shows that the SOC of ESS is less than the ESS limit. ESS will discharge when the electricity rate is high, and if ESS has more stored energy than the expected ESS minimum level given in Equation (9). The ESS stored electricity at time t is shown in Equations (7)–(9). There are storage losses due to the charging and discharging effect. To detect the limitation, the efficiency of the battery is computed in Equation (10):

$$B^{eff} = B^{eff}(t-1) + j \times \eta^{ESS} \times ESS_t^{charging} - \frac{j.ESS_t^{discharging}}{\eta^{ESS}}, \tag{10}$$

where B^{eff} has a storage capacity of kWh at t. The time slots are denoted as j, and η^{ESS} denotes the storage efficiency rate.

4. Proposed Schemes

To achieve reliable management of energy and systematic operations of smart grid via DSM, a scheme of a power system that has a smart building with a number of residential households and a single utility is presented. The consumers' electricity demands are satisfied via power grid or ESS. All consumers within the residential household have distinct electricity consumption behavior, which is related to the different load profiles. Smart meters are installed in all of the residential household for easy calculation of the electricity consumption. The smart meters ensure bidirectional communication between each residential households and utility for price sharing as well as the quantity of electricity needed to meet the load demands. We divide a day into 24 h operational time slots, for each time slot denotes one hour. We further categorize smart household appliances into shiftable and non shiftable operational mode based upon their functions.

Figure 1 shows the graphical representation of the proposed scheme that is used as the basis for developing the optimization scheme. It comprises of RES and integrated power utility that is required to meet the residential loads. The individual power grid and RES act as a node. The delivery of energy to the residential household and the ESS that are utilized during high load demands that are controlled by the optimization scheme. The residential load demand is satisfied from the power grid, where direct utilization of RES and ESS depend on the electricity price in that particular hour. However, RES and ESS are instantly used to deliver energy to the residential household. In this manner, energy demand concentrated on the power grid is drastically reduced. In addition, incorporating the RES, ESS and HEMS are beneficent at reducing the heavy load on the power grid with respect to high demand.

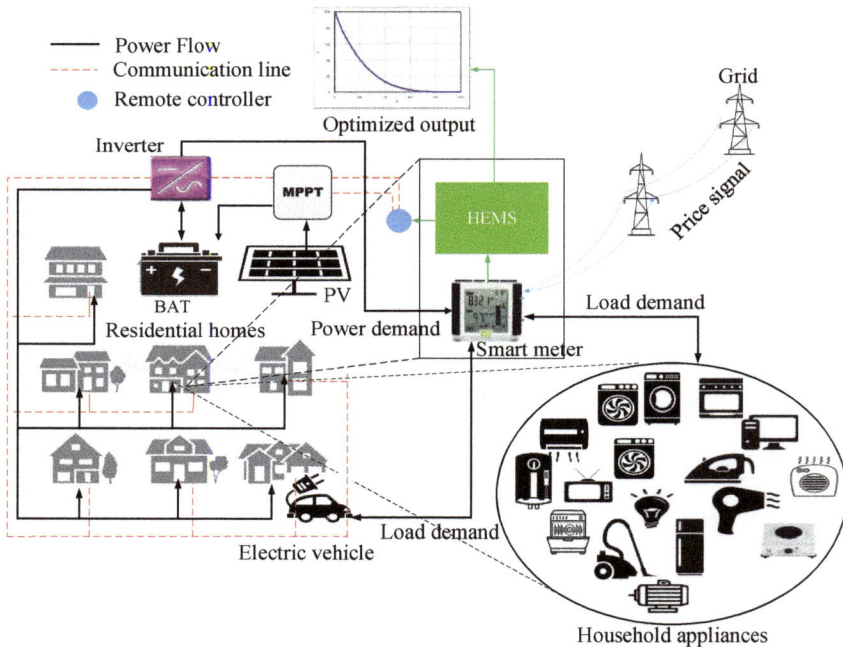

Figure 1. Propose system model.

Figure 2 shows the proposed Earliglow based HEMS architecture. The electricity price signals (CPP and ToU) are obtained from the utility, which are used for electricity billing. The entire architecture is divided into two phases. Phase 1 estimates the output energy of PV that depends on several temperatures and the effect of irradiation on the PV modular. The generated energy of PV is not solely dependent on irradiation but on its surrounding temperatures. A maximum power point tracker (MPPT) is mostly used on the PV array to achieve energy output at any irradiation level. Due to cloud cover and other isolation constraints, we take a random variable of irradiation. The different irradiations and temperatures are used to calculate the maximum output energy of PV using the equation in [33].

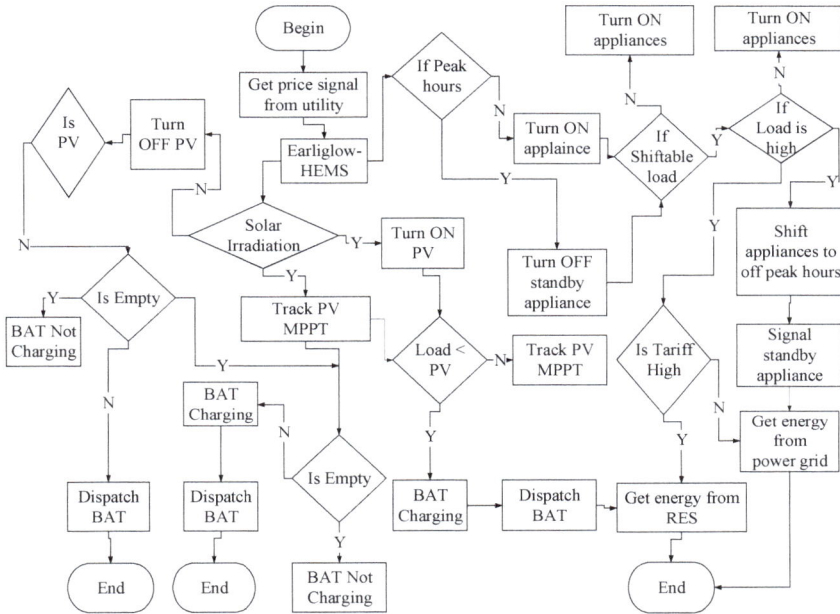

Figure 2. Proposed Earliglow based home energy management system (HEMS) architecture.

Once the irradiation and temperature are calculated for each hour, the PV is Turn ON. If the maximum energy output of PV is greater than the load demand of the household, thus the battery starts charging. On the other hand, if the household load is greater than the output energy of PV and the RES output is not enough to supply the household load, then the battery starts discharging. The inverter is used as the electric circuitry to convert direct current to alternative current for giving energy to the residential household. The voltage input/output, frequency and the overall energy handling depend on the system design. However, the output from the inverter is less than its input. The battery as earlier described is used for energy storage, which also enhances the energy supply to the residential household.

Phase 2: the actual load scheduling is performed. In this phase, the proposed Earliglow algorithm is implemented as the optimization scheme. The ideal solution derived after a series of iterative steps is converted into a binary format (0 and 1) which forms the decision variables (Section 4.1). A 24-h scheduling plan is performed with a one hour time interval. The time periods are divided into on-peak (12–8) p.m. and off-peak (12–8) a.m. hours for the ToU signal, while we consider the on-peak hours as the critical event associated with the CPP price signal. The electricity cost, consumption and load balancing are calculated using Equations (4)–(6). The household appliances are sorted in descending order of their load consumption per hour. This is done in order to compare the next load from the rest of the loads. Let t_i denotes the event at time i and t_j be the event following t_i at time j. If an appliance with maximum load wants to be shifted within t_i and t_j, then swap between t_i and t_j, the appliance with maximum load is shifted to the new swapped event as presented in Algorithm 1.

Algorithm 1: Proposed optimal scheduling technique [26]

A: set of appliances; /* *app*: number of appliances in *A*, *t*: time of scheduling event *T**/;
t= 0; /*Set appliance initial scheduling time to 0*/;
foreach *time slots* $t_i \in T$ **do**

  Determine the initial energy consumption $P^{\text{consumption}}$; /*Energy consumption must not be
    zero*/;
  Calculate the electricity cost, consumption and PAR using Equations (4)–(6);
  Sort household appliance in descending order of their energy consumption; /*appliance is
    now sorted */;
  a=*A* ; /* *a* is set equal to the sorted set of *A**/;
  for *a=1 to app* **do**

    pointer= TRUE;
    for *i=1 to app-1* **do**

      $j = i + 1$; /* t_j: event following t_i in *T* */;
      For all maximum energy consumption between t_i and t_j; /*analyze in 2 phases*/;
      if t_i *is not current peak hours* **then**
        | Set t_i as not current peak hours; /* current peak hours: either ToU or CPP*/
      end
      if t_j *is not current peak hours* **then**
        | Set t_j as not current peak hours; /* current peak hours: either ToU or CPP */
      end
      Compute energy consumption of *A* at $t(i,j)$ and $t(j,i)$; /* compute energy
        consumption at distinct time*/;
      Pick appliance with the maximum consumption load;
    end
    if *appliance with the maximum consumption is in need of shifting between* t_i *and* t_j **then**

      exchange (t_i, t_j);
      $T = T + T(t_j)$; /*scheduling time of t_j in current time slot */;
      pointer=FALSE;
    else
      | $T = T + T(t_i)$; /* scheduling time of t_i in current time slot*/
    end
  end
  if *pointer=TRUE* **then**

    Stop;
    No shifting; /*so all time slots are in order*/;
    a is already shifted; return time slot *t*;
  end
end

Figure 3 provides an illustration of the household load shifting activity. The change in residential load demands is characterized by the size of load reduction (S), duration of load reduction Δt_1 as well as duration of the load recovery, which is either greater or smaller than the load recovery (W). Subsequently, the load losses between S and load recovery (K) give rise to the load demand in the recovery time which may vary over load reduction. In a situation when load losses are not experienced (i.e., ($K = 1$)), the extra amount of power consumed in the recovery time will be equal to the energy consumed during the load reduction time. The disparity between (Δt_1) and (Δt_2) show that the recovery time does not correlate with the reduction time.

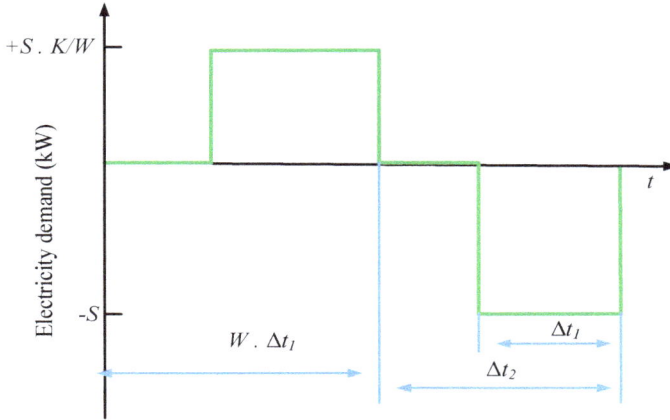

Figure 3. Illustration of a load shifting operation by a household appliance. S is the size of load reduction; K is the load recovery; W is the duration of K, greater or smaller than K; t: time.

Algorithm 1 presents the optimal scheduling technique. The execution steps of the proposed optimal scheduling techniques depend on the complexity at lines (8–24) in the for-loop. The complexities of other lines (i.e., 10–22) are more than these lines. However, complexity of line 8 is influenced by line 10. All of the sub-lines 10–22 take $O(a)$, while line 10 can iterate maximally with a iterations, and hence its cost is $O(a)$. Lines 8–24 perform at the maximum iteration cost of $O(app^2)$. Then, the overall T process takes the cost of $O(T^3)$.

4.1. Jaya Algorithm

In the Jaya algorithm [34], the objective function is presented as V_t and is minimized at any iteration t. Suppose that there is a k number of decision variables $(j = 1, 2, \ldots, k)$ and the number of the individual solution is presented as l. The individual with the ideal value is presented as V_t^* in the whole individual solution. While the individual with the worse value is presented as V_t^{**} in the whole individual solution. If in the jth variable we obtain the $V_{j,l,t}$ value for the lth individual during the tth iteration, then we present a modified value using Equation (11):

$$V'_{j,l,t} = V_{j,l,t} + r_{1,j,l}(V_{j,t}^* - |V_{j,l,t}|) - r_{2,j,l}(V_{j,t}^{**} - |V_{j,l,t}|),\tag{11}$$

where $V_{j,t}^*$, is the value of variable l for the ideal individual and $V_{j,t}^{**}$ is value of the variable l for the worst individual. $V'_{j,l,t}$ is the updated value of $V_{j,l,t}$, where $r_{1,j,l}$ and $r_{2,j,l}$ are two random numbers for jth variable during tth iteration in the range of $[0, 1]$. The term $r_{1,j,l}(V_{j,t}^* - |V_{j,l,t}|)$ indicates the tendency of solution to move towards the ideal solution, whereas the term $r_{2,j,l}(V_{j,t}^{**} - |V_{j,l,t}|)$ indicates the tendency of the solution to avoid the worst solution. $V'_{j,l,t}$ is selected if it gives the ideal function value. All selected function values at each iteration are preserved and used as the input for the next iteration. Parameters used for simulation are shown in Table 3. Algorithm 2 illustrates the proposed Jaya based HEMS.

Table 3. Algorithms parameters. *Max* and *Min* denote the maximum and minimum population bound.

Jaya-Parameter	Value	SBA-Parameter	Value	EDE-Parameter	Value
Population size	30	Population size	30	Population size	30
No. decision variables	16	No. decision variables	16	No. decision variables	16
Min	0.1	Min	0	Min	0.1
Max	0.9	Max	1	Max	0.9
		No. of runner	50	Scaling factor	0.5
		Length of root	5	Crossover probabilities	0.3, 0.6, 0.9
Maximum Iteration	100	Maximum Iteration	100	Maximum Iteration	100

Algorithm 2: Proposed Jaya based HEMS [12]

Initialize Min, Max, and termination criterion *MaxIter*;
iter = 0;
Generate a random population;
while *iter* < *MaxIter* **do**
    | *iter* = *iter* + 1;
    | Evaluate the objective value of each one of the population;
    | Get new solution;
    | Find the local search;
    | Begin mutation strategy;
    | End mutation strategy;
    | Accept new solution if it gives best objective value;
end
Perform schduling of household using the Algorithm 1;

4.2. SBA

Several plant intelligence-based inspired optimization algorithms have been proposed in [35]. The strawberry plant is generally known for its characteristic aroma, bright red color, juicy texture and its sweetness. This plant propagates via a runner in order to survive. If the plant found itself in a good spot on the ground, with plenty of soil nutrients, water and sunlight. Then, the plant produces several short runners, which reproduces a new strawberry plant and occupies the neighborhood as much as possible. On the contrary, if the strawberry plant is not on the spot of a good ground (i.e., there is no sunlight, poor nutrient and water), it will try to survive by sending fewer smaller runners to further exploit the far neighborhood, hence finding a better spot for its offspring. Since sending a longer runner is a huge investment for the plant that is in a poor spot, we therefore assumed that the quality of the spot is reflected on the growth of the plant; then, we build our optimization based on these notions: the strawberry that falls on a good spot of the ground propagates well by reproducing many short runners. Those that fall on the poor spot tends to reproduce a few long runners. The parameters of SBA are shown in Table 3. Algorithm 3 describes the entire operations for the SBA. The $f(x) \in [0,1]$ is the fitness function. Therefore, the distance between each runner and the number of the runners is computed using Equation (12):

$$N(x) = \frac{1}{2}(\tanh(4 \times f(x) - 2) + 1) \tag{12}$$

by default, the number of runner is proportional to its fitness and is computed using Equation (12):

$$n_i = [K, N_i r], \tag{13}$$

where K is the maximum number of runner and n_i denotes the number of the runners generated by the solution at iteration i after sorting. $N_i r$ mapped the fitness to its solution i, where $r \in [0,1]$

denotes the number generated randomly for each candidate in each generation. To ensure all solutions, generate a runner for the best candidate. The $f(x) \equiv 0$, fitness must generate at least K maximum runner. The distance of each runner is inversely proportional to its growth and it is computed using Equation (14):

$$d_{i,j} = 2(1 - N_i) \times (r - 0.5),\qquad(14)$$

where j denotes the size of search space. Every $d_{i,j}$ is in the range [0,1]. $d_{i,j}$ determines the growth of runner. The computed distance will be used for updating the i candidate solutions, which depends on the limit of x_j^{T} given in Equation (15):

$$x_j^{\mathsf{T}} = x_j + (s_j - q_j) \times d_{i,j}.\qquad(15)$$

The x_j^{T} ensures that the solution lies within the limit $[s_j, q_j]$, where s_j, q_j denote the upper and lower bound, respectively.

Algorithm 3: SBA based HEMS

Objective function $f(x), x \in [0,1]$ generate a population $P\{p_1, \ldots, p_m\}, g = 1, MaxIter :$ maximum iteration, $drunner :$ number of runner, $droot :$ length of root ;

for *g=1 to MaxIter* **do**

  compute$\{N_i = f(\{p_i\}), \forall p_i \in P\}$ sort P in descending order of N

   create a new population P_{new};

  foreach *{$p_i, i = 1 \ldots, m\}$* **do**

    Best m only;

    drunner and *droot* are proportional to the fitness N_i ;

    $P_{\text{new}} = P_{\text{new}} \cup drunner$;

    Append to population, death occurs by omission above ;

  end

  $p = P_{\text{new}}$; {new population}

  Return p, the population solutions

   Perform scheduling of household load using the Algorithm 1;

end

4.3. Earliglow Algorithm

Every meta-heuristic optimization algorithms have their advantages and disadvantages. Several algorithms have a parameter tuning problem that can be resolved by trial and error. Generally, the smaller the number of tuning parameters, the more superior the algorithm is. Other limitations include an accuracy problem and high sensitivity to the initial guess while finding the global solution using low probability. In addition, achieving solutions outside the region defined by boundary values of variables. However, some algorithms may be efficient in solving one problem and yet not suitable for another. SBA has the following limitations. Firstly, SBA duplicates every number of computational individual for each iteration. Lastly, local and global searches are done simultaneously where each computational individual is subjected to large and small movement from the start to end. In addition, Jaya algorithm falls into local minimum as the number of iteration increases. To ensure the better performance of both algorithms, a new algorithm known as Earliglow which has both flavors of the existing techniques, SBA and Jaya algorithm is proposed. Based on our assumption, the named Earliglow is chosen due to its features to ripen its fruits sooner (accuracy) than other strawberry variants and is also resistant to many known strawberry diseases (fast convergence). The detailed procedures of Earliglow are illustrated in Figure 4.

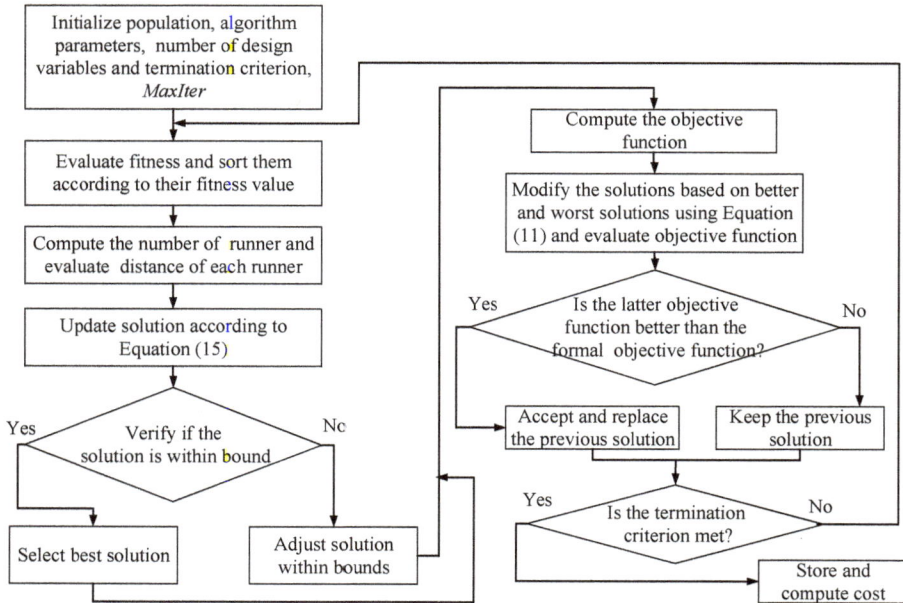

Figure 4. Proposed Earliglow algorithm.

In the Earliglow algorithm, SBA algorithm is performed first on the random population using Equations (12)–(15) due to the problem of adjusting the individual solution within the population bounds, high memory usage, and since SBA duplicates every number of computational individual for each iteration. The Jaya algorithm is implemented to reduce the number of iterations. If the solution falls outside the given range of the defined bounds, the solution is modified on the basis of ideal and worse individual solution. For this purpose, a mutation policy is adopted as well as trimming of the population using Equation (16). This policy updates individual solutions and ensures that the solution moves towards achieving the ideal solution while circumventing the worse solution. In addition, the number of iterations is further reduced leading to a global solution within a minimal execution time:

$$
\begin{aligned}
x(x(:,i) < \mathrm{mini}(i), i) &= \mathrm{mini}(i), \\
x(x(:,i) > \mathrm{maxi}(i), i) &= \mathrm{maxi}(i),
\end{aligned}
\tag{16}
$$

where mini and maxi are the minimum and maximum population bound of the ith individual population, x.

4.4. Enhanced Differential Evolution Algorithm (EDE)

The differential evolution (DE) algorithm was first introduced by Storn and Price in 1995 [36], since then it has become the state-of-the-art algorithm for solving most optimization problems. This algorithm uses the concept of mutation, selection and crossover. Adjusting the common known parameters like population size, mutation, scaling factor and crossover rate. In this paper, we implement the enhanced version of DE where the performance is enhanced by adjustment of population size and scaling factor. The crossover rate is implicitly used during the crossover process. The parameters of EDE are shown in Table 3.

5. Simulations and Discussions

In order to evaluate the performance of the proposed appliance scheduling scheme, we simulate the hourly energy use of the set of household appliances. The parameters in terms of length of operation time (LOT) and power ratings of appliance are presented in Table 2. Simulations are performed in three cases: (I) without HEM, (II) with HEM, and (III) HEM with RES. We also discussed the hourly load behavior of household appliances and finally the feasible regions (FR) in terms of electricity cost, load and waiting time. Peak load shifting is achieved by load adjustment of appliances with feasible schedules. Shiftable appliance usage was systematically changed by turning them ON and OFF.

In this paper, consumers are charged based on the ToU and CPP pricing schemes. The ToU provides varying price structure for off-peak, mid peak and on-peak; it also provides an incentive to consumers who engage in the DR program for shifting their load to off-peak hours. In this pricing scheme, different time periods have different electricity charging rates. On the other hand, the CPP pricing scheme creates different price structure for different seasons and events of a year. The electricity cost is charged for this period with a specified rate given by the utility. Figure 5a presents the ToU and CPP signals received from the utility and send to consumers.

5.1. Case I (without HEM)

The total electricity cost of without HEM using CPP and ToU is presented in Figure 5b, whereas the hourly electricity cost of consumers without HEM is illustrated in Figure 5c,d respectively.

The consumers have no HEMS architecture in their households and thus take electricity from the power grid when required. Figure 5e,f show the different CPP and ToU pricing scheme hourly appliance consumption of energy from the main grid.

5.2. Case II (with HEM)

Four types of HEM schemes are presented to the smart consumers. The HEM scheme based on Earliglow allows few appliances to operate during on-peak hours. The achievement of Earliglow based HEM scheme is allowing optimal household appliance scheduling in the 24-h time period as presented in Figure 5e,f, respectively. Furthermore, the electricity cost paid to utility against this consumption is presented in Figure 5c,d, respectively. It is proven from these figures that HEMS not only shift load from on-peak to off-peak hours; however, find and shift load to the respective hours when electricity cost is minimal.

The smart consumers that implement HEM scheme based on Jaya, EDE and SBA utilize the energy optimally and stabilize the household load by moving loads from on-peak to off-peak hours by considering the different consumers' setting and constraints. The behavior of load consumption after participating in the DR program is presented in Figure 5e,f.

Likewise, Figure 5c–f show the load in kWh and electricity cost in cents/h using the proposed four algorithms. The load has been scheduled from h_{8-14} to off-peak h_{18-10} for CPP, whereas the load has been scheduled from h_{8-14} to off-peak h_{16-10} for ToU. Thus, it minimizes the electricity cost and intense peak created because of load shifting where consumers depend on grid energy (Figure 6e).

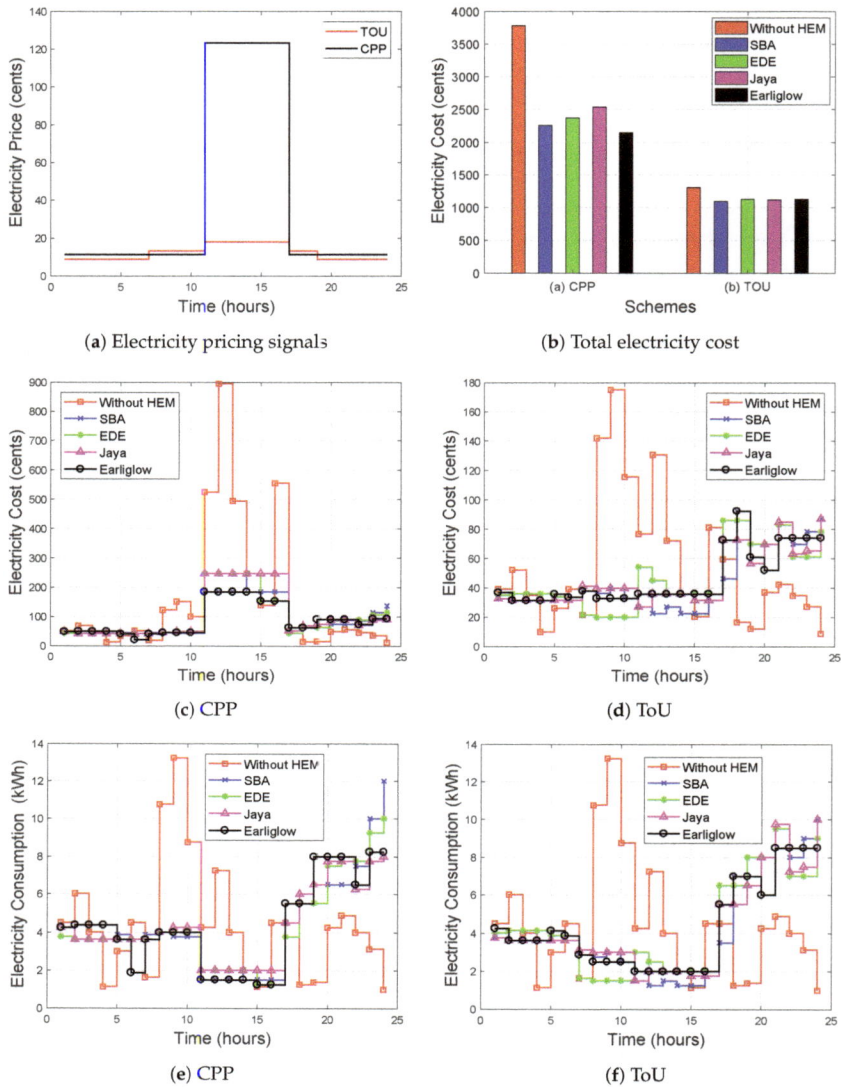

(a) Electricity pricing signals

(b) Total electricity cost

(c) CPP

(d) ToU

(e) CPP

(f) ToU

Figure 5. (a) electricity pricing signals; (b) total electricity cost; (c) hourly electricity cost using critical peak price (CPP); (d) hourly electricity cost using time-of-use (ToU); (e) hourly electricity consumption without renewable energy sources (RES) using CPP; (f) hourly electricity consumption without RES using ToU.

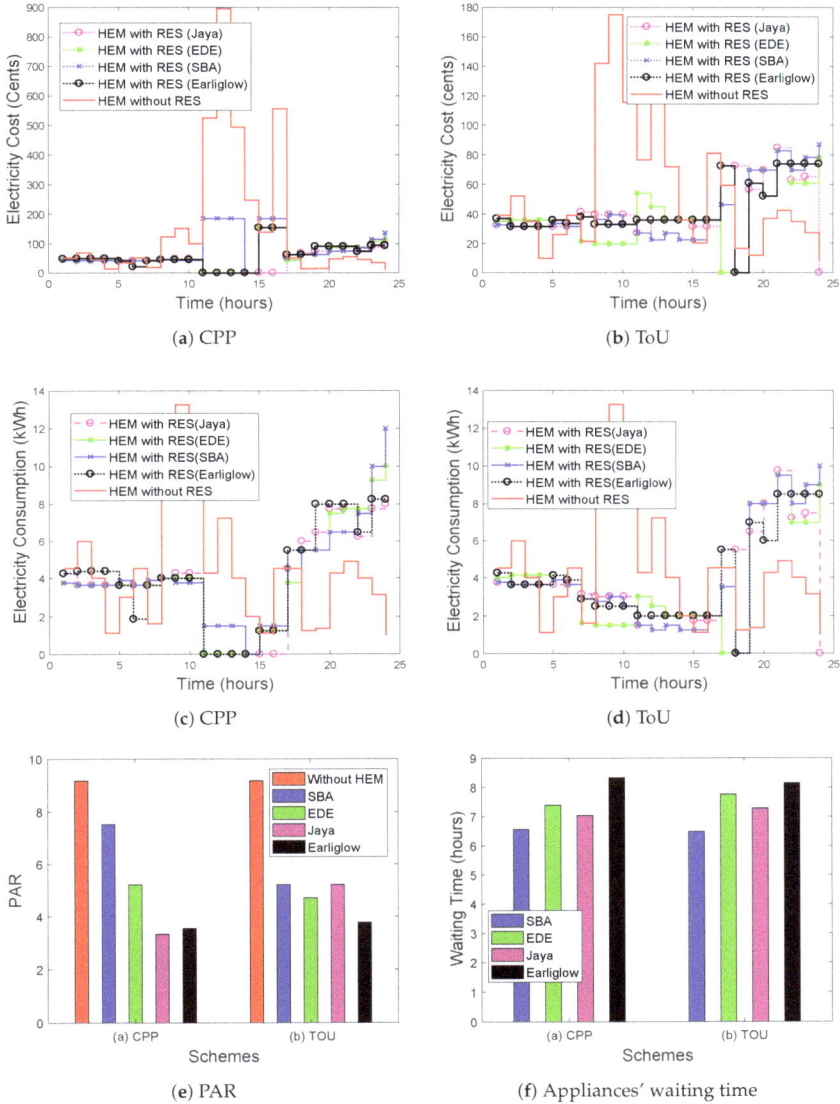

Figure 6. (**a**) hourly electricity cost with RES using CPP; (**b**) hourly electricity cost with RES using ToU; (**c**) hourly electricity consumption with RES using CPP; (**d**) hourly electricity consumption with RES using ToU; (**e**) PAR; (**f**) appliances' waiting time.

After achieving this improvement, the consumer is encouraged to engage in the effective management of energy through their household appliances scheduling. The total daily cost savings with and without HEM are presented in Table 4. Similarly, the total daily energy savings with and without HEM are also presented in Table 5.

In fact, Tables 4 and 5 present the comparisons of the schemes used in this paper. Distinctly, the proposed scheme is efficient in the management of the household load. In addition, the penetration of RES improves it practicality for smart consumers.

Table 4. Summary of the results for electricity cost in (cents).

Technique	Electricity Cost		Electricity Cost Savings (%)		Computation Time (s)	
Pricing scheme	CPP	ToU	CPP	ToU	CPP	ToU
without HEM	3787.00	1309.10	-	-	-	-
Earliglow without ESS	2149.00	1128.00	43.25	13.83	0.1185	0.1202
Earliglow with ESS	1408.60	1035.60	62.80	20.89	0.1185	0.1202
Jaya without ESS	2541.00	1119.90	32.90	14.45	0.0755	0.0806
Jaya with ESS	1060.20	1032.90	72.00	21.09	0.0755	0.0806
SBA without ESS	2261.00	1092.20	38.07	18.21	0.8673	0.3709
SBA with ESS	2014.20	999.79	45.41	25.35	0.8673	0.3709
EDE without ESS	2373.00	1125.10	37.33	16.49	2.2728	0.3803
EDE with ESS	1385.80	953.51	56.88	24.63	2.2728	0.3803

Table 5. Summary of the results for energy consumption in (kWh).

Technique	Energy Consumption		Energy Consumption Savings (%)		Computation Time (s)	
Pricing scheme	CPP	ToU	CPP	ToU	CPP	ToU
without HEM	105.00	105.00	-	-	-	-
Earliglow without ESS	105.00	105.00	-	-	0.1185	0.1202
Earliglow with ESS	98.00	98 00	6.66	6.66	0.1185	0.1202
Jaya without ESS	105.00	105.00	-	-	0.0755	0.0806
Jaya with ESS	93.00	95 00	11.42	9.52	0.0755	0.0806
SBA without ESS	105.00	105.00	-	-	0.8673	0.3709
SBA with ESS	103.00	98 00	1.90	6.66	0.8673	0.3709
EDE without ESS	105.00	105.00	-	-	2.2728	0.3803
EDE with ESS	97.00	92 00	7.61	12.38	2.2728	0.3803

5.3. Case III (HEM with RES)

In this case, the consumer employs the ToU and CPP pricing scheme and further uses the stored RES energy optimally to reduce electricity cost. The household considered in this scenario has the combination of HEM (Earliglow, Jaya, SBA and EDE) and RES generation with a storage system. The HEMS utilises the RES stored energy wherever the electricity cost of the utility is enormous and move the load from the power grid to the RES energy and therefore decreases the cost of electricity by a meaningful amount.

The achievement of all HEM schemes towards the optimal consumption of power grid depends on stored RES which is based upon the PV energy and the total area of the PV array. This means that energy can be used directly during peak hours; otherwise, it can be stored in batteries.

Huge peaks during the off-peak hours have been prevented by using the RES stored energy. During the on-peak hours, the consumers do not fully depend on the energy from the power grid and decide to use RES stored energy. In this manner, the cost of electricity and the huge peaks are drastically minimized. In addition, it will provide grid stability (Figure 6e).

5.4. Performance Trade-Off Made by Optimization Schemes

HEMS allows consumers to optimally shift their appliances from on-peak to off-peak hours. Thus, the trade-off between cost of electricity and waiting time exists because of load shifting. The HEMS estimates the waiting time for appliances and utilizes the energy from the power grid efficiently. This enables consumer to pay a lower electricity cost and also gets the best satisfaction. Moreover, the consumer is encouraged through a reduced electric billing. Likewise, the HEMS scheduled load in order that delay is created within the operable limit. Hence, a minimum waiting time during scheduling is accepted.

Apart from RES, smart consumers are allowed to use energy optimally from the power grid and also the RES generation energy at their disposal. This class of smart consumers has HEMS that allows them to consume electricity at a low cost. The pricing schemes enable smart consumers to optimally utilize grid energy and also energy stored from RES. In conclusion, the comparisons are taken based on cost, cost savings, consumption and consumption savings.

In Table 4, consumers without HEM pay high electricity bills for the same energy consumed by the consumers with HEM; thus, the consumer with HEM get the maximum benefit from the pricing schemes. Meanwhile, the waiting time of the consumers without HEM is little bit higher; however, they pay the maximum cost of electricity to the utility using the proposed pricing schemes. Thus, there is a trade-off between consumer's waiting time and the cost of electricity. On the other hand, consumers with HEM pay a minimal electricity bill as compared to consumers without HEM.

The load scheduling behavior of an Earliglow based HEM scheme is smooth without creating peaks during on-peak hours. This enhances the household appliance's operations. As shown in Figure 5e,f, the Earliglow scheduled load in a more advanced manner and maintains the completion of the appliance's operation. Meanwhile, the other algorithms (Jaya, EDE and SBA) demonstrate the load movement to off-peak hours efficiently.

From Figure 5e,f, it is shown that consumers having Jaya, EDE and SBA schemes may not be able to utilize the energy stored optimally as compared to Earliglow because of high load peaks (Figure 6e).

In conclusion, the results from the simulation confirmed the performance of Earliglow based HEM over the other HEM (Jaya, EDE and SBA) in terms of peak reduction, electricity (cost and savings), using CPP, consumption (load and savings) using the ToU scheme and computational time is presented in Tables 4 and 5.

5.5. Hourly Load Behavior of Household Appliances

Figure 7 shows the consumers without HEM for hourly household appliance behavior. The figure demonstrates that the majority of the consumer household appliances are concentrated within the on-peak hours. The consumers then implement the proposed HEMS to optimally shift load from on-peak to off-peak hours as presented in Figure 8 to illustrate the individual household appliance's energy consumption behavior for each hour with respect to the two pricing schemes.

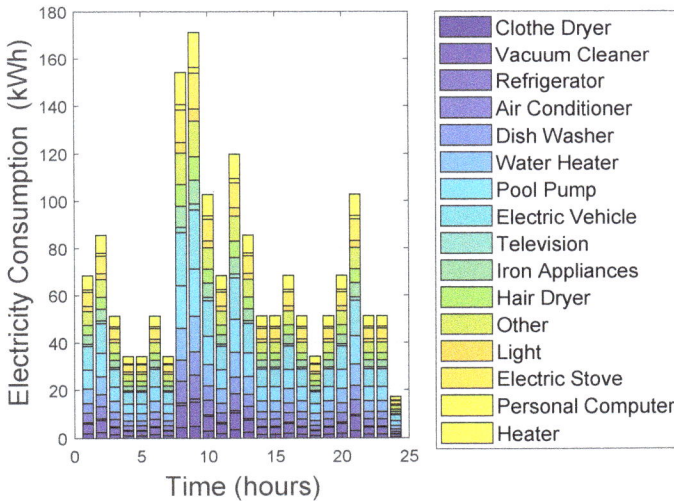

Figure 7. Hourly electricity consumption for each appliance without HEM.

Using the CPP scheme: in Figure 8a, Jaya based HEMS shift more loads to the first h_{10} and h_{19-24} than the other hours. Thus, the loads shifting during these hours consist of high energy consuming appliances. In Figure 8b, most appliances are shifted to the first h_{10} and h_{17-24} of the day for SBA based HEMS. These are high load profile appliances like pool pump, hair dryer, iron appliance, electric stove, water heater, dishwasher and air conditioner. Finally, in Figure 8c, most appliances are shifted to the first h_{1-10} and h_{15-24} in a day for Earliglow based HEMS than the other hours.

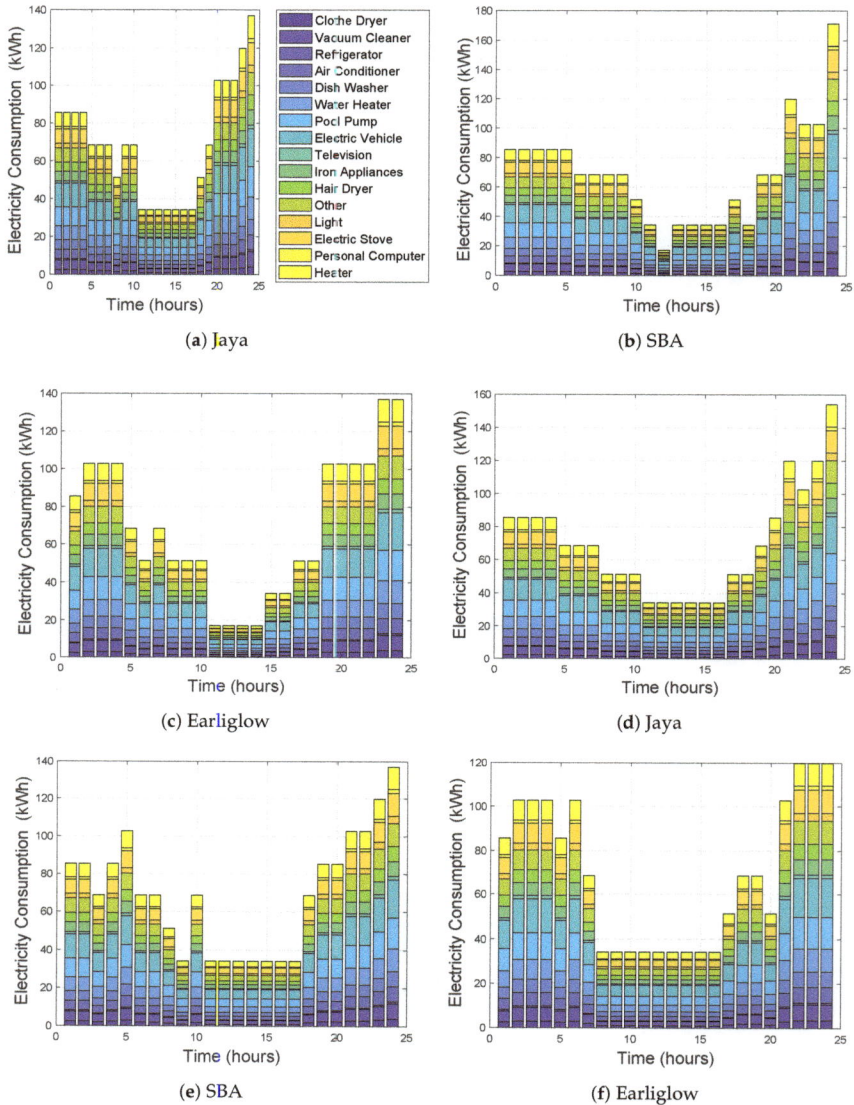

Figure 8. Hourly electricity consumption for each appliance with HEM. (**a**) hourly electricity consumption for each appliance with HEM for Jaya using CPP; (**b**) hourly electricity consumption for each appliance with HEM for SBA using CPP; (**c**) hourly electricity consumption for each appliance with HEM for Earliglow using CPP; (**d**) hourly electricity consumption for each appliance with HEM for Jaya using ToU; (**e**) hourly electricity consumption for each appliance with HEM for SBA using ToU; (**f**) hourly electricity consumption for each appliance with HEM for Earliglow using ToU.

Using the ToU scheme: in Figure 8d, more appliances are shifted to the h_{2-10} and h_{21-24} of the day for Jaya based HEMS than the rest hours. From these results, most of the high energy consuming loads are moved to off-peak from on-peak hours. This reduces the electricity cost and may generate peaks. While SBA based HEMS in Figure 8e, SBA based HEMS shifts most of its appliances to the first h_{1-6}

and h_{20-24} with high impact on the h_{23-24}. Finally, Figure 8f shows that Earliglow based HEMS shifts most of its appliances to the first h_{1-7} and h_{17-24} of the day, especially for high load profile appliances like air conditioner, dishwasher, pool pump, iron appliances, electric stove and electric vehicle.

5.6. FR for Electricity Cost and Energy Load

A search space is the set of all possible points which satisfy the objective constraints, and it is known as the FR [37]. In this paper, ToU price signal for all households falls within (8.70, 2317.40) cents. Similarly, ToU pricing scheme hourly energy consumption for all households in all cases falls within (1.00, 13.25) kW. In CPP, the price signal ranges from (11.40, 11,854.00) cents for all household appliances and the without HEM scheme ranges from (1.00, 13.25) kW for all the possible cases. As shown in Figure 9, the electricity cost in the FR for each pricing scheme must be less than or equal to the maximum without HEM hourly electricity cost of 2317.40 and 11,845 cents, respectively. The following are the formulated possible cases:

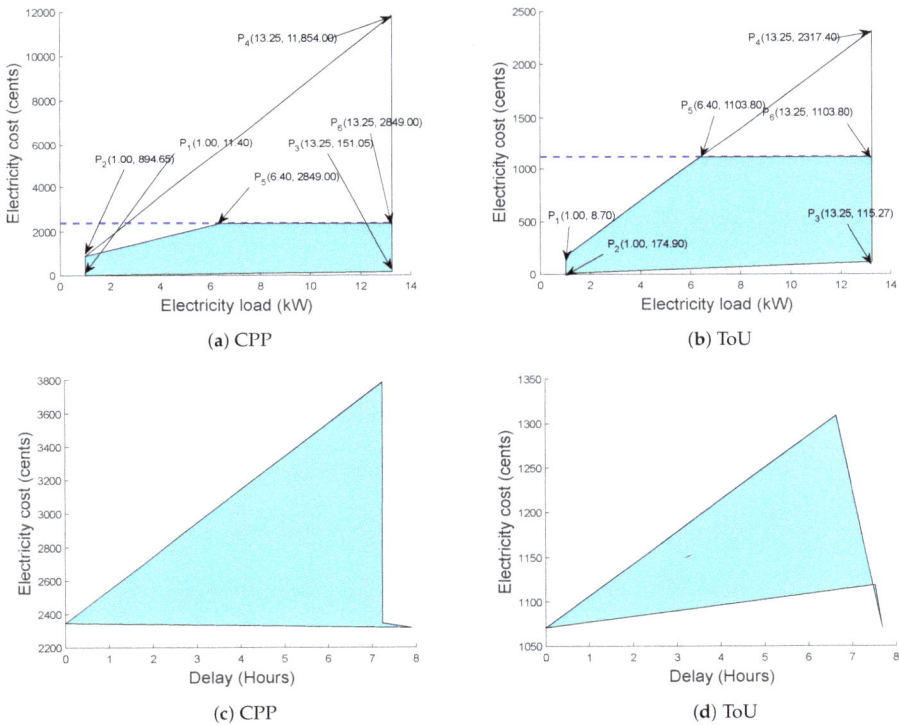

Figure 9. (a) FR for cost and electricity consumption using CPP; (b) FR for cost and electricity consumption using TOU; (c) FR for cost and waiting time using CPP; (d) FR for cost and waiting time using ToU.

The constraints obtained from the possible formulated cases in Table 6 and 7 are explained below.

1. The hourly cost of electricity for each load must fall within the lowest and highest electricity cost without HEM.
2. The hourly cost of electricity for each load must be less than the hourly electricity cost without HEM.
3. The entire hourly load must fall within the lowest and highest combined energy without HEM.

Table 6. Showing the possible formulated cases for CPP.

Cases	Load (kW)	Price (Cents)	Cost (Cents)
Minimum load, minimum price	1.00	11.40	11.40
Minimum load, maximum price	1.00	894.65	894.65
Maximum load, minimum price	13.25	11.40	151.05
Maximum load, maximum price	13.25	894.65	11,854.00

Table 7. Showing the possible formulated cases for ToU.

Cases	Load (kW)	Price (Cents)	Cost (Cents)
Minimum load, minimum price	1.00	8.70	8.70
Minimum load, maximum price	1.00	174.90	174.90
Maximum load, minimum price	13.25	8.70	115.27
Maximum load, maximum price	13.25	174.90	2317.40

Based on the defined constraints, the electricity cost of scheduled loads must fall within the lower and upper limit of load without HEM hourly electricity cost. Notably, 13.25 kW is maximum hourly load without HEM which must be greater than or equal to the scheduled loads for all schemes. The FR in Figure 9a,b, describes the relationship between electricity cost and energy load. We denote all points with $P_1, P_2, P_3, P_4, P_5, P_6$, where (P_1, P_2, P_3, P_4) is the set of points in which the without HEM load boundary lies. To calculate these points, we multiply the minimum and maximum loads with the minimum and maximum electricity price signals collected from the utility. The ToU and CPP achieve maximum electricity cost per time slot without HEM of 2317.40 cents and 11,854.00 cents, respectively. The cost for scheduled load should not be more than the electricity cost of without HEM. The FR covers all points $(P_1, P_2, P_3, P_5, P_6)$. The FR covers all point $(P_1, P_2, P_3, P_5, P_6)$. This assumes the realistic electricity cost reduction while using HEM.

5.7. FR for Cost and Waiting Time

Waiting time is measured as the amount of time the household appliances stay idle before they are turned ON. Figure 9c,d show the FR of electricity cost versus the waiting time of each household appliance for the two pricing schemes. The shaded region indicates the points where the problem constraints are satisfied. The formulated possible case tells the average waiting time for each scheduled load per time slot and their corresponding cost. The above-formulated case is used to derive the constraints of the FR as explained below:

1. The hourly waiting time should not be more than the maximum average waiting cost.
2. The cost of hourly waiting time must be within the minimum and maximum waiting time.
3. The total hourly waiting time must not exceed the minimum and maximum total waiting time.

From Figure 9c, the cost of 2210.00 cents shows the household appliance waiting time is zero, whereas, in Figure 9d, the cost of 1080.00 cents shows zero waiting time of household appliances. From the simulation results, the maximum electricity cost for the two pricing schemes indicates the maximum delay.

6. Conclusions and Future Work

This paper proposes a model for SG that handles the emerging advancement in technology of smart households and the power grid. It is also established that integrating the RES and the proposed optimization algorithm with its solution optimally addresses the multi-objective scheduling problem. The performance of the proposed models shows that, unlike Jaya, EDE and SBA based HEMS, Earliglow based HEMS reduces the electricity cost by 43.20%, 13.83% for CPP and ToU, respectively. The proposed Earliglow based HEMS achieves a reduced PAR up to 61.23% as compared to Jaya with PAR reduction up to 63.54%, whereas SBA and EDE have PAR reduction up to 17.97% and 43.61%

for the CPP scheme. Similarly, Earliglow based HEMS reduces the PAR up to 58.84% as compared to Jaya and SBA that each have PAR reduction of 43.03%, while EDE achieves PAR reduction up to 48.59% for ToU schemes. The achievements of Earliglow based HEMS over the counterparts are as follows. Firstly, Earliglow based HEMS shows that power system stability and reliable grid operation are enhanced through the efficient scheduling and utilization of the ESS. Lastly, Earliglow uses the advantages of both Jaya and SBA, which resolves the limitations of the two algorithms. Additionally, ESS is included to ensure reliable operation of the power grid as well as minimize the electricity cost. FRs illustrate the effect of household appliance scheduling on electricity cost with power consumption and consumer waiting time.

The proposed optimization method is scalable to any infrastructure capacity of the electricity system. Moreover, the proposed optimization method is efficient in achieving a global optimum solution within a small amount of execution time.

In the future, we will consider micro-grid and HEMS for minimization of the load burden on the power grid and reduction of electricity cost for the household users. We are also concerned in the coordination among the RES and ESS to utilize the renewable and sustainable energy resources. This coordination of RES in a residential area will not only enable the exchange of surplus renewable energy among the micro-grid but also minimize the load on the power grid.

Author Contributions: All authors equally contributed.

Acknowledgments: The present research has been conducted by the Research Grant of Kwangwoon University in 2018.

Conflicts of Interest: The authors declare no conflict of interest.

References

1. Gungor, V.C.; Sahin, D.; Kocak, T.; Ergut, S.; Buccella, C.; Cecati, C.; Hancke, G.P. Smart grid technologies: Communication technologies and standards. *IEEE Trans. Ind. Inform.* **2011**, *7*, 529–539. [CrossRef]
2. El-Hawary, M.E. The smart grid—State-of-the-art and future trends. *Electr. Power Compon. Syst.* **2014**, *42*, 239–250. [CrossRef]
3. Ahmed, M.S.; Mohamed, A.; Khatib, T.; Shareef, H.; Homod, R.Z.; Ali, J.A. Real time optimal schedule controller for household energy management system using new binary backtracking search algorithm. *Energy Build.* **2017**, *138*, 215–227. [CrossRef]
4. Basit, A.; Sidhu, G.A.S.; Mahmood, A.; Gao, F. Efficient and autonomous energy management techniques for the future smart households. *IEEE Trans. Smart Grid* **2017**, *8*, 917–926.
5. Abubakar, I.; Khalid, S.N.; Mustafa, M.W.; Shareef, H.; Mustapha, M. Application of load monitoring in appliances' energy management—A review. *Renew. Sustain. Energy Rev.* **2017**, *67*, 235–245. [CrossRef]
6. Li, H.; Zeng, P.; Zang, C.; Yu, H.; Li, S. An Integrative DR Study for Optimal Home Energy Management Based on Approximate Dynamic Programming. *Sustainability* **2017**, *9*, 1248. [CrossRef]
7. Marzband, M.; Ghazimirsaeid, S.S.; Uppal, H.; Fernando, T. A real-time evaluation of energy management systems for smart hybrid home Microgrids. *Electr. Power Syst. Res.* **2017**, *143*, 624–633. [CrossRef]
8. Ahmad, A.; Khan, A.; Javaid, N.; Hussain, H.M.; Abdul, W.; Almogren, A.; Alamri, A.; Niaz, I.A. An optimized home energy management system with integrated renewable energy and storage resources. *Energies* **2017**, *10*, 549. [CrossRef]
9. Javaid, N.; Javaid, S.; Abdul, W.; Ahmed, I.; Almogren, A.; Alamri, A.; Niaz, I.A. A hybrid genetic wind driven heuristic optimization algorithm for demand side management in smart grid. *Energies* **2017**, *10*, 319. [CrossRef]
10. Aslam, S.; Iqbal, Z.; Javaid, N.; Khan, Z.A.; Aurangzeb, K.; Haider, S.I. Towards Efficient Energy Management of Smart Buildings Exploiting Heuristic Optimization with Real Time and Critical Peak Pricing Schemes. *Energies* **2017**, *10*, 2065. [CrossRef]
11. Rao, R.V.; Waghmare, G.G. A new optimization algorithm for solving complex constrained design optimization problems. *Eng. Optim.* **2017**, *49*, 60–83. [CrossRef]

12. Samuel, O.; Javaid, N.; Aslam, S.; Rahim, M.H. JAYA optimization based energy management controller for smart grid: JAYA optimization based energy management controller. In Proceedings of the IEEE 2018 International Conference on Computing, Mathematics and Engineering Technologies (iCoMET), Sukkur, Pakistan, 3–4 March 2018; pp. 1–8.

13. Hussain, H.M.; Javaid, N.; Iqbal, S.; Hasan, Q.U.; Aurangzeb, K.; Alhussein, M. An Efficient Demand Side Management System with a New Optimized Home Energy Management Controller in Smart Grid. *Energies* **2018**, *11*, 190. [CrossRef]

14. Al-Ali, A.R.; Zualkernan, I.A.; Rashid, M.; Gupta, R.; Alikarar, M. A smart home energy management system using IoT and big data analytics approach. *IEEE Trans. Consum. Electron.* **2017**, *63*, 426–434. [CrossRef]

15. Joo, I.-Y.; Choi, D.-H. Distributed Optimization Framework for Energy Management of Multiple Smart Homes With Distributed Energy Resources. *IEEE Access* **2017**, *5*, 15551–15560. [CrossRef]

16. Shafie-khah, M.; Siano, P. A Stochastic Home Energy Management System considering Satisfaction Cost and Response Fatigue. *IEEE Trans. Ind. Inform.* **2017**, *14*, 629–638. [CrossRef]

17. Javaid, N.; Ullah, I.; Akbar, M.; Iqbal, Z.; Khan, F.A.; Alrajeh, N.; Alabed, M.S. An intelligent load management system with renewable energy integration for smart homes. *IEEE Access* **2017**, *5*, 13587–13600. [CrossRef]

18. Paterakis, N.G.; Erdinc, O.; Bakirtzis, A.G.; Catalão, J.P.S. Optimal household appliances scheduling under day-ahead pricing and load-shaping demand response strategies. *IEEE Trans. Ind. Inform.* **2015**, *11*, 1509–1519. [CrossRef]

19. Muralitharan, K.; Sakthivel, R.; Shi, Y. Multiobjective optimization technique for demand side management with load balancing approach in smart grid. *Neurocomputing* **2016**, *177*, 110–119. [CrossRef]

20. Du, Y.F.; Jiang, L.; Li, Y.Z.; Counsell, J.; Smith, J.S. Multi-objective demand side scheduling considering the operational safety of appliances. *Appl. Energy* **2016**, *179*, 864–874. [CrossRef]

21. Javaid, N.; Naseem, M.; Rasheed, M.B.; Mahmood, D.; Khan, S.A.; Alrajeh, N.; Iqbal, Z. A new heuristically optimized Home Energy Management controller for smart grid. *Sustain. Cities Soc.* **2017**, *34*, 211–227. [CrossRef]

22. Du, Y.; Jiang, L.; Li, Y.; Wu, Q.H. A robust optimization approach for demand side scheduling under energy consumption uncertainty of manually operated appliances. *IEEE Trans. Smart Grid* **2016**, *9*, 743–755. [CrossRef]

23. Li, M.; Li, G.; Chen, H.; Jiang, C. QoE-Aware Smart Home Energy Management Considering Renewables and Electric Vehicles. *Energies* **2018**, *11*, 2304. [CrossRef]

24. Veras, J.; Silva, I.; Pinheiro, P.; Rabêlo, R.; Veloso, A.; Borges, F.; Rodrigues, J. A Multi-Objective Demand Response Optimization Model for Scheduling Loads in a Home Energy Management System. *Sensors* **2018**, *18*, 3207. [CrossRef] [PubMed]

25. Oprea, S.-V.; Bâra, A.; Reveiu, A. Informatics solution for energy efficiency improvement and consumption management of householders. *Energies* **2018**, *11*, 138. [CrossRef]

26. Samuel, O.; Javaid, N.; Ashraf, M.; Ishmanov, F.; Afzal, M.K.; Khan, Z.A. Jaya based Optimization Method with High Dispatchable Distributed Generation for Residential Microgrid. *Energies* **2018**, *11*, 1513. [CrossRef]

27. Zhang, X. Optimal scheduling of critical peak pricing considering wind commitment. *IEEE Trans. Sustain. Energy* **2014**, *5*, 637–645. [CrossRef]

28. Celebi, E.; Fuller, J.D. Time-of-use pricing in electricity markets under different market structures. *IEEE Trans. Power Syst.* **2012**, *27*, 1170–1181. [CrossRef]

29. Liserre, M.; Sauter, T.; Hung, J.Y. Future energy systems: Integrating renewable energy sources into the smart power grid through industrial electronics. *IEEE Ind. Electron. Mag.* **2010**, *4*, 18–37. [CrossRef]

30. Che, L.; Zhang, X.; Shahidehpour, M.; Alabdulwahab, A.; Abusorrah, A. Optimal interconnection planning of community microgrids with renewable energy sources. *IEEE Trans. Smart Grid* **2017**, *8*, 1054–1063. [CrossRef]

31. Tushar, M.H.; Assi, C. Volt-VAR control through joint optimization of capacitor bank switching, renewable energy, and home appliances. *IEEE Trans. Smart Grid* **2017**, *9*, 4077–4086. [CrossRef]

32. Lund, H.; Andersen, A.N.; Østergaard, P.A.; Mathiesen, B.V.; Connolly, D. From electricity smart grids to smart energy systems—A market operation based approach and understanding. *Energy* **2012**, *42*, 96–102. [CrossRef]

33. Li, J.; Wei, W.; Xiang, J. A simple sizing algorithm for stand-alone PV/wind/battery hybrid microgrids. *Energies* **2012**, *5*, 5307–5323. [CrossRef]

34. Rao, R.J. A simple and new optimization algorithm for solving constrained and unconstrained optimization problems. *Int. J. Ind. Eng. Comput.* **2016**, *7*, 19–34.

35. Akyol, S.; Alatas, B. Plant intelligence based metaheuristic optimization algorithms. *Artif. Intell. Rev.* **2017**, *47*, 417–462. [CrossRef]

36. Price, K.; Storn, R.M.; Lampinen, J.A. *Differential Evolution: A Practical Approach to Global Optimization*; Springer Science & Business Media: Berlin, Germany, 2006.

37. Soyster, A.L. Convex programming with set-inclusive constraints and applications to inexact linear programming. *Oper. Res.* **1973**, *21*, 1154–1157. [CrossRef]

energies

MDPI

Article

Integration of Measurements and Time Diaries as Complementary Measures to Improve Resolution of BES

Jakob Carlander [1,*], Kristina Trygg [2] and Bahram Moshfegh [1,3]

1   Division of Building, Energy and Environment Technology, Department of Technology and Environment, University of Gävle, 80176 Gävle, Sweden; bahram.moshfegh@liu.se
2   Technology and Social Change, Linköping University, 58183 Linköping, Sweden; kristina.trygg@liu.se
3   Division of Energy Systems, Department of Management and Engineering, Linköping University, 58183 Linköping, Sweden
*   Correspondence: Jakob.Carlander@hig.se

Received: 23 April 2019; Accepted: 27 May 2019; Published: 30 May 2019

Abstract: Building energy simulation (BES) models rely on a variety of different input data, and the more accurate the input data are, the more accurate the model will be in predicting energy use. The objective of this paper is to show a method for obtaining higher accuracy in building energy simulations of existing buildings by combining time diaries with data from logged measurements, and also to show that more variety is needed in template values of user input data in different kinds of buildings. The case studied in this article is a retirement home in Linköping, Sweden. Results from time diaries and interviews were combined with logged measurements of electricity, temperature, and CO_2 levels to create detailed occupant behavior schedules for use in BES models. Two BES models were compared, one with highly detailed schedules of occupancy, electricity use, and airing, and one using standardized input data of occupant behavior. The largest differences between the models could be seen in energy losses due to airing and in household electricity use, where the one with standardized user input data had a higher amount of electricity use and less losses due to airing of 39% and 99%, respectively. Time diaries and interviews, together with logged measurements, can be great tools to detect behavior that affects energy use in buildings. They can also be used to create detailed schedules and behavioral models, and to help develop standardized user input data for more types of buildings. This will help improve the accuracy of BES models so the energy efficiency gap can be reduced.

Keywords: building energy simulation; occupant behavior; energy performance; indoor climate; retirement home; user input data

1. Introduction

In the European Union (EU), buildings are responsible for approximately 40% of energy use and 36% of CO_2 emissions [1]. The building sector is also responsible for about 60% of electricity consumption, and about one-third of the total energy use in the building sector can be related to non-domestic buildings [2]. In 2016, the total energy use in the building sector in Sweden was 80.5 TWh, and approximately 27% of this energy use was from public buildings [3]. When renovating or constructing new buildings, it is important to have an idea on how the energy use will be affected by the users. One way to do this is with building energy simulation (BES). However, BES models rely on a variety of different input data, and the more accurate the input data are, the more accurate the model will be in predicting energy use. In many cases, when it comes to whole-building simulation, there are significant discrepancies between simulated results and actual energy use of the real buildings [4]; this is called the "energy performance gap" [5]. However, whole-building simulation is often held as the best approach when it comes to analyzing performance in the building industry [6]. Many times, these differences come from the behavioral patterns of the residents, which are hard to predict and to simulate [7]. In Sweden, it is now standard to use template values and schedules for user behavior. The template values used in Sweden are usually from Sveby´s reports on standardized user input data, which are available for residential housing, office buildings, and schools [8–10]. Sveby stands for "standardize and verify energy performance in buildings", and it is a branch overlapping program that produces instruments that aid and standardize energy use. However, these template values and schedules were only developed for residential buildings in the form of apartment buildings and detached homes, and not for other types of residential buildings such as retirement homes or homes for people with special needs. In Reference [11], the authors made a review article where they studied research articles which dealt with the impact of occupant behaviors on building energy analysis. They concluded that most research in the field studied residential and office buildings, while a small number studied commercial and educational buildings, and sparse attention was given to recreational and healthcare facilities. As this paper shows, the available template values from Sveby do not always provide a good prediction of user behavior and, therefore, the simulations lack in accuracy compared to real buildings.

Another way to simulate user behavior and its impact on energy use in a building is to use behavioral models. Many behavioral models that has been developed uses data from large national surveys based on time diaries [12–15]. However, these behavioral models were also only developed for residential housing. In Reference [12], the authors presented a method for generating realistic occupancy schedules and electricity load profiles for United Kingdom (UK) households, where they used data from a large time-use survey (TUS) on how people used their time, which was conducted in the UK in 2000. The data used included location of participants and if they were active (not asleep) for each ten-minute diary period. The data did not contain any information on electricity use; however, according to the article, electricity use is highly connected to active occupants. Their conclusions were that the simulated output and original TUS data correlate very closely, and that the technique of building transaction matrices from such data in order to generate synthetic data series is very effective and computationally efficient. The authors of Reference [13] presented a method to generate load profiles for household electricity and domestic hot-water (DHW) use from time-use data. The profiles were generated from a detailed dataset on the time use for everyday activities in Swedish households, and the results were compared to electricity and hot-water profiles from recently performed measurements. They said that their model makes realistic reproductions of electricity demand for individual households and generates well-corresponding load distributions when compared to available measurement data. "The overall energy-use pattern found in measured data is well described by the model, while magnitudes sometimes deviate" [13]. In Reference [14], the authors developed a high-spatial-resolution model of energy use in residential buildings using data from national TUS data from the United States (US) residential sector. Their model was more detailed then previous models in the way that it was used down to a zone level in a building rather than an entire building. One conclusion they drew is that, if energy simulation tools can result in better decisions in energy-efficient renovations of single-unit structures where owners have full control over their property (in the US, this is approximately 65 million properties), it can have substantial societal impact. In Reference [15], the authors also made a probabilistic occupancy model for residential buildings, and tested it against Belgian TUS data. However, none of these studies or behavioral models were made for specific types of buildings; rather, they were made from large datasets and only from regular residential buildings. To the best knowledge of the authors of this article, such behavioral models are not widely used compared to the use of template values in the building industry.

The aim of this article is to show a method on how to improve the resolution of BES models on existing buildings by integrating measurements of electricity use, indoor temperature, and CO_2 levels with time diaries. The aim is also to show how occupant behavior in a retirement home can differ from occupant behavior according to template values for regular residential buildings, and how these differences can affect energy use in a building. It shows the need for the development of template values to use in BES models for more types of buildings than currently available. The article will also contribute to the knowledge on occupant behavior in retirement homes, which according to Reference [11] is sparse.

2. Method and Case Description

In the present study, a model of an existing building was created in the general simulation software IDA-ICE (IDA Indoor Climate and Energy) version 4.8 [16]. Input data for the model were obtained through interviews and time diaries, logged measurements, blueprints, and onsite observations. IDA-ICE is a dynamic whole-building energy simulation software that was released in 1998, in which building energy use, indoor climate, and heating and cooling loads can be simulated. The validity of IDA-ICE was tested several times throughout the years with good results [16–18].

2.1. Theory and Related Research

2.1.1. Validation of BES Models

There are three basic approaches for validating BES models. The first one is analytical validation, where the model is compared to a given and exact solution, the second approach is peer model validation, where the model is compared to peer models with the same input data, and the third approach is empirical/realistic validation, where the model is compared to empirically collected data [19,20]. When performing a realistic validation, one can use metering and auditing data from actual residential and commercial buildings to compare the model [19]. User behavior is usually also included in these models, but setting occupant behavior schedules is difficult due to the variable nature of occupants [19]. In their conclusion, Sanquist and Ryan [19] stated that realistic validation cases need to be included in building energy validation procedures to improve the accuracy of building energy models. Sanquist and Ryan [19] also stated that there were not any major improvements in the methods used to model occupant behavior, and that the use of stochastic and other detailed behavior models could improve predictions of energy models, but at a much greater cost than current schedule-based occupant models. There were not many attempts at empirical validation of occupied buildings [21]. A few examples on studies that used empirical validation include References [21–24]. La Fleur, Moshfegh, and Rohdin [21] measured electricity, indoor temperature, and CO_2 level in two apartments, both before and after renovation. The aim of their study was to present numerical predictions, validation, and evaluation of energy use and indoor climate for the building before and after renovation. Good agreement was reported, both in annual heat demand and indoor temperature, between simulation results and measurements. La Fleur, Moshfegh, and Rohdin [21] concluded that assumptions on user behavior have significant impact on energy-saving potential. In the article by Liu et al. [22], two almost identical buildings built in the 1970s in Sweden were studied. One building was retrofitted during 2009-2010, and the other was not yet retrofitted. In the study, a mixed-methods approach was used, i.e., conducted measurements and simulations on the buildings, as well as handing out questionnaires with questions about the indoor environment to the residents of both buildings. The electricity use and indoor temperature were measured. Based on the electricity measurements, schedules for electricity use were created and compared with the predicted results of indoor temperature and heat demand; good agreement was achieved between them. A similar approach was conducted by Reference [23], as they validated their BES model against detailed measurements of electricity at the appliance level in a passive house, with good agreement. Some of the differences in simulated and real results in building simulation models can, as previously stated by Reference [21] amongst others, be due to behavior of the building's occupants. Large offsets between simulated results and actual energy use, due to differences between actual and simulated behavior, can be a problem [7]. Gauthier and Shipworth [25] described a couple of different behavioral responses connected with thermal discomfort according to the following factors: increasing/decreasing clothing insulation level (clo), operative temperature, and increasing frequency, duration, and/or amplitude of localized behavioral responses such as consuming hot food and/or liquids, changing location to another room in the dwelling, opening and closing of curtains and/or windows, and changing body position. Some of these responses have a direct impact on energy use in the dwelling, such as opening and closing of windows [26]. A lot of behavior models were designed for use in energy simulations, and many of them use data from national time-use surveys, which are based on data from time diaries [12–15], as described above. These models rarely took opening of windows and doors into account; however, other models were developed for this purpose (e.g., Reference [26]). A method on how to model and validate BES models was proposed by Raftery et al. [4]. The proposed approach was evidence-based, which means that changes to the input parameters could only be made according to available evidence under clearly defined priorities. This was done to minimize the differences between real and simulated energy use for buildings [4]. The proposed method follows a certain sequence of steps that should be done in order to get as good results as possible. The first step is preparation, where an initial model is constructed, and historical weather data and calibration data are gathered, in addition to setting acceptance criteria for the model. Calibration data should be gathered and used according to the following hierarchy: (1) data-logged measurement; (2) spot or short-term measurement; (3) direct observation; (4) operator and personnel interviews; (5) operation documents; (6) commissioning documents; (7) benchmark studies and best practice guides; (8) standards, specifications, and guidelines; (9) design stage information. The acceptance criteria are values of, for instance, energy use and zonal indoor temperatures during a certain period of time, whereby the model should match to be called validated. In the second step, the model is updated with the information gathered, and the simulation is run. The model is tested against the acceptance criteria, and, if it checks out, the model is validated. If it does not meet the acceptance criteria, an iterative process is started where new measurements or data are obtained, and the model is updated and tested until it meets the acceptance criteria. According to Reference [4], the proposed methodology was based on some of the best techniques from the reviewed literature of their study, and it combined these with an evidence-based approach that used version control software to track the calibration process.

Standards, specifications, guidelines, and template values, such as Sveby's standardized user input data, come in at place eight out of nine in the hierarchy proposed by Reference [4] for input data in building energy simulations. Despite this, standards such as user input data for housing, office buildings, and schools from Sveby are consistently used when new building energy performance is simulated in the design phase of a project. They are also used when renovations and/or retrofits are done in existing buildings. This is mainly because more accuracy is not required by the Swedish Boverket (National Board of Housing, Building, and Planning).

2.1.2. Time Diaries and Interviews

Time diaries are a method developed from the time-geographical perspective, and are commonly used in everyday life studies on energy use with a time-geographical approach (e.g., References [27–29]). With this method, it is possible to describe and analyze the relationship between activity, location, and movements of individuals. On the individual level, a rhythmic pattern may occur, which shows the central role of the dwelling, the workplace, and, for shorter visits, the places where friends and relatives live and where shops are located. It is also of interest to understand how often or for how long a social activity occurs, and how these social activities relate to other events that are involved in structuring an individual's daily, weekly, or monthly pattern (e.g., how a coffee break with a friend in the apartment affects energy use). The difference between a time diary and an ordinary diary is that it is a written biography that describes a period of an individual's life more or less systematically. Even though the time diary often has a certain aim related to the study objectives, and the researchers specify what they want the respondent to reflect upon, in the end, it is the respondent who decides what will be written and what will be left out [30]. Asking people to write diaries will make people more aware of the practices they are involved in (e.g., how many times they open the window (routines)). A time diary offers a chance to reflect upon their everyday life, which might get them to remember things and why they are doing things in a specific way. This can also help the respondents explain why they do certain things. Today, a time diary might be in paper or digital format. In a digital format time diary, people record their activities on their mobile phone or computer and then share those electronically [31]. This is less time-consuming for both the respondent and the researcher.

2.1.3. Energy Affecting Behavior in Swedish Households

According to Reference [10], the general population in Sweden has the following routines and behavior which affect energy: they are at home for about 14 h per day and use annually around 30 kWh/m^2 of household electricity, with about 4 kWh/m^2 of energy losses due to airing and 25 kWh/m^2 due to DHW use.

2.2. The Case Study

Comparable with many countries around the world, the population of Sweden is aging [32,33]. The city block consists of tenants with different types of rental forms and, in the building, both housing and care homes (dementia housing) and service dwellings are present. Some tenants rent directly from the private property owner, while other tenants rent through the municipality (service and dementia housing and special housing). Most people who live in the housing complex are age 65 or older. Older people in Sweden are more reluctant to change housing than younger age groups [34,35]. In Sweden, aging in place was a political goal for a long time; hence, most older people in Sweden live in their ordinary homes with the assistance of home and medical care when needed. How many people live in assisted-living accommodation and how many have assistance in home can be seen in Table 1.

Table 1. Number of people with assistance in their own home and number of people living in assisted-living accomodation in Sweden (men and women; 2016) [32].

Age	People with Assistance in Home	People in Assisted-Living Accommodation
65+	228,654	106,002
65–79	68,536	84,843
80+	160,118	21,159

The city block that was the case for this study contains 99 apartments, of which 35 are assisted-living accommodation for elderly people. Elderly people have their own apartments with a safety alarm, and two healthcare providers (private and public) operate in the building, as well as a nurse. The building complex consist of five different buildings, where it is possible to walk between three of them without having to go outside (they are in the same building body); the other two buildings are detached three-story buildings with 15 apartments in each, evenly distributed on the three stories. No renovation was done since the block was built in 1983, with only minor repairs taking place. This type of housing includes a canteen where lunch is served, and there are also leisure activities in which all elderly people in the municipality can take part. Most of the people living in the block are at least 65 years old. Tenants in this specific building complex rent an apartment directly from the property owner, and the people in need of assisted living rent their apartment from the municipality. There is also a residence for elderly people with dementia in a separate part of the building complex. The building that was modeled and simulated was one of the two detached buildings. The building and the simulation model can be seen in Figure 1. Technical data of the modeled building, as well as data of its location, can be found in Table 2. This specific city block is geographically located in Östergötland, a county in the south of Sweden that has a continental climate and belongs to the northern part of the temperate zone.

Table 2. Technical data of the modeled building.

Location Data

Country	Sweden
City	Linköping
Longitude	15.53 east (E)
Latitude	58.4 north (N)
Annual mean temperature	6.8 °C

Building

Model floor area	1366.2 m² (A_{temp})
Floor area apartments	1131.7 m²
Model volume	3370.1 m²
Model ground area	427.2 m²
Model envelope area	1770 m²
Window/envelope	5.5%
Average U-value	0.42 W/m²K
Number of apartments	15
Common area (Stairwell, entrance and storage)	234.5 m²
Time Constant	201 h
Winter Outdoor Design Temperature	−13 °C
Q_{tot}	1036 W/K

Building Envelope

Envelope Part	Area (m²)	U (W/m²K)	U × A (W/K)
Walls above ground	775.38	0.25	197.54
Roof	427.88	0.18	75.86
Floor towards ground	427.20	0.45	191.56
Windows	97.81	1.91 (incl. frame)	186.81
Doors	38.06	0.25	9.61

Windows

Direction	Area (m²)	Glazing	g-factor
N	21.32	3-pane	0.68
E	26.97	3-pane	0.68
South (S)	23.63	3-pane	0.68
West (W)	25.89	3-pane	0.68

Construction

External wall 1	Brick 87 mm	Air gap 23 mm	Light insulation 170 mm	Gypsum 13 mm
External wall 2	Wood 20 mm	Air gap 16 mm	Light insulation 120 mm	Concrete 160 mm
External wall 3	Wood 20 mm	Air gap 16 mm	Light insulation 120 mm	Gypsum 13 mm
External wall 4	Brick 87 mm	Air gap 23 mm	Light insulation 120 mm	Concrete 160 mm
Internal wall	Concrete 160 mm			
Internal floor	Concrete 200 mm	Floor coating 5 mm		
External slab	Concrete 250 mm	Floor coating 5 mm		
Roof 1	Wood 25 mm	Light insulation 250 mm	Concrete 200 mm	
Roof 2	Wood 25 mm	Light insulation 10 mm	Aluminum 1 mm	

Figure 1. (**a**) Building that was modeled. (**b**) Model of the building in IDA-ICE (IDA Indoor Climate and Energy).

2.3. Data Collection

The data for this study were obtained through interviews and time diaries, logged measurements, blueprints, and onsite observations. The data were largely collected between 2014 and 2016, based on a pilot study carried out in 2013.

2.3.1. Conducting Time Diaries and Interviews

All tenants received an invitation to participate in the study (except those in the residence for dementia) by participating in interviews and/or by writing time diaries for a week. The interviews each lasted for at least one hour. Interviews were used to understand social phenomena and to complement the time diaries. For the interviews and time diaries, a total of 16 tenants agreed to participate. Ultimately, 11 of them participated in both the interviews and the time diaries. The tenants that agreed to participate kept a time diary for one week in which they wrote time, activity, place, with whom, which electrical appliances were used, and other comments. Two of the subjects lived with their spouse and also wrote their spouse's activities in the time diary. Table 3 presents general information on the subjects that participated in both time diaries and interviews.

Table 3. Gender, age, civil status, and working status of the occupants in the present study.

Description	Facts
Gender	Female (8), Male (5)
Age	58–94 (mean 85, median 78)
Civil status	Single (9), married (4: two couples)
Retired/working	11/2

2.3.2. Field Measurements

At the same time as the tenants wrote time diaries, measurements of CO_2 levels, relative humidity, indoor temperature, and electricity use were logged in their apartment. Temperature was measured with three loggers, one in the kitchen, one in the living room, and one in the bedroom. In some apartments, there were more than one bedroom, as a result, the temperature logger was set up in the master bedroom or the one that was used as bedroom. Relative humidity was logged in the kitchen and CO_2 levels were logged either in the living room or the hallway. Electricity use was logged with an EliQ optical eye at five-minute intervals for each apartment. The logged measurements and time diaries were done between the period of 17 November and 2 February, two apartments at a time. Logged measurements were done in 12 of the 15 apartments in the building, and 11 of these tenants also kept a time diary; the tenants in the three remaining apartments did not wish to participate in the study. An energy mapping was also performed for the city block to allocate the different uses of electricity, energy for heating of domestic hot water (DHW), and energy for space heating. Some data could only be obtained for the entire block, such as energy for space heating and DHW use. Electricity meters of 15 apartments were, therefore, read once a week for two months, with approval from the tenants. The same was done with the electricity meters measuring the facility electricity, which includes electricity for the lighting of stairwells, corridors, common areas, and outside lighting, electricity for operation of automatic doors, elevators, and laundry rooms, and electricity used for a central ventilation system in the common area and offices of the block. Apart from the energy mapping, the tightness of the building envelope was measured by using the blower door technique in one apartment. Readings, measurements, and used measurement equipment can be found in Table 4, and each measuring equipment's accuracy can be found in Table 5. On-site observations were made throughout the building complex, as it was useful in terms of understanding how the building was used by the tenants and the people working there, as well as also to check that the construction matched the provided blueprints of the building. Photographs were also taken to enrich the data material (see Reference [30]). Which data were used and for what purpose can be seen in Table 6.

Table 4. Measurements at the city block and the modeled building, time resolution, and time span, as well as which measurement equipment was used.

Measuring Type	Measuring Equipment	Area	Time Resolution	Timespan
District heating	Meter at the city block's heating central	Entire city block (one measuring point)	weekly	2 months
Domestic hot-water use and temperature	Meter at the city block's heating central	Entire city block (one measuring point)	weekly	2 months
Domestic cold-water use	Meter at the city block's heating central	Entire city block (one measuring point)	weekly	2 months
Facility electricity	Electricity meter	Entire city block (six measuring points)	weekly	2 months
Apartment electricity	Electricity meter	Ekholmsvägen 106 (15 apartments)	weekly	2 months
Total electricity	Eliq	Entire city block (incoming electricity for each facility and for the central ventilation; 5 measuring points in total)	Every 5 minutes	1 week
Momentary total electricity	Fluke 41B Power harmonics analyzer + Universal Technic current clamps	Entire city block (incoming electricity for each facility and for the central ventilation; 5 measuring points in total)	-	-
Momentary temperature	Swema3000	Stairwells, ventilation outtakes	-	-
Logged temperature	Tinytag Plus2	12 apartments	Every 5 minutes	1 week
Logged apartment electricity	Eliq	12 apartments	Every 5 minutes	1 week
Relative humidity	Tinytag View2 Temp and RH logger	12 apartments	Every 5 minutes	1 week
CO_2	Tinytag CO_2	Dining hall + 12 apartments	Every 5 minutes	1 week
Energy use from apartment ventilation	Everflourish power meter EMT707CTL	Ventilation unit in one apartment	1 week	1 week
Momentary ventilation flows	Swema3000	Outlets on roof + measured on central ventilation system	-	-
Building envelope tightness	Retrotec DM32 Blower door	1 apartment	-	-

Table 5. Accuracy and area of use for measuring equipment used in the study. N/A—not applicable.

Equipment	Measures	Equipment Accuracy	Manufacturer
EliQ optical eye	Household electricity use	N/A	Eliq, Gothenburg, Sweden
Fluke 41B Power harmonics analyser	Facility electricity	±1% for active watts (VA) + probe specs	Fluke, Everett, WA, USA
Universal Technic current clamps	Facility electricity	N/A	Universal Technic, Paris, France
Swema3000	Temperature and air speed	±0.1 °C, ±0.03 m/s	Swema, Stockholm, Sweden
Tinytag Plus2	Temperature	±0.35 °C	Intab, Stenkullen, Sweden
Tinytag View2	Temperature and relative humidity (RH)	±0.4 °C, RH ±3%	Intab, Stenkullen, Sweden
Tinytag CO_2 logger	Concentration of carbon dioxide in air	±3%	Intab, Stenkullen, Sweden
Everflourish power meter EMT707CTL	Household electricity	N/A	Everflourish, Friedrichsthal, Germany
Retrotec Blower Door 3100 with DM2 digital pressure gauge	Building envelope airtightness	±5% flow rate accuracy, ±1% pressure reading accuracy (or ±0.15 Pa)	Retrotec, Everson, WA, USA

Table 6. Part of the process in which the gathered data were used. BES—building energy simulation.

Step in Process	Construction of Model	Creating Detailed User Schedules	Validating BES Model
Used data	Blueprints of the building and on-site observation of construction, measured ventilation airflows, temperature, number of occupants in each apartment, and measurements of building envelope tightness.	Time diaries, interviews, and household electricity use	Logged measurements of CO_2 levels, indoor temperature, and household electricity use

2.4. Modeling and Validation of BES Models for This Study

Two models of the same building were created in this study: one reference model, BES-ref (building energy simulation reference model), which used template values for occupancy, airing, and electricity use, and one model where detailed schedules for occupancy, airing, and electricity were used, BES-v.2 (building energy simulation model version 2). The modeling of both BES-ref and BES v.2 was inspired by the method developed by Reference [4]. An initial model was built according to blueprints made available from the housing company, and the construction in the blueprints was confirmed with on-site observations where possible.

2.4.1. Model Input Data and Creation of Detailed Schedules

Input data for BES-ref and BES-v.2 can be seen in Table 7. Input data in the form of schedules for occupancy (absent from 7:00 a.m. to 5:00 p.m.) and electricity use (electricity use between 6:00 and 8:00 a.m. and between 3:00 and 11:00 p.m.) in BES-ref were set according to Sveby's [10] user input data, and losses due to airing were added after simulations according to Sveby's [10] user input data. For both BES-ref and BES-v.2, temperature was set according to logged measurements from 12 apartments, while ventilation air flows were set according to measurements at the inlets and outlets, DHW use was set according to the mean use per square meter for the city block, the heat exchanger efficiency was approximated for each apartment by analyzing the electricity use of the air handling unit (AHU), and the building airtightness was set according to measurements with the blower door technique. For BES-v.2, input data in the form of schedules for occupancy, electricity use, and airing were created by combining information from time diaries and interviews, with logged measurements of electricity use, indoor temperature, and CO_2 levels. In cases where there were multiple people living in an apartment, several occupants were used in the simulations, each with their own unique occupancy schedule. An example on how the detailed schedules were created is presented below. The example follows the creation of detailed schedules for one day and one apartment. The first thing one needs to do is analyze the logged measurements and check the time diaries for reasons on why drops and peaks occur in temperature, CO_2 levels, and electricity use. In Figure 2, we can see the logged measurements from apartment 5 on the second floor for 19 November, and Table 8 shows a transcription of the time diary from the same day written by the tenant living in that specific apartment.

The first interesting thing in the measurements for CO_2 levels and temperature occurred around 8:30 a.m., as can be seen in Figure 2. To see the reason for the drop in CO_2 and temperature, we can look at the entries from the time diary in Table 8. The tenant wrote that she put out linen on the balcony and left the balcony door open from 8:30-8:50 a.m. This is the most probable cause for the drop in both temperature and CO_2 level. To represent this in the model, an opening of the balcony door from 8:30 to 8:50 a.m. was entered into the opening schedule.

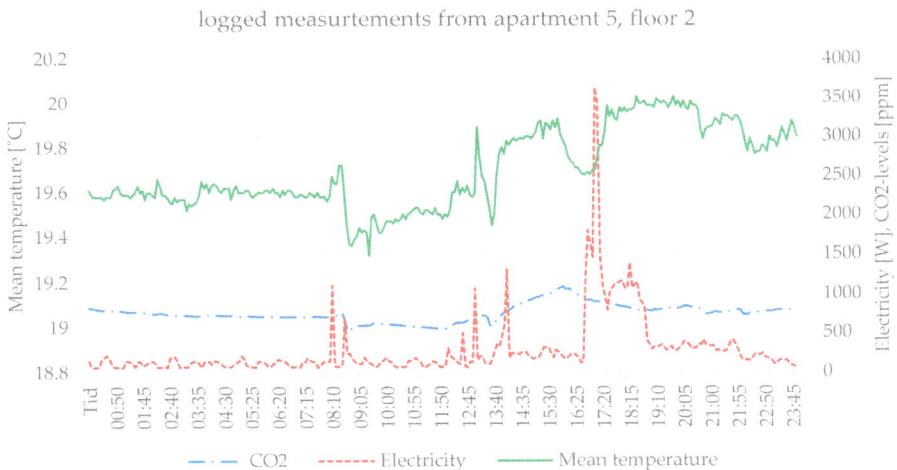

Figure 2. The logged measurements from apartment 5, second floor, on 19 November. The green (full) curve shows temperature, the blue (dotted and dashed) curve shows CO_2 level, and the red (dashed) curve shows electric power.

Table 7. Input data for BES-ref and BES-v.2.

Input data	Floor 1			Floor 2			Floor 3		
	BES-ref	BES-v.2	Difference	BES-ref	BES-v.2	Difference	BES-ref	BES-v.2	Difference
Apartment 1									
Tenant electricity use/year (kWh)	2489	1098	−56%	3000	918	−69%	3000	1098	−63%
Number of people in apartment	2	2	-	1	1	-	2	2	-
Heat exchanger efficiency (%)	60	60	0	60	55	−5.00	60	60	0.00
Temperature set-point (°C)	21.45	21.5	0.05	21.2	21.3	0.10	21.3	21.35	0.05
AHU schedule	Always on	Always on	-	Always on	Always on	-	Always on	Always on	-
Zone area (m²)	83	83	-	100	100	-	100	100	-
Apartment 2									
Tenant electricity use (kWh)	2311	1859	−20%	2311	1858	−20%	2311	1849	−20%
Number of people in apartment	1	1	-	2	2	-	2	2	-
Heat exchanger efficiency (%)	60	55	−5.00	60	55	−5.00	60	50	−10.0
Temperature set-point (°C)	19.8	19.9	0.10	19.6	19.8	0.20	21	21	0.00
AHU schedule	Always on	Always on	-	Always on	Always on	-	Always on	Always on	-
Zone area (m²)	77.5	77.5	-	77.5	77.5	-	77.5	77.5	-
Apartment 3									
Tenant electricity use (kWh)	2311	1714	−26%	2311	645	−72%	2311	1629	−30%
Number of people in apartment	1	1	-	2	2	-	1	1	-
Heat exchanger efficiency (%)	60	45	−15.00%	60	60	0.00	60	60	0.00
Temperature set-point (°C)	19.6	19.6	0.00	20.3	20.3	0.00	21.5	21.6	0.10
AHU schedule	Always on	Always on	-	Always on	Always on	-	Always on	Always on	-
Zone area (m²)	77.5	77.5	-	77.5	77.5	-	77.5	77.5	-
Apartment 4									
Tenant electricity use (kWh)	1982	1389	−30%	1982	1848	−7%	1982	456	−77%
Number of people in apartment	1	1	-	2	2	-	1	1	-
Heat exchanger efficiency (%)	60	60	0.00	60	50	−10.0	60	60	0.00
Temperature set-point (°C)	21.5	21.6	0.10	20.3	21	0.00	20.3	20.4	0.10
AHU schedule	Always On	Always On	-	Always On	Always On	-	Always On	Always On	-
Zone area (m²)	66	66	-	66	66	-	66	66	-

Table 7. *Cont.*

Input data	Floor 1			Floor 2			Floor 3		
Apartment 5									
Tenant electricity use (kWh)	1562	2852	83%	1978	789	−60%	1978	744	−62%
Number of people in apartment	1	1	-	1	1	-	1	1	-
Heat exchanger efficiency (%)	60	60	0.00	60	60	0.00	60	50	−10.0
Temperature set-point (°C)	18.5	18.75	0.25	18.7	18.6	−0.10	20	20.1	0.10
AHU schedule	Always on	Tenant-specific	-	Always on	Always on	-	Always on	Always on	-
Zone area (m²)	53	53	-	66	66	-	66	66	-
Schedules for all apartments	BES-ref	BES-v.2		BES-ref	BES-v.2		BES-ref	BES-v.2	
Tenant electricity use schedule	6:00-8:00 a.m. and 3-11 p.m.	Tenant-specific		6:00-8:00 a.m. and 3-11 p.m.	Tenant-specific		6:00-8:00 a.m. and 3-11 p.m.	Tenant-specific	
occupancy schedule	absent 7:00 a.m. to 5:00 p.m.	Tenant-specific		absent 7:00 a.m. to 5:00 p.m.	Tenant-specific		absent 7:00 a.m. to 5:00 p.m.	Tenant-specific	
Ventilation air flows (L/s.m²)	0.4 in, 0.42 out	0.4 in, 0.42 out		0.4 in, 0.42 out	0.4 in, 0.42 out		0.4 in, 0.42 out	0.4 in, 0.42 out	
Domestic hot water use (L/m².year)	345	345		345	345		345	345	
Airing	4 kWh/m²·Year	Tenant-specific		4 kWh/m²·Year	Tenant-specific		4 kWh/m²·year	Tenant-specific	

Table 8. Transcribed time diary from tenant in apartment 5, second floor, on 19 November. The effects of entries on the gray background are clearly visible in the logged measurements in Figure 2.

Time	Activity	Place	With Whom	Electrical Appliances	Comments
07:30 a.m.	Wakes up, turns on radio	Bedroom	Alone	Radio	
08:15 a.m.	Turns off radio, gets out of bed	Bedroom	Alone	Radio	
08:16 a.m.	Turns on radio, feeds the cats, makes tea	Kitchen	Alone	Water boiler 4 min, radio	
08:22 a.m.	Showers and gets dressed	Bathroom	Alone		
08:30 a.m.	Hangs out linens on the balcony	Balcony/living room	Alone		
08:30–08:50 a.m.	Opens balcony door				Cats wanted to be out
08:40 a.m.	Empties litter box	Bathroom	Alone	Lamp	
08:45 a.m.	Makes breakfast, eats breakfast	Kitchen	Alone	Lamp, microwave (4 min)	
09:20 a.m.	Brushes teeth	Bathroom	Alone	Electrical toothbrush	
09:30 a.m.	Takes in the linens and makes the bed	Balcony/bedroom	Alone		
09:35 a.m.	Turns off the radio	Kitchen		Radio	
09:35 a.m.–12:00 p.m.	Not at home	Out			
12:00 p.m.	Home again, turns on radio, washes hands, puts groceries in fridge	Kitchen	Alone	Radio	
12:10 p.m.	Makes copies and laminates	Bedroom/kitchen	Alone	Copying machine, laminator	
12:20 p.m.	At computer, talks on the phone	Bedroom	Alone	Computer, cell phone	
12:40 p.m.	Toilet	Bathroom	Alone		
12:42 p.m.	Heats and eats food, does dishes	Kitchen	Alone	Microwave (3 min)	
1:05 p.m.	Empties litter box, vacuums	Bathroom	Alone	Lamp, vacuum cleaner	
1:10 p.m.	Vacuums	Kitchen, living room, bedroom	Alone	Vacuum cleaner	
1:20 p.m.	Talks on the phone	Kitchen	Alone	Cell phone	
1:25 p.m.	Turns on lamps in windows	Kitchen, living room	Alone	Lamps	
1:26–1:40 p.m.	Lights lantern	Balcony	Alone		Balcony door open; cats outside
1:40 p.m.	Calls hair salon	Kitchen	Alone	Cell phone	

Table 8. *Cont.*

Time	Activity	Place	With Whom	Electrical Appliances	Comments
1:45 p.m.	Has a guest at home, has coffee, talks, gets help with computer		Friend	Computer, water boiler (4 min)	Very nice with good company
4:05 p.m.	Toilet	Bathroom	Alone		
4:10 p.m.	Turns off radio and kitchen lamp	Kitchen	Alone		Nice with a walk
4:10 p.m.	Goes for a walk	Out	Alone		
5:00 p.m.	Turns on radio, starts to make dinner	Kitchen	Alone	Radio, stove (17–19 min)	
5:15 p.m.	Copies	Bedroom	Alone	Copying machine	
5:25 p.m.	Does the dishes	Kitchen	Alone	Lamp	
5:40 p.m.	Goes through the mail	Kitchen	Alone	Lamp	
5:50 p.m.	Cleans toilet	Bathroom	Alone		
6:00 p.m.	Turns off radio, turns on floor lamp, watches television (TV)	Living room	Alone	TV, radio, lamp	
7:00 p.m.	Eats, does dishes, cleans stove	Kitchen	Alone		
7:00–7:45 p.m.	Charges cell phone			Cell phone	
8:05 p.m.	Watches TV	Living room	Alone	TV	
8:10 p.m.	Answers an e-mail	Bedroom	Alone	Computer	
8:15–10:00 p.m.	Watches TV	Living room	Alone	TV	Puts on thick socks Cats wanted to be out
8:40–8:46 p.m.	Opens balcony door				
10:00 p.m.	Empties litter box, washes hands	Bathroom	Alone		
10:03 p.m.	Puts out lantern	Balcony	Alone		Balcony door open 10:03–10:09
10:04 p.m.	Makes the bed	Bedroom	Alone		
10:10 p.m.	Cuddles with cat	Kitchen	Alone		
10:20 p.m.	Brushes teeth	Bathroom	Alone	Electrical toothbrush	
10:25 p.m.	Goes through some papers	Kitchen	Alone		
10:59 p.m.	Gets ready for bed	Bathroom	Alone		
11:15 p.m.	Turns off all lights and goes to bed		Alone		Finally in bed

At around 9:30 a.m., we can see that the CO_2 level slightly decreased, and it coincided with the tenant leaving the apartment and returning around 12:00 p.m., which is written in the time diary. The occupancy schedule was updated according to measurements and time diary entries. When the tenant returned around 12:00 p.m., one can see that both the temperature and CO_2 levels started to increase; then, there was a drop around 1:30 p.m., which once again can be traced to the opening of the balcony door. Another observation was that the CO_2 level increased even more and it peaked at around 4:00 p.m. This can be traced to the tenant having a friend over, which she also wrote in her time diary. This was modeled by inserting an extra occupant into the zone, who was only present from 1:45–4:00 p.m. on Wednesday. The peaks in electricity use can be traced to the tenant using the microwave, water boiler, and stove. The base load is most likely from the fridge and also the ventilation system, which is connected to the tenant's electricity since each apartment has its own ventilation system (this conclusion was based on the fact that the base load was approximately the same in each apartment). The electricity schedule was created by reading the maximum amount of power and inserting equipment with the same maximum power into the zone in the model. The schedules were made for one week. If there were drops in temperature in the logged measurements, but no entry in the time diaries of any specific behavior that could be linked to a drop in temperature, these drops were considered to be due to airing. In cases where occupancy could not be determined by time diaries, it was determined by drops and peaks in the logged measurements of CO_2 level.

2.4.2. Validation and Calibration of BES-ref and BES-v.2

The process of calibrating the schedules for electricity use, airing, and occupancy was then done. The schedule for electricity use was altered slightly after each simulation so that the simulated graph eventually mimicked the graph from the measurements as closely as possible, and the total amount of used electricity for each day was compared to the total simulated electricity use for the same day after each simulation, making sure that they matched. In the same way, the airing and occupancy schedules were slightly altered after each simulation until the simulated temperature and CO_2 graphs corresponded with the measured temperature and CO_2 graphs in a satisfying way (see Figure 3). The simulated mean temperature for the week was also checked against the measured mean temperature to make sure that they corresponded. The validation criteria for mean indoor temperature was ±0.1 °C for both BES-ref and BES-v.2, while the criteria for highest and lowest indoor temperature was ±0.1 °C for BES-v.2 if it was feasible. For electricity use, it was ±5% kWh, and, for the start and end time of drops and peaks in temperature, CO_2 levels, and electricity use, it was ±30 min. The same procedure was performed for each apartment and its corresponding zone in the model.

As can be seen in Figure 3, the CO_2 levels in the simulations are somewhat lower than the measurements, but the curves have a very similar shape. This was the same for all apartments, both in BES-ref and BES-v.2, and is probably due to a higher background level of CO_2 in the measurements than that which was used in the simulations. This, however, does not interfere much with the results since the objective in this study was not to compare simulated CO_2 levels to measurements, but rather to compare them to other simulated results, and the CO_2 measurements were used as a guide to see that the occupancy and airing schedules were done properly.

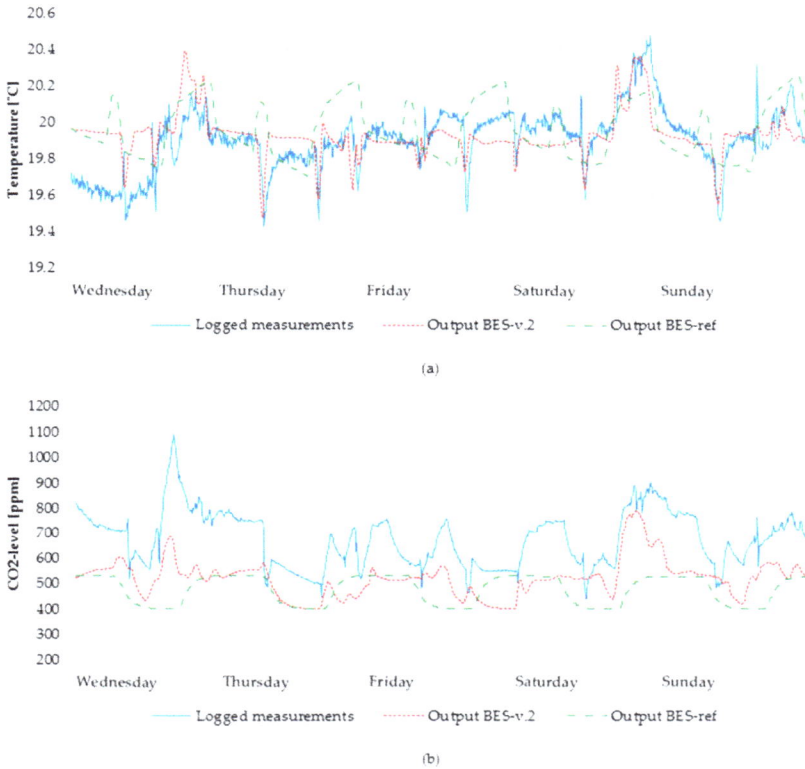

Figure 3. Logged measurements compared to simulated output from apartment 5 on the second floor, for both reference building energy simulation (BES-ref) and building energy simulation version 2 (BES-v.2), from Wednesday 19 November to Sunday 23 November: (**a**) mean temperature; (**b**) CO_2 levels.

The validation of the BES models was done with empirical validation. BES-ref was validated against mean indoor temperature from the measurements, and BES-v.2 was validated against temperature and CO_2 graphs from the measurements, making sure that the graphs from the simulations followed the graphs from the measurements as described above. The validation and calibration process used in this study can be seen in Figure 4.

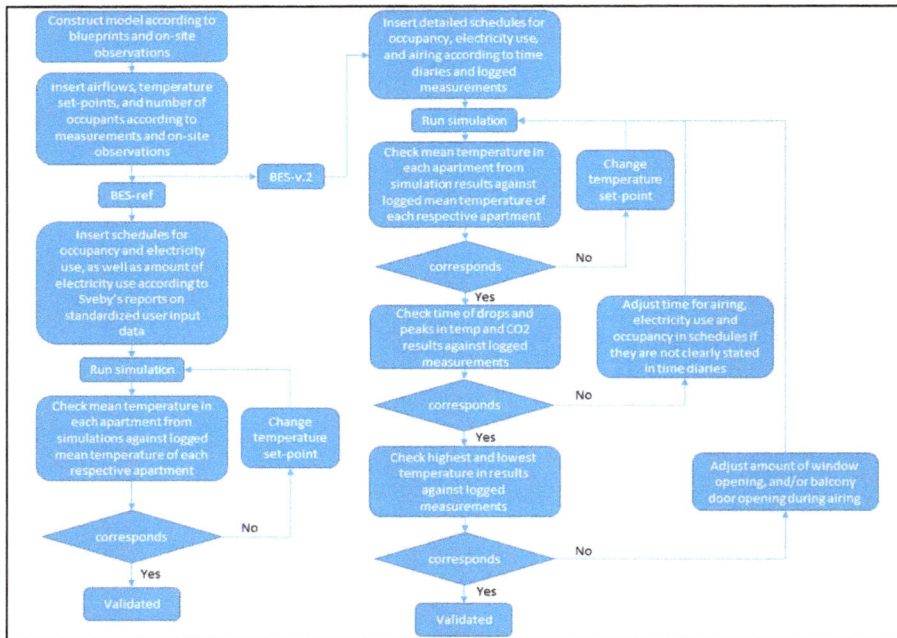

Figure 4. The validation process for BES-ref and BES-v.2 models.

2.4.3. Comparison of the Two Models

After both models were validated, several simulations were conducted, and the two models were compared at three different levels: (1) building level, (2) apartment level, and (3) room level. This meant that all apartments were originally modeled as one zone; however, to compare at room level, one apartment was later modeled with separate zones for each room (see Figure 5). The parameters that were compared at each level can be seen in Table 9.

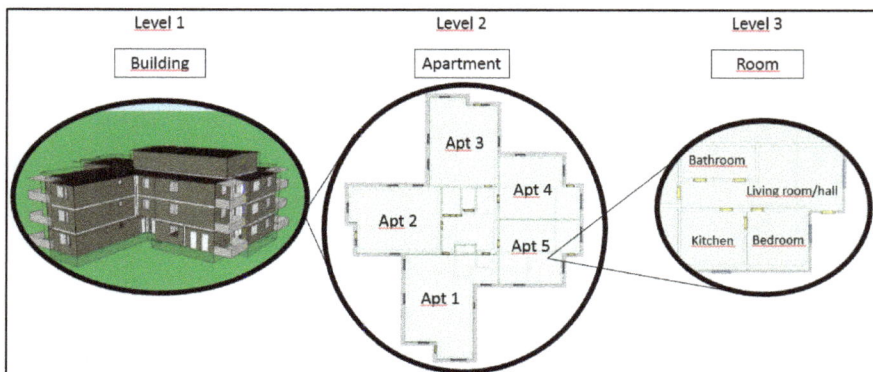

Figure 5. The three different levels on which simulated results between BES-ref and BES v.2 were compared.

Table 9. This table describes which parameters were compared between reference BES (BES-ref) and BES version 2 (BES-v.2) at each level.

Parameter	Level 1 Building	Level 2 Apartment	Level 3 Room
Energy supplied by water radiators (District Heating)	X	X	X
Air Handling Unit heating (electrical)	X	X	X
Energy losses due to airing	X	X	X
Indoor temperature		X	X
CO_2 level		X	X
PPD (Predicted Percentage Dissatisfied)		X	
Simulation Period			
Full-year simulation	X	X	X
Simulation of measuring period		X	X

X: Parameter that is compared between the models.

3. Results and Discussion

3.1. Measurements, Interviews, and Time Diaries

The mean temperature in the 12 apartments varied between 19.5 °C in the coldest apartment to 22.1 °C in the warmest apartment during the measuring period. This may seem odd; however, in interviews, some of the tenants stated that they used to turn down their radiators since they liked it a bit cooler inside. Domestic hot-water use was calculated from the total use for the city block to 345 L/m^2 and year, which led to a total energy use for heating of water as 34.9 MWh in the modeled building. From all 12 apartments, logged measurements of indoor temperature, CO_2 levels, relative humidity, and electricity use were collected for a week, and results from 19 November in apartment 5 on the second floor can be seen in Figure 2 (see Figure 5 for construction plan).

The level of detail in the time diaries varied between the different apartments, and one day from one of the most detailed time diaries was transcribed, as can be seen in Table 8. This is the time diary written by the tenant of the apartment from which the measurements in Figure 2 were collected. It is also from the same day, 19 November, as the results in Figure 2.

From the time diaries and the measurements, clear patterns could be seen for each household. Even though the patterns differed between the households, some clear differences compared to the general population of Sweden, henceforth called the general population, could be seen. For instance, many of them stated during the interviews and/or wrote in their time diaries that they aired their apartments a couple of times a day to get fresh air in. This could be seen in the measurements as drops in temperature and CO_2 levels. This behavior is something that seems to be connected to something they "used to do" in their old house or apartment. As stated earlier, most of the tenants are over 65 years old and moved to this city block upon getting older; thus, this is probably a behavior that they grew up with, since it was common practice in Sweden to ventilate homes in this way. They also stayed at home for a longer time each day than the general population. The tenants are at home for approximately 20 h/day in this case, which is a difference of 6 h/day compared to Sveby´s user input data, which say that occupancy should be entered as 14 h/day. This was expected and does not seem strange since it is a retirement home and most of the tenants do not work anymore. If the household electricity use, which was read weekly for two months, is scaled up to be the same use for an entire year, this will also show large differences compared to the general population. The tenants in this study had an electricity use of approximately 19 kWh/m^2·year, while Sveby´s standard user input data say 30 kWh/m^2·year for the general population. The reasons for this might be many, but the most probable cause is that, since this is an elderly population, they do not seem to own as many electrical appliances as the general population. According to Sveby´s reports of user input data [36], electricity use can differ by around 30% between the summer and winter. This is, among other things, due to the large difference in daylight and the fact that people spend more time indoors during winter, since Sweden is situated in the northern part of the temperate and polar climate zone. This means that the difference between the occupants in the study and the regular population might differ even more since they probably have an even lower electricity use during the summer.

3.2. Comparing BES-ref to BES-v.2

3.2.1. Building Level

At building level, the simulated values of energy for space heating, electricity use, and energy loss due to airing were compared between BES-ref and BES v.2. The results show an increase in energy supplied by water radiators, district heating (DH), with 20%, an increase in air handling unit (AHU) heating (electrical) with 85%, a decrease in tenant electricity use with 39% (this is an input value, see Table 7), and an increase in energy loss due to airing with 99%from BES-ref to BES v.2 (see Table 10).

Table 10. Simulated values for a whole-year simulation at building level from BES-ref and BES v.2, and difference in percentage between them (unit KWh).

Parameter	BES-ref	BES-v.2	Difference BES-ref to BES-v.2
Energy supplied by water radiators (DH)	48,545	58,345	20%
AHU heating (electrical)	3432	6357	85%
Total energy use for space heating	51,976	64,702	24%
Energy loss due to airing	5464	10,865	99%
Household electricity (input)	34,092	20,659	−39%

The two things that really stand out are losses due to airing and the increase in heating from the AHU. The reasons for the increase in electrical heating from the AHU can be traced back to the difference in input data between the two models. In many apartments, the efficiency of the heat exchanger was lowered after measurements showed a higher electricity baseload in these apartments that could only be traced to the AHU. In apartment 1 on the second floor, the heater in the AHU was set a lot higher than in the remaining apartments, which gave the tenant a temperature of 20 °C in the supply air from the AHU (this was entered into BES-v.2); this also led to a higher electricity use from the AHU. The difference in the amount of energy losses from airing might seem odd at a first glance; however, taking into account that, according to time diaries and interviews, they seem to air a lot more than the general population, it does not seem as odd anymore. However, this type of airing should not be needed in a building with an exhaust and supply air ventilation system with heat recovery. In a previous study of the same building block made by Carlander and Tullsson [37], where they conducted a survey using questionnaires of indoor climate, they got evidence pointing out that the AHUs do not work as well as they are supposed to, which might also contribute to more excessive airing than usual. The difference could also be due to the fact that the schedules could only be validated for one specific week in each apartment, but the schedules were then also used for full-year simulations; thus, it is possible that the tenants air less during colder days and more during warmer days. However, since the measurements and time diaries were all done during the heating season, the authors are confident that the tenants air quite a lot even during this period. The third interesting thing is the household electricity use. In BES-v.2, the input value was 39% less than in BES-ref. The value for BES-v.2 was acquired by reading the electricity meters of each apartment, and through the logged measurements. The input value of household electricity for BES-ref was taken from Sveby´s reports on standardized user input data. As can be seen, the standardized user input data are much higher than the actual use in this case. Electricity use can be 30% higher than average during the winter and 30% less than average during the summer according to Reference [10]. If this applies for the tenants in this case, it means even less internal gains during the summer and, therefore, probably even higher energy use for space heating. The tenants are, however, home a lot more than the average population. This means higher internal gains from occupancy than what you get from using template values; however, in the end, there was still a difference in total energy use for space heating of 24% between the two models, where BES-v.2 had the highest use.

3.2.2. Apartment Level

When analyzing the models on an apartment level, quite big differences can be seen between the two models (BES-ref and BES-v.2), as well as between the apartments themselves, especially in BES-v.2. The results from the five-day simulation periods can be seen in Table 11, and the results for the full-year simulation can be seen in Table 12. The apartment with the biggest difference in total energy use for space heating between BES-ref and BESv.2 was apartment 5 on the second floor, which is the apartment with the tenant who made the most detailed time diary. The difference in total energy use for space heating in this apartment was 170% according to the simulations. This seems to be due to the fact that she is one of the tenants with the lowest indoor temperature, which means that the BES-ref model of this apartment hardly used any energy for space heating at all (only 6.6 kWh including losses due to airing in the five-day simulation period). However, when airing, occupancy, and schedules for electricity use according to logged measurements and the time diary were added, the energy use for space heating increased with 329%. This is a product of the tenant using about 60% less electricity, and airing quite a lot more in BES-v.2 than in BES-ref. However, this tenant is not the one that seems to be doing the most airing. In the apartment with the highest losses due to airing, the difference between BES-ref and BES-v.2 was 991%. This, however, might not be as strange as it sounds, as the tenants in this apartment wrote in their time diary that they always had a window open during the night when they were sleeping, which can be seen quite clearly in the temperature measurements from their apartment. A distinct pattern of dropping temperature during the nights can be seen in Figure 6, which corresponds to the tenants' entry of opening windows in their time diary. Figure 6 also shows the simulated temperature curves from the models BES-ref and BES-v.2. The temperature curve from BES-v.2 follows the logged measurement curve in a good way, while the curve from BES-ref has much less fluctuation in temperature compared to the logged measurements and BES-v.2.

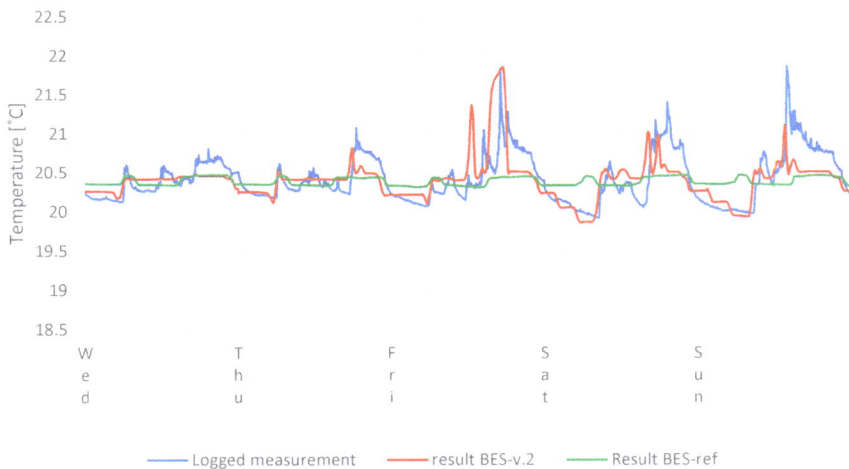

Figure 6. Temperature curves from logged measurements, BES-ref, and BES-v.2 for the period of 21–25 January. The blue curve shows the logged measurements, the red curve shows results from BES-v.2, and the green curve shows results from BES-ref.

Table 11. Results at the apartment level from simulations using the two models (BES-ref and BES-v.2). The simulations were done for the same period as the logged measurements, and the time diaries were done for each apartment.

Parameters	Floor 1			Floor 2			Floor 3		
	BES-ref	BES-v.2	Difference	BES-ref	BES-v.2	Difference	BES-ref	BES-v.2	Difference
Apartment 1	Simulation period 10 to 14 December 2014			Simulation period 14 to 18 January 2015			Simulation period 10 to 14 December 2014		
Energy supplied by water radiators (DH) (kWh)	127.5	137.4	8%	83	90.3	9%	101.3	115.8	14%
AHU heating (electrical) (kWh)	3.4	3.6	6%	4.9	11.4	133%	4.2	4.2	0%
Total energy use for space heating (kWh)	130.9	141	8%	87.9	101.7	16%	105.5	120	14%
Energy loss due to airing (kWh)	4.6	4	−13%	5.5	5.9	7%	5.5	1.4	−75%
Indoor temperature (°C)	22.0–22.1	21.8–22.1	-	21.8–22.0	21.0–22.0	-	21.9–22.0	21.9–22.1	-
Maximum CO_2 (ppm)	670.7	672.7	2.0	512.5	594.9	82.4	626.3	627.4	1.1
Maximum PPD (%)	13.5	13.9	0.4	13.3	14.6	1.3	12.3	12.3	0.0
Apartment 2	Simulation period 21 to 25 January 2015			Simulation period 21 to 25 January 2015			Simulation period 10 to 14 December 2014		
Energy supplied by water radiators (DH) (kWh)	83.6	123.7	48%	40.8	77.5	90%	82.6	94.6	15%
AHU heating (electrical) (kWh)	14	19.7	41%	14.3	19.1	34%	4.1	13	217%
Total energy use for space heating (kWh)	97.6	143.4	47%	55.1	96.6	75%	86.7	107.6	24%
Energy loss due to airing (kWh)	4.3	46.9	991%	4.3	38.3	791%	4.3	12.8	198%
Indoor temperature (°C)	20.4–20.5	19.9–21.9	-	20.3–20.5	19.9–22.5	-	21.6–21.7	21.2–21.8	-
Maximum CO_2 (ppm)	545.3	686.5	141.2	69.7	686.4	616.7	692.1	692.4	0.3
Maximum PPD (%)	23.5	25.5	2.0	21.5	23.0	1.5	13.7	14.4	0.7
Apartment 3	Simulation period 3 to 7 December 2014			Simulation period 19 to 23 November 2014			Simulation period 28 January to 1 February 2015		
Energy supplied by water radiators (DH) (kWh)	74	77	4%	45.4	69.2	52%	105.8	136.4	29%
AHU heating (electrical) (kWh)	9.1	22	142%	4.1	4.3	5%	6.5	6.4	−2%
Total energy use for space heating (kWh)	83.1	99	19%	49.5	73.5	48%	112.3	142.8	27%
Energy loss due to airing (kWh)	4.3	4.2	−2%	4.3	15	249%	4.3	25	481%
Indoor temperature (°C)	20.2–20.4	20.1–20.6	-	20.9–21.2	20.8–21.5	-	22.0–22.2	22.0–22.3	-
Maximum CO_2 (ppm)	545.1	689	143.9	691.3	689.6	−1.7	546	579.1	33.1
Maximum PPD (%)	23.8	24.8	1.0	15.6	16.8	1.2	13.3	13.0	−0.3

Table 11. *Cont.*

Parameters	Floor 1			Floor 2			Floor 3		
	BES-ref	BES-v.2	Difference	BES-ref	BES-v.2	Difference	BES-ref	BES-v.2	Difference
Apartment 4	Simulation period 21 to 25 January 2015			Simulation period 10 to 14 December 2014			Simulation period 14 to 18 January 2015		
Energy supplied by water radiators (DH) (kWh)	105	138.9	32%	48.4	57.8	19%	46.2	76.5	66%
AHU heating (electrical) (kWh)	8.4	8.5	1%	3.5	11.1	217%	5.1	5.1	0%
Total energy use for space heating (kWh)	113.4	147.4	30%	51.9	68.9	33%	51.3	81.6	59%
Energy loss due to airing (kWh)	3.6	27.5	664%	3.6	10.2	183%	3.6	14.7	308%
Indoor temperature (°C)	22.0–22.1	21.8–22.1	-	21.6–21.8	21.2–21.9	-	20.9–21.1	20.8–21.1	-
Maximum CO_2 (ppm)	570.9	606.3	35.4	737.6	742.1	4.5	570.6	570.7	0.1
Maximum PPD (%)	14.4	15.0	0.6	12.7	13.7	1.0	16.2	16.6	0.4
Apartment 5	Simulation period 28 January to 1 February 2015			Simulation period 19 to 23 November 2014			Simulation period 3 to 7 December 2014		
Energy supplied by water radiators (DH) (kWh)	7.6	20.8	174%	6.6	28.3	329%	48.5	89.4	84%
AHU heating (electrical) (kWh)	9	3.1	-66%	6.1	6	-2%	7	14.6	109%
Total energy use for space heating (kWh)	16.6	23.9	44%	12.7	34.3	170%	55.5	104	87%
Energy loss due to airing (kWh)	2.9	13.5	366%	3.6	11.6	222%	3.6	24.4	578%
Indoor temperature (°C)	19.3–20.2	19.3–19.9	-	19.5–20.6	19.4–20.4	-	20.6–20.8	20.5–20.8	-
Maximum CO_2 (ppm)	613	2252	1639.0	566.3	714.3	148.0	566.6	566.8	0.2
Maximum PPD (%)	27.7	26.9	-0.9	22.7	22.3	-0.4	18.6	19.5	0.9

Table 12. Results from yearly simulations at the apartment level for both models (BES-ref and BES-v.2).

Parameters	Floor 1			Floor 2			Floor 3		
Apartment 1	BES-ref	BES-v.2	Difference	BES-ref	BES-v.2	Difference	BES-ref	BES-v.2	Difference
Energy supplied by water radiators (DH) (kWh)	4856.8	5386.6	11%	3177.5	3279.1	3%	3417.3	3993.8	17%
AHU heating (electrical) (kWh)	188.7	194.7	3%	240.8	1695.4	604%	227.9	230.1	1%
Total energy use for space heating (kWh)	5045.5	5581.3	11%	3418.3	4974.5	46%	3645.2	4223.9	16%
Energy loss due to airing (kWh)	332	188.9	-43%	400	217.3	-46%	400	60.2	-85%
Indoor temperature (°C)	21.9–31.0	21.6–29.5	N/A	21.6–33.0	18.4–31.2	N/A	21.8–33.7	21.8–31.9	N/A
Maximum CO_2 (ppm)	682.4	682.3	0%	517.8	610.1	18%	637.8	637.7	0%
Apartment 2									
Energy supplied by water radiators (DH) (kWh)	2896	4158.6	44%	1055.1	1952.4	85%	2859.3	3119.7	9%
AHU heating (electrical) (kWh)	289.9	442.3	53%	282.3	432.5	53%	197.4	490.6	149%
Total energy use for space heating (kWh)	3185.9	4600.9	44%	1337.4	2384.9	78%	3056.7	3610.3	18%
Energy loss due to airing (kWh)	310	1842.6	494%	310	798.9	158%	310	498.7	61%
Indoor temperature (°C)	20.3–30.5	17.7–30.6	N/A	20.2–32.5	18.8–32.7	N/A	21.5–33.0	20.8–31.9	N/A
Maximum CO_2 (ppm)	551.3	696.3	26%	704.9	698.2	-1%	705.6	705	0%
Apartment 3									
Energy supplied by water radiators (DH) (kWh)	2845	3129.3	10%	1786.6	2667.5	49%	3668.4	4557.8	24%
AHU heating (electrical) (kWh)	307.6	780.7	154%	242.4	253.5	5%	169.2	167.5	-1%
Total energy use for space heating (kWh)	3152.6	3910	24%	2029	2921	44%	3837.6	4725.3	23%
Energy loss due to airing (kWh)	310	441.9	43%	310	797	157%	310	782.7	152%
Indoor temperature (°C)	20.1–30.7	19.9–30.1	N/A	20.8–32.7	20.6–31.2	N/A	21.9–33.3	21.5–31.7	N/A
Maximum CO_2 (ppm)	551.1	702.3	27%	704.6	702.9	0%	552.5	649	17%
Apartment 4									
Energy supplied by water radiators (DH) (kWh)	4068.7	4981.8	22%	1741.4	1893.5	9%	1692.1	3006.3	78%
AHU heating (electrical) (kWh)	150.1	151.5	1%	169.7	418.6	147%	208.3	214	3%
Total energy use for space heating (kWh)	4218.8	5133.3	22%	1911.1	2312.1	21%	1900.4	3220.3	69%
Energy loss due to airing (kWh)	264	734	178%	264	403.6	53%	264	1154	337%
Indoor temperature (°C)	21.8–30.6	21.4–29.4	N/A	21.5–32.7	20.9–31.6	N/A	20.8–33.0	20.5–30.9	N/A
Maximum CO_2 (ppm)	577.1	614.6	6%	753	756.9	1%	578.8	577.8	0%
Apartment 5									
Energy supplied by water radiators (DH) (kWh)	457.5	641.4	40%	425.9	587	38%	1492.3	3009.6	102%
AHU heating (electrical) (kWh)	244.4	68.86	-72%	277.7	317.1	7%	229.1	499.3	118%
Total energy use for space heating (kWh)	701.9	710.26	1%	703.6	904.1	28%	1721.4	3508.9	104%
Energy loss due to airing (kWh)	212	404.2	91%	264	771.9	192%	264	1269.3	381%
Indoor temperature (°C)	19.2–30.6	18.9–29.6	N/A	19.3–33.1	16.3–31.2	N/A	20.5–33.5	20.1–31.4	N/A
Maximum CO_2 (ppm)	622.7	3120	401%	574.8	791.7	38%	575.1	573.3	0%

The average difference in total energy use for space heating in the 15 apartments for the five-day simulations was 47%, and, for a full-year simulation, it was 37%. In general, the biggest differences between the models can be found in the amount of energy loss due to airing, both in full-year simulations and in the simulations of the validation periods. Since this was also the case in the building-level simulations, no further discussion on this takes place here. To be able to see how well the two models, BES-ref and BES-v.2, could replicate the logged measurements of temperature during the measurement period, a statistical analysis was done by calculating the mean bias error (MBE), normalized mean bias error (NMBE), and coefficient of variance (CV) of the root-mean-square error (RMSE) according to Reference [38]. Table 13 shows the results of MBE, NMBE, and CV (RMSE), when comparing simulated results of temperature to logged measurements. As can be seen in Table 13, BES-v.2 had a lower MBE, NMBE, and CV (RMSE) in all cases, which shows that BES-v.2 is better for predicting indoor temperature than BES-ref. According to ASHRAE Guideline 14, the MBE for hourly data of energy use should not exceed 10, and the CV (RMSE) should not be more than 30. In this case, the data are from every five minutes and do not represent energy use; instead, the temperature is compared. However, all values for both models were under the recommended values from ASHRAE Guideline 14 [39].

Table 13. Mean bias error (MBE), normalized mean bias error (NMBE), and coefficient of variation (CV) of the root-mean-square error (RMSE) for logged measurements of temperature and the simulated results for BES-ref and BES-v.2.

Apartment and Floor	Logged vs. BES-ref			Logged vs. BES-v2		
Floor 1	MBE	NMBE	CV (RMSE) (%)	MBE	NMBE	CV (RMSE) (%)
Apartment 1	−0.068	−0.31	11.7	−0.0077	−0.035	1.33
Apartment 2	0.028	0.14	5.19	0.012	0.059	2.26
Apartment 3	0.076	0.38	14.3	0.071	0.35	13.2
Apartment 4	0.051	0.23	8.75	−0.015	−0.067	2.55
Apartment 5	0.078	0.40	15.2	−0.054	−0.28	10.49
Floor 2						
Apartment 1	0.032	0.15	5.54	0.021	0.098	3.71
Apartment 2	0.028	0.14	5.19	−0.0099	−0.048	1.84
Apartment 3	−0.079	−0.38	14.2	−0.008	−0.038	1.46
Apartment 4	0.0084	0.039	1.48	0.0061	0.028	1.07
Apartment 5	−0.056	−0.28	10.8	−0.023	−0.11	4.36
Floor 3						
Apartment 1	−0.068	−0.31	11.7	−0.0077	−0.035	1.33
Apartment 2	0.0084	0.039	1.48	0.0067	0.31	1.18
Apartment 3	−0.065	−0.30	11.2	−0.012	−0.055	2.11
Apartment 4	−0.059	−0.28	10.6	0.021	0.10	3.80
Apartment 5	−0.14	−0.68	25.7	0.040	0.20	7.53

3.2.3. Room Level

For the room-level simulations, the apartment with the most detailed time diary was chosen. The apartment was then modeled as four zones: living room (including hall), kitchen, bedroom, and bathroom together with a walk-in closet. The doors inside the apartment were set to never close since this seemed to be the case when visiting the tenant. Table 14 shows the comparison between BES-ref and BES-v.2 in apartment 5 on the second floor.

Table 14. Comparison of apartment 5, second floor, at the room level between BES-ref and BES-v.2 for the validation period and for a full-year simulation.

Apartment 5, Floor 2	Simulation Period 19/11-23/11			Full Year Simulation		
	BES-ref	BES-v.2	Difference	BES-ref	BES-v.2	Difference
Living room						
Energy supplied by water radiators (DH) (kWh)	2.4	15.9	563%	190.7	564.8	196%
Energy loss due to airing (kWh)	1.7	2.6	53%	122	242.9	99%
Indoor temperature (°C)	19.6-20.1	19.6-20.2	-	19.4-32.1	18.7-30.9	-
Maximum CO_2 (ppm)	579.6	846.8	267.2	597.5	883.6	286.1
Maximum PPD (%)	21.7	21.6	-0.10	87.8	66.7	-21.06
Kitchen						
Energy supplied by water radiators (DH) (kWh)	1.44	0	-100%	100.6	0	-100%
Energy loss due to airing (kWh)	0.84	0.2	-76%	61.1	45.1	-26%
Indoor temperature (°C)	19.6-20.1	19.6-21.1	-	19.4-32.6	19.0-32.0	-
Maximum CO_2 (ppm)	559	821.5	821.5	569.6	912.8	343.2
Maximum PPD (%)	N/A	22.4	-	N/A	79.5	-
Bedroom						
Energy supplied by water radiators (DH) (kWh)	5.41	8.7	61%	215.7	265.8	23%
Energy loss due to airing (kWh)	0.81	4.5	456%	59.4	172.3	190%
Indoor temperature (°C)	19.5-19.8	18.9-19.9	-	19.4-32.2	17.6-31.2	-
Maximum CO_2 (ppm)	546.8	793.1	246.3	557.4	807.6	250.2
Maximum PPD (%)	N/A	25.7	-	N/A	72.7	-
Bathroom						
Energy supplied by water radiators (DH) (kWh)	0.55	0	-100%	40	3.4	-92%
Energy loss due to airing (kWh)	0.55	0.1	-82%	39.9	27.1	-32%
Indoor temperature (°C)	20.0-20.5	20.1-20.2	-	19.7-31.9	19.7-30.3	-
Maximum CO_2 (ppm)	579.5	834.1	254.6	596.4	881.7	285.3
Maximum PPD (%)	N/A	18.5	-	N/A	60.6	-
Apartment total						
Energy supplied by water radiators (DH) (kWh)	9.8	24.6	151%	547	834	52%
AHU heating (electrical) (kWh)	7.5	7.7	3%	332.6	345.3	4%
Total energy use for space heating (kWh)	13.4	32.3	87%	879.6	1179.3	34%
Energy loss due to airing (kWh)	3.9	7.4	90%	282.4	487.4	73%
Indoor temperature (°C)	19.5-20.5	18.9-21.1	-	19.4-32.6	17.6-32.0	-
Maximum CO_2 (ppm)	579.6	846.8	267.2	597.5	912.8	315.3
Maximum PPD (%)	21.7	25.7	4.02	87.8	79.5	-8.32

When comparing the models at the room level, there can also be seen large differences between the results. The living room is the room where the most difference can be seen in energy supplied by the radiators. Even though it does not have the highest losses due to airing, it does seem to use a lot of energy trying to heat up the other rooms in the apartment. This, in combination with airing through the balcony door (which in this apartment is very frequent, and the door is situated in the living room), and the fact that the occupant in BES-ref is only active in the living room, contributing to higher internal gains in that model, should be the most probable cause of this. In the kitchen, no energy is supplied by the water radiators during simulations for the validation period. This is because the tenant spends quite a lot of time in the kitchen, as well as using the stove and other household appliances there, which is all in the detailed schedules in BES-v.2, which results in all energy coming from internal gains. The bedroom is the room where the most airing losses occur in BES-v.2, which is probably due to the fact that the tenant sometimes has the window open during the nights.

3.3. General Discussion

La Fleur, Moshfegh, and Rohdin [21] concluded that assumptions on user behavior have a significant impact on energy-saving potential when renovating or retrofitting a building, and this study shows how much difference there can actually be due to assumptions of user behavior. The largest differences in energy use in this study between using standardized user input data and behavioral schedules based on data collected from the actual building were due to airing and electricity use. It seems as though the standardized user input data are a bit too generalized and cannot actually be used in this case where the studied object is a retirement home. Even so, when designing a building for elderly care, one is supposed to use template values for regular housing in the calculations and/or simulations, as was done in BES-ref. The results show that schedules created from using time diaries and logged measurements bring the simulated results much closer to reality than just using template values. There was, however, quite a difference between the level of detail in the time diaries. The time diary shown in this article was the most detailed one, but some of the tenants almost only wrote whether they were home or not, and some of them wrote activities with no time. It is, therefore, of great importance, when conducting a study with time diaries, that the participants are told what sort of activities they should write down and to make sure they understand the importance of the level of detail in their time diaries. Since logged measurements could only be done in two apartments at the same time in this study due to a lack of equipment, it means that the schedules for the apartments were validated during different weeks, and only for one week during the heating season. In future studies, one should try to use time diaries and logged measurements at least four times during a year, if the building is situated in the temperate climate zone, i.e., one week in each season of the year (winter, spring, summer, and autumn), to see if there are any changes in the user behavior and to be able to make even more detailed schedules which vary during full-year simulations. Since this is a case study, conducted on a single retirement home, it might be hard to generalize the findings of this study to all retirement homes. However, the authors believe that most of the behavioral patterns are probably quite similar in homes of elderly people and retirement homes in Sweden. More studies on this type of housing are, however, required to be able to generalize and to create template values for retirement homes or housing for elderly people. One thing that also needs to be accounted for is that people that are over 70 today might behave a lot differently from people that will turn 70, for example, in 30 years, since they probably will have other accustomed behaviors. It should also be pointed out that user behavior is not the only thing that can affect the energy performance gap; the building parameters themselves such as U-values, air flow in the ventilation system, temperature set-point, etc. are of great importance to have a model that can predict energy use well. In this study, however, the main focus was on user behavior and the difference in energy use when using standardized user input data compared to using schedules based on gathered data from the actual building. The building parameters that were measured were indoor temperature to determine temperature set-point, leakage through the building envelope with blower door technology, ventilation air flow, and temperature, while U-values for the walls and windows were not measured. The U-values are based on construction blueprints and product specifications of the used materials. This can of course affect how the different user input data affect the simulated energy use and could, therefore, give either higher or lower discrepancies between the two cases (BES-ref and BES-v.2).

4. Conclusions

Time diaries and interviews, together with logged measurements, can be great tools to detect behavior that affects energy use in buildings; as far as the authors are concerned, this was not done simultaneously before. The time diaries provide a way of determining what actually causes variations in logged measurements. This can greatly benefit researchers studying user behavior and energy use in buildings and/or people that work with BES models. The reason is that it eliminates a lot of assumptions about the user behavior and/or why variations occur in measurements. This can also be good when deciding on which measures should be taken if the building is being retrofitted, since it gives a clear view on how the building is being used. They can also be used to create detailed schedules and behavioral models for BES models so that they are better at predicting actual energy use. However, using and creating these schedules from scratch is time-consuming and costly, which means that this method would probably not be viable in industry. Because of this, we believe that standardized user input data are still necessary, but there needs to be more variety in these standards. As shown in this article, user data for regular residential housing did not work well when used in this particular retirement home. Therefore, we believe that more studies of this nature need to be done on these sorts of buildings and also different kinds of buildings, both public and domestic. Thus, when designing new buildings or planning to renovate or retrofit old ones, there will be standardized user input data for that exact type of building. The largest offsets between using standardized user input data and input data from actual logged measurements and time diaries in this case could be seen in energy losses due to airing, and the amount of household electricity use. The tenants in this study aired a lot more and used a lot less electricity than the general population living in residential houses, according to Reference [10], which affected the energy use quite substantially. Based on the present findings, an update to the source hierarchy is also suggested for input data to BES models described by Raftery et al. [4]. Since time diaries represent a sort of in situ measurement (but not technical, since it is also behavioral), it should be in second place in the source hierarchy created by Raftery et al. [4], together with spot and short-term measurements. With time diaries, short-term measurements can be greatly expanded, since one can actually determine why drops and peaks occur in the different measurements. This would create the following source hierarchy: data-logged measurements, spot or short-term measurements and *time diaries*, direct observation, operator and personnel interviews, operation documents, commissioning documents, benchmark studies and best practice guides, standards, specifications, and guidelines, and design stage information.

Author Contributions: Conceptualization, J.C. and B.M.; data curation, J.C.; funding acquisition, B.M.; investigation, J.C. and K.T.; methodology, J.C.; project administration, B.M.; supervision, B.M.; writing—original draft, J.C. and K.T.

Funding: This work was funded by the KK foundation (KK-stiftelsen, Stiftelsen för Kunskaps och Kompetensutveckling), grant number 20150133, and the Swedish Energy Agency (Energimyndigheten), grant number 37492-1.

Acknowledgments: The authors of this article would like to thank Jakob Rosenquist, previously employed at Linköping University's department of energy system, for the great help while conducting measurements at the studied site. We would also like to thank Patrik Rohdin for helping with the IDA-ICE simulations and for initiating the project together with Kajsa Ellegård at Linköping University, Sweden. Also, huge thanks goes out to all the residents and staff at Räknestickan Retirement Home for their participation in the study and for letting us into their homes and their workplace to do our measurements.

Conflicts of Interest: The authors declare no conflicts of interest.

References

1. European Commission Buildings. Available online: https://ec.europa.eu/energy/en/topics/energy-efficiency/energy-performance-of-buildings (accessed on 29 May 2019).
2. Gynther, L.; Lappillone, B.; Pollier, K. Energy Efficiency Trends and Policies in the Household and Tertiary Sectors. An Analysis Based on the ODYSSEE and MURE Databases. Available online: http://www.odyssee-mure.eu/publications/br/energy-efficiency-trends-policies-buildings.pdf (accessed on 29 May 2019).
3. *Energimyndigheten flerbostadshus och lokaler 2016*; Energimyndigheten: Eskilstuna, Sweden, 2016.
4. Raftery, P.; Keane, M.; O'Donnell, J. Calibrating whole building energy models: An evidence-based methodology. *Energy Build.* **2011**, *43*, 2356–2364. [CrossRef]
5. van den Brom, P.; Meijer, A.; Visscher, H. Performance gaps in energy consumption: household groups and building characteristics. *Build. Res. Inf.* **2017**, *46*, 1–17. [CrossRef]
6. Clarke, J.A.; Joe, A. *Energy Simulation in Building Design*; Butterworth-Heinemann: Oxford, UK, 2001; ISBN 9780750650823.
7. Andersen, R.V.; Olesen, B.W.; Toftum, J. Simulation of the Effects of Occupant Behaviour on Indoor Climate and Energy Consumption. In Proceedings of the Clima 2007: 9th REHVA world congress: WellBeing Indoors, Helsinki, Finland, 10–14 June 2007.
8. Svebyprogrammet. *Brukarindata Undervisningsbyggnader*; Sveby: Stockholm, Sweden, 2016.
9. Sveby. *Brukarindata Kontor: Version 1.1*; Sveby: Stockholm, Sweden, 2013.
10. Sveby. *Brukarindata Bostäder*; Sveby: Stockholm, Sweden, 2012.
11. Delzendeh, E.; Wu, S.; Lee, A.; Zhou, Y. The impact of occupants' behaviours on building energy analysis: A research review. *Renew. Sustain. Energy Rev.* **2017**, *80*, 1061–1071. [CrossRef]
12. Richardson, I.; Thomson, M.; Infield, D. A high-resolution domestic building occupancy model for energy demand simulations. *Energy Build.* **2008**, *40*, 1560–1566. [CrossRef]
13. Widén, J.; Lundh, M.; Vassileva, I.; Dahlquist, E.; Ellegård, K.; Wäckelgård, E. Constructing load profiles for household electricity and hot water from time-use data-Modelling approach and validation. *Energy Build.* **2009**, *41*, 753–768. [CrossRef]
14. Chiou, Y.-S.; Carley, K.M.; Davidson, C.I.; Johnson, M.P. A high spatial resolution residential energy model based on American Time Use Survey data and the bootstrap sampling method. *Energy Build.* **2011**, *43*, 3528–3538. [CrossRef]
15. Aerts, D.; Minnen, J.; Glorieux, I.; Wouters, I.; Descamps, F. A method for the identification and modelling of realistic domestic occupancy sequences for building energy demand simulations and peer comparison. *Build. Environ.* **2014**, *75*, 67–78. [CrossRef]
16. Sahlin, P.; Eriksson, L.; Grozman, P.; Johnsson, H.; Shapovalov, A.; Vuolle, M. Whole-building simulation with symbolic DAE equations and general purpose solvers. *Build. Environ.* **2004**, *39*, 949–958. [CrossRef]
17. Kropf, S.; Zweifel, G. *Validation of the Building Simulation Program IDA-ICE According to CEN 13791 "Thermal Performance of Buildings—Calculation of Internal Temperatures of a Room in Summer Without Mechanical Cooling—General Criteria and Validation Procedures"*; Fachhochschule Zentralschweiz: Luzern, Schweiz, 2001.
18. Achermann, M. *Validation of IDA ICE, Version 2.11.06 With IEA Task 12—Envelope BESTEST*; Fachhochschule Zentralschweiz: Luzern, Schweiz, 2000.
19. Ryan, E.M.; Sanquist, T.F. Validation of building energy modeling tools under idealized and realistic conditions. *Energy Build.* **2012**, *47*, 375–382. [CrossRef]
20. Coakley, D.; Raftery, P.; Keane, M. A review of methods to match building energy simulation models to measured data. *Renew. Sustain. Energy Rev.* **2014**, *37*, 123–141. [CrossRef]
21. La Fleur, L.; Moshfegh, B.; Rohdin, P. Measured and predicted energy use and indoor climate before and after a major renovation of an apartment building in Sweden. *Energy Build.* **2017**, *146*, 98–110. [CrossRef]
22. Liu, L.; Rohdin, P.; Moshfegh, B. Evaluating indoor environment of a retrofitted multi-family building with improved energy performance in Sweden. *Energy Build.* **2015**, *102*, 32–44. [CrossRef]
23. Molin, A.; Rohdin, P.; Moshfegh, B. Investigation of energy performance of newly built low-energy buildings in Sweden. *Energy Build.* **2011**, *43*, 2822–2831. [CrossRef]
24. Rohdin, P.; Molin, A.; Moshfegh, B. Experiences from nine passive houses in Sweden—Indoor thermal environment and energy use. *Build. Environ.* **2014**, *71*, 176–185. [CrossRef]

25. Gauthier, S.; Shipworth, D. Behavioural responses to cold thermal discomfort. *Build. Res. Inf.* **2015**, *43*, 355–370. [CrossRef]
26. Andersen, R.V.; Toftum, J.; Olesen, B.W. Simulation of the effects of window opening and heating set-point behaviour on indoor climate and building energy performance. In Proceedings of the Healthy Buildings 2009, Syracuse, NY, USA, 13–17 September 2009.
27. Isaksson, C.; Ellegård, K. Anchoring energy efficiency information in households' everyday projects: Peoples' understanding of renewable heating systems. In *Energy Efficiency*; Springer: Berlin/Heidelberg, Germany, 2015; pp. 353–364.
28. Ellegård, K.; Palm, J. Visualizing energy consumption activities as a tool for making everyday life more sustainable. *Appl. Energy* **2011**, *88*, 1920–1926. [CrossRef]
29. Hellgren, M. *Energy Use as a Consequence of Everyday Life*; Linköping University: Linköping, Sweden, 2015; ISBN 9789176859100.
30. Clifford, N.; Cope, M.; Gillespie, T.; French, S. *Key Methods in Geography*, 3rd ed.; Sage Publications: Thousand Oaks, CA, USA, 2016; ISBN 9781446298602.
31. Vrotsou, K.; Bergqvist, M.; Cooper, M. PODD: A Portable Diary Data Collection System. In Proceedings of the 2014 International Working Conference on Advanced Visual Interfaces, Como, Italy, 27-29 May 2014; pp. 381–382.
32. SCB. Statistikdatabas för Äldreomsorg. Available online: https://www.socialstyrelsen.se/statistik/ statistikdatabas/aldreomsorg (accessed on 16 May 2019).
33. United Nations, UN. *Transforming our World: The 2030 Agenda for Sustainable Development*; United Nations: New York, NY, USA, 2015.
34. Abramsson, M.; Andersson, E.V.A. Changing preferences with ageing—housing choices and housing plans of older people. *Hous. Theory Soc.* **2016**, *33*, 217–241. [CrossRef]
35. Abramsson, M.; Andersson, E.V.A.K. Residential mobility patterns of elderly—leaving the house for an apartment. *Hous. Stud.* **2012**, *27*, 582–604. [CrossRef]
36. Sveby. *Brukarindata för energiberäkningar i bostäder*; Sveby: Stockholm, Sweden, 2009.
37. Carlander, J.; Tullsson, F. *Utredning av energi och inomhusklimat samt åtgärdsförslag för kvarteret Räknestickan 1*; Linköping University: Linköping, Sweden, 2014.
38. Ruiz, G.R.; Bandera, C.F. Validation of calibrated energy models: Common errors. *Energies* **2017**, *10*, 1587. [CrossRef]
39. ASHRAE. *Guideline 14-2002: Measurement of Energy and Demand Savings*; ASHRAE: Atlanta, GA, USA, 2002.

Article

The Influence of Energy Renovation on the Change of Indoor Temperature and Energy Use

Anti Hamburg * and Targo Kalamees

Nearly Zero Energy Research Group, Tallinn University of Technology, Tallinn 19086, Estonia;
targo.kalamees@taltech.ee
* Correspondence: anti.hamburg@taltech.ee; Tel.: +372-53-419-274

Received: 26 September 2018; Accepted: 12 November 2018; Published: 16 November 2018

Abstract: The aim of the renovation of apartment buildings is to lower the energy consumption of those buildings, mainly the heating energy consumption. There are few analyses regarding those other energy consumptions which are also related to the primary energy need for calculating the energy efficiency class, including the primary energy need of calculated heating, domestic hot water (DHW), and household electricity. Indoor temperature is directly connected with heating energy consumption, but it is not known yet how much it will change after renovation. One of the research issues relates to the change of electricity and DHW usage after renovation and to the question of whether this change is related to the users' behavior or to changes to technical solutions. Thirty-five renovated apartment buildings have been analyzed in this study, where the data of indoor temperature, airflow, and energy consumption for DHW with and without circulation and electricity use in apartments and common rooms has been measured. During research, it turned out that the usage of DHW without circulation and the usage of household electricity do not change after renovation. Yet there is a major increase in indoor temperature and DHW energy use in buildings that did not have circulation before the renovation. In addition, a small increase in the use of electricity in common areas was discovered. This study will offer changes in calculations for the energy efficiency number.

Keywords: indoor temperature after renovation; electricity use; DHW energy use; user behavior; standard use

---

1. Introduction

Buildings are responsible for approximately 40% of energy consumption in the European Union countries. Final energy use in Estonia is 33.0 TWh/a and the share of buildings is 50% [1]. The Energy Performance of Buildings Directive (EPBD) [2], the Energy Efficiency Directive (EED) [3], and the Renewable Energy Directive (RED) [4] define a framework for long-term improvements in the energy performance of Europe's building stock.

To decrease energy use, EU Member States shall establish a long-term renovation strategy to support the renovation of the national stock, into a highly energy efficient and decarbonized building stock by 2050, facilitating the cost-effective transformation of existing buildings into nearly zero-energy buildings (nZEB) [2]. D'Agostino et al. [5] provide an overview of the status of implementation of nZEBs in Europe and showed that building retrofit is one of the biggest challenges that Europe is facing.

Energy renovation is one of the most effective and cost-efficient ways to improve indoor climate and achieve energy savings. Indoor climate and energy modeling have estimated the savings potential to be in the range of 40–80% of energy use [6–9]. Modeling has usually been done on the standard use of buildings [10]. In reality, the use of user-related energy can be different compared with the standard use because of the density of occupants or the number of apartments in a building [11]. The use of standardized user profiles for modeling is good for comparing similar buildings and to work out the building stock level. To work out cost effective energy renovation measures for specific buildings, this peculiarity has to be taken into account. That is why it is important to investigate user-related indoor climate and energy consumptions before renovation and to compare that with standard use energy.

The rebound effect has been investigated by Sorrell [12]. He has found that most governments are seeking solutions to improve energy efficiency to fulfill their energy policy goals. But measured energy savings generally turn out to be appreciably lower. He postulates that one explanation could be that improvements in energy efficiency encourage a higher use of those services which are provided by the energy supply. This situation where the calculated energy savings are not being achieved due to behavioral responses has come to be known as the energy efficiency 'rebound effect'. In some cases this rebound effect is high enough to lead to an overall increase in energy consumption, an outcome termed as 'backfire' [13]. In general, the rebound effect is not taken in to account in energy efficiency calculations, which may lead to an overestimation of the future energy savings [12]. The occupants' behavior has also been identified as one of the reasons for the energy performance gap in other studies [14,15]. The systematic review of the literature on occupant and building energy performance by Zhang et al. [16] estimated that the occupant behavior-related energy-saving potential could be in the range of 10–25% for residential buildings. Menezes et al. [17] highlighted the need for a better understanding of occupancy behavior patterns and the use of more realistic input parameters in energy models; needed to bring the predicted figures closer to reality.

This study investigates indoor climate and energy consumption, which is connected with occupant behavior before and after renovation. Energy renovated apartment buildings in Estonia are used as an example. The research questions of the study are the following:

- Whether and how much does energy renovation influence indoor climate and human related energy use?
- How well do real indoor climate parameters correspond to the standard use of a building before and after the renovation?
- Is it appropriate to use a different standard use for the energy certification process for apartment buildings?

2. Methods

2.1. Studied Buildings

In Estonia, the majority of apartment buildings that have been constructed between WWII and 1990 have the same typical problems: high energy-consumption levels, insufficient ventilation (natural ventilation without any outdoor air inlets), uneven indoor temperatures, and insufficient thermal comfort levels [18–20]. From the year 2010, more than 1000 apartment buildings have undergone renovation, the majority of them supported by Fund KredEx. The energy renovation of 663 apartment buildings resulted in average energy savings of 43% [21]. The main challenge was to achieve the same level of heating energy consumptions as estimated by modeling before renovation [22].

The energy use and indoor climate were investigated in 35 apartment buildings (Table 1).

The average number of apartments in one building was 27 (varied between 12 and 72, standard deviation is 17), average heated area was 1757 m^2 (varied between 550 m^2 and 5030 m^2, standard deviation is 1046). Average occupancy in one apartment was 2.2 persons (varied between 1.1 and 3.3, standard deviation is 0.5) and the average area per person was 31 m^2/person (varied between 16 m^2/person and 55 m^2/person, standard deviation is 7.7).

Table 1. Studied buildings. DHW: domestic hot water.

Code	No. of Apartments	Heated Net Area, m²	No. of People	Ventilation	DHW Circulation before/after Renovation	Additional Insulation, cm/Thermal Transmittance (U W/(m²·K))		
						Walls	Roof	Windows
1.1	25	1665	47	Exhaust fan	−/+	+20/0.16	+30/0.10	≤1.1
1.2	18	1673	45	Exhaust fan	−/+	+15/0.18	+45/0.10	≤1.6
1.3	18	1592	44	Exhaust fan	+/+	+15/0.18	+30/0.12	≤1.5
Target: EPC "D", PE ≤ 180 kWh/(m²·a) (DHW with electrical boilers). 40% grant.								
2.1	12	1029	40	Central AHU	−/−	+15–20/0.21	+23/0.13	≤1.4
2.2	18	1490	27	Central AHU	−/−	+15–20/0.20	+30/0.11	≤1.3
2.3	18	1508	40	Central AHU	−/−	+15/0.24	+21/0.15	≤1.1
2.4	24	1370	41	Central AHU	−/−	+15/0.20	+30/0.12	≤1.3
2.7	18	1180	40	Central AHU	−/−	+15/0.21	+40/0.09	≤1.1
Target: EPC "C" PE ≤ 150 kWh/(m²·a) (with central Air Handling Unit (AHU)). 40% grant.								
2.5	18	1306	45	Central AHU	−/+	+15/0.20	+28/0.11	≤0.9
2.6	18	1306	35	Central AHU	−/+	+15/0.21	+28/0.12	≤1.1
2.8	18	886	25	Central AHU	−/+	+15/0.21	+35/0.09	≤1.1
2.9	12	903	24	Central AHU	+/+	+15/0.20	+28/0.12	≤1.3
Target: EPC "C" PE ≤ 150 kWh/(m²·a) (with exhaust air heat pump). 40% grant.								
2.10	55	3378	89	Exhaust fan	+/+	+20/0.16	+25/0.16	≤1.1
2.11	32	1505	96	Exhaust fan	+/+	+15/0.21	+30/0.12	≤0.9
2.12	50	3904	130	Exhaust fan	+/+	+20/0.19	+35/0.15	≤1.1
Target: Heating energy saving 30% (with natural ventilation and extra outdoor air inlets (FAI)). 15% grant.								
15.1	60	3163	150	NAT	−/−	+10/0.38	+15/0.20	≤1.8
15.2	36	1718	61	NAT+FAI	+/+	+15–20/0.21	+0/0.4	≤2.0
15.3	60	2959	150	NAT	+/+	+0–10/0.75	+23/0.15	≤2.0
15.4	24	1737	60	NAT+FAI	+/+	+15/0.21	+20/0.17	≤1.8
15.5	40	3075	100	NAT	+/+	+0–10/0.75	+10/0.25	≤2.0
Target: Heating energy saving 40% (with natural ventilation (NAT) and extra outdoor air inlets (FAI)). 25% grant.								

Table 1. *Cont.*

Code	No. of Apartments	Heated Net Area, m²	No. of People	Ventilation	DHW Circulation before/after Renovation	Additional Insulation, cm/Thermal Transmittance (*U* W/(m²·K))		
						Walls	Roof	Windows
25.1	12	777	27	NAT+FAI	−/−	+15/0.21	+25/0.13	≤1.6
25.2	40	2623	80	NAT+FAI	+/+	+10–15/0.30	+25/0.13	≤1.4
25.3	60	3519	150	NAT+FAI	+/+	+15/0.21	+20/0.17	≤1.6
25.4	12	550	24	NAT	−/−	+15/0.21	+25/0.13	≤1.6
25.5	16	1903	38	NAT+FAI	−/−	+10–15/0.28	+30/0.11	≤1.6
Target: Heating energy saving 50% (supply-exhaust room units (SERU)). 35% grant.								
35.1	18	1064	40	SERU	−/+	+10–15/0.30	+13/0.20	≤1.4
35.2	18	1285	44	SERU	−/+	+15/0.21	+13/0.20	≤1.6
35.7	18	1026	34	SERU	+/+	+5–15/0.28	+23/0.15	≤1.6
35.9	12	940	30	SERU	−/−	+15–20/0.20	+20/0.17	≤1.6
Target: Heating energy saving 50% (with exhaust air heat pump). 35% grant.								
35.3	21	1527	60	Exhaust fan	/−	+15/0.21	+25/0.15	≤1.6
35.4	18	1041	40	Exhaust fan	−/+	+15/0.21	+23/0.16	≤1.6
35.5	18	1162	40	Exhaust fan	+/+	+10/0.28	+23/0.16	≤1.6
35.6	15	1151	38	Exhaust fan	+/+	+15/0.21	+23/0.16	≤1.6
35.8	72	5030	200	Exhaust fan	+/+	+15/0.21	+23/0.16	≤1.6
Target: Heating energy saving 50% (with central Air Handling Unit (AHU)). 35% grant.								
35.10	15	561	16	Central AHU	−/−	+15/0.21	+10/0.25	≤1.6

An example of a building before (a) and after (b); a renovation is shown in the following Figure 1.

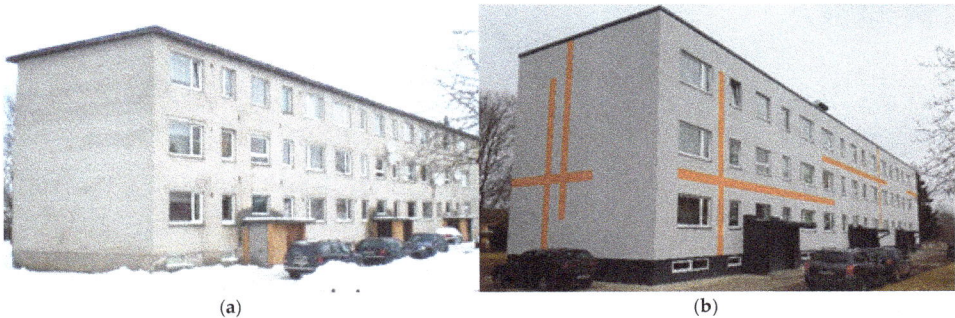

(a)   (b)

Figure 1. An example of a building (**a**) before and (**b**) after the renovation.

All 35 buildings have district heating for space heating. The heating system was renovated in all of the buildings: a hydronic radiator with thermostat valves (TRV) was installed in all apartment buildings, (before renovation, the existing one pipe system didn't have TRV). In ten buildings, the performance of natural ventilation was improved by adding outdoor air inlets. In 11 buildings, centralized exhaust ventilation (without ventilation heat recovery (VHR)) was installed. In eight buildings, the exhaust ventilation was equipped with an exhaust air heat pump (EXHP) for heat recovery. Supply and exhaust ventilation with heat recovery was installed in 14 buildings: four apartment buildings had supply-exhaust room units (SERU) and ten buildings had central air handling units (AHU).

In 11 buildings, DHW was heated by electrical boilers, located in apartments, as before renovation. In nine apartment buildings, the DHW heating by local electric boilers was changed into a central system heated by district heating after renovation (installing DHW and DHW circulation pipes). In all other buildings, district heating for DHW was used before and after the renovation. In all those buildings where DHW is heated by district heating there also exists DHW circulation, (Table 1 shows where DHW circulation was in use before renovation and how the situation is after renovation).

2.2. Evaluating Energy Consumption before and after Renovation

Energy audits before renovation were done for each building by professional energy auditors. Energy audits are documents which show the energy consumption of a building for different requirements and how to renovate the building to decrease energy usage. There were no special standards or guides for auditing in existence during that period in Estonia. The majority of energy auditors were educated through special courses and most of auditors used the same audit methodology and form. From year 2015, a new energy audit procedure was developed by Fund Kredex [23]. The information about energy consumption (electricity, space heating together with ventilation air heating (heat) and domestic hot water (DHW)) and indoor temperature before renovation was taken from an energy audit. Energy consumption after renovation was measured and data was collected from building managers. In apartment buildings with district heating, where heat for space heating and DHW was measured together, the heat for DHW was calculated based on the assumption that 40% of the total water used is hot water [24] and the difference between the temperatures is 50 °C. Circulation heat loss was calculated by using the difference between theoretical (energy consumption from water use and temperature difference) and measured energy use for DHW during the summer months.

2.3. Indoor Climate Measurements

We measured indoor temperature and ventilation airflow as the most important parameters to guaranteeing thermal comfort and indoor air quality. Measurements were conducted in all buildings in at least 3–4 apartments (altogether 120 apartments) after the renovation during the heating period between the beginning of December until the end of February, (buildings coded from 15.1 to 35.10 during the period December 2013 until February 2014, and coded 1.1 to 2.12 during the period December 2016 until February 2017).

Temperatures were measured at fifteen-minute intervals. The temperature was measured with portable data loggers (EVIKON E6226, measurement range −10–50 °C with an accuracy of ±0.6 °C) (Evikon MCI OÜ, Tartu, Estonia). The data loggers were located on the separating walls mainly in master bedrooms.

Airflow was measured in apartments twice, generally at the beginning of December and again at the end of February. In all apartments we measured exhaust air outlet airflow. The criteria for the selection of apartments was that they should be located on different floors and that in the selected apartments there should be living more persons than there are bedrooms. Ventilation airflow was measured with a Testo 435 hot wire anemometer sensor (measurement range 0–20 m/s, with an accuracy ±0.03 + 5% m/s) (Testo SE & Co. KGaA, Lenzkirch, Germany) together with a volume flow funnel Testovent 410 (Ø 340 mm).

In every apartment, where indoor temperature and ventilation airflow were measured, we collected data regarding the appropriateness of the indoor temperature via a questionnaire (5 step scale: rather cool, slightly cool, neutral, slightly warm, and rather warm). Also, we asked a question on how they feel temperature after renovation (5 step scale: warmer, slightly warmer, neutral, slightly cooler, and cooler). In most buildings the ventilation system has been renovated. That is why we asked also how they evaluated ventilation air quality (5 step scale: fresh, rather fresh, neutral, rather stuffy, and stuffy).

Thermal comfort was calculated based on ISO 7730 standard [25] by using Excel based tool [26]. Air temperature and relative humidity values were taken from measurements from all 120 apartments. The surface temperature of external wall (1/5 from all surface area) was calculated based on its thermal resistance (taken from design documentation) and typical surface resistance (0.13 $m^2 \cdot K/W$). For other input parameters (clothing = 1.0 clo, activity level = 1.2 met, and air velocity = 0.1 m/s) we used values recommended in EN 15,251 standard [27] for indoor climate category Indoor climat calss (ICC) II.

2.4. Standard Use of Buildings and Performance Gap

Pursuant to an Estonian regulation [28], the standard use of a building (indoor climate, water and electricity use, and heat gains) for indoor climate and energy modeling of an apartment building are the following:

- Indoor temperature during heating period: 21 °C;
- Ventilation airflow: 0.42 $L/(s \cdot m^2)$ for apartments with a local air handling unit and 0.5 $L/(s \cdot m^2)$ for apartments with central air handling unit. The minimum requirement for renovation is 0.35 $L/(s \cdot m^2)$;
- The use of DHW:520 $L/(m^2 \cdot a)$, i.e., 30 $kWh/(m^2 \cdot a)$;
- The use of electricity for appliances, lighting, and circulation pumps is 30 $kWh/(m^2 \cdot a)$.

The performance gap is calculated as a relative difference between the measured and standard use values according to Equation (1):

$$\text{Performance gap} = \frac{100 \times (\text{Measured value} - \text{Standard use})}{\text{Measured value}}\% \quad (1)$$

3. Results

3.1. Indoor Climate

Before renovation, the indoor temperature during the heating period was 20.8 °C on average, which is slightly lower than the standard value [28] for energy simulations (21 °C). After renovation, the indoor temperature was higher than the standard value in almost all buildings: 22.4 °C on average (varied between 19.4 °C and 24.5 °C), Figure 2a, i.e., 1.6 °C higher than before renovation, on average. In Figure 2a, on the right Figure 2b, we can see that after renovation the room temperature is 1.4 °C on average (relative difference 6%) higher than the value for standard use.

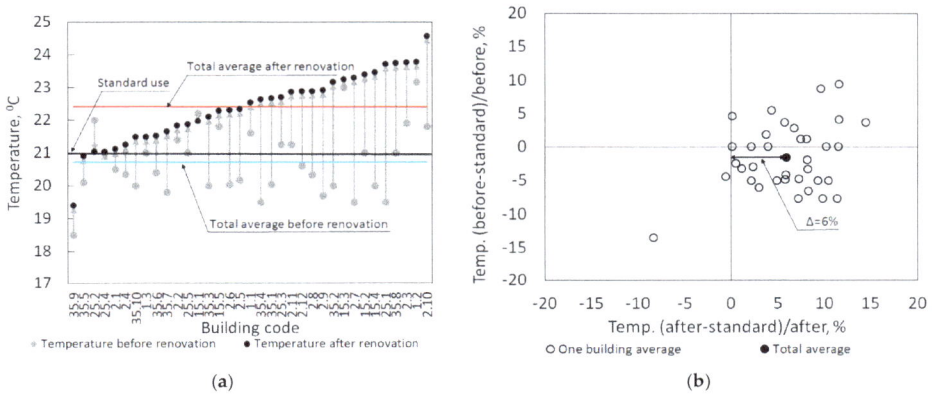

Figure 2. (**a**) Indoor temperature before and after renovation; (**b**) Indoor temperature performance gap from standard.

Based on the questionnaire, occupants were satisfied with the indoor temperature. 78% from 120 occupants answered that indoor temperature was comfortable (Figure 3). Only 11% of the occupants said that the temperature is lightly or rather warm.

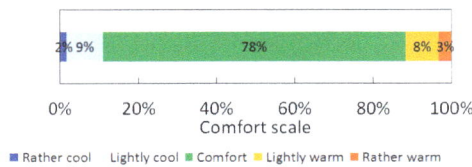

Figure 3. Occupant satisfaction with the room temperature in apartments.

The lower and higher calculated Predicted Mean Vote (PMV) (values are -0.66 and 0.67 and maximum Predicted Percentage of Dissatisfied (PPD) value is 14.4%. From 120 apartments 10 are outside from neutral thermal comfort ($-0.5 < PMV < 0.5$) zone. Based on calculations 89.8% of apartments inside of comfort zone are satisfied. Based on this we can conclude that there was not large difference between the reported satisfaction and the satisfaction calculated based on measurements (Figure 4).

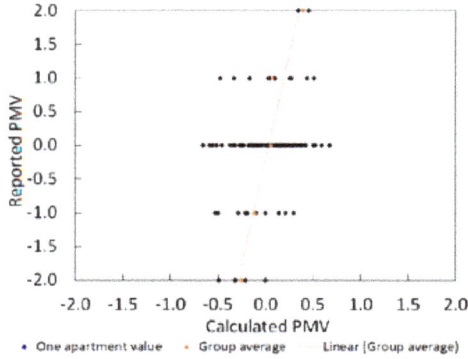

Figure 4. Calculated PMV index and PMV index by questionnaire.

Also, in Figure 5 we can see that 68% of occupants understand that the indoor temperature has increased after renovation. Only 8% of occupants said that the temperature has decreased.

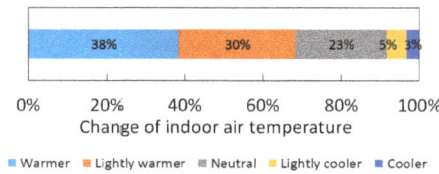

Figure 5. Occupant evaluation on the change of room air temperature after renovation.

The average ventilation air change rate of old Estonian apartments with natural ventilation before renovation was 0.24 h^{-1} and 0.17 L/(s·m^2) [20]. The ventilation airflow after renovation of 0.36 h^{-1}, 0.25 L/(s·m^2) (varied between 0.05 h^{-1} and 0.86 h^{-1}, 0.03 L/(s·m^2) and 0.60 L/(s·m^2)) on average was much less than the standard value [28] for energy simulations 0.5–0.6 h^{-1}; 0.35–0.42 L/(s·m^2) (Figure 6).

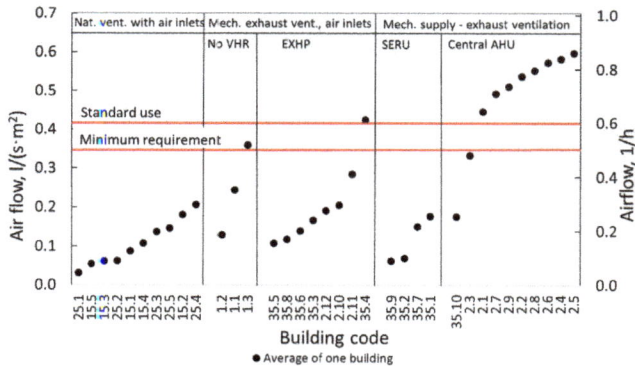

Figure 6. Ventilation airflow after renovation in studied buildings.

In our study we asked how the occupants rated also ventilation air quality after renovation. Based on the results of the measurements it can be said that airflows in most of building can be improved, but the questionnaire showed (Figure 7) that 56% of occupants feel that air is rather fresh after renovation.

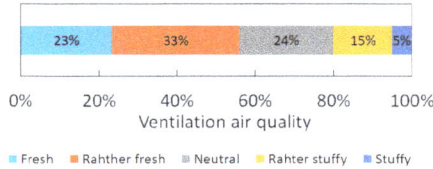

Figure 7. Occupant evaluation of ventilation quality.

3.2. Domestic Hot Water Use

The average DHW use in studied buildings was, on average, 31 L/(pers.·d) before renovation and 28 L/(pers.·d) after renovation (without circulation losses 24 L/(pers.·d) and 22 kWh/(m^2·a) correspondingly). DHW use with circulation losses was in all buildings, on average, 31 kWh/(m^2·a) before renovation and 33 kWh/(m^2·a) after renovation. We divided houses in three groups depending on DHW circulation. Table 2 features DHW energy use before and after renovation. Buildings with DHW circulation have an average DHW use of 38 kWh/(m^2·a) after renovation and without circulation, 21 kWh/(m^2·a). In buildings where circulation was installed during the renovation, the average increase of energy consumption for DHW was 13.4 kWh/(m^2·a) (Figure 8a).

Table 2. The influence of DHW energy consumption on circulation and renovation.

DHW before and after Renovation		DHW Circulation after Renovation	
		Yes	No
DHW circulation before renovation	Yes	Before renovation: 42 kWh/(m^2·a) After renovation: 39 kWh/(m^2·a)	-
	No	Before renovation: 24 kWh/(m^2·a) After renovation: 37 kWh/(m^2·a)	Before renovation: 21 kWh/(m^2·a) After renovation: 21 kWh/(m^2·a)

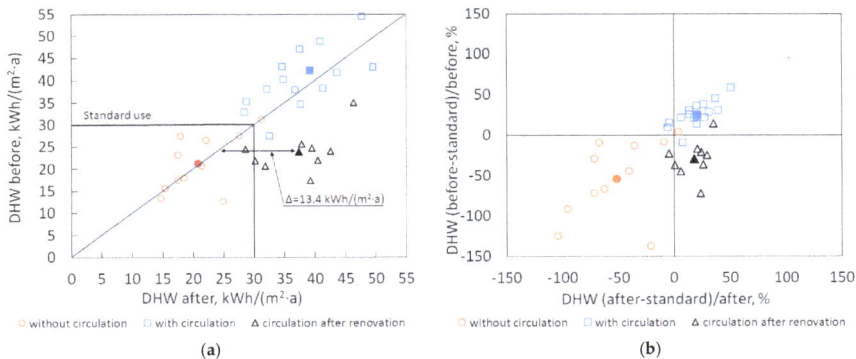

Figure 8. (**a**) DHW use before and after renovation; (**b**) DHW performance gap from standard (one building parameter is with hole and group average is filled).

Figure 8b shows the gap between the measured and standard use of DHW. Almost all buildings where there was no DHW circulation before and after the renovation used less DHW energy compared to the standard use. The relative difference between the measured energy and standard use was 54% before renovation and 52% after renovation. On the other hand, buildings with DHW circulation had a higher DHW energy use compared with standard use: before renovation 26% and after renovation 20%. Hence, independently from the availability of DHW, the energy for DHW decreased a little. The main difference in the change in DHW use was apparent in buildings where DHW circulation was installed during renovation: energy for DHW increased 56%.

In the regulations, DHW use is defined as water use per heated area. In reality, an area does not use the water; it is the occupants in the building who do it. To analyze what is the better DHW use presenting unit—L/(pers.·d) or kWh/(m²·a), we measured energy use with average DHW usage per person (28 L/(pers.·d)) and with standard usage (30 kWh/(m²·a)) with and without DHW circulation (Figure 8a). We can see that in most cases, DHW use without circulation compared with standard use per heated area is lower; the average gap from the standard use in all buildings is −48% (Figure 9a). The gap between the standard use (kWh/(m²·a)) is −140% to +4%; from DHW use per person (L/(pers.·d)), it is between −61 and 40%. When we take into account DHW circulation, then we can see that the average use from standard use per heated area moves to the positive side and when hot water circulation is considered, then the average difference with standard use after renovation is +19%, which is between −5 and +50% (Figure 9b).

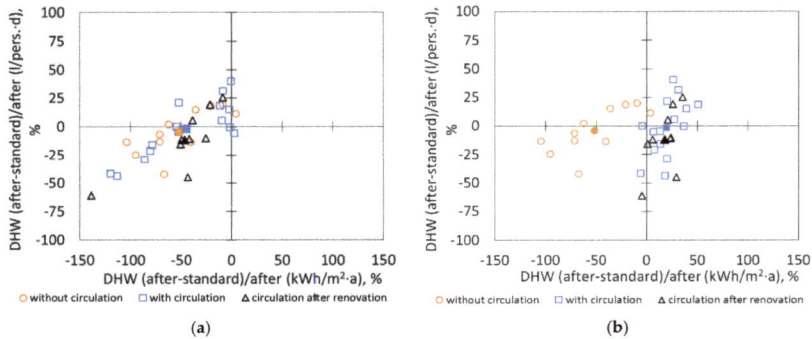

Figure 9. (a) DHW use gap from average usage per person (L/(pers.·d)) and use gap from standard use per heated area (kWh/(m²·a)) without DHW circulation and (b) with DHW circulation.

3.3. Household Electricity

The renovation did not influence the average use of household electricity (apartments + common spaces): before renovation, it was 30.1 kWh/(m²·a), and after renovation, approximately the same, 29.5 kWh/(m²·a) (Figure 10a). In general, we see that the renovation did not change the use of electricity that much. The gap between the standard use, which has been taken without electricity use for ventilation (30 kWh/(m²·a)), is, on average, −3% before renovation (between −54 until 35%) and after renovation −4% (between −29 until 30%) (Figure 10b).

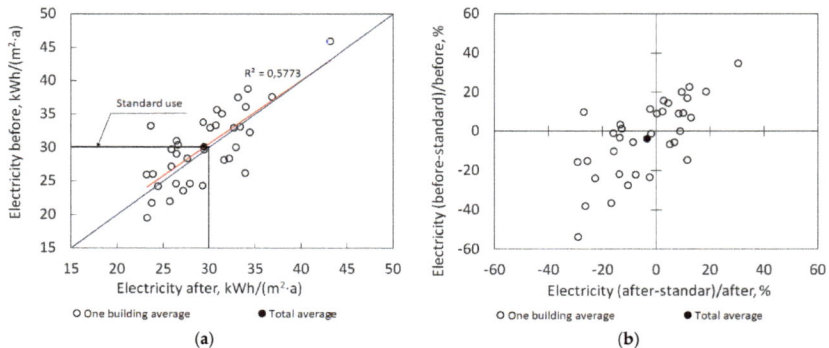

Figure 10. (a) Electricity use before and after renovation; (b) Electricity performance gap from standard use.

The use of electricity in common spaces (includes circulation pumps for DHW and heating, and electricity for central ventilation units) in all buildings was, after renovation, 0.9 kWh/(m²·a) higher (Figure 11a) than before renovation. The increase of the use of electricity in common spaces was significantly higher ($p = 0.001$) in buildings with central AHU compared with buildings with other ventilation types. Figure 11a, shows that in buildings with a central AHU, the average electricity use increased from 1.6 kWh/(m²·a) before renovation to 4.9 kWh/(m²·a) after renovation. Figure 11b, shows that after the renovation, airflow in these buildings was also higher than in other buildings (average 0.5 L/(s·m²)). An increase in the use of electricity in general spaces after the renovation was very small in buildings with other ventilation systems.

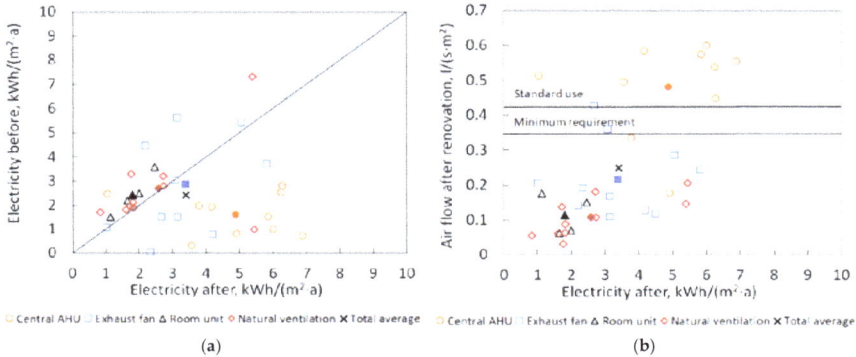

Figure 11. (**a**) Electricity use in common spaces (including pumps and ventilators) before and after renovation; (**b**) Electricity use in common spaces (including pumps and fans) after renovation compared with airflow.

4. Discussion

Indoor temperature was, on average, 1.6 °C higher after renovation (22.4 °C), which is 1.4 °C higher than the value used for indoor climate and energy modeling. If thermostatic valves were installed during the renovation, inhabitants now had the possibility to regulate their living temperature. This could be a reason for higher indoor temperatures. After renovation, the building is well insulated and should use less energy for space heating. As the heating bill is now not so high for occupants, they enjoy a higher temperature. This phenomenon can be described by the rebound effect. Higher room temperatures after renovation have been shown in other studies [29–32]. Higher room temperature also causes higher heating energy consumption. Földveary et al. [33] showed that a room temperature increase of 1 °C increases the heating energy consumption in energy efficient buildings by 16.8%. Based on the questionnaire, occupant satisfaction about indoor temperatures was good. Some difference existed between the reported and the calculated PMV based on measurements values in the rage outside of neutral zone. Occupants reported very severe conditions than we may calculate based on measurements. This may be caused on different clothing and activity levels and there always exist some unsatisfied persons [34].

This situation is much better than previous cross-sectional studies about the building's technical condition and occupant behavior have shown. Kalamees at al. [35] showed the main problems related to building physics, indoor climate, HVAC systems, and energy efficiency. Typical indoor climate related problems have been stuffy air, uneven temperature in different rooms, problems with temperature regulation possibility, etc.

Based on our questionnaire, occupants were satisfied with the indoor temperature even though the temperature was more than 1 °C higher than that used for energy modeling. To achieve realistic estimates for energy use after renovation, we suggest increasing the room temperature to 22 °C.

It is proposed that an individual heating metering system in apartments could motivate occupants to avoid a too high room temperature. Hamburg et al. [36,37] showed that instead of lowering the room temperature, occupants started decreasing the ventilation airflow and neighboring heating in well-insulated buildings.

Ventilation airflow was lower than designed in buildings with natural ventilation, mechanical exhaust ventilation, and supply-exhaust room units. In apartments with outdoor air inlets, drafts occur during the cold period. Therefore, occupants start closing the ventilation air inlets, thereby also decreasing exhaust airflow. In apartments with room-based supply and exhaust ventilation units, the drawbacks of using designed airflow are a high noise level, low pressure drop, operation management, and inefficient heat recovery. To achieve the designed airflows, we recommend using, in the renovation of residential buildings, central supply and exhaust ventilation units with heat recovery or apartment-based supply and exhaust ventilation units with heat recovery that showed a satisfactory performance in detached houses in a cold climate [38]. Based on questionnaire only 20% of occupants were dissatisfied with indoor air quality even when required ventilation airflows were not guaranteed after renovation. This shows that occupants adapted to the worsened air quality.

We measured that the use of DHW was similar with other Estonian apartment buildings [39,40] but higher than in other countries: the EU average is 25 kWh/(m^2·a), Sweden 29 kWh/(m^2·a), and Norway 30 kWh/(m^2·a) [10]. Our study showed a difference in the use of energy in buildings with and without DHW circulation. A difference in the energy use for DHW with and without circulation shows the need to calculate DHW circulation losses separately. Cali [14] has also showed that DHW distribution losses can be very high. We recommend calculating DHW circulation separately from DHW to get comparable values with standard use.

The use of electricity in buildings showed a good match between the use before and after the renovation. This shows that it does not influence occupant behavior too much. Liu [41] showed that household electricity can increase after renovation, but this was related to new installations. When comparing the use of household electricity with standard use, we can see a large variation between buildings. The relative difference varied between −54% until +35% but average difference between after and before renovation is 3.1 kWh/(m^2·a). In three buildings the electricity use difference was more than 5 kWh/(m^2·a). In the same buildings the difference in electricity use was also apparent for a three-year period before the renovations.

The installation of mechanical ventilation increased the use of electricity due to electric fans. The increase was significantly higher in buildings with a central air-handling unit. Compared with other ventilation systems, the higher values were due to the better performance of ventilation, as the ventilation airflow was much lower than required in buildings with other ventilation systems. Even though the electricity use increases when installing mechanical ventilation, the total energy balance is positive in cold climate conditions. Many studies have shown that installation of mechanical ventilation with heat recovery in cold climates is cost-effective in total [39,40,42].

5. Conclusions

Our study room temperature increased after the renovation. Temperature after the renovation is, on average, 1.6 °C higher than before the renovation, which shows a rebound effect during the renovation. Even though the indoor temperature was higher compared to the standard use; occupants were satisfied with the temperature. To achieve a realistic estimation for energy use after the renovation, we suggest increasing the room temperature in simulations to 22 °C.

The current study confirmed that the current standard electricity and DHW use in Estonian energy-modeling regulations are correct. We showed that installation DHW circulation significantly influences the energy use for DHW ($p \leq 0.001$). We recommend in the future separating DHW energy use for heating and circulation energy use. The electricity usage before and after renovation depends in most cases only whether a central AHU is installed or not and on the ventilation airflow.

Ventilation airflow was lower than designed in buildings with natural ventilation, mechanical exhaust ventilation, and supply-exhaust room units. In the majority of buildings with central supply and balanced ventilation with heat recovery, ventilation airflow was as designed. To achieve required airflows, we recommend using, in the renovation of residential buildings, central or apartment-based supply and exhaust ventilation units with heat recovery.

Our study also showed that the behavior of people is more or less the same as it was before renovation. Even for energy performance certification, the standard use of buildings is unavoidable; for cost-efficient energy renovation measures we recommend taking into account building-specific user profiles.

In future studies it will be important to analyze DHW circulation losses more deeply, as our study showed that in renovated apartment buildings which are using less energy, distribution losses have an impact on energy efficiency. As after renovation the total energy use decreases, all deviation from target values makes large relative difference for more energy efficient buildings. As user behavior become more and more important topic in constructing new and renovating existing energy efficient buildings, it is important to analyze occupants behavior more deeply.

Author Contributions: Analyses of the measured data was carried out by A.H. The research principles of the study were developed together with T.K.

Funding: This research was funded by the Estonian Centre of Excellence in Zero Energy and Resource Efficient Smart Buildings and Districts, ZEBE, grant number TK146, funded by the European Regional Development Fund, and by the Estonian Research Council, with Institutional research funding grant IUT1–15.

Acknowledgments: Authors would also like to thank Fund Kredex for cooperation and financial support for our research work.

Conflicts of Interest: The authors declare no conflict of interest.

References

1. Kurnitski, J.; Kuusk, K.; Tark, T.; Uutar, A.; Kalamees, T.; Pikas, E. Energy and investment intensity of integrated renovation and 2030 cost optimal savings. *Energy Build.* **2014**, *75*, 51–59. [CrossRef]
2. Directive (EU) 2018/844 of the European Parliament and of the Council of 30 May 2018 Amending Directive 2010/31/EU on the Energy Performance of Buildings and Directive 2012/27/EU on Energy Efficiency. *Off. J. Eur. Union* **2018**. Available online: https://eur-lex.europa.eu/legal-content/EN/TXT/?uri=OJ:L:2018:156:TOC (accessed on 15 November 2018).
3. EED. Directive 2012/27/EU of the European Parliament and of the Council of 25 October 2012 on Energy Efficiency, Amending Directives 2009/125/EC and 2010/30/EU and Repealing Directives 2004/8/EC and 2006/32/EC. *Off. J. Eur. Union* **2012**. Available online: https://eur-lex.europa.eu/legal-content/EN/TXT/?uri=OJ:L:2012:315:TOC (accessed on 15 November 2018).
4. RED, Directive 2009/28/EC of the European Parliament and of the Council of 23 April 2009 on the Promotion of the Use of Energy from renewable sources and amending and subsequently repealing Directives 2001/77/EC and 2003/30/EC. *Off. J. Eur. Union* **2009**. Available online: https://eur-lex.europa.eu/legal-content/EN/TXT/?uri=OJ:L:2009:140:TOC (accessed on 15 November 2018).
5. D'Agostino, D.; Zangheri, P.; Castellazzi, L. Towards Nearly Zero Energy Buildings in Europe: A Focus on Retrofit in Non-Residential Buildings. *Energies* **2017**, *10*, 117. [CrossRef]
6. Pombo, O.; Allacker, K.; Rivela, B.; Neila, J. Sustainability assessment of energy saving measures: A multi-criteria approach for residential buildings retrofitting—A case study of the Spanish housing stock. *Energy Build.* **2016**, *116*, 384–394. [CrossRef]
7. Paiho, S.; Pinto, I.S.; Jimenez, C. An energetic analysis of a multifunctional façade system for energy efficient retrofitting of residential buildings in cold climates of Finland and Russia. *Sustain. Cities Soc.* **2015**, *15*, 75–85. [CrossRef]
8. Thomsen, K.E.; Rose, J.; Mørck, O.; Jensen, S.Ø.; Østergaard, I.; Knudsen, H.N.; Bergsøe, N.C. Energy consumption and indoor climate in a residential building before and after comprehensive energy retrofitting. *Energy Build.* **2016**, *123*, 8–16. [CrossRef]

9. Kuusk, K.; Kalamees, T. nZEB Retrofit of a Concrete Large Panel Apartment Building. *Energy Procedia* **2015**, *78*, 985–990. [CrossRef]
10. Kurnitski, J.; Ahmed, K.; Hasu, T.; Kalamees, T.; Lolli, N.; Lien, A.; Jan, J. Nzeb Energy Performance Requirements in Four Countries vs. European Commission Recommendations. In Proceedings of the REHVA Annual Meeting Conference, Brussels, Belgium, 23 April 2018; pp. 1–8.
11. Ahmed, K.; Pylsy, P.; Kurnitski, J. Monthly domestic hot water profiles for energy calculation in Finnish apartment buildings. *Energy Build.* **2015**, *97*, 77–85. [CrossRef]
12. Sorrell, S. Energy Substitution, Technical Change and Rebound Effects. *Energies* **2014**, *7*, 2850–2873. [CrossRef]
13. Sorrell, S. *The Rebound Effect: An Assessment of the Evidence for Economy-Wide Energy Savings from Improved Energy Efficiency*; UKERC: London, UK, 2007; ISBN 1-903144-0-35.
14. Calì, D.; Osterhage, T.; Streblow, R.; Müller, D. Energy performance gap in refurbished German dwellings: Lesson learned from a field test. *Energy Build.* **2016**, *127*, 1146–1158. [CrossRef]
15. Mohareb, E.; Hashemi, A.; Shahrestani, M.; Sunikka-Blank, M. Retrofit Planning for the Performance Gap: Results of a Workshop on Addressing Energy, Health and Comfort Needs in a Protected Building. *Energies* **2017**, *10*, 1177. [CrossRef]
16. Zhang, Y.; Bai, X.; Mills, F.P.; Pezzey, J.C.V. Rethinking the role of occupant behavior in building energy performance: A review. *Energy Build.* **2018**, *172*, 279–294. [CrossRef]
17. Desideri, U.; Yan, J.; Menezes, A.C.; Cripps, A.; Bouchlaghem, D.; Buswell, R. Predicted vs. actual energy performance of non-domestic buildings: Using post-occupancy evaluation data to reduce the performance gap. *Appl. Energy* **2012**, *97*, 355–364.
18. Ilomets, S.; Kuusk, K.; Paap, L.; Arumägi, E.; Kalamees, T. Impact of linear thermal bridges on thermal transmittance of renovated apartment buildings. *J. Civ. Eng. Manag.* **2017**, *23*, 96–104. [CrossRef]
19. Ilomets, S.; Kalamees, T.; Vinha, J. Indoor hygrothermal loads for the deterministic and stochastic design of the building envelope for dwellings in cold climates. *J. Build. Phys.* **2017**. [CrossRef]
20. Mikola, A.; Kalamees, T.; Kõiv, T.-A. Performance of ventilation in Estonian apartment buildings. *Energy Procedia* **2017**, *132*, 963–968. [CrossRef]
21. Kuusk, K.; Kalamees, T. Estonian Grant Scheme for Renovating Apartment Buildings. *Energy Procedia* **2016**, *96*, 628–637. [CrossRef]
22. Hamburg, A.; Kalamees, T. Improving the indoor climate and energy saving in renovated apartment buildings in Estonia. In Proceedings of the 9th International Cold Climate HVAC 2018, Kiruna, Sweden, 12–15 March 2018.
23. Jõesaar, T.; Hamburg, A. *Korterelamute Energiaauditite Koostamise Juhend (Guideline for Energy Audits of Apartment Buildings)*; SA KredEx: Tallinn, Estonia, 2015.
24. Toode, A.; Kõiv, T.-A. Investigation of the Domestic Hot Water Consumption in the Apartment Building. *Proc. Est. Acad. Sci. Eng.* **2005**, *11*, 207–214.
25. ISO 7730. *Moderate Thermal Environments—Determination of the PMV and PPD Indices and Specification of the Conditions for Thermal Comfort*; ISO: Geneva, Switzerland, 1994.
26. Da Silva, M.C.G. *Spreadsheets for Calculation of Thermal Comfort Indices PMV and PPD*; University of Coimbra: Coimbra, Portugal, 2014. [CrossRef]
27. EN 15251. *Indoor Environmental Input Parameters for Design and Assessment of Energy Performance of Buildings Addressing Indoor Air Quality, Thermal Environment, Lighting and Acoustics*; CEN: Brussels, Belgium, 2007.
28. 7 RT I, 19.01.2018, MKM määrus nr. 58, Hoonete Energiatõhususe Arvutamise Metoodika (Minister of Economic Affairs and Communications Regulation nr. 58, Methodology for Calculating the Energy Performance of Buildings). 2018. Available online: https://www.riigiteataja.ee/akt/119012018007 (accessed on 15 November 2018).
29. Branco, G.; Lachal, B.; Gallinelli, P.; Weber, W. Predicted versus observed heat consumption of a low energy multifamily complex in Switzerland based on long-term experimental data. *Energy Build.* **2004**, *36*, 543–555. [CrossRef]
30. La Fleur, L.; Moshfegh, B.; Rohdin, P. Measured and predicted energy use and indoor climate before and after a major renovation of an apartment building in Sweden. *Energy Build.* **2017**, *146*, 98–110. [CrossRef]
31. Földváry, V.; Bukovianska, H.P.; Petráš, D. Analysis of Energy Performance and Indoor Climate Conditions of the Slovak Housing Stock before and after its Renovation. *Energy Procedia* **2015**, *78*, 2184–2189. [CrossRef]

32. Broderick, Á.; Byrne, M.; Armstrong, S.; Sheahan, J.; Coggins, A.M. A pre and post evaluation of indoor air quality, ventilation, and thermal comfort in retrofitted co-operative social housing. *Build. Environ.* **2017**, *122*, 126–133. [CrossRef]

33. Földváry, V.; Bekö, G.; Langer, S.; Arrhenius, K.; Petráš, D. Effect of energy renovation on indoor air quality in multifamily residential buildings in Slovakia. *Build. Environ.* **2017**, *122*, 363–372. [CrossRef]

34. Fabbri, K. Thermal comfort evaluation in kindergarten: PMV and PPD measurement through datalogger and questionnaire. *Build. Environ.* **2013**, *68*, 202–214. [CrossRef]

35. Kalamees, T.; Ilomets, S.; Arumägi, E.; Kuusk, K.; Liias, R.; Kõiv, T.-A.; Õier, K. Research demand of old apartment buildings in Estonia. In Proceedings of the IEA Annex 55 (RAP-RETRO), Working Meeting, Holzkirchen, Germany, 15–16 April 2010.

36. Hamburg, A.; Kalamees, T. Method to divide heating energy in energy efficient building without direct measuring. *Energy Procedia* 2017. [CrossRef]

37. Hamburg, A.; Thalfeldt, M.; Kõiv, T.; Mikola, A. Investigation of heat transfer between neighbouring apartments. In Proceedings of the 9th International Conference "Environmental Engineering", Vilnius, Lithuania, 22–23 May 2014.

38. Kurnitski, J.; Eskola, L.; Palonen, J. Ventilation in 102 Finnish single-family houses. In Proceedings of the 8th REHVA World Congress High Tech, Low Energy: Experience the Future of Building Technologies, Lausanne, Sweden, 9–12 October 2005; p. 6.

39. Kuusk, K.; Kalamees, T.; Maivel, M. Cost effectiveness of energy performance improvements in Estonian brick apartment buildings. *Energy Build.* **2014**, *77*. [CrossRef]

40. Arumägi, E.; Kalamees, T. Analysis of energy economic renovation for historic wooden apartment buildings in cold climates. *Appl. Energy* **2014**, *115*, 540–548. [CrossRef]

41. Liu, L.; Rohdin, P.; Moshfegh, B. Evaluating indoor environment of a retrofitted multi-family building with improved energy performance in Sweden. *Energy Build.* **2015**, *102*, 32–44. [CrossRef]

42. Alev, Ü.; Allikmaa, A.; Kalamees, T. Potential for financial- and energy saving of detached houses in Estonia. *Energy Procedia* **2015**, *78*, 907–912. [CrossRef]

![energies logo] *energies*

MDPI

Article

In Situ Measurements of Energy Consumption and Indoor Environmental Quality of a Pre-Retrofitted Student Dormitory in Athens

Nikolaos Barmparesos *, Dimitra Papadaki, Michalis Karalis, Kyriaki Fameliari and Margarita Niki Assimakopoulos

Department of Applied Physics, Faculty of Physics, University of Athens, Building Physics 5, University Campus, 157 84 Athens, Greece: dpapadaki@phys.uoa.gr (D.P.); sph1200071@uoa.gr (M.K.); fameliari-k@hotmail.com (K.F.); masim@phys.uoa.gr (M.N.A.)
* Correspondence: nikobar@phys.uoa.gr; Tel.: +30-210-7276845

Received: 15 May 2019; Accepted: 8 June 2019; Published: 11 June 2019

Abstract: In the following years all European Union member states should bring into force national laws on the energy performance of buildings. Moreover, university campus dormitories are buildings of great importance, due to their architectural characteristics and their social impact. In this study, the energy performance along with the indoor environmental conditions of a dormitory of a university has been analysed. The in situ measurements included temperature, relative humidity, concentrations of carbon dioxide, total volatile organic compounds, and electrical consumption; lastly, the energy signature of the whole building was investigated. The study focused on the summer months, during which significantly increased thermal needs of the building were identified. The ground floor was found to be the floor with the highest percentage of thermal conditions within the comfort range, and the third floor the lowest. Lastly, a significant correlation between electrical consumption and the outdoor temperature was presented, highlighting the lack of thermal insulation. Overall, it was clear that a redesign of the cooling and heating system, the installation of a ventilation system, and thermal insulation are essential for improving the energy efficiency of this building.

Keywords: energy efficiency; student dormitories; Indoor Environmental Quality (IEQ), Pro-GET-onE H2020; in situ measurements; monitoring measurements; energy signature

---

1. Introduction

During the last decades the subjects of energy efficiency and indoor air quality of different types of building have gained the attention of the scientific community. In the European Union (EU), the Member States are intensively trying to improve energy efficiency in all end-use sectors, on the one hand by increasing the usage of renewable energy sources (RES), and on the other, by minimizing the environmental threats caused by energy consumption of fossil fuels, thus, supporting energy security [1]. Moreover, the EU directive on the energy performance of buildings [2] states that all EU members should bring into force national laws, regulations, and administrative provisions for setting minimum requirements on the energy performance of new and existing buildings that are subject to major renovations. For that reason, the Greek Regulation on the Energy Efficiency of Buildings [3] sets specific limitations and minimum requirements for energy efficiency for the design and the construction of different types of building (residential, educational, cultural, etc.).

Numerous scientific publications set the main target as energy retrofitting of buildings, such as decreasing the negative impact on the ambient environment, creating thermal comfort zones for tenants, and controlling the consumption of energy and material resources effectively [4–12]. However, research by Szodrai et al. reported that several environmental parameters within the indoor environment of a building, should also be taken into account before the implementation of deep energy retrofitting actions [13]. Recent experiments have demonstrated that indoor climate has either improved or deteriorated as a result of additional energy efficiency measures in a building's envelope [14,15]. Specifically, for the case of offices or educational buildings, which usually contain large glazed areas, the indoor air temperature may rise to very high levels, reducing the thermal comfort conditions for tenants. On the contrary, additional improvement of air tightness could have a negative impact on indoor environmental quality (IEQ) [16–20]. For similar building types, different strategies of ventilation have also been suggested, with the personalized ventilation system reported as the most appropriate solution in order to combine significant energy savings together with acceptable levels of IEQ [21–26].

Along with IEQ, an additional parameter that strongly influences the levels of energy consumption within a building is the economic profile and the behaviour of tenants. Several research studies demonstrated results of an experimental campaign in 124 households in China, which showed that on average, education on energy-conscious behaviour for tenants could result in reduction of household electricity consumption by more than 10% [27,28]. Moreover, other studies reported that personal exposure indoors is also correlated with the social status of the occupants in different parts of the world, such as the United States, France, Germany, Japan, and South Africa. They concluded that within lower income dwellings, concentrations of air pollutants, such as carbon dioxide (CO_2), total volatile organic compounds (TVOC), fine particulate matters ($PM_{2.5}$), carbon monoxide (CO), and formaldehyde (CH_2O) were found to be higher due to frequent smoking, use of cleaning products, disinfectants, and sprays, and wooden stoves [29–34]. As people spend a significant proportion of time within their houses, exposure to the air pollutants mentioned above can have a serious impact on their health [35,36].

Different studies have investigated the impact of energy retrofitting on IEQ in buildings by comparing the results from field measurements or data modelling before and after renovation actions [37–39]. All researchers highlighted the development of a state-of-the-art ventilation system as the optimum solution in order to achieve adequate levels of IEQ along with energy efficiency. That is because in some cases occupants tend to frequently keep the window shut after renovation actions, and thus, even though the thermal comfort conditions had been improved, concentrations of different indoor air pollutants were found to be increased. Characteristically, short term increases in $PM_{2.5}$ and CH_2O concentrations immediately after the retrofit process have led to a long-term decrease. Previous studies [40,41] reported that concentrations of CH_2O and nitrogen dioxide (NO_2) increased in some instances. A research study conducted on fifteen social houses in Ireland highlighted the importance of securing good IEQ levels within the building after energy retrofitting in order to protect the health of the inhabitants [42]. Similarly, in another case study, thirty-five renovated apartments in Estonia were inspected, studied, and tested. The researchers report that indoor air temperature was found to be higher (by 1.6 °C on average) after the renovation even though occupants were satisfied. In addition, they suggested that central or apartment-based air supply and exhaust ventilation units with heat recovery as an optimal ventilation system [43].

Moreover, the scientific community is focusing on energy efficiency and IEQ measurements for student housing. University campus dormitories are buildings of great importance, firstly because of their architectural and technical characteristics (such as large glazed spaces, common use areas, numerous apartments, etc.), and secondly because of their social impact (as they host students). Lastly, the impact of energy savings of public buildings is significant, but also the decrease of the operating costs positively influence the quality of life of the users. Different studies in Italy [44], Serbia [45], the United States [46–48], and China [49] have remarked on the importance of renovating such buildings in order to achieve better energy efficiency and living conditions for the occupants, reduce CO_2 emissions, and motivate more students to inhabit them. A novel holistic methodology to design the refurbishment of educational buildings in Mediterranean regions is described in the research of Assimakopoulos et al. conducted in 2018 [50]. Particularly in Greece, five different scenarios have been tested concerning the reduction of energy consumption for a dormitory at the University of Crete campus. This research concluded that the installation of a photovoltaic roof panel led to satisfactory results (62% energy savings) for the target of near-zero energy performance [51].

In this study, the energy performance and the evaluation of the IEQ conditions of a dormitory within the campus of the University of Athens is presented. An experimental campaign (field measurements) took place in order to demonstrate the energy profile of the building's current state. A deep retrofitting of the building is planned in 2019 under the framework of "Proactive synergy of inteGrated Efficient Technologies on buildings' Envelopes" (Pro-GET-onE -Horizon 2020 Grant Agreement number 723747). Pro-GET-onE is based on the integration of different technologies to achieve a multi-benefit approach through the closer integration between energy and non-energy related benefits, promoting a holistic vision in order to achieve the highest performance of buildings in terms of energy requirements, safety, and socio-economic sustainability [52]. The present research focuses on the investigation of the levels of energy consumption and IEQ conditions within the student housing during different monitoring periods based on already established and further developed protocols, highlighting the specific needs on which the renovation actions should focus on. This work will help building designers in decision making processes towards sustainable design in these types of buildings, taking into account both energy efficiency methods and also IEQ aspects.

2. Materials and Methods

2.1. Building Site

The selected case study, named "Dormitory of the University of Athens—Building B" (B FEPA), is a student dormitory that consists of 138 single rooms for students. It is a property of the National and Kapodistrian University of Athens, located in the University campus of Zografou, a suburb in Athens, Greece. All available technical details have been provided by the technical service of the University of Athens. Figure 1a,b shows a global view of the site, where it can be seen that the building has 4 floors (ground level included) and a basement. Latitude and Longitude of the site are, respectively, 37.97° and 23.76°. The building is located next to a busy road (Taxilou Street) with an altitude of 153 m above sea level. The building was constructed in 1986. It has been in continuous operation since then. The net height of each room is 2.40 m, while in the basement floor the net height is 2.60 m. The main entrance is on the northwest side.

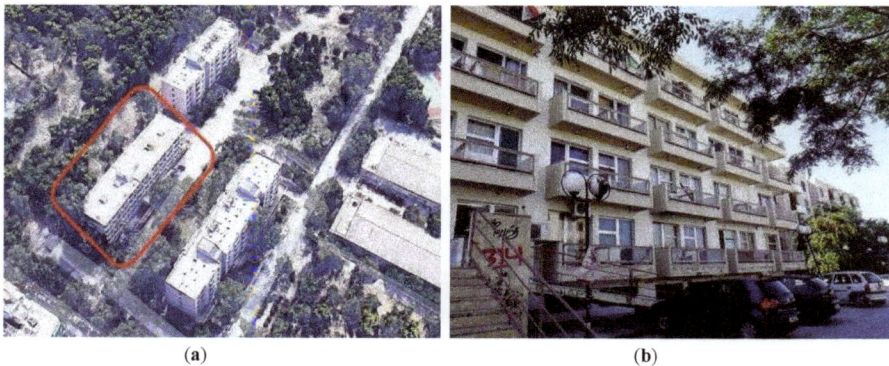

| (a) | (b) |

Figure 1. Arial view of the location (a) and (b) the façade of the building

Additionally, the building has a rectangular shape. The gross building area is around 3642 m^2 with 138 bedrooms, which in many cases have a surface area of 9.50 m^2. The building is accessible from a central staircase or two lifts. The net conditioned building volume is equal to 6960 m^3. Table 1 summarizes the main geometrical characteristics of each floor; each area represents the net liveable area.

Table 1. Main geometrical characteristics of each block.

Location	Area Elevation	Elevation	Number of Bedrooms	Planimetric Dimensions
Basement	742 m^2	2.60 m	0	$56.59 \times 13.36 \text{ m}^2$
Ground-floor	725 m^2	2.40 m	36	$56.59 \times 15.37 \text{ m}^2$
First floor	725 m^2	2.40 m	36	$56.59 \times 15.37 \text{ m}^2$
Second floor	725 m^2	2.40 m	36	$56.59 \times 15.37 \text{ m}^2$
Third floor	725 m^2	2.40 m	36	$56.59 \times 15.37 \text{ m}^2$

The building mainly consists of a reinforced concrete structure. The bearing body of the building is formed by two pillars, separated by a joint. Separation is probably due to the large overall length of the building, so as to limit individual lengths below 40 m. Moreover, the two sub-frames have been formed with a similar grid and cross-sectional dimensions. The surfaces of the two vectors per floor are: (a) around 290 m^2; and (b) around 430 m^2. External walls have an overall thickness of 0.25 m, consisting of plaster (2.5 cm) on both sides and brick (double wall without insulation). Mortars and beams of thickness of 0.30 m do not have external insulation. The basement is made with the use of 3 cm of marble and 20 cm of concrete, while the roof is composed, from the outer side to inner side, of: asphalt cover (6 mm), perlite-bitumen bonded layer (3 cm), concrete (20 cm), and plaster (2.5 cm). Windows and glazed doors are made of single glass with an aluminium frame (5 cm width). They can be divided into four types whose dimensions are specified, including the frame:

- Type 1: 1.30 m × 1.10 m and Type 2: 1.00 m × 2.30 m in single rooms;
- Type 3: 5.70 m × 2.30 m balcony door in common use zones;
- Type 4: 2.16 m × 0.6 m in the basement.

For heating purposes, the building is equipped with a centralized boiler (natural gas), which provides the thermal vector fluid to in-room radiators. The thermal power plant is installed in the basement floor. In the beginning of 2017, the boilers that used oil were replaced with natural gas boilers, one with nominal power of 850,000 kcal/h ($\approx$988.6 kW) and another one with nominal power of 630,000 kcal/h ($\approx$732.7 kW), with nominal efficiency of around 94%. The heating system is turned on for 7 h each day (cycle of 2 + 2 + 3 h) for around 5 months, depending on the year. The local emitters are very old static radiators. There are no thermostatic valves or zone thermostats. Moreover, the pipes delivering hot water are not isolated. This decreases the efficiency of the system, as there are high distribution losses from the basement floor across each floor. There is no central air conditioning system, and only few rooms (warden's room and living room on the ground floor) have autonomous split systems. There is no mechanical ventilation system—only natural ventilation is provided through the external frames.

Furthermore, for lighting needs, the dormitory has, in total, installed power of 26.5 kW for the rooms and 18.7 kW for shared floors, with the addition of 12.5 kW for an external lighting system. It is also important to note that the building includes the following functions per floor:

- Basement: technical rooms and warehouses;
- Ground floor: central entrance, staircase and 2 elevators, seating area, 30 single rooms, public bathroom (for men), common kitchen;
- 1st, 2nd, 3rd floors: 36 single rooms and shared kitchens for men and women, living rooms.

There is a call center area on the ground floor. A TV set is installed in the ground floor lounge. Electrical equipment consists of 6 washing machines per floor and 1 large refrigerator per floor in the kitchen.

2.2. Experimental Campaign and Monitoring Protocols

Guideline 14-2002, created by a committee of the American Society of Heating, Refrigerating, and Air-Conditioning Engineers (ASHRAE) members, addresses the determination of energy savings by comparing before and after energy use measurements. The basic method involves the projection of energy use or demand patterns of the pre-retrofit (baseline) period into the post-retrofit period. Typical adjustments to the baseline energy use or demands include weather, occupancy, and system variables. Savings represent the amount of energy use between the projected baseline and the post-retrofit consumption and are calculated using the following formula [53]:

Savings = (Baseline energy use or demand projected for Post-Retrofit conditions) − (Post-Retrofit energy use or demands)

The monitoring process followed in this study is designed based on the ASHRAE Guideline 14 on Measurement of Energy and Demand Savings, a protocol which was designed for the following six performance categories: energy use, water use, thermal comfort, indoor air quality, lighting, and acoustics, measuring post-retrofit energy use and comparing that to the measured pre-retrofit use, which will take place in 2020, after the retrofit of the student housing (within Pro-GET-onE project) is finished. The following investigations have been done:

(1) Building environmental parameters (including thermal and IEQ conditions)

The measurements of IEQ took place from April 2018 to March 2019 in order to investigate the annual behaviour of the building. The equipment consisted of portable continuous recording Tongdy sensors simultaneously measuring temperature (T), relative humidity (RH), and concentrations of CO_2 and TVOC. The CO_2 sensor had a recording range from 0 to 2000 ppm, and accuracy of $\pm$ 40 ppm at 25 °C and TVOC ranging from 1 to 30 ppm with accuracy of 1 ppm. In addition, the measuring ranges for temperature and RH were 0 to 50 °C and 0 to 95% (non-condensing). These sensors operate under conditions of 0 to 50 °C and 5 to 95%, respectively. All parameters were recorded on a 24 h basis at 15-min intervals. Quality assurance for the equipment was performed on several occasions during the experiment and all of the instruments were calibrated according to the manufacturers' standards. Four sensors were placed within the commonly used areas of the building on each floor. Thus, the influence of different occupancy patterns on the results could be investigated. Furthermore, in order to ensure that differences between mean values (depending on the floor) are statistically significant, a Kruskal-Wallis (non-parametric) test ($p < 0.05$) was implemented. Each case includes four grouping variables (ground, first, second, and third floor) and one parameter (T, RH, CO_2, or TVOC). For all cases, the *p*-value was found to be close to zero, depicting a statistically significant difference between parameters among the floors.

(2) Meteorological parameters

Moreover, the following meteorological parameters were recorded from nearby stations of the National Observatory of Athens on an hourly basis: air temperature and relative humidity, provided by the National Observatory of Athens Institute of Environmental Research.

(3) Energy use

All energy data refer to electricity consumption of the building, as data from the direct hot water (DHW) system are not available. Electrical consumption was measured constantly by Landis+Gyr E650 meters [54] installed in the central power supply network of the University Campus. These data were acquired from the "Students Achieving Valuable Energy Savings 2" (SAVES2) project, H2020 Grant Agreement number 754203. All data were transferred to an EMM100 system [55] via S_0 output pulse according to international protocol IEC 62056-21. The uploaded measurements (from April 2018 to January 2019) were obtained from an online platform (hourly or daily) in order to assess the electrical consumption patterns of the dormitory under investigation during different time periods. Figure 2a,b illustrates the sensors used for all types of measurements. The impact of different weather conditions on the energy behaviour of the building was also investigated, defined as the "energy signature" of the building. Specifically, the mathematical identification of electrical consumption (E in kWh) and ambient temperature (Tout in °C) is expressed in the form of Equation (1):

$$E = C_0 + C_1 \cdot Tout \tag{1}$$

Two simple linear regression models have been used for cooling (from April to October) and heating (from November to January) periods, respectively, in order to highlight the influence of seasonality on the results.

Figure 2. Experimental equipment for measurements of (**a**) IEQ and (**b**) electrical consumption within the building of B FEPA.

3. Results and Discussion

3.1. Indoor Environmental Quality

3.1.1. Thermal Conditions

In order to investigate the thermal conditions of the B FEPA building, temperature and RH measurements were implemented within the common spaces of each floor. All data were collected from April 2018 to March 2019. Ambient temperature and relative humidity data was provided by the National Observatory of Athens from a local meteorological station [56].

The thermal behaviour of each floor differs, and therefore it is important to look into the thermal conditions of each floor separately. Table 2 summarizes the results of temperature and RH levels at every experimental point.

On the ground floor, the highest internal mean temperature was found in August (27.93 °C) and the lowest in January (21.36 °C). One may also notice that standard deviations of temperature during cold months (November to March) are slightly higher, as expected, due to sudden decrease of temperature levels caused by increased air infiltration as the main entrance opens more frequently. However, fluctuations of indoor temperature within this floor are not significant, especially compared to the respective outdoors. Based on the analysis of the aforementioned data, it is the only floor where the mean temperature levels remain lower than the respective outside level during summer, while it retains relatively higher temperature levels during the cold months in comparison with the other floors. The frequent air exchange with the ambient environment does not significantly affect monthly average values. This result is due to the increased presence of people in the living room of the ground floor and the usage of extra electrical heaters for some hours during the day. The ground floor is also favoured, since it is situated close to the central heating system (the two boilers located in the basement). This, combined with the enhanced thermal stability provided by the rest of the floors, minimizes thermal losses and means this space is (relatively) unaffected by the outdoor conditions. The highest mean levels of RH are observed in October (50.96%), while the lowest during March

(36.14%). Standard deviations are relatively lower during winter, due to closed windows and regular functioning of the building's central heating system. The highest standard deviation (7.48%), reported in October, demonstrates that the influence of the external environment is significant during this month. The windows remain open more often when external levels of RH are relatively high (68.37%).

For the first floor, the maximum mean value is observed in July (29.03 °C) and the lowest is in December (18.10 °C). Standard deviations follow the same pattern as in the ground floor. It is also important to notice that the difference between the maximum mean temperature and the minimum mean temperature is higher on this floor (almost 9 °C) compared to the underlying (ground) floor. As a building's level increases, thermal insulation decreases; therefore, the heat losses are more significant. Based on the results of RH levels, it is obvious that throughout the year, the most humid month for this floor is October (55%) and the least humid is February (36.38%). These results are in alignment with the ground floor. Because of higher ambient temperature levels, the occupants tend to open the windows frequently for ventilation and cooling purposes, allowing the outdoor humid air of October to infiltrate into the building.

The mean temperature level on the second floor follows the same pattern as the aforementioned floors. The warmest month is found to be July with 29.46 °C and the coldest is January with 19.43 °C. The difference between the maximum and minimum mean temperature for this floor is 10 °C (almost 1 degree higher than the respective difference of the first floor). Regarding relative humidity levels, October was found to be, once again, the month with the highest mean RH (56.82%), and March is the month with the lowest (38.86%).

Table 2. Mean values and standard deviations of temperature (°C) and relative humidity (%) for each floor, for all experimental periods (April 2018–March 2019).

		Experimental Site				
Month	Parameter	Ground Floor	1st Floor	2nd Floor	3rd Floor	Outdoors
April	T (°C)	24.69 ± 1.07	22.00 ± 1.99	23.01 ± 1.91	24.94 ± 2.09	21.68
	RH (%)	37.12 ± 4.84	40.79 ± 6.34	38.86 ± 5.73	34.92 ± 4.58	51.22
May	T (°C)	25.58 ± 1.38	23.85 ± 2.29	24.64 ± 1.39	26.12 ± 1.40	23.13
	RH (%)	46.48 ± 7.25	50.74 ± 11.35	47.18 ± 7.84	43.54 ± 7.03	60.73
June	T (°C)	26.97 ± 1.16	27.16 ± 1.76	27.05 ± 1.94	28.83 ± 1.57	25.99
	RH (%)	43.30 ± 8.56	44.64 ± 9.91	44.91 ± 11.23	41.07 ± 8.87	57.25
July	T (°C)	27.86 ± 0.82	29.03 ± 1.81	29.46 ± 1.58	32.11 ± 1.36	28.67
	RH (%)	44.6 ± 5.57	47.42 ± 8.61	46.28 ± 8.09	39.07 ± 6.55	58.13
August	T (°C)	27.93 ± 0.70	28.91 ± 1.35	28.93 ± 1.22	31.47 ± 0.92	28.14
	RH (%)	40.02 ± 5.07	42.30 ± 7.55	41.73 ± 7.44	36.41 ± 6.32	53.59
September	T (°C)	26.54 ± 1.35	26.94 ± 2.25	26.13 ± 2.62	28.72 ± 2.59	24.85
	RH (%)	41.69 ± 6.76	44.56 ± 9.05	45.70 ± 9.62	39.39 ± 0.90	56.16
October	T (°C)	23.7 ± 0.72	22.00 ± 1.46	21.79 ± 0.85	23.43 ± 0.89	19.60
	RH (%)	50.96 ± 5.42	55.00 ± 9.10	56.82 ± 8.77	50.97 ± 8.44	68.37
November	T (°C)	22.87 ± 1.60	19.66 ± 2.12	21.14 ± 1.53	19.49 ± 1.92	15.90
	RH (%)	46.46 ± 6.36	53.31 ± 8.36	51.51 ± 6.95	53.79 ± 7.85	70.97
December	T (°C)	22.57 ± 1.05	18.10 ± 3.46	19.79 ± 1.89	18.81 ± 1.86	10.89
	RH (%)	37.86 ± 4.38	44.80 ± 8.23	40.68 ± 6.08	43.87 ± 3.32	71.39
January	T (°C)	21.36 ± 1.90	20.22 ± 3.39	19.43 ± 2.24	19.49 ± 1.47	9.66
	RH (%)	37.90 ± 5.42	38.85 ± 8.53	42.19 ± 6.91	44.58 ± 6.29	72.49
February	T (°C)	22.11 ± 1.51	21.64 ± 2.19	20.07 ± 1.62	19.08 ± 2.04	10.12
	RH (%)	37.04 ± 4.27	36.38 ± 5.00	41.67 ± 5.95	41.97 ± 3.74	71.03
March	T (°C)	24.30 ± 1.25	23.13 ± 2.03	22.33 ± 1.83	22.14 ± 1.52	13.51
	RH (%)	36.14 ± 3.13	36.51 ± 4.90	38.55 ± 4.87	38.94 ± 4.33	66.56

The third floor demonstrated some interesting results. On average, the hottest month was found to be July (32.11 °C) and the coldest December (18.81 °C). The mean internal temperature during the summer months exceeded the respective outdoor temperature by 5–6 °C. For the same period, temperature levels within this floor were measured to be higher compared to the rest of the building by approximately 3–4 °C. Despite the continuous inflow of outdoor air through the open windows, the indoor temperature remained elevated. The third floor, as the highest of the building, is directly exposed to solar radiation, which is in abundance during summer due to the Earth's angle. The thermal mass of the building is able to absorb the extra energy during the day and gradually yield it afterwards. In addition, during the heating period temperature levels within this floor are comparable to the first and second floors because of central heating. As for RH levels, November was found to be, on average, the most humid month (with internal levels of 53.79%), while April was the least humid (34.92%). RH values of the cold period are comparable to the warmer months. The third floor presents a high dependency on the ambient climate conditions, especially for warm months. Increased solar radiation during summer causes internal overheating due to insufficient thermal insulation of the roof (directly exposed to solar radiation) and inadequate ventilation rates.

The results are also presented graphically, demonstrating a significant difference between internal and external temperature levels for different periods. Figure 3 illustrates that during the cooling period (from April to October), the indoor and outdoor temperature values are comparable. It should be noted that the building is naturally ventilated with the absence of central heating, ventilation, and an air conditioning (HVAC) system or possible night ventilation. Therefore, the opening of windows and doors is the only way to cool the building, especially the common areas (such as corridors). As a result, air exchange rate between indoor and outdoor environments during the cooling period is higher, and thus, the thermal condition is almost similar. On the other hand, for the heating period (from November to March), the inflow of cold and humid air that causes discomfort to the occupants is avoided as the windows stay mainly closed for most of the time. In addition, the central heating system (natural gas-radiators) operates for 7 h per day, preserving indoor temperature at relatively high levels. Figure 4 shows similar results for RH levels. The indoor mean RH values for each floor are close to the respective outdoor values during the cooling period, although during the heating period the difference is significant. Interestingly, the monthly internal mean RH peaks in between the end of the cooling period (October) and the start of the heating period (November) follow the pattern of the respective external levels. As mentioned above, during those months ambient climatic conditions are not severe and tenants tend to open the windows more frequently. On the contrary, during the cold months, even if outdoor RH increases, the opposite behaviour is observed for indoor RH in all floors. Closed windows, as well as the operation of a central heating system, mitigate internal levels of RH. In both figures, the dotted lines represent temperature and RH limit ranges in which adequate thermal comfort conditions for the tenants are achieved, based on the standards reported in the available literature [57,58]. The limit ranges for internal temperature and RH levels are the following:

- 23–26 °C: refers to the warm period, from April to October.
- 20–23 °C: refers to the cold period, from November to March.
- 30–60%: refers to the entire year.

Additionally, Figure 5 illustrates the daily behavior of temperature levels within different floors when the external air temperature reached its highest recorded value. The selected date of interest (May 7, 2018) was found to be the warmest day during the entire experimental period, and therefore, the examination of the building's thermal behavior on that specific day should be examined. One may notice that the ground floor presented the lowest levels of indoor air temperature and the third floor presented the highest. All measured values were found to overcome the proposed limit range of comfort. Specifically, from 12:00 to 17:00, all floors appear to be cooler in respect to the ambient environment. Nevertheless, during night hours (from 0:00 to 7:00 and from 20:00 to 23:00), all floors (except the ground floor) retained higher temperature levels than outdoors. The delayed thermal response of the building is presumably because of its high heat capacity.

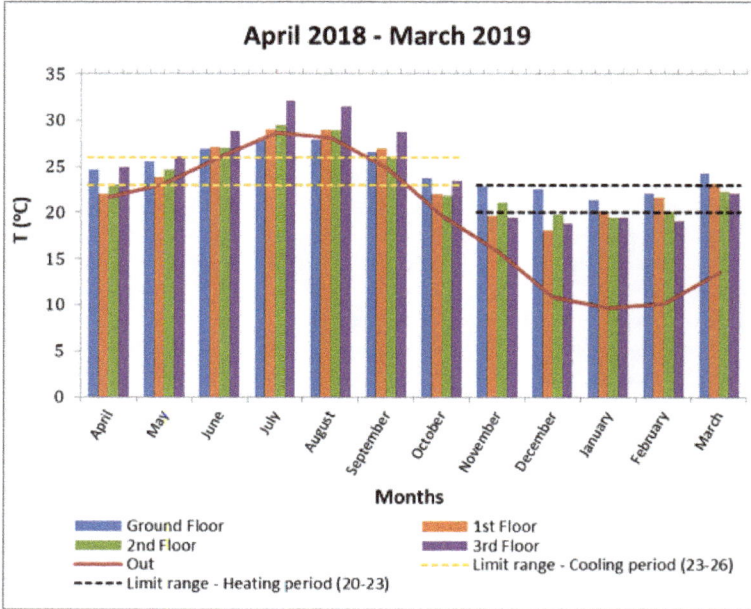

Figure 3. Monthly average T (°C) levels within the B FEPA building along with the respective limits.

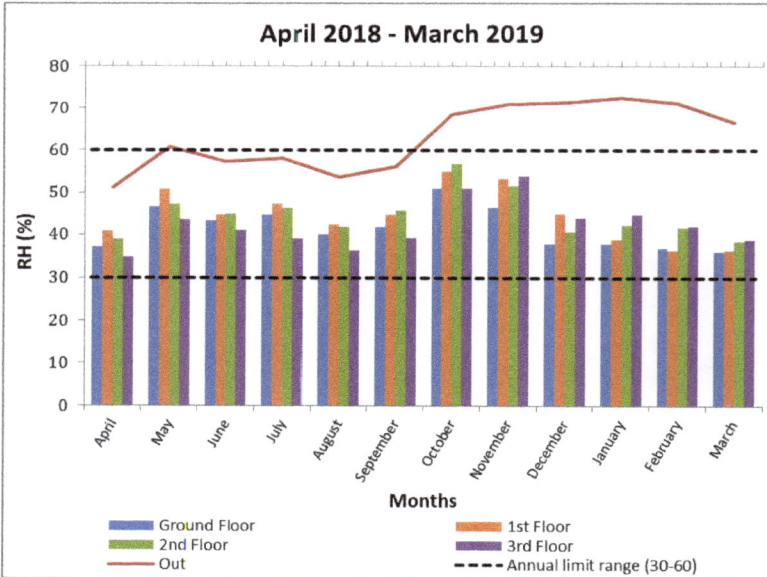

Figure 4. Monthly average RH (%) levels within the B FEPA building along with the respective limits.

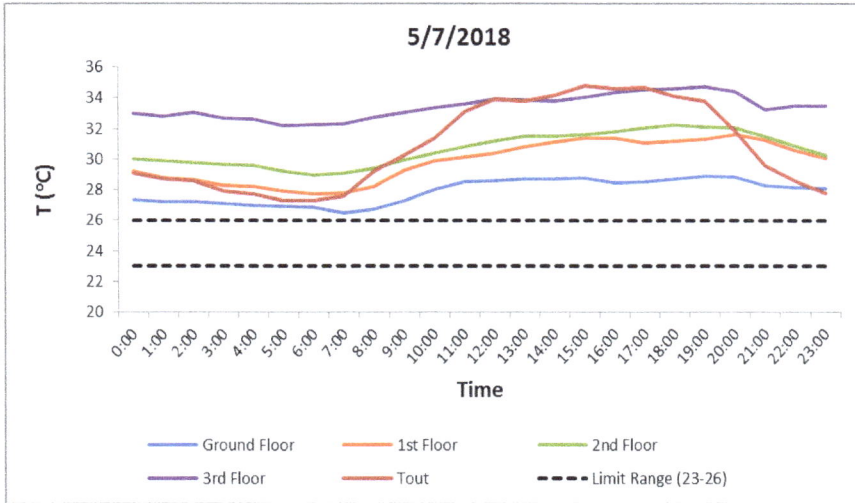

Figure 5. Daily behaviour of T (°C) levels within the B FEPA building along with the respective limits for a selected day of interest (May 7, 2018).

In order to assess the percentage within the range of the thermal comfort conditions, percentages of hourly measurements that were recorded within the recommended limit ranges of temperature and RH were calculated. The range of indoor RH within the respective limit is fairly wide, and thus, the percentage of values exceeding the limit is not significant. However, measurements of temperature levels showed some interesting results. Table 3 illustrates that in all floors, thermal conditions are considered within the range of thermal comfort for tenants only for 34.62% of the total measurements. This result, even though it is considered general, highlights the poor thermal behaviour of this building and the need for deep retrofitting actions. Indoor thermal conditions were found to be improved in February (51.11%), during the heating period, and in April (53.57%) and May (53.68%) amid the cooling period. The time period when thermal conditions are admittedly insufferable is definitely summer. The percentage within the comfort range decreases below 20% in June and is calculated from 0 to 5.77% during July and August, respectively. It should be noted that for these two warm months, none of the temperature data collected on the third floor were found to be within the recommended limit range. This floor is considered to be within the range of thermally comfortable conditions for only 28.97% of the total measurements, which is the lowest estimated percentage among the whole building. The highest percentage (43.26% in total) was found in the ground floor, which is expected to have the most satisfied tenants, especially during the heating period.

Table 3. Percentages of measured indoor temperature values within the recommended limit ranges.

Month	B FEPA Building (%)	Ground Floor (%)	1st Floor (%)	2nd Floor (%)	3rd Floor (%)
April	53.57	87.15	33.15	45.78	48.19
May	53.68	53.18	47.53	70.22	43.79
June	18.63	17.78	24.03	28.75	3.97
July	2.19	0.69	5.77	2.28	0.00
August	1.08	0.54	2.02	1.75	0.00
September	24.76	35.69	25.81	25.86	11.68
October	47.42	84.62	29.65	3.60	71.83
November	44.56	55.62	40.69	49.22	32.69
December	35.64	56.45	27.38	49.09	9.64
January	39.51	49.06	37.37	38.1	33.53
February	51.11	62.65	60.42	52.38	29.00
March	43.32	15.71	34.29	60.00	63.3
Total	34.62	43.26	30.68	35.59	28.97

3.1.2. Air Pollution

Indoor air pollution levels were also investigated in the common spaces of the B FEPA building. For that reason, concentrations of CO_2 and TVOC were measured from the ground to the third floor. It has to be noted that indoor CO_2 is mainly derived from human exhalation and that in some cases it is used as an indicator of adequate ventilation [59]. Additionally, the study of Yrieix et al. notes that the main internal emission sources of TVOC are building materials, paints, furnishings, and smoking [60]. A critical limit of exposure to indoor CO_2 is 1000 ppm, according to ASHRAE Standard 62-2001 [61], while the limit range of tolerance for internal TVOC is from 15 to 20 ppm, as Raatikainen et al. reported [62].

Concentrations of CO_2 were found to be very low during the experimental period, as they did not exceed the respective limit of exposure for any observation. As expected, the ground floor demonstrated the highest average (474 ppm) and maximum (799 ppm) concentrations, due to the increased occupancy of the shared space and numerous people passing by continuously (entering and exiting the building) (Table 4). However, the measured values within all floors are comparable, demonstrating non-significant differences. Standard deviations were also found to be decreased in all floors (from 28 to 48 ppm), demonstrating a relatively stable pattern of CO_2 within the building. This behaviour can also be noticed in the results of coefficient of variation, ranging from 5.88% in the first floor to 10.41% in the third floor, respectively. The low occupancy levels in common, areas along with the frequent window openings (due to natural ventilation), mitigate CO_2 fluctuations. Similarly, TVOC concentrations were found to be significantly decreased without surpassing the respective limit range. Table 5 shows that once again, the ground floor demonstrates the highest mean concentration (5.38 ppm). As mentioned above, it is the space with the most frequent human activity (such as smoking) compared to the other floors. However, it should be noted that the values of coefficient of variation, especially from the first to the third floor, are extremely high (92.47% to 113.62%), indicating a large dispersion of data. Thus, in these microenvironments, mean values of TVOC are not representative. An additional explanation for the low levels of TVOC within the building is that no refurbishing actions have been implemented recently (absence of fresh paint, new building materials, and furnishings).

Table 4. Descriptive statistics for CO_2 in each floor, for all experimental periods (April 2018–March 2019).

Statistics		Ground Floor (ppm)	1st Floor (ppm)	2nd Floor (ppm)	3rd Floor (ppm)
Mean		474	442	452	461
Standard Deviation		28	26	46	48
Coefficient of Variation		5.91%	5.88%	10.18%	10.41%
Minimum		414	401	405	395
Maximum		799	688	698	763
	25	454	425	422	433
Percentiles	50 (median)	470	434	434	445
	75	489	450	463	468

Table 5. Descriptive statistics for TVOC in each floor, for all experimental periods (April 2018–March 2019).

Statistics		Ground Floor (ppm)	1st Floor (ppm)	2nd Floor (ppm)	3rd Floor (ppm)
Mean		5.38	<1	2.79	1.47
Standard Deviation		2.31	<1	3.17	1.66
Coefficient of Variation		42.94%	92.47%	113.62%	112.93%
Minimum		1.54	<1	<1	<1
Maximum		22.85	22.01	28.22	13.38
	25	3.72	<1	<1	<1
Percentiles	50 (median)	5.08	<1	<1	<1
	75	6.59	<1	3.99	<1

Furthermore, the behaviour of extreme instant observations (outliers) for both air pollutants has been examined for all experimental periods. Figure 6 depicts that in all floors, none of the CO_2 observations exceed the limit of 1000 ppm, even during the time of high occupancy. On the contrary, Figure 7 illustrates that from the ground to the third floor a large number of instant maximum values of TVOC appeared. Individual extreme measurements were reported during specific events that took place in the dormitory when numerous people were present. However, from the same figure it is clear that only a negligible number of outliers are found to be higher than the upper limit of 20 ppm. It is obvious that for this building, indoor levels of CO_2 and TVOC are found to be relatively low.

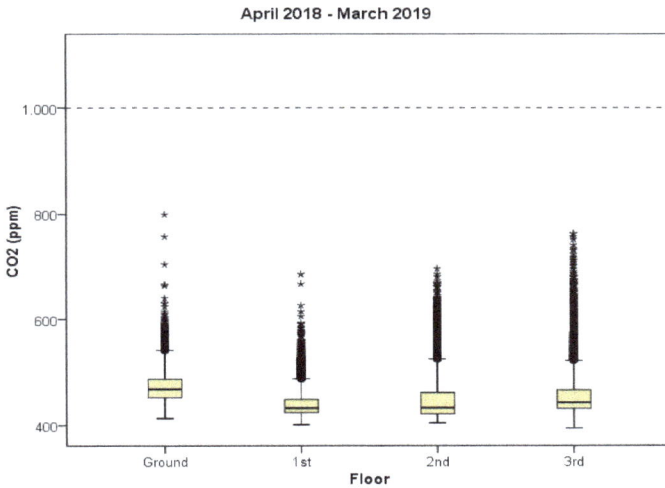

Figure 6. Boxplots of CO_2 within each floor, for all experimental periods (April 2018–March 2019), along with the respective limit of exposure.

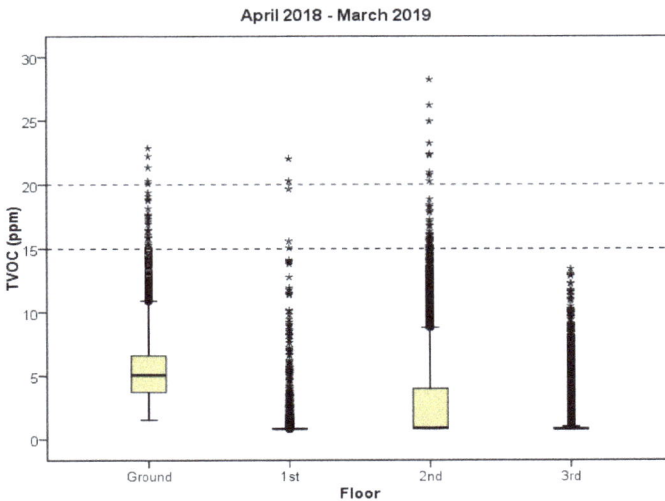

Figure 7. Boxplots of TVOC within each floor, for all experimental periods (April 2018–March 2019), along with the respective limit of exposure.

3.2. Energy Assessment

3.2.1. Electrical Consumption

Energy consumption of the B FEPA building depends mainly on the habits of the residents and the ambient meteorological conditions. For the hot days of summer there is only one air conditioning unit, installed in the common space of the ground floor, and some portable fans or air conditioners that inhabitants may use. It should be noted that these results refer only to the electrical consumption, corresponding to all electrical appliances (fridges, cooking stoves, cooking plates, laundry machines, personal appliances), all lighting needs, and some cooling needs (from a single air conditioning unit). From Figure 8, it is clear that during summer, energy consumption was found to be higher due to the typically high temperatures of the Greek summer. However, the same graph demonstrates that June is the month with the highest consumption, reaching 18,853.47 kWh in total. The respective measurements for July and August demonstrated lower levels of consumption. That is because residents (students) tend to leave after the end of the examination period (approximately in the beginning of July), and thus, the personal use of electrical devices is decreased. The consumption during summer remains higher than during the spring months. As expected, June presented the highest consumption levels, as the building has full capacity.

During autumn, local ambient temperature levels are quite moderate, and thus, the electrical consumption is low for September–October (9873 and 8169.13 kWh, respectively). However, the amounts of electrical consumption increase gradually for the following months (winter period) when the outdoor temperature decreases severely. An additional reason for this behaviour is the constant use of portable electrical heaters by the occupants. Even though the data for the cold period (November to March) are relatively limited, it is worth mentioning that December demonstrated lower consumption levels (12,985.17 kWh) than July (17,648.91 kWh). This result (as in July) is also affected by the absence of a large amount of tenants, due to Christmas holidays the last two week of the month.

Generally, autonomous HVAC systems seem to play a key role in the electrical consumption of the examined building. Characteristically, during July and August, even though the inhabitants are mainly away from the dormitory, the consumption levels are significantly high. Tenants that remain in the building during this period constantly use portable fans or air conditioning units, increasing the total amount of electricity. On the contrary, during October and November, when there is no need for cooling but the building has maximum occupancy, the electrical consumption levels appear to be significantly lower. It is clear that not only a redesign of the cooling/heating system is essential but also a reinforcement of the thermal insulation, as the building is prone to thermal losses.

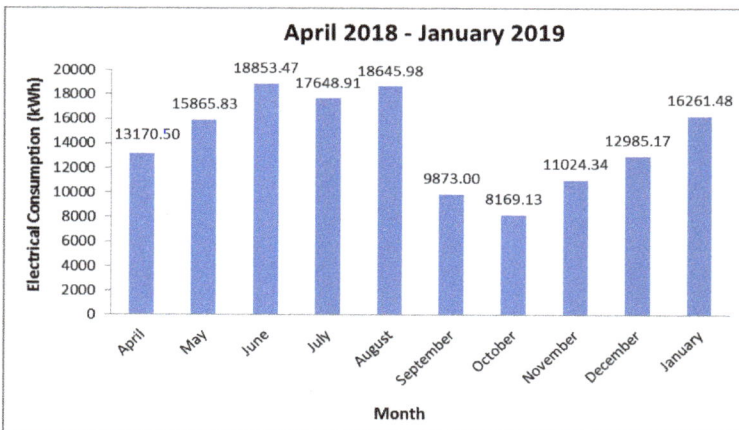

Figure 8. Monthly electrical consumption of B FEPA from April 2018 to January 2019.

3.2.2. Energy Signature

Energy behaviour of old buildings is usually correlated with external climatic conditions [63]. For the case of B FEPA, Figure 9 illustrates different levels of electrical consumption along with the respective outdoor temperature levels in order to investigate possible patterns. A higher energy demand during early morning and late evening hours is depicted. For that reason, a thorough investigation of the energy signature of the building (for the specific experimental period is considered mandated). The term "energy signature", also called thermal performance line, is considered the best fit correlating energy consumption with external climatic conditions [64–66]. The approach of obtaining different energy signatures (heating, DHW, electrical) is a system identification approach, also known as inverse modelling [67]. Measured energy consumption and weather data are used simultaneously in order to perform the regression (most commonly least squares method).

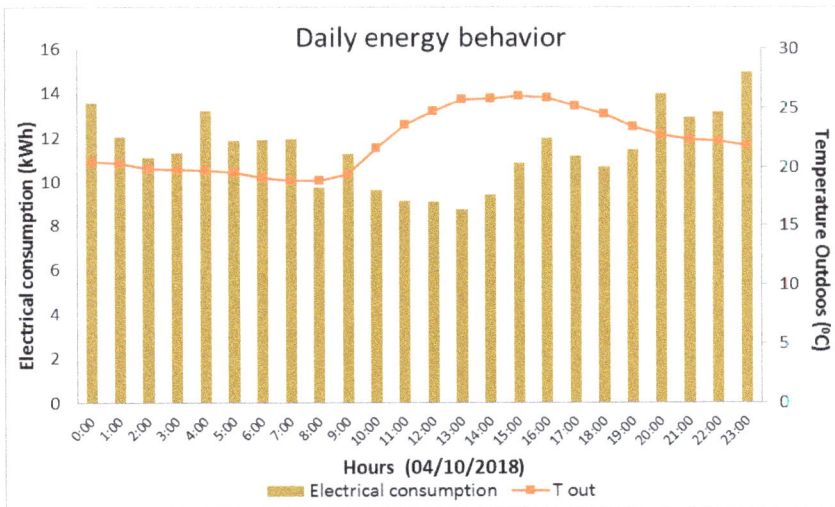

Figure 9. Hourly electrical consumption along with ambient temperature levels.

In order to calculate the energy signature for the B FEPA building, the total electrical consumption and the average outdoor temperature for all experimental periods were examined. The energy signature is separated into two time periods—the cooling period (from April to September) and heating period (from October to January). The results show that when the outdoor temperature is low during winter, the energy consumption is increased. The coefficient of determination (R^2) is equal to 0.4436 and the correlation coefficient (R) is found to be 0.6660. It is clear that 44.36% of the variability of electrical consumption is explained by the changes of ambient temperature. On the contrary, during the cooling period, when the temperature increases, the electrical consumption is relatively influenced. Characteristically, R^2 and R are 0.1902 and 0.4361, while the slope was found to be lower compared to the heating season. More specifically, in Figure 10, one may observe that approximately above 25 °C, the dependency of electrical consumption from the ambient temperature increases. However, it is important to note a deviation from this pattern for three specific days of June, when the dormitory festival was organized, and thus, plenty of people were in the building and large amounts of electric power was consumed because of numerous activities (parties, film projections, etc.). The days that this behaviour was observed are the following:

- June 16   E = 743.60 kWh,   Tout = 23.38 °C
- June 17   E = 746.16 kWh,   Tout = 22.33 °C

- June 18   E = 745.85 kWh,   Tout = 22.67 °C

Figure 11 illustrates the results for the heating period. The electrical consumption increases as outdoor temperature decreases. The same figure also shows that there are very few cases that diverge. The same results were obtained in a research study by Arregi et al., which refers to the energy signature of a university building in the United Kingdom. The authors highlighted that the pre-retrofitted building followed a seasonal pattern with energy consumption strongly correlated with the respective external temperature and that occupancy and usage patterns of the building also influenced the results. In conclusion, the two figures demonstrated a significant correlation between electrical consumption and the outdoor temperature, especially during the cold period [68]. This is an indicator that the indoor environment of the building is relatively vulnerable to outdoor conditions, mainly due to insufficient thermal insulation. It has to be noted that electrical consumption also demonstrated a strong correlation with indoor air temperature, but relatively weak correlations with the rest of the variables (RH, CO_2, and TVOC).

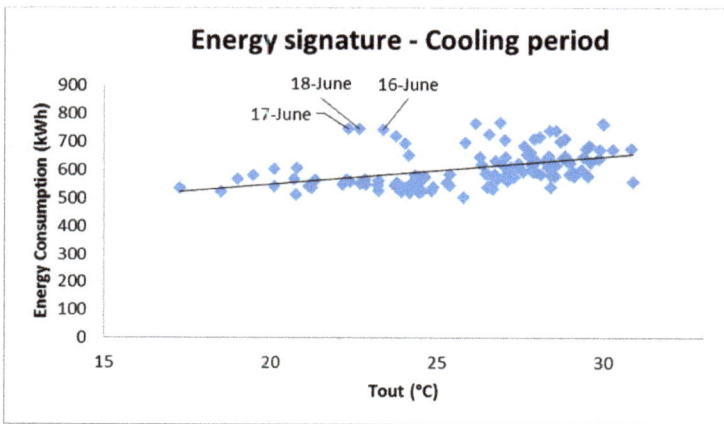

Figure 10. Energy signature of the B FEPA building for the cooling period (April–October 2018).

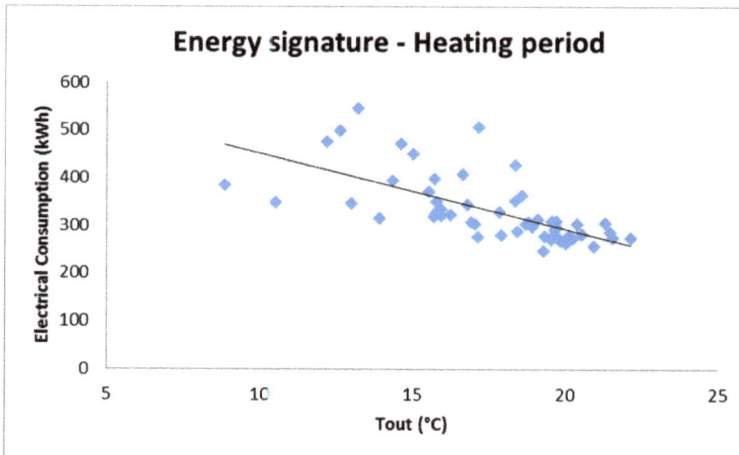

Figure 11. Energy signature of the B FEPA building for the heating period (November 2018–January 2019).

4. Conclusions

In this study, the energy performance along with the IEQ conditions of a dormitory within the campus of the University of Athens have been analyzed. The dormitory in question was selected as the pilot case of the Pro-GET-onE (Horizon 2020 G.A. n. 723747) project, in which a deep retrofit is planned to take place in late 2019. In fact, through Pro-GET-onE H2020, a 3D building information model through BIM modelling is being developed and will be available and presented within the European community at a later stage of the project.

A series of field measurements took place in order to demonstrate the thermal and energy profile of the building's current state, along with the IEQ performance during different monitoring periods. The needs on which the renovation actions should focus on, which is the main goal of this paper, are highlighted.

The most significant conclusions drawn from the experimental campaign are the following:

- Thermal conditions within the whole building cannot be considered comfortable for tenants, as high percentages of temperature measurements were found to surpass the proposed limits. This phenomenon is enhanced during summer months, with high ambient temperatures. Installation of a central air conditioning unit or an alternative cooling system is considered mandatory.
- The third floor is strongly influenced by the ambient climate conditions, especially during warm months, when solar radiation causes internal overheating. This depicts problematic thermal insulation of the roof, which needs to be addressed.
- The investigation of the IEQ regime did not demonstrate high concentrations of air pollutants. Low occupancy numbers, along with adequate natural ventilation of the common use areas, were found to maintain CO_2 and TVOC at decreased levels, within all experimental points. However, similar measurements should be carried out after the refurbishment in order to validate this result.
- Energy metering showed that the examined building is generally vulnerable to thermal loses. This is because electrical consumption was found to be significantly correlated with ambient climatic conditions. It is noticeably increased during extreme outdoor temperature levels. This is an additional result that demonstrates that redesign of the cooling and heating system, as well as reinforcement of the building's thermal insulation, are the key role actions that need to be assessed for the energy retrofitting of this building.

Author Contributions: All authors equally contributed to the paper.

Funding: This project has received funding from the European Union's Horizon 2020 Innovation action under grant agreement No. 723747 (Pro-GET-onE).

Acknowledgments: The authors wish to acknowledge the Technical Service department of the National and Kapodistrian University of Athens for their contribution to the technical specifications of the student housing. Furthermore, the authors would like to acknowledge SAVES2 project, a H2020 project under the grant agreement no. 754203, for the acquisition of the electrical data for the B FEPA student housing, and also the company *Ether Applications Ltd.* for the installation of the electrical meters. Lastly, Pro-GET-onE (Horizon 2020 G.A. n. 723747) and the National Observatory of Athens Institute of Environmental Research are highly appreciated.

Conflicts of Interest: The authors declare no conflict of interest.

References

1. Balaras, C.A.; Gaglia, A.G.; Georgopoulou, E.; Mirasgedis, S.; Sarafidis, Y.; Lalas, D. European residential buildings and empirical assessment of the Hellenic building stock, energy consumption, emissions and potential energy savings. *Build. Environ.* **2007**, *42*, 1298–1314. [CrossRef]
2. EU. On the Energy Performance of Buildings. In *Directive 2002/91/EC of the European Parliament and of the Council*; Official Journal of the European Communities: Brussels, Belgium, 2002.
3. YPEKA. *EK407/B/9.4.2010, "Regulation on Energy Performance in the Building Sector—KENAK"*; YPEKA: Athens, Greece, 2010.
4. Baranova, D.; Sovetnikov, D.; Borodinecs, A. The extensive analysis of building energy performance across the Baltic Sea region. *Sci. Technol. Built Environ.* **2018**, *24*, 982–993. [CrossRef]

5. Zhaoa, D.; McCoyb, A.; Duc, J. An empirical study on the energy consumption in residential buildings after adopting green building standards. *Procedia Eng.* **2016**, *145*, 766–773. [CrossRef]
6. Huang, Y. Energy saving technology measures of green building. *Adv. Mater. Res.* **2012**, *347–353*, 4114–4117. [CrossRef]
7. Huang, Z. Discussion on energy saving technology and energy saving way of green building. *Int. Conf. Intell. Transp.* **2015**, *1*, 313–315.
8. Santamouris, M.; Sfakianaki, A.; Pavlou, K. On the efficiency of night ventilation techniques applied to residential buildings. *Energy Build.* **2010**, *42*, 1309–1313. [CrossRef]
9. Santamouris, M. Cooling the buildings: Past, present, and future. *Energy Build.* **2016**, *128*, 617–638. [CrossRef]
10. Vatin, N.; Nemova, D.; Kazimirova, A.; Gureev, K. Increase of energy efficiency of the building of kindergarten. *Adv. Mater. Res.* **2014**, *953–954*, 1537–1544. [CrossRef]
11. Naumov, A.; Tabunshchikov, D.; Kapko, M.; Brodach, M. Research of the Microclimate formed by the Local DCV. *Energy Build.* **2015**, *90*, 1–5. [CrossRef]
12. Deng, Y.; Feng, Z.; Fang, J.; Cao, S.J. Impact of ventilation rates on indoor thermal comfort and energy efficiency of ground-source heat pump system. *Sustain. Cities Soc.* **2018**, *37*, 154–163. [CrossRef]
13. Szodrai, F.; Kalmár, F. Simulation of Temperature Distribution on the Face Skin in Case of Advanced Personalized Ventilation System. *Energies* **2019**, *12*, 1185. [CrossRef]
14. Zalejska-Jonsson, A.; Wilhelmsson, M. Impact of perceived indoor environment quality on overall satisfaction in Swedish dwellings. *Build. Environ.* **2013**, *63*, 134–144. [CrossRef]
15. Thomsen, K.E.; Rose, J.; Christen Mørck, O.; Jensen, S.Ø.; Østergaard, I.; Knudsen, H.N.; Bergsøe, N.C. Energy consumption and indoor climate in a residential building before and after comprehensive energy retrofitting. *Energy Build.* **2016**, *123*, 8–16. [CrossRef]
16. Kalmár, F. Interrelation between glazing and summer operative temperatures in buildings. *Int. Rev. Appl. Sci. Eng.* **2016**, *7*, 51–60. [CrossRef]
17. Kalmár, F. Summer operative temperatures in free running existing buildings with high glazed ratio of the facades. *J. Build. Eng.* **2016**, *6*, 236–242. [CrossRef]
18. Zemitis, J.; Borodinecs, A.; Frolova, M. Measurements of moisture production caused by various sources. *Energy Build.* **2016**, *127*, 884–891. [CrossRef]
19. Prasauskas, T.; Martuzevicius, D.; Kalamees, T.; Kuusk, K.; Leivo, V.; Haverinen-Shaughnessy, U. Effects of Energy Retrofits on Indoor Air Quality in Three Northern European Countries. *Energy Procedia* **2016**, *96*, 253–259. [CrossRef]
20. Crump, D.; Dengel, A.; Swainson, M. *Indoor Air Quality in Highly Energy Efficient Homes—A Review*; Report n° NF18; NHBC Foundation: Milton Keynes, UK, 2009; Available online: http://www.zerocarbonhub.org/sites/default/files/resources/reports/Indoor_Air_Quality_in_Highly_Energy_Efficient_Homes_A_Review_NF18.pdf (accessed on 22 April 2019).
21. Melikov, A.K. Advanced air distribution. *ASHRAE J.* **2011**, *53*, 73–78.
22. Huang, Y.; Wang, Y.; Liu, L.; Nielsen, P.V.; Jensen, R.L.; Yang, X. Performance of constant exhaust ventilation for removal of transient high-temperature contaminated airflows and ventilation-performance comparison between two local exhaust hoods. *Energy Build.* **2017**, *154*, 207–216. [CrossRef]
23. Gao, R.; Wang, C.; Li, A.; Yu, S.; Deng, B. A novel targeted personalized ventilation system based on the shooting concept. *Build. Environ.* **2018**, *135*, 269–279. [CrossRef]
24. Xu, C.; Nielsen, P.V.; Liu, L.; Jensen, R.L.; Gong, G. Impacts of airflow interactions with thermal boundary layer on performance of personalized ventilation. *Build. Environ.* **2018**, *135*, 31–41. [CrossRef]
25. Xu, C.; Liu, L. Personalized ventilation: One possible solution for airborne infection control in highly occupied space? *Indoor Built Environ.* **2018**, *27*, 873–876. [CrossRef]
26. Chludzińska, M. The effect of front pattern perforation shape on thermal sensations of occupants in personalized ventilation systems. *Build. Environ.* **2019**, *151*, 140–147. [CrossRef]
27. Yu, Z.; Fung, B.C.M.; Haghighat, F.; Yoshino, H.; Morofsky, E. A systematic procedure to study the influence of occupant behavior on building energy consumption. *Energy Build.* **2011**, *43*, 1409–1417. [CrossRef]
28. Ouyang, J.; Hokao, K. Energy-saving potential by improving occupants' behavior in urban residential sector in Hangzhou City, China. *Energy Build.* **2009**, *41*, 711–720. [CrossRef]

29. Brown, T.; Dassonville, C.; Derbez, M.; Ramalho, O.; Kirchner, S.; Crump, D.; Mandin, C. Relationships between socioeconomic and lifestyle factors and indoor air quality in French dwellings. *Environ. Res.* **2015**, *140*, 385–396. [CrossRef]

30. Hiscock, R.; Bauld, L.; Amos, A.; Fidler, J.A.; Munafò, M. Socioeconomic status and smoking: A review. *Ann. N. Y. Acad. Sci.* **2012**, *1248*, 107–123. [CrossRef]

31. Klepeis, N.E.; Bellettiere, J.; Hughes, S.C.; Nguyen, B.; Berardi, V.; Liles, S.; Obayashi, S.; Hofstetter, C.R.; Blumberg, E.; Hovell, M.F. Fine particles in homes of predominantly low-income families with children and smokers: Key physical and behavioral determinants to inform indoor-air-quality interventions. *PLoS ONE* **2017**, *12*, e0177718. [CrossRef]

32. Calì, D.; Osterhage, T.; Streblow, R.; Müller, D. Energy performance gap in refurbished German dwellings: Lesson learned from a field test. *Energy Build.* **2016**, *127*, 1146–1158. [CrossRef]

33. Shinohara, N.; Tokumura, M.; Kazama, M.; Yonemoto, Y.; Yoshioka, M.; Kagi, N.; Hasegawa, K.; Yoshino, H.; Yanagi, U. Indoor air quality and thermal comfort in temporary houses occupied after the Great East Japan Earthquake. *Indoor Air* **2014**, *24*, 425–437. [CrossRef]

34. Makonese, T.; Bradnum, C.M. Design and performance evaluation of wood-burning cookstoves for low-income households in South Africa. *J. Energy S. Afr.* **2018**, *29*, 1–12. [CrossRef]

35. Franchi, M.; Carrer, P.; Kotzias, D.; Rameckers, E.; Seppänen, O.; Van Bronswijk, J.; Viegi, G.; Gilder, J.; Valovirta, E. Working towards healthy air in dwellings in Europe. *Allergy* **2006**, *61*, 864–868. [CrossRef] [PubMed]

36. Steinemann, A.; Wargocki, P.; Rismanchi, B. Ten questions concerning green buildings and indoor air quality. *Build. Environ.* **2017**, *112*, 351–358. [CrossRef]

37. Collignan, B.; Le Ponner, E.; Mandin, C. Relationships between indoor radon concentrations, thermal retrofit and dwelling characteristics. *J. Environ. Radioact.* **2016**, *165*, 124–130. [CrossRef] [PubMed]

38. Coombs, K.C.; Chew, G.L.; Schaffer, C.; Ryan, P.H.; Brokamp, C.; Grinshpun, S.A.; Adamkiewicz, G.; Chillrud, S.; Hedman, C.; Colton, M.; et al. Indoor air quality in green-renovated vs. non-green low-income homes of children living in a temperate region of US (Ohio). *Sci. Total Environ.* **2016**, *554–555*, 178–185. [CrossRef] [PubMed]

39. Hamilton, I.; Milner, J.; Chalabi, Z.; Das, P.; Jones, B.; Shrubsole, C.; Davies, M.; Wilkinson, P. Health effects of home energy efficiency interventions in England: A modelling study. *Br. Med. J. Open* **2015**, *5*, e007298. [CrossRef] [PubMed]

40. Frey, S.E.; Destaillats, H.; Cohn, S.; Ahrentzen, S.; Fraser, M.P. Characterization of indoor air quality and resident health in an Arizona senior housing apartment building. *J. Air Waste Manag. Assoc.* **2014**, *64*, 1251–1259. [CrossRef] [PubMed]

41. Noris, F.; Adamkiewicz, G.; Delp, W.W.; Hotchi, T.; Russell, M.; Singer, B.C.; Spears, M.; Vermeer, K.; Fisk, W.J. Indoor environmental quality benefits of apartment energy retrofits. *Build. Environ.* **2013**, *68*, 170–178. [CrossRef]

42. Broderick, Á.; Byrne, M.; Armstrong, S.; Sheahan, J.; Coggins, A.M. A pre and post evaluation of indoor air quality, ventilation, and thermal comfort in retrofitted co-operative social housing. *Build. Environ.* **2017**, *122*, 126–133. [CrossRef]

43. Hamburg, A.; Kalamees, T. The Influence of Energy Renovation on the Change of Indoor Temperature and Energy Use. *Energies* **2018**, *11*, 3179. [CrossRef]

44. Cascone, S.; Sciuto, G. Recovery and reuse of abandoned buildings for student housing: A case study in Catania, Italy. *Front. Archit. Res.* **2018**, *7*, 510–520. [CrossRef]

45. Furundzic, A.K.; Kosoric, V.; Golic, K. Potential for reduction of CO_2 emissions by integration of solar water heating systems on student dormitories through building refurbishment. *Sustain. Cities Soc.* **2012**, *2*, 50–62. [CrossRef]

46. Mostafavi, N.; Farzinmoghadam, M.; Hoque, S. Envelope retrofit analysis using eQUEST, IESVE Revit Plug-in and Green Building Studio: A university dormitory case study. *Int. J. Sustain. Energy* **2015**, *34*, 594–613. [CrossRef]

47. Tabatabaee, S.; Weil, B.; Aksamija, A. Negative life-cycle emissions growth rate through retrofit of existing institutional buildings. In Proceedings of the ARCC Conference, Chigaco, IL, USA, 6–9 April 2015; pp. 212–221.

48. Karp, A.; McCauley, M.; Byrne, J.; Byrne, J. The value of adding ambient energy feedback to conservation tips and goal-setting in a dormitory. *Int. J. Sustain. High Educ.* **2016**, *17*, 471–488. [CrossRef]

49. Ning, H.R.; Wang, Z.J.; Zhang, X.X.; Ji, Y.C. Adaptive thermal comfort in university dormitories in the severe cold area of China. *Build. Environ.* **2016**, *99*, 161–169. [CrossRef]

50. Assimakopoulos, M.N.; De Masi, R.F.; Papadaki, D.; Ruggiero, S.; Vanoli, G.P. Energy audit and performance optimization of a residential university building in heating dominated climates of Italian backcountry. *TEMA Technol. Eng. Mater. Archit. [S.L.]* **2018**, *4*, 19–33.

51. Petidis, I.; Aryblia, M.; Daras, T.; Tsoutsos, T. Energy saving and thermal comfort interventions based on occupants' needs: A students' residence building case. *Energy Build.* **2018**, *174*, 347–364. [CrossRef]

52. Ferrante, A.; Mochi, G.; Predari, G.; Badini, L.; Fotopoulou, A.; Gulli, R.; Semprini, G. A European Project for Safer and Energy Efficient Buildings: Pro-GET-onE (Proactive Synergy of inteGrated Efficient Technologies on Buildings' Envelopes). *Sustainability* **2018**, *10*, 812. [CrossRef]

53. ANSI/ASHRAE Guideline 14-2002. *Measurement of Energy and Demand Savings*; American Society of Heating, Refrigeration and Air Conditioning Engineers, Inc.: Atlanta, GA, USA, 2002.

54. Landis+Gyr E65C, Industrial & Commercial Meter. Available online: https://www.landisgyr.eu/product/landisgyr-e650/ (accessed on 12 April 2019).

55. Smart System for Energy Monitoring and Management, Ether Technological Applications Ltd. Available online: http://www.ether.gr/en (accessed on 18 April 2019).

56. Lagouvardos, K.; Kotroni, V.; Bezes, A.; Koletsis, I.; Kopania, T.; Lykoudis, S.; Mazarakis, N.; Papagiannaki, K.; Vougioukas, S. The automatic weather stations NOANN network of the National Observatory of Athens: Operation and database. *Geosci. Data J.* **2017**, *4*, 1–16. [CrossRef]

57. ANSI/ASHRAE Standard 55-2004. *Thermal Environmental Conditions for Human Occupancy*; American Society of Heating, Refrigerating and Air-Conditioning Engineers, Inc.: Atlanta, GA, USA, 2004.

58. Pegas, P.; Alves, C.; Evtyugina, M.; Nunes, T.; Cerqueira, M.; Franchi, M.; Pio, C.; Almeida, S.; Freitas, M. Indoor air quality in elementary schools of Lisbon in spring. *Environ. Geochem. Health* **2011**, *33*, 455–468. [CrossRef]

59. Shendell, D.G.; Prill, R.; Fisk, W.J.; Apte, M.G.; Blake, D.; Faulkner, D. Associations between classroom CO2 concentrations and student attendance in Washington and Idaho. *Indoor Air* **2004**, *14*, 333–341. [CrossRef] [PubMed]

60. Yrieix, C.; Dulaurent, A.; Laffargue, C.; Maupetit, F.; Pacary, T.; Uhde, E. Characterization of VOC and formaldehyde emissions from a wood based panel: Results from an inter-laboratory comparison. *Chemosphere* **2010**, *79*, 414–419. [CrossRef] [PubMed]

61. ANSI/ASHRAE Standard 62-2001. *Ventilation for Acceptable Indoor Air Quality*; American Society of Heating, Refrigerating and Air Conditioning Engineers, Inc.: Atlanta, GA, USA, 2001.

62. Raatikainen, M.; Skön, J.P.; Turunen, M.; Leiviskä, K.; Kolehmainen, M. Evaluating Effects of Indoor Air Quality in School Buildings and Students' Health: A Study in Ten Schools of Kuopio, Finland. In Proceedings of the 2nd International Conference on Environment, Energy and Biotechnology (PCBEE), Kuala Lumpur, Malaysia, 8–9 June 2013.

63. Zhao, H.; Magoules, F. A review on the prediction of building energy consumption. Renewable and Sustainable. *Energy Rev.* **2012**, *16*, 3586–3592.

64. Hammarsten, S. A critical appraisal of energy-signature models. *Appl. Energy* **1987**, *26*, 97–110. [CrossRef]

65. Westergren, K.E.; Högberg, H.; Norlén, U. Monitoring energy consumption in single-family houses. *Energy Build.* **1999**, *29*, 247–257. [CrossRef]

66. Belussi, L.; Danza, L. Method for the prediction of malfunctions of buildings through real energy consumption analysis: Holistic and multi-disciplinary approach of energy signature. *Energy Build.* **2012**, *55*, 715–720. [CrossRef]

67. Fumo, N.; Biswas, M.A.R. Regression analysis for prediction of residential energy consumption. *Renew. Sustain. Energy Rev.* **2015**, *47*, 332–343. [CrossRef]

68. Arregi, B.; Garay, R. Regression analysis of the energy consumption of tertiary buildings. *Energy Procedia* **2017**, *122*, 9–14. [CrossRef]

Article

A Data-Driven Approach for Enhancing the Efficiency in Chiller Plants: A Hospital Case Study

Serafín Alonso *, Antonio Morán, Miguel Ángel Prada, Perfecto Reguera, Juan José Fuertes and Manuel Domínguez

Grupo de Investigación en Supervisión, Control y Automatización de Procesos Industriales (SUPPRESS),
Esc. de Ing. Industrial e Informática, Universidad de León, Campus de Vegazana s/n, 24007 León, Spain;
a.moran@unileon.es (A.M.); ma.prada@unileon.es (M.A.P.); prega@unileon.es (P.R.); jj.fuertes@unileon.es (J.J.F.);
mdomg@unileon.es (M.D.)
* Correspondence: saloc@unileon.es; Tel.: +34-987261694

Received: 14 January 2019; Accepted: 26 February 2019; Published: 2 March 2019

Abstract: Large buildings cause more than 20% of the global energy consumption in advanced countries. In buildings such as hospitals, cooling loads represent an important percentage of the overall energy demand (up to 44%) due to the intensive use of heating, ventilation and air conditioning (HVAC) systems among other key factors, so their study should be considered. In this paper, we propose a data-driven analysis for improving the efficiency in multiple-chiller plants. Coefficient of performance (COP) is used as energy efficiency indicator. Data analysis, based on aggregation operations, filtering and data projection, allows us to obtain knowledge from chillers and the whole plant, in order to define and tune management rules. The plant manager software (PMS) that implements those rules establishes *when* a chiller should be staged up/down and *which* chiller should be started/stopped according different efficiency criteria. This approach has been applied on the chiller plant at the Hospital of León.

Keywords: energy efficiency; HVAC systems; chiller plants; chiller performance; COP; data-driven analysis

---

1. Introduction

Energy consumption in large buildings, such as hotels, museums, hospitals, commercial buildings, etc., represents more than 20% of the global energy consumption in developed countries [1]. The reasons behind such a high consumption are the addition of new building services, the increase of comfort levels, the additional time spent by people inside buildings, and the proliferation of heating, ventilation and air conditioning (HVAC) systems, among others [2]. Four types of energy consumption can be distinguished in those buildings: electricity, heating, hot water, and cooling [3]. Cooling loads usually have a seasonal behaviour and, in some buildings, they are not the most noteworthy, so their study is often disregarded. Cooling loads, however, represent an important percentage of the overall energy demand (up to 44%) in utility buildings with special facilities, such as hospitals [4]. For instance, hospitals keep a minimum level of cooling load during the whole year to guarantee the operation of hospital services: refrigeration of surgeries, scanners, magnetic resonance imaging systems and data centers, among other key facilities. Furthermore, cooling loads have a direct influence on the electricity demand, since chillers and their auxiliary elements (pumps, fans, cooling towers, etc.) are electric systems [4]. Therefore, both cooling load and chiller performance should be considered and analyzed in large buildings, in order to achieve energy efficiency. Additionally, many buildings require a reliable and secure cooling supply, so aspects such as monitoring, automatic management and assets maintenance play also an important role [5].

To achieve energy efficiency in a multiple-chiller plant, it is recommended to study, first, the individual chiller performance and, later, the overall plant performance [6]. The building management systems (BMS) acquire and store a great amount of real data, which can be analyzed and exploited to extract the implicit knowledge. So far, the vast amount of data was rarely translated into useful knowledge about potential energy performance improvements, due to its extreme complexity or a lack of effective data analysis techniques [7]. However, the novel advances in the data science allow us to address a complex data analysis.

Therefore, a data-driven analysis should be carried out in order to acquired knowledge about the plant [8]. The implicit knowledge which is discovered (analyzing past data from the chillers, from the whole plant and from the environment) can be added to the management strategies. It can be converted into rules to be used in an expert module with the final aim of enhancing energy efficiency [9]. A periodic data analysis of the chiller plant can help us to achieve a better understanding and to monitor efficiency, aiming to upgrade and tune the management rules and to implement more efficient up/down sequencing strategies.

The contribution of this paper is the proposal of a comprehensive methodology for improving the efficiency in multiple-chiller plants. This methodology is based on a data analysis of the operation of the chillers and the overall plant, using real data instead of simulations. The proposed data analyses highlight relevant information by applying aggregation, filtering and data projection. Using the knowledge extracted specifically from the plant, control parameters of the chillers can be adjusted and management rules can be defined or tuned. The aim is to achieve an efficient management of the plant, without the need of incorporating cutting-edge controllers, since the management rules obtained through the proposed approach can be easily deployed in existing controllers. The proposed approach is applied on the real chiller plant at the Hospital of León.

This paper is structured as follows: Section 2 reviews the previous related work. The data-driven approach to improve energy efficiency in chiller plants is proposed in Section 3. Then, the multiple-chiller plant at the Hospital of León is described in detail in Section 4. Section 5 explains the application of proposed methodology to that plant. Next, results on chiller and plant efficiencies are presented in Section 6. Finally, conclusions are drawn in Section 7.

2. Related Work

Reviewing the literature, some examples of research on data mining for improving energy efficiency in buildings can be found. These works focus on the use of data mining techniques to extract relationships and patterns of interest from a large dataset [7]. However, many works rely on simulations, using software as TRNSYS, EnergyPlus, etc. [10]. Data analytics on a detailed measured building performance can help us to identify and estimate energy savings and then to inform the decision making system [11].

Other research focuses on the use of machine learning techniques for forecasting the energy efficiency and consumption in the building and, afterwards, comparing the predicted values with the nominal ones in order to detect possible deviations [12]. Simulated data are generally based on a physical model of the system, often used to build prediction models [13]. On the other hand, prediction models of the HVAC systems have been also built using real data [14]. The third type of approach found in the literature is a grey box model, which merges the qualities of both the physics-based and data-driven models [15].

With regard to the measurement of energy efficiency, several indicators (EEI) can be used in the data analysis [16], ranging from the COP (Coefficient of Performance) or EER (Energy Efficiency Ratio) to more sophisticated indicators such as SCOP (Seasonal Coefficient of Performance), SEER (Seasonal Energy Efficiency Ratio) and IPLV (Integrated Part Load Value), which consider seasonal chiller operations and capacity modulation. Other research defines and uses specific EEIs [9]. Nevertheless, the computation of COP is quite simple from measurements of electricity consumption and cooling load, so this indicator is often used to characterize the chiller efficiency and the overall performance (including the performance of chillers, pumps, fans, refrigeration towers, etc.) [6]. The COP value depends on the chiller technology and on the surrounding conditions [17–19].

Smart buildings should incorporate the feedback from the data analysis in the management and control system in order to optimize the use of energy in different conditions [20,21]. The structure of the control system is usually based on a hierarchical multilevel concept [22,23], with a coordinator layer over the local control units. For instance, a hierarchical cascade control strategy for energy management of intelligent buildings is used in [24]. The plant management software (PMS) is in charge of operating the plant (together with the BMS) with the minimum energy consumption. For that, PMS implements efficient chiller sequencing strategies which decide *when* a chiller should be staged up/down and *which* chiller should be started/stopped, considering a cooling load, weather conditions, chiller load capacities, etc. [25–27]. Note that reducing condensing temperature leads also to an increase of the chiller performance [28,29]. The aim of the PMS is to maximize the overall COP, by adjusting the capacity of the plant to the fluctuating cooling load. Therefore, a PMS becomes essential to improve energy efficiency in multiple-chiller plants [30].

Most PMSs rely on complex optimization methods [31–33], which require a high computational effort and make the deployment on existing controllers so difficult that often the rules of these PMSs can be only tested on simulated plants. Other commercial software uses relational control, based on the equal marginal performance principle [34]. The aim of relational control is to achieve optimal energy efficiency of the plant, requiring each chiller to be operated in relation to the operation of the others.

Rule-based management strategies, together with performance monitoring tools and model-based predictive control, have been outlined as outstanding methods for intelligent HVAC control to enhance energy efficiency [35]. Rule-based management enables the translation of best practices, experience and knowledge of HVAC control engineers into a set of rules, which can be applied to operate the plant. Other control methods and optimization techniques developed in the HVAC field have been reviewed in [36].

3. Methodology for Enhancing Efficiency in Multiple-Chiller Plants

In this paper, we propose a data-driven approach to define, upgrade and tune the rules of a PMS and, in consequence, to improve energy efficiency in multiple-chiller plants (see Figure 1). The data analysis provides, on the one hand, information about individual chillers and, on the other hand, information about the chiller plant. The aim of chiller data analysis is to enhance individual chiller efficiency, whereas the objective of plant data analysis is to enhance overall plant efficiency. The knowledge about individual chiller performance is used for two purposes: to adjust internal chiller parameters and to define or tune rules implemented in the PMS. For that, conditional rules (*If-Then-Else*) allow us to decide *when* a chiller should be staged up/down, whereas sorting rules based on multi-criteria rankings allow us to choose *which* chiller should start/stop (the fittest one in each situation). Finally, the knowledge extracted from the plant lets us update management rules. Our approach is based on a hierarchical multilevel control system, requiring the implementation of the coordination level (an expert system with the management rules) and some configuration actions in the local units (chiller control boards).

The proposed approach requires several iterative analyses in order to achieve an optimal efficiency in the plant, but it has the advantage of low computational cost. For that reason, the application of the proposed approach can include new data from subsequent years, providing an incremental improvement in efficiency to the plant.

In a first step, we propose acquiring data from chillers and carry out an individual data-driven analysis. Using the extracted knowledge about chillers, internal chiller parameters are adjusted to improve their operation. Moreover, knowledge of each chiller is used to implement the management rules in the chiller PMS. Unlike complex optimization methods, heuristic rules can be implemented using simple programming structures and executed by any conventional building management system, without requiring extra computational resources.

In the second step, data from the whole plant are also collected and analyzed in order to obtain global knowledge about the plant. This step allows us to redefine and tune the rules or to add/delete sorting criteria regularly in the PMS.

Prior to their analysis, data must be collected from the BMS logs. Chiller-wise and overall plant COPs, which can be computed using chiller power and cooling load, are used as energy efficiency indicators. Data from other variables, such as chiller load ratio, condensing pressure and temperature, compressor current, outdoor temperature, etc., can also be considered for the analysis.

The data analysis of chillers and plant performance is based on highlighting relevant information by applying aggregation operations on measures, subject to some attributes [37]. Expression (1) represents, in a general way, those operations:

$$Aggregation_{[method]}(efficiencyVariable)\Pi_{[projectionVariable]} \cdot \sigma_{[dataSelection]} \tag{1}$$

In our approach, average and counting samples are used as aggregation methods. Measures (the object of analysis) are energy efficiency indicators such as chiller and plant COPs, power demand and cooling load. Attributes are outdoor temperature, chiller load ratio, type of chiller, number of chillers running, year, month, day of year, weekday, hour, etc. Finally, data can be selected either from a specific chiller or from the plant. The attributes listed above can be also used to filter the data. For example, data from an specific weekday (Monday), hour interval (0–8 h), outdoor temperature limit (*OutdoorTemp* < 10 °C), etc. Expression (2) summarizes some operations which can be carried out in different data analyses:

$$Agg\begin{bmatrix} average \\ count \end{bmatrix}\begin{pmatrix} COP \\ CoolingLoad \\ PowerDemand \end{pmatrix}\Pi\begin{bmatrix} Year \\ DayOfYear \\ Month \\ Weekday \\ Hour \\ ChillerLoad \\ OutdoorTemp. \\ ChillersOn \\ ChillerType \\ \dots \end{bmatrix}\sigma\begin{bmatrix} Chiller\ 1-n \\ Plant \\ Anyfilter \end{bmatrix}. \tag{2}$$

Data projections can be completed using visualization techniques to incorporate extra information [38]. For that purpose, additional attributes can be coded with properties such as size, texture, color, shape or weight, given a projection Π.

Other additional variables (condensing pressure and temperature, compressor current, etc.) can be also involved in the data analysis of each chiller. In this case, the addition of their information in terms of simple time-series plots can help us to monitor the chiller behavior.

From the knowledge acquired about each chiller performance, modifications in its internal configuration can be suggested. Note that local chiller control is implemented by manufacturers and they often only allow us to modify schedules and setpoints within a specific range. Parameters such as output water temperature, control zone, rate of changes, slide percentage, delays, etc. can be modified in order to improve chiller efficiency.

Data-driven analysis also provides information about how attributes influence the chiller and plant efficiencies. That knowledge is converted into management rules which are implemented in an expert module of the PMS. In this sense, the extracted knowledge is used to setup chillers, in order to update the set of conditional rules or to change their sorting criteria, resulting in appropriate plant management strategies. The staff in charge of energy management in the building should be involved in performing the data analyses and defining or tuning the management rules. They could provide their expertise and approve the management rules before their deployment on controllers.

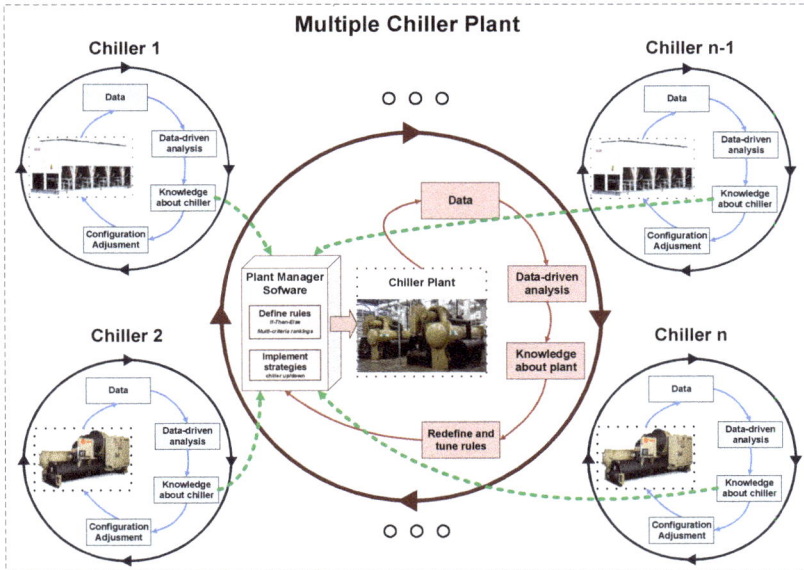

Figure 1. Methodology for extracting knowledge and enhancing efficiency in Chiller Plants.

The strategies to decide when chiller stages up/down can be implemented using basic programming structures, executable by any building management system (see Expression (3)):

$$If \begin{bmatrix} Condition1 \\ Condition2 \\ ... \\ ConditionC \end{bmatrix}, Then \begin{bmatrix} Action1 \\ Action2 \\ ... \\ ActionA \end{bmatrix}, Else \begin{bmatrix} Opposite\ action1 \\ Opposite\ action2 \\ ... \\ Opposite\ action A \end{bmatrix}. \quad (3)$$

Basic logic operations (AND "&", OR "|", NOT "!"), comparison operators (" ==", " <", " >") and math functions ("+", "−", "∗", "/") can be used to define complex relationships among variables (see Expression (4)). Furthermore, nonlinearity strategies based on Fuzzy logic could be also applied using these conditional rules [39]:

$$\begin{bmatrix} Condition1\ "Operator"\ Condition2\ "Operator" ... ConditionC \end{bmatrix}. \quad (4)$$

These strategies constitute an expert module that determines the actions the PMS (Chiller up/down, up/down disable and no change) performs on the chiller in operation, according to the set of rules and sensor data in every situation. As an example, these management rules could be expressed according to Expression (5):

$$If \begin{bmatrix} CoolingLoad < 1000KW \\ OutdoorTemp > 20°C \\ ChillerLoad > 0.9 \\ (Month >= Jun)\&(Month < Sept) \\ ... \\ (Hour >= 8am)\&(Hour < 15am) \end{bmatrix}, Then \begin{bmatrix} ChillerUp \\ ChillerDown \\ UpDisable \\ DownDisable \\ NoChange \end{bmatrix}. \quad (5)$$

The strategies to select which chiller starts/stops are defined. They can be implemented using multi-criteria rankings. The criteria must be established beforehand: for example, total running hours, count starts, efficiencies, priorities, chiller loads, etc. can be used. For each criterion, a sorting order has to be chosen (ascending or descending order). The proposed procedure is to create one table for each criterion "*c*" with *n* rows (as many as chillers) and two columns (the chiller index and the values of the corresponding criterion). Next, these tables are sorted, obtaining the chiller rankings for each criterion. Finally, a suitability table is obtained by weighting all previous rankings. According to all criteria, the most suitable chiller to start/stop among the available ones will be on the top of the table. All criteria can be weighted either equally (weights = $1/c$) or differently (even excluding some criteria with zero weight, provided that the sum of all weights is 1). The following expression describes the sorting rules, based on multi-criteria ranking:

> **for each c in criteria**
> > $Rankings[c] = sort\ Table[c]\ by\ value\ (Order);$
> > $FitTable+ = Weights[c] \cdot Rankings[c];$
> **end**

4. Description of the Chiller Plant at the Hospital of León

The chiller plant at the Hospital of León can be divided into a chilled water production system and a distribution system. In the production system, air-cooled and water-cooled chillers can be found. The distribution system comprises the decoupling bypass pipe and a variable chilled-water flow system. Figure 2 displays the general structure of the water flow system. Note that the colors of the streams try to represent the temperature of the water through the pipes during normal operation of the plant.

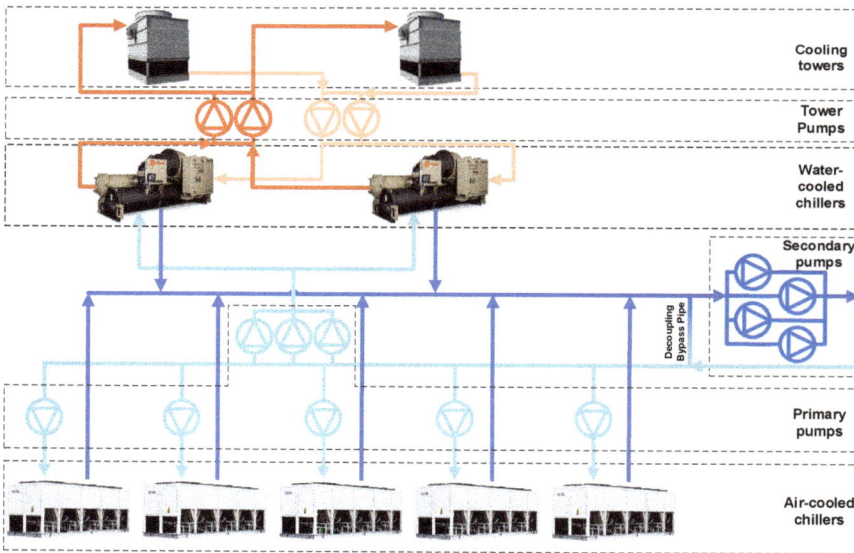

Figure 2. Chiller plant at the Hospital of León.

4.1. Production System

The main components of the chilled water production system at the Hospital of León are two groups of chillers: air-cooled and water-cooled chillers. Additionally, valves, sensors and pumps are needed to manage the chiller operation.

4.1.1. Air-Cooled Chillers

The plant comprises five identical air-cooled chillers. Each chiller needs a primary pump to force water flow through the evaporator and an on/off valve to avoid water flow when the chiller is not running.

- Chiller: The model is a Petra APSa 400-3, with a maximum cooling capacity of 400 tons (approximately 1407 KW). It includes three identical and independent refrigeration circuits (of 469 KW each one). Each one is composed of a screw compressor, an electronic expansion valve (EEV), and three individual condensers in V form. A common evaporator is used for the three circuits. The compressor has a maximum displacement of 791 m³/h of R134a refrigeration gas, its capacity can be regulated between 50–100% of the maximum value by means of two auxiliary loading and unloading valves, and it is driven by a three-phase induction motor (400 V; 109 KW). The condensers use 16 fans of 1.5 KW, driven by ABB ACH550 variable speed drives. The control board is provided by Micro Control Systems. The nominal chiller COP is 4.3 (with Condensing temperature: 40 °C; Evaporating temperature: 0 °C).
- Primary pump: It is a Grundfos TDP 150-200/4 whose function is to maintain a water flow through the evaporator, depending on the cooling load. It is driven by a three-phase induction motor (400 V; 15 KW). A variable speed drive (Danfoss VLT HVAC FC 102), drives the pump, although the traditional star-delta starting is also available.
- On/Off valve: It is a Siemens Acvatix SQL33, installed on the return pipe to block water flow through the evaporator when the corresponding chiller is stopped.
- Sensors: Several sensors are installed on pipes, associated with the chillers. An on/off flow sensor is used to confirm whether chilled water flows or not (Siemens QVE1900). Moreover, a differential pressure sensor (Siemens QBE61.3-DP10) is connected to measure the flow. In addition, the temperature of output chilled water is measured (with a Siemens QAE2120.010).

4.1.2. Water-Cooled Chillers

The plant comprises two identical water-cooled chillers. Three primary pumps are used to force water flow through the evaporator, whereas an on/off valve cuts water flow when chiller is stopped. Furthermore, in this case, a cooling tower and tower pumps are required for condensing.

- Chiller: The model is a Trane CVGF650 with a maximum cooling capacity of 650 tons (approximately 2286 KW). It comprises only one refrigeration circuit with a centrifugal compressor, an electronic expansion valve (EEV) and tubular heat exchangers (for evaporator and condenser). The compressor works using R134a refrigeration gas and its capacity can be regulated between 50–100% of the maximum value, by changing the angle of turbine blades. It is driven by a three-phase induction motor (400 V; 367 KW). The nominal chiller COP is 6.23 (Condensing temperature: 40 °C; Evaporating temperature: 0 °C).
- Primary pumps: There are three Grundfos NK 125-250/247/A/BAQE pumps that are started, depending on the cooling load, in order to maintain a water flow through the evaporator. They are driven by a three-phase induction motors (400 V; 18.5 KW), with star-delta starting.
- On/Off valve 1: It is a Bernard OA15, installed on the return pipe to block water flow through the evaporator when corresponding chiller is stopped.
- Cooling tower: The Baltimore Aircoil Company S-3654-NM cooling tower enables the control of condensing temperature, transferring heat to the environment when condensing water evaporates. An axial fan, driven by a three-phase induction motor (400 V; 18.5 KW) and managed by a variable speed drive (Moeller DF6-340-22K), helps the heat exchange. Moreover, it incorporates three resistors of 5 KW to avoid water freezing.
- Tower pumps: There are two groups of two Grundfos NK 150-315/307/BAQE pumps each, one to propel condensing water to the cooling tower and the other to maintain a water flow through the condenser. They are driven by three-phase induction motors (400 V; 30 KW), with star-delta starting.

- On/Off valve 2: A Bernard AS25, which allows for selecting which cooling tower will be used for condensing.
- Sensors: Several sensors are installed on pipes, associated with the chillers. An on/off flow sensor is used to confirm whether chilled and condensing water flows or not (Johnson Controls F61SB-9100). In addition, the temperatures of input and output chilled water are measured (Johnson Controls TS-9101-8224).

Table 1 summarizes the nominal chiller data in the plant at the Hospital of León.

Table 1. Chiller data overview.

	Water-Cooled Chillers	Air-Cooled Chillers
Number	2	5
Compressors	1	3
Cooling power [kW]	2286	1407
Electric power [kW]	367	327
Primary Pump [kW]	18.5	15
Fans [kW]	18.5	24
Cooling Tower Pump [kW]	30	-
Compressor COP	6.23	4.30
Overall COP	5.27	3.84

4.2. Distribution System

The design of the distribution system enables variation of chilled water flow using a set of four secondary pumps (Grundfos NK 125-400/375/BAQE) driven by four three-phase induction motors (400 V; 55 kW) with their corresponding variable speed drives (Danfoss VLT 6000 HVAC and Schneider Electric Altivar 61). The aim of this system is to adjust the building load to the instantaneous cooling demand. In addition, pressure variations due to differences between cooling generation and consumption are alleviated by the decoupler piping placed between supply and return pipes. Additionally, a cooling meter and supply/return temperature sensors allow us to measure demanded cooling energy.

5. Application of the Proposed Approach to a Chiller Plant

The proposed methodology has been applied to the multiple-chiller plant at the Hospital of León. As mentioned above, that plant comprises two different chiller groups: five air-cooled chillers (ACC1–ACC5) and two larger water-cooled chillers (WCC1, WCC2). The first aim of our approach is to analyze the operation of chillers in order to acquire knowledge (detect failures or malfunctions, extract patterns, find external influences, etc.). The final aim is to analyze the operation of the plant in order to monitor its efficiency and tune or upgrade control rules, if necessary. Thus, our approach requires past data of the chiller and plant to carry out the analyses. For that reason, data were collected from the BMS, including variables, such as COP, cooling load, power demand, chiller load ratio, condensing pressure and temperature, compressor current, outdoor temperature, etc. Data were gathered every 1 min during a one-year period, so 525,600 samples were obtained for each variable. The chiller plant is located in León, a city with continental climate where cooling loads have a clear seasonal nature. Therefore, data from a whole year should cover that seasonal behavior. Nevertheless, the addition of more data in subsequent years would improve the coverage.

5.1. Data-Driven Analyses and Knowledge about the Chillers

Using data from chillers, an analysis was carried out for each one, focusing on the chiller behavior and its operation with regard to external conditions such as chiller load ratio or outdoor temperature.

Prior to the study, maintenance staff inspected the main chiller control and protection elements (valves, solenoids, relays, sensors, fuses, etc.), with the aim of repairing them, if required. Faults in control relays, broken fuses, damaged solenoids, blocked valves, earth defects or wrong wirings can provoke abnormal chiller operation. Once faults in chiller elements are detected and corrected, the analyses were performed. In this case, time series plots were used for checking the R134a refrigeration cycle with its main parameters (gas suction, discharge temperature and pressure). Sharp oscillations were discovered in the control signal acting on condensing fans of ACC1. The condensing pressure was higher than normal, being affected by those variations. Thus, this high pressure caused electricity peak demands, adversely affecting the chiller efficiency Moreover, it provoked damages on air-cooled chiller elements—for example, the solenoid of some loading and unloading valves broke down frequently, fan bearings suffered strong strains, compressors were working in extreme conditions affecting the compression ratio, etc. The remaining air-cooled chillers showed similar behavior, so condensing control setpoints were verified in order to reduce condensing pressure (below 900 KPa) and smooth the control signal. Variables involved in the R134a refrigeration cycle of water-cooled chillers were in a normal range.

First of all, chiller operation at partial loads was analyzed. Studying the chiller performance at partial loads is very important since rarely chillers run at their nominal load (only a few hours per day). Air-cooled chillers can modulate cooling capacity between 0.22 and 1.0 of nominal value, whereas water-cooled chillers can regulate capacity from 0.5 to 1.0. It should be remarked that the nominal capacity of water-cooled chillers is 1.6 times higher than that of the air-cooled chillers (2286 KW versus 1407 KW). Some aggregation operations are applied to each chiller data set according to the following expression:

$$Agg_{Average(COP)} \Pi_{Hour;\ ChillerLoad} \sigma_{ACC\ 1-5;\ WCC\ 1-2}.$$

Figure 3 shows the relationship between COP and chiller load for two kinds of chillers (ACC1 and WCC2). In Figure 3a, it can be observed that ACC1 COP remains constant with regard to the load (around 3.7). However, WCC2 COP has an exponential relationship with load, with a maximum value of 5.2 and a minimum value of 2.5, when the chiller runs at the half capacity (see Figure 3b). Therefore, water-cooled chillers operation should be avoided when the cooling load is lower than 1143 KW for a long time. In this case, air-cooled chillers with a better load partition are preferable, since their capacity can be regulated up to 310 KW, keeping a constant performance.

Next, the chiller operation regarding to outdoor temperature is analyzed. For that purpose, aggregation operations are applied to each chiller data set according to the following expression:

$$Agg_{Average(COP)} \Pi_{DayOfYear,\ OutdoorTemp.} \sigma_{ACC\ 1-5;\ WCC\ 1-2}.$$

As can be seen in Figure 4, outdoor temperature influences negatively the air-cooled chiller performance, whereas COP for water-cooled chillers increases with outdoor temperature. Observing Figure 4a, ACC1 COP decreases slightly when temperature increases, since the heat exchange with air becomes difficult. However, daily average COP is always greater than 3. Looking at Figure 4b, it can be pointed out that daily average COP can reach high values, greater than 3. Nevertheless, lower COP values can be achieved when outdoor temperature is below 12 °C. This is due to water-cooled chillers run at a low load ratio with low cooling loads. Thus, air-cooled chillers should run those days when average temperature is below 12 °C. Water-cooled chillers should run on days with an average temperature above 16 °C. In the range 12–16 °C, all chillers can operate and other patterns should be considered.

ACC1 Hourly Avg. COP

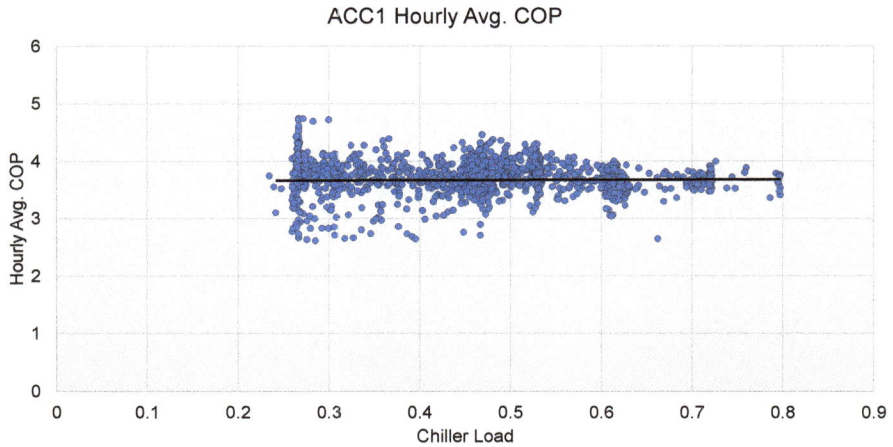

(a) Air-cooled chillers.

WCC2 Hourly Avg. COP

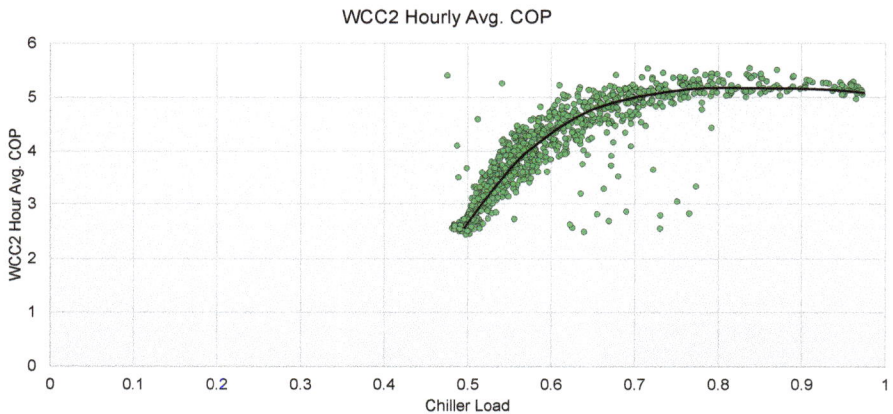

(b) Water-cooled chillers.

Figure 3. Evolution of COP with chiller load.

Summarizing what was learned about the chillers, it can be stated that air-cooled chillers should be started with low outdoor temperatures, when cooling load is low. On the other hand, water-cooled chillers should be run with high cooling loads, ensuring that the chiller load ratio is quite high. Condensing control, especially on air-cooled chillers, should be adjusted since high condensing pressures and strong oscillations have been detected.

After the chiller analyses, it can be necessary to adjust the internal configuration of chillers. The maintenance staff was advised to adjust some parameters, especially the configuration of air-cooled chillers. The main changes in the internal configuration of the chillers were focused on slightly increasing the deadband in temperature control, improving the chiller response in the presence of short peak cooling loads, minimizing the operation cycles of capacity control valves, reducing the proportional action in condensing control, balancing refrigeration circuits and avoiding unsafe compressor runnings. Note that manufacturers do not allow us to modify remotely some internal parameters and they are only accessible through the local panel with the right access privileges. Table 2 summarizes the parameters modified in air-cooled chillers after chiller analyses.

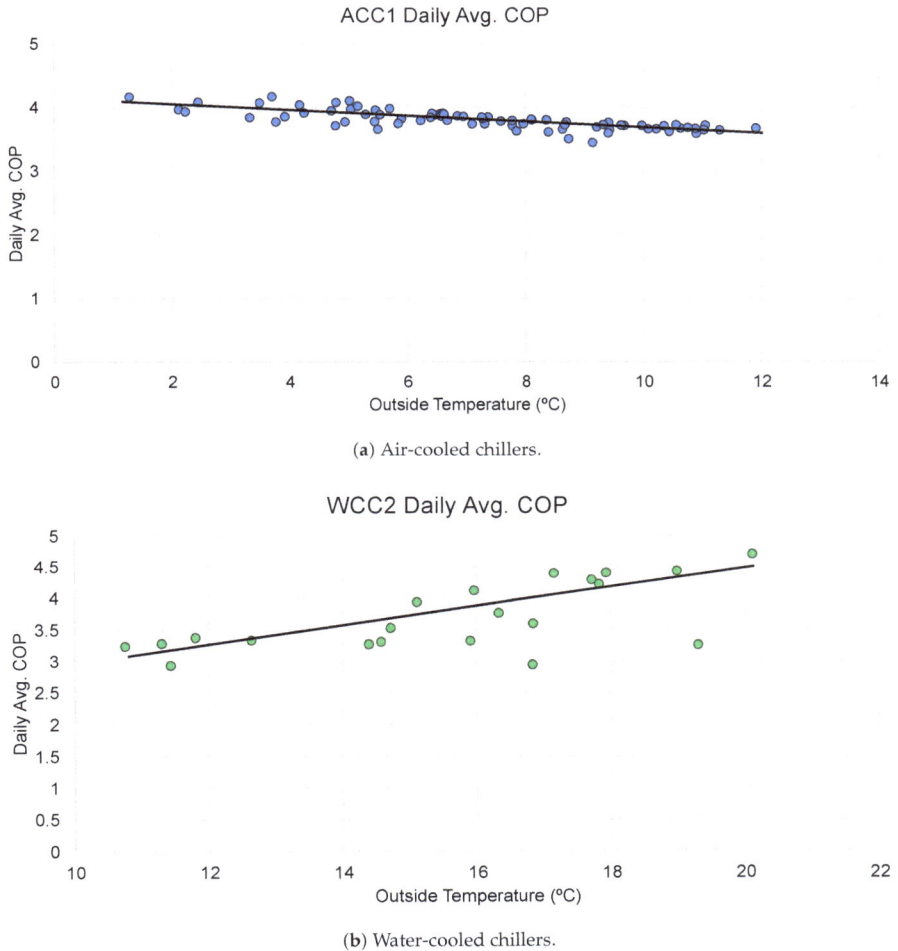

(a) Air-cooled chillers.

(b) Water-cooled chillers.

Figure 4. Evolution of COP with outdoor temperature.

The configuration of water-cooled chillers was also examined, but, in this case, most of the internal parameters remained unchanged. Just a few parameters were adjusted to be in coherence with the water flow system. As in air-cooled chillers, the temperature setpoint was also decreased due to the unbalance of primary and secondary chilled water flows. It causes a recirculation excess of return chilled water through the decoupling bypass pipe.

Table 2. Main changes in the air-cooled chiller configuration.

Parameter	Value Before	Value After
Water setpoint [°C]	7	6
Control zone [°C]	±0.5	±1.5
Delay [s]	30	300
Rate of Change [°C/s]	±0.1	±0.6
Slide [%]	50–100	65–100
Adjust [%]	1–5	2–10
Deadband [A]	±1	±3
Load and unload pulses [s]	0.2	0.8
Pulse delay [s]	1	5
Condensing control range [KPa]	758–896	758–1241
Unsafe suction warning [KPa]	138	34

5.2. Converting Knowledge into Management Strategies

Once knowledge about the chillers was extracted and operators upgraded the internal configuration of each one, our efforts focused on designing and implementing efficient chiller management strategies, which can be incorporated into a PMS. Note that chiller operation is automated and data are stored by BMS.

5.2.1. Architecture of the Plant Manager Software

An automatic existing system was modified for chiller plant management at the Hospital of León (see Figure 5). The system is based on an *ad hoc* software which implements management strategies. The PMS has been developed as a Software as a Service (SaaS) application, so that web clients can easily access a software manager using a standard web browser to manage the chiller plant.

The BMS communicates with two controllers (one for air-cooled chillers and the other for water-cooled chillers) using the BACnet IP protocol. The BMS receives all signals from both controllers (chillers, valves, pumps, cooling towers, etc.) for monitoring purposes and sends them start/stop commands and setpoints. The PMS decides which chiller should be staged up/down and when, and communicates with BMS using also BACnet IP.

The first controller (AS by Schneider Electric) is used to implement control of air-cooled chillers and auxiliary elements (valves, primary pumps, etc.). Additional I/O modules capture signals from field elements and weather sensors. This device also works as a Modbus gateway to communicate with chiller cards and retrieve internal variables. The AS controller stores data in local logs and gradually transfers them to the BMS. The second controller (FEC by Johnson Controls), with integrated I/O modules, is used to implement control of water-cooled chillers and their auxiliary elements (valves, primary pumps, tower pump, cooling tower, fan, etc.) and to read building cooling meter. The FEC controller stores data in local log files and gradually transfers them to BMS too. Additionally, a protocol interface card (Trane PIC BAS-SVX08D-E4) has been installed in each water-cooled chiller to gain access to their internal variables and parameters. This card provides data using Modbus RTU protocol. A gateway (Com'X 510 by Schneider Electric) is necessary to convert Modbus RTU to Modbus TCP. On the other hand, power meters measure the main variables of electricity supply to both groups of chillers. They are also connected to Modbus RTU networks to communicate with BMS.

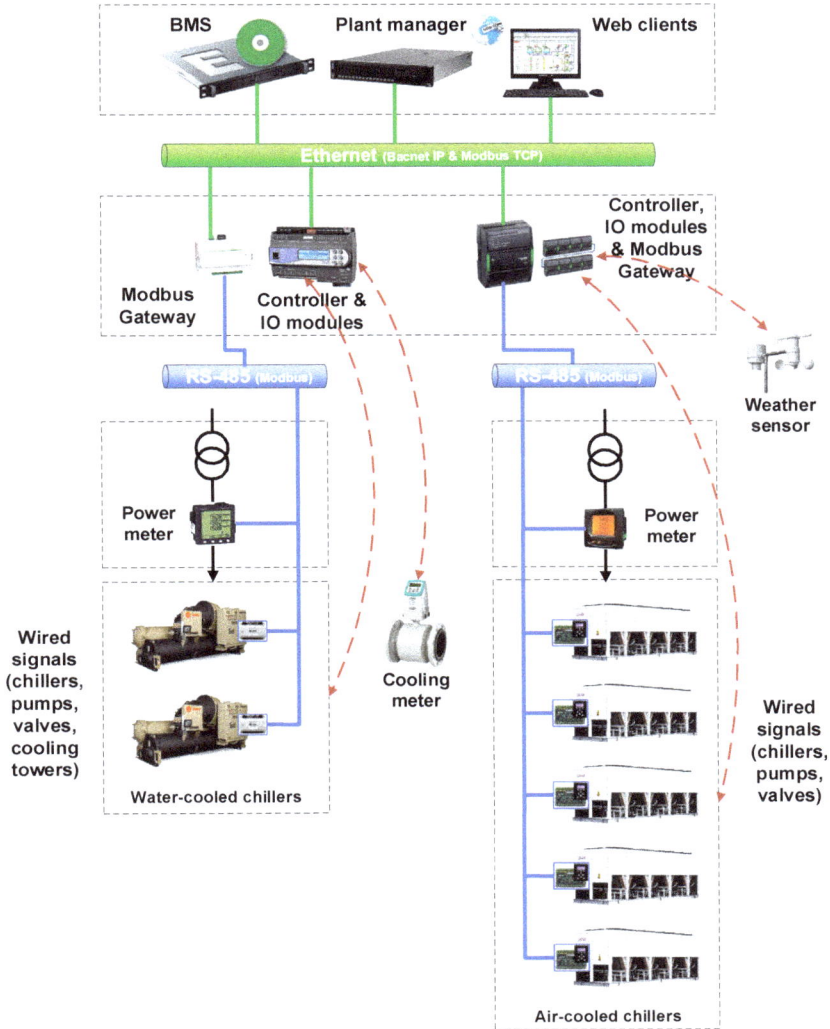

Figure 5. Architecture of the plant manager software.

5.2.2. Defining Rules

Using knowledge about the chillers, several sets of rules are defined, tuned or upgraded. These rules are the core of an expert module that uses them to manage the plant efficiently. The following aspects are taken into account in the rule definition:

- Reliability and security of supply of chilled water (since a hospital has critical systems such as the refrigeration of surgeries, magnetic resonance systems, scanners, a data center, HVAC systems in patient rooms, etc.).
- Maximization of the plant operation by choosing the most efficient chiller (or combination of chillers) available to meet the required cooling load.
- Maximization the chiller efficiency by enforcing high chiller loads.
- Reduced maintenance of the chiller plant (the manager should balance running time and start counts for all chillers).

- Adjustment of the plant operation according to weather conditions (mainly, outdoor temperature).
- Automatic and manual operation modes of the chiller plant. The manager should consider user preferences, either giving priority to a chiller or enabling/disabling an specific chiller.

The basis of sequencing method is the measurement of the cooling load and the computation of the current operating capacity, based on all chiller load ratios. If cooling load is greater than chiller capacity during a short period of time, a new chiller will be started. Otherwise, a running chiller will be stopped. Prior to proceeding, it is checked that the chiller load is decreasing. Note that the plant requires, at least, one chiller running in order to provide the base cooling load to the building. In other cases, a schedule could be considered to stage up/down chillers. A chiller will be also stopped when consecutive running time exceeds the rotation setpoint (168 h). Additional rules considering supply and return chilled water temperatures are also defined in order to ensure cooling supply to the building. These rules contain extreme thresholds and will be activated in exceptional situations, keeping chilled water temperature in range [6–10] °C. Note that simple rules can be combined using logic operations in order to build advanced rules. Table 3 summarizes the main rules used to stage up/down chillers.

Table 3. Stage up/down rules set.

Conditions	Actions	Opposite Actions
If	*Then*	*Else*
(CoolingLoad > ChillerCapacity) & & (ChillerLoad > 0.9)	ChillerUp	-
(CoolingLoad < ChillerCapacity) & & (ChillerLoad < 0.7) & (NoChillersOn > 1)	ChillerDown	-
NoChillersOn == 0	ChillerUp	-
RelativeRunningHours > RotationSp	ChillerDown	-
(SupplyTemp > 10 °C) | (ReturnTemp > 14 °C)	ChillerUp	-
(SupplyTemp < 6 °C) | (ReturnTemp < 10 °C)	ChillerDown	-

Some rules taking into consideration the physical environment are defined. The plant contains different types of chillers and their efficiencies are influenced by external conditions as data analysis revealed. Therefore, it is required to decide which type of chiller should be running in each situation. In this way, outdoor temperature is used for creating such delimiting rules. Average temperature is computed each day and used to predict the conditions of the following day. If daily average outdoor temperature of the previous day was higher than 16 °C, then the chiller selected to stage up/down will be a water-cooled chiller (WCCy; y = 1–2). If daily average outdoor temperature of the day before is lower than 12 °C, then the chiller selected to stage up/down will be an air-cooled chiller (ACCx; x = 1–5). In the range [12–16] °C, other rules will be taken into account. An overview of delimiting rules can be seen in Table 4.

Table 4. Delimiting rule set.

Conditions	Actions	Opposite Actions
If	*Then*	*Else*
(AvgOutdoorTemp > 16 °C)	WCCyUp	WCCyDown
(AvgOutdoorTemp < 12 °C)	ACCxUp	ACCxDown

Exclusion rules are required in order to determine when to disable chiller up/down commands due to either alarms or planned maintenance tasks. User can enable/disable a chiller from the web interface. On the other hand, if power demand exceeds 1300 KW in the plant, a new chiller starting is blocked and a warning is triggered (a user check is required to allow staging up a new chiller). That avoids peak power demands and inefficiencies in the plant provoked by anomalous operations.

Moreover, stopping a chiller is disabled as long as the idle capacity is less than the load ratio of the chiller to stop. An overview of exclusion rules can be seen in Table 5.

Table 5. Exclusion rule set.

Conditions	Actions	Opposite Actions
If	*Then*	*Else*
ChillerEnable == 1	ChillerUp/Down	Up/Down Disable
ChillerAlarm == 1	Up/Down Disable	ChillerUp/Down
PowerDemand > 1300 KW	UpDisable	ChillerUp
IdleCapacity < StopChillerLoad	Down Disable	ChillerDown

Finally, rules for sorting chillers are defined with the aim of determining which chiller starts/stops. First, different criteria are established and, later, the corresponding chiller rankings are obtained. Table 6 summarizes the sorting rules.

Table 6. Criteria and order for sorting chillers.

	Criterion	Asc./des. Order
	Chiller up?	
1	Chiller Priority (P)	Ascending
2	Chiller COP (COP)	Descending
3	Total Running Hours (TRH)	Ascending
4	Start Count (SC)	Ascending
5	Hours From Last Stop (HLS)	Descending
	Chiller down?	
11	Chiller Load (CLR)	Ascending
12	Relative Running Hours (RRH)	Descending
13	Chiller COP (COP)	Ascending

In the stage up sequence, criteria such as chiller priorities, efficiencies, total running hours, hours from last stop and start counts are used. A weighting method is applied, balancing all criteria and obtaining a weighted ranking. The chiller on the top of that ranking should have a high position in the partial rankings, matching most of the criteria. For example, the next chiller to start should have a high priority, noteworthy efficiency, lower total running hours and start counts and it should not have been stopped recently, i.e., the chiller should have a high position in all rankings. In this case, the same weight is applied for each criterion. However, different values could be used:

$$w_1 \cdot Ranking_1 + w_2 \cdot Ranking_2 + w_3 \cdot Ranking_3 + w_4 \cdot Ranking_4 + w_5 \cdot Ranking_5.$$

In the same way, the criteria for the staging down sequence are relative running hours, chiller loads and efficiencies, which are weighted to decide the chiller to stop. Note that sorting chillers according to the COP is now performed in the opposite order. For instance, the next chiller to stop should have the lowest efficiency and load ratio and should also have been running for many consecutive hours:

$$v_{11} \cdot Ranking_{11} + v_{12} \cdot Ranking_{12} + v_{13} \cdot Ranking_{13}.$$

5.2.3. Plant Management Strategies

Using the previous sets of rules, efficient management strategies are implemented on the *ad hoc* software. Mainly, two strategies determine chiller sequencing: "chiller up" and "chiller down" (see Figure 6). The "chiller up" strategy monitors the cooling load (measured by a cooling meter) as well as the supply and return chilled-water temperatures. If the cooling demand increases and the running chillers are fully loaded, a new chiller is started. Depending on supply and return chilled water temperatures, the chiller is started immediately or after a short delay (trying to absorb transitory load fluctuations). The criteria described in the previous section are applied in order to decide which chiller starts. Furthermore, it is checked the presence of alarms (such as the internal faults in chillers and auxiliary elements) and the user activations (a chiller can be disabled due to maintenance tasks). The prediction of the outdoor temperature is also considered in order to take advantage of weather conditions. Finally, the manager sends commands up to BMS and it provides start commands to the selected chiller and its auxiliary elements, such as valves, pumps, etc., and verifies possible alarms in the starting sequence.

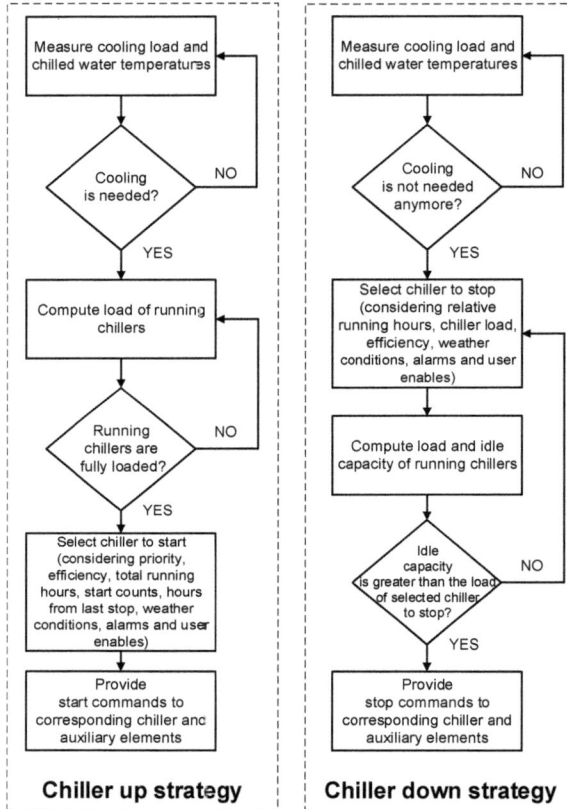

Figure 6. Chiller up and down management strategies.

The "chiller down" strategy also monitors the cooling load and supply/return chilled water temperatures. If the manager detects that cooling is not needed anymore (because of a decrease in cooling load and a sudden drop in chilled water temperature), one of the running chillers should be stopped. The criteria described in the previous section are applied in order to decide which chiller stops. Before making that decision, the software must verify that the remaining running chillers can absorb the cooling load of the chosen chiller. For that purpose, the manager computes the difference between the idle capacity of running chillers and the load of the chiller that is stopped. Finally, if idle capacity is slightly greater than the chiller load, the manager sends down a command to BMS and it provides stop commands in sequence to the selected chiller and its auxiliary elements.

These management strategies based on rules are used by the system in automatic mode (default). However, the system can be managed in manual mode, according to the staff expertise. The staff can vary setpoints, delays, thresholds and chiller activations.

5.3. Data-Driven Analysis and Knowledge about the Plant

Once the management strategies are deployed in the plant and new data are collected, an analysis has to be carried out to monitor plant efficiency. That study focuses on the plant operation and efficiency with regard to cooling load, external conditions, chiller sequencing, etc.

First of all, cooling demand is analyzed (see Figure 7) applying the following expression:

$$Agg_{Average(CoolingLoad)} \Pi_{Month;\ Weekday;\ Hour} \sigma_{Plant}.$$

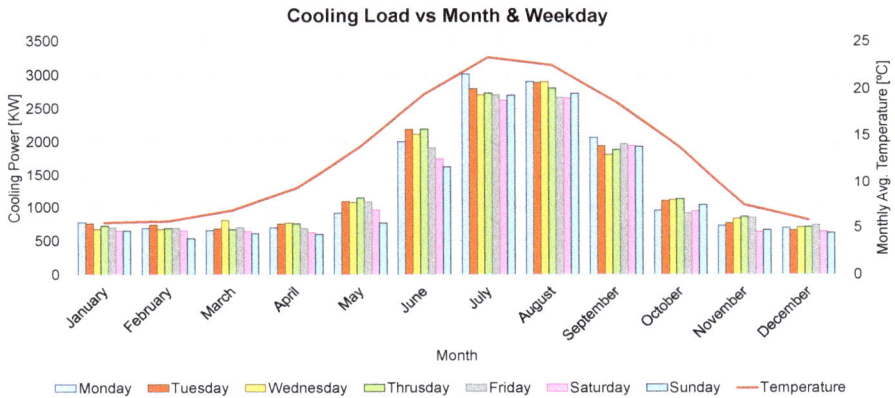

(a) Cooling demand vs Month, Weekday and Outdoor Temperature.

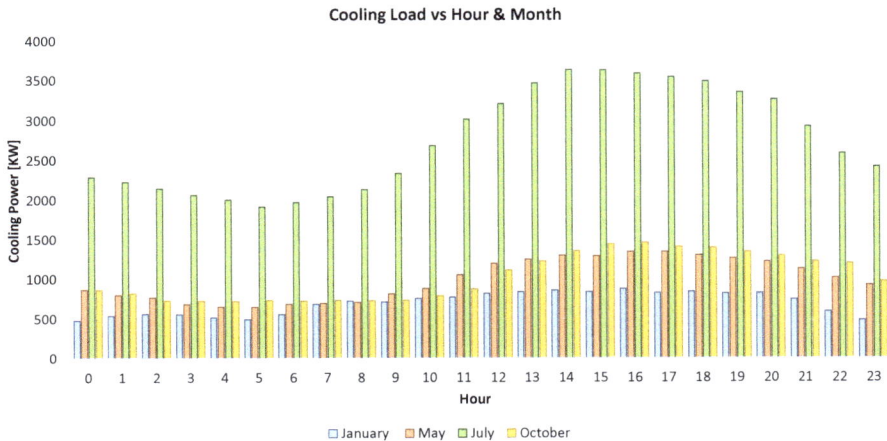

(b) Cooling demand vs Hour and Month.

Figure 7. Analysis of cooling demand.

Figure 7a shows the average cooling load for each month and weekday. It can be seen that cooling demand has a seasonal evolution, exceeding 2000 KW in Summer and not exceeding 1000 KW in

Winter. Therefore, water-cooled chillers should be run in Summer to cover the cooling demand. On the contrary, they could operate in Winter, but at half capacity, achieving lower COP. Due to lower nominal power and better load partition, air-cooled chillers are more efficient in that season. In Figure 7a, the average outdoor temperature is also represented. It can be seen that outdoor temperature influences directly on cooling load due to demand of HVAC systems. In May and October, free-cooling operation in those systems can be observed since cooling load does not follow the temperature evolution.

Figure 7b displays the average cooling load for each hour of the day and months. Only four representative months (January, May, July and October) have been represented only for simplicity. In winter (January), cooling load is quite flat during all the day, whereas, in Summer (July), it is flat at nights, becoming steep from 10 h on and decreasing after 22 h. In May (Spring) and October (Autumn), the day profile of cooling load is quite similar.

Next, the influence of outdoor temperature on plant COP is analyzed using the expression:

$$Agg_{Average(COP)}\Pi_{Day;\ OutdoorTemp;\ ChillerType;\ NoChillers}\sigma_{Plant}.$$

The daily average COP value was plotted with respect to the daily average temperature (see Figure 8). Note that the size of points represents the type and the number of chillers running: the smallest points correspond to one air-cooled chiller and the largest ones are from two water-cooled chillers. Three zones can be distinguished in the graph (below 10 °C, above 20 °C and between 10 °C and 20 °C). Lower temperatures imply low cooling loads (little use of HVAC systems), so air-cooled chillers (smaller points) are more appropriate, since condensing refrigerant using cold air is quite efficient. In contrast, higher temperatures entail high cooling loads (strong use of HVAC systems) and, therefore, water-cooled chillers (larger points) are more convenient because they have better performance and higher nominal capacity (so a smaller number of chillers is required to cover high cooling loads). It can be also seen that the plant COP would slightly decrease if air-cooled chillers run with temperatures above 20 °C. Similar results (plant COP reduction) would be obtained if water-cooled chillers run with temperatures below 10 °C. Between 10 °C and 20 °C, some chiller combinations can be observed, depending on other factors considered by the management strategies.

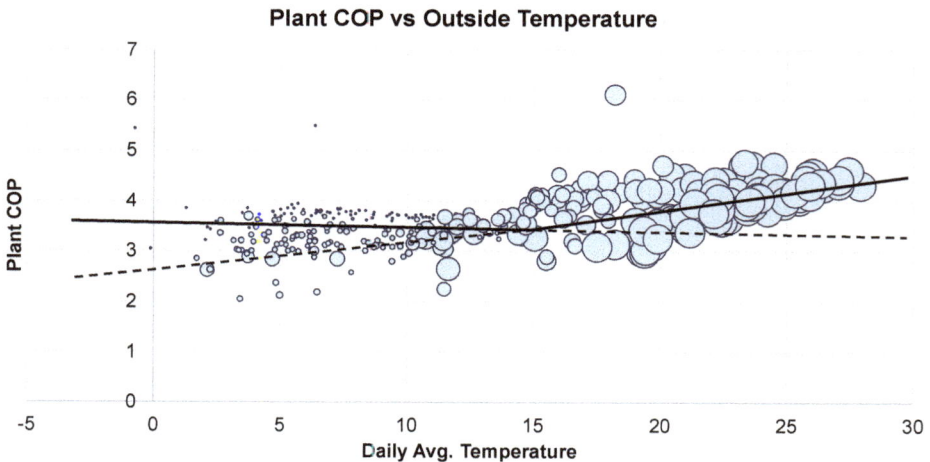

Figure 8. Plant COP vs. Daily average outdoor temperature.

Analyzing the plant performance with regard to the cooling load allows us to verify the chiller sequencing management strategies. In this sense, the daily average COP was plotted with respect to the cooling load of the building (see Figure 9). The used aggregation was:

$$Agg_{Average(COP)}\Pi_{CoolingLoad; ChillerType; NoChillers}\sigma_{Plant}.$$

As in the previous figure, the size of points represents the type and the number of chillers running. It can be seen that, the higher the cooling load is, the better the plant performance is. High cooling loads are typically covered by two water-cooled chillers, with better individual performance. On the contrary, low cooling loads are provided by one air-cooled chiller, with finer load partition. Medium cooling loads can be covered by several chiller combinations (two air-cooled chillers, one water-cooled chiller or one of each).

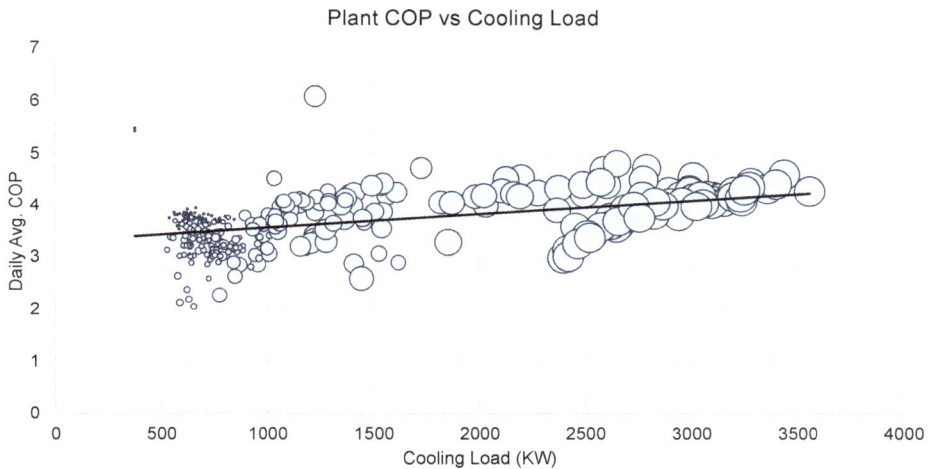

Figure 9. Plant COP vs. Cooling Load.

Studying the contribution of each chiller to the total cooling load of the building can also help us to verify the efficacy of chiller sequencing strategies (see Figure 10). The manager tries to consider external conditions and to balance several criteria (total running hours, start counts, priorities, efficiencies, etc.), choosing the fittest ones in order to cover the cooling load. It can be seen in Figure 10a that chiller sequencing has a seasonal behavior (as cooling load). In winter, cooling is provided by air-cooled chillers, whereas, in summer, cooling is generated by water-cooled chillers. During both periods, cooling is quite stable, so chillers only alternate their operation when the relative running hours exceed the rotation setpoint. For example, only ACC1 and ACC4 were running in January, and only WCC1 and WCC2 produced cooling in August. In spring and autumn, the manager combines the operation of air-cooled and water-cooled chillers, since cooling varies daily. Therefore, the manager has to send up/down commands daily to the chillers in order to cover fluctuating cooling load. Focusing on October (see Figure 10b), it can be confirmed that up to three chillers were started during a day (October 6th). This causes the rise of start counts, a crucial step in the sequencing, and thus an increasing possibility of appearing faults.

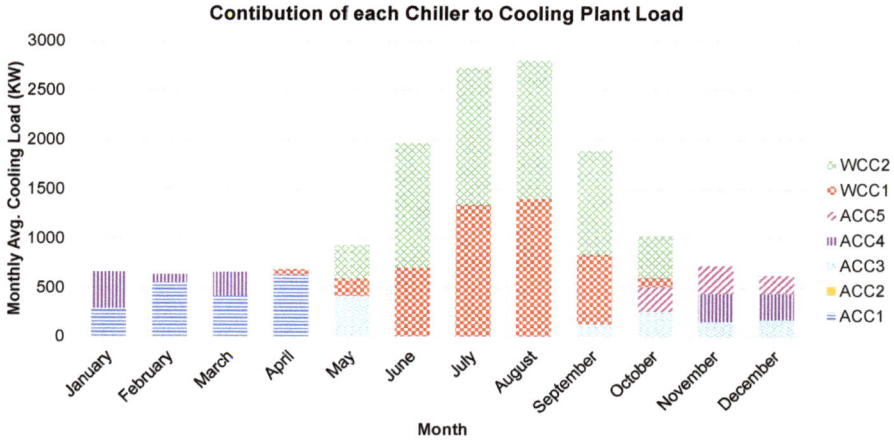

(a) Distribution of cooling production.

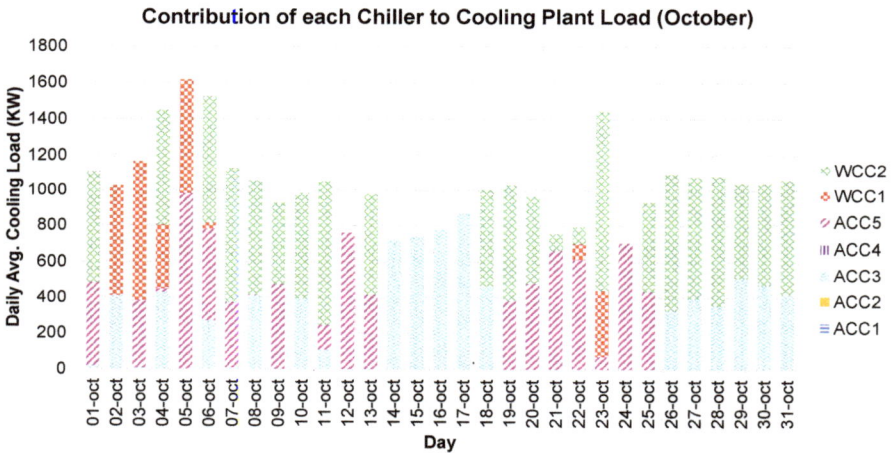

(b) Distribution of cooling production in October.

Figure 10. Contribution of each chiller to cooling load of the building.

Summarizing the knowledge about the plant, it can be observed that cooling load at the Hospital of León has a noticeable seasonal and daily behavior, so a different operation can be established each season. Outdoor temperature can be used to predict abnormal days with regard to the current season. Two hours (10 h and 22 h) of daily profile can be used as indicators of the expected cooling load (hours in which the cooling load usually increases or decreases, respectively). In order to avoid the increase of start counts in spring and autumn, chiller down command can be delayed until the idle capacity is a bit higher.

Updating Rules

Rules set can be updated in three ways:

- **Tuning the threshold.** The original rule is maintained, but the threshold which triggers this rule is modified. For example, as long as the capacity of running chillers does not drop below 0.6 (not 0.7), a running chiller will not be stopped, avoiding the increase in the number of start counts.
- **Upgrading the complete condition.** The rule is redefined completely, either establishing new premises or combining individual ones. For instance, outdoor temperature allows us to adapt the seasonal and daily operation. Furthermore, cooling load at certain hours allows us to estimate its evolution.
- **Tuning criteria and weights.** Sorting rules can be changed either by adding/deleting a criterion or adjusting the weights. Staff can vary weights from the web interface, for instance, to consider only one criterion (its weight is 1 and the remaining ones are 0) or to weight some criteria more than others (being the sum of weights 1). However, during this work, criteria have not been modified.

An overview of updated rules can be observed in Table 7.

Table 7. Updated rules set.

Conditions	Actions	Opposite Actions
If	*Then*	*Else*
(CoolingLoad < ChillerCapacity) & & (ChillerLoad < 0.6) & (NoChillersOn > 1)	ChillerDown	-
Winter == 1	ACCxUp	ACCxDown
Summer == 1	WCCyUp	WCCyDown
((Spring == 1) | (Autumn == 1)) & & (AvgOutdoorTemp < 14 °C)	ACCxUp	ACCxDown
((Spring == 1) | (Autumn == 1)) & & (AvgOutdoorTemp > 14 °C)	WCCyUp	WCCyDown
(CoolingLoad > 1500 KW) & (Hour < 10 h)	WCCyUp	ACCxDown
(CoolingLoad < 900 KW) & (Hour > 22 h)	ACCxUp	WCCyDown

6. Results and Discussion

After the application of the proposed approach, some efficiency enhancements have been obtained in the multiple-chiller plant at the Hospital of León. Below, these results for the individual chillers (ACC1–5, WCC1–2) and overall plant are presented. For that purpose, COPs corresponding to one year after changes were collected in order to compare them with the previous COPs before changes. Note that annual average outdoor temperature was very similar in both periods (11.2 °C versus 11 °C). Data acquired using a sampling time of one minute during a one-year period, i.e., 1,052,640 samples, are used in the projections.

6.1. Chiller Efficiencies

First, results on individual chiller efficiency are presented and discussed. Efficiency was monitored for each chiller according to the following expression:

$$Agg_{Count(COP)} \Pi_{Year} \sigma_{ACC\ 1-5;\ WCC\ 1-2}.$$

The study focuses on ACC1 and WCC2 chillers, since they were running the greatest number of hours in that period.

Figure 11 shows the histograms of ACC1 and WCC2 COP corresponding to before and after chiller modifications. At a glance, it can be seen that COP indicator was increased for both chillers. For ACC1 (see Figure 11a), the average COP was increased from 3.18 to 3.76, resulting a noteworthy enhancement of 18.35%. Mainly, it was possible due to two reasons. On the one hand, changes on condensing control setpoints caused the reduction of condensing pressure (200 KPa approx.) and, as a result, the power demand decreased. On the other hand, changes on condensing control parameters (mainly proportional action) ensured a more stable operation of air-cooled chillers. In case of WCC2 (see Figure 11b), the average COP was increased from 4.08 to 4.13, resulting in a slight enhancement of 1.22%. It was probably due to the adjustment of the chilled water setpoint and the reparation of cooling towers.

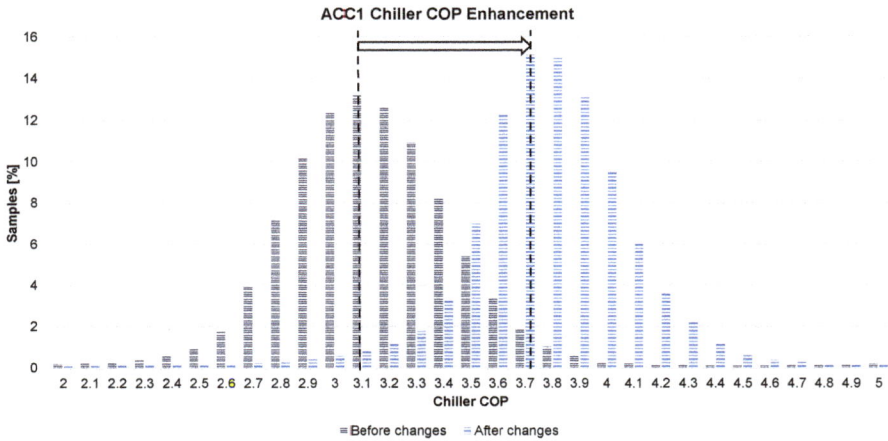

(a) Air-cooled chillers (ACC1).

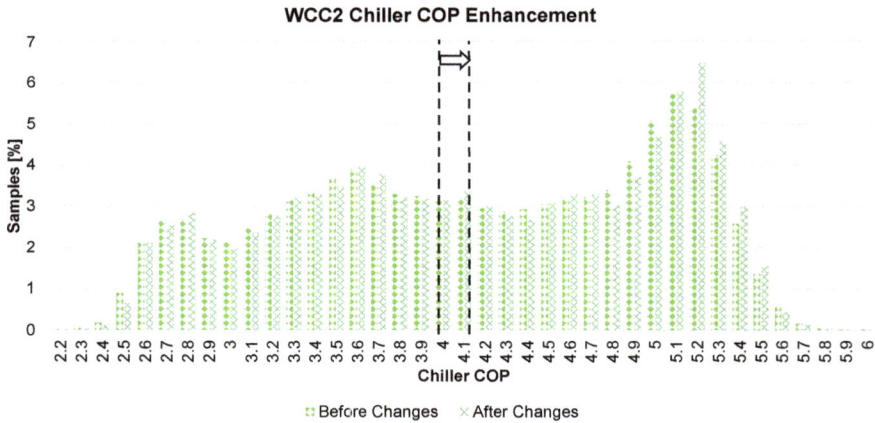

(b) Water-cooled chillers (WCC2).

Figure 11. Histogram of chiller COP before and after control changes.

COPs of identical chillers have been compared with each other (see Figure 12) in order to detect deviations and inefficient operations. COP histograms corresponding to five air-cooled chillers (ACC1–5) are represented on Figure 12a. This comparison reveals a clear difference between ACC1 and ACC4 performance. At a glance, it can be observed that the COP of ACC4 is lower than the one of ACC1 (the histogram moved to the left), whereas the COP of ACC3 and ACC5 are located in the middle and they have a very similar performance. The noticeable difference between ACC1 and ACC4 performance could be due to the experiments carried out in order to test a new decoupled condensing control strategy in ACC4. Refrigerant charge was also checked, verifying that, for example, the pressure of R134a in ACC3 was lower than the nominal value. Furthermore, oscillating chiller operation could have affected the compressor ratio and the efficiency of other chiller elements, such as condensing fans. In fact, some fans have been replaced in ACC1. Regarding water-cooled chillers (see Figure 12b), no significant difference can be appreciated (apart from WCC2 run longer than WCC1), so both chillers have a similar efficiency indicator.

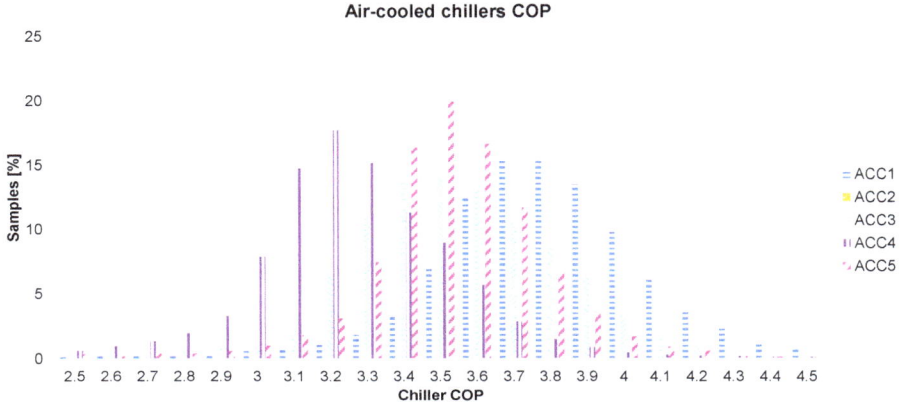

(a) Air-cooled chillers.

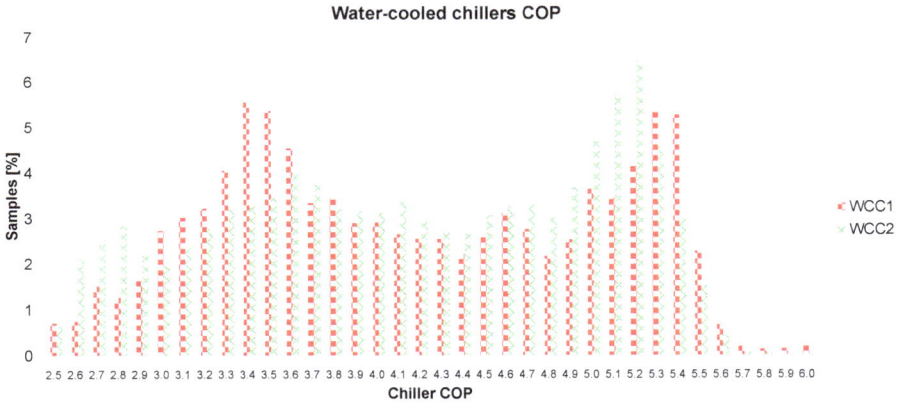

(b) Water-cooled chillers.

Figure 12. Comparison of chillers in terms of COP.

Table 8 summarizes efficiency results for all chillers. Note that high efficiency enhancements have been obtained for air-cooled chillers compared with water-cooled chillers. ACC2 was not running during the studied period due to severe electrical faults.

Table 8. Efficiency results for all chillers.

Chiller	COP Before	COP After	Variation [%]
ACC1	3.18	3.76	+18.35
ACC2	-	-	-
ACC3	3.04	3.48	+14.28
ACC4	2.80	3.17	+12.99
ACC5	2.91	3.39	+16.30
WCC1	4.15	4.19	+0.96
WCC2	4.08	4.13	+1.22

6.2. Plant Efficiency

The results on chiller plant efficiency are presented and discussed. Efficiency was monitored for overall plant according to the following expressions:

$$Agg_{Count(COP)}\Pi_{Year}\sigma_{Plant},$$

$$Agg_{Average(COP)}\Pi_{Month;\ Weekday}\sigma_{Plant}.$$

Figure 13 displays both histograms, before and after the changes. Considering all the exposed modifications and new management strategies, the plant performance has been increased from 3.2 to 3.65, i.e., an efficiency enhancement of 12.33%. It can be pointed out that the plant COP before changes remained a long time around 2.6.

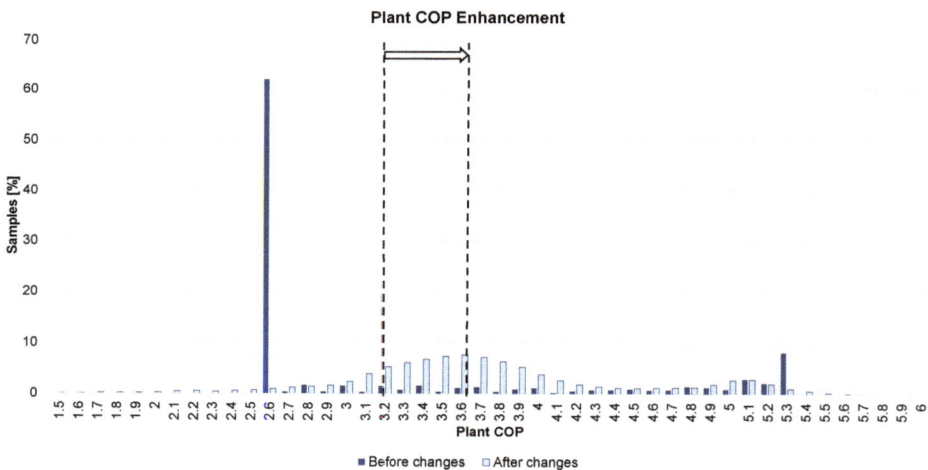

Figure 13. Histogram of plant COP before and after chiller plant changes.

Before applying the proposed approach and deploying efficient management rules, air-cooled chillers were running a short time compared to water-cooled chillers, due to unsteady operations and electricity peak demands. Thus, water-cooled chillers were used longer, even with low cooling demands (below 1143 KW), in order to guarantee cooling supply. In this case, the chiller load ratio was the lowest possible, i.e., 0.5, often being the cooling production higher than the demand. This caused a decrease of chilled water temperature and, sometimes, a temporal chiller stop for overproduction. On the other side, there were several COP values around 5.3, corresponding to high cooling loads in summer, when water-cooled chillers run at maximum load, and a no air-cooled chiller is started in order to avoid electricity peak demands. The distribution of the plant COP after the changes is much more uniform, varying between 3.6 (influenced by air-cooled chillers operation) and 5.1 (influenced by water-cooled chillers operation). Figure 14 shows the average plant COP for each month and weekday of studied period (after changes). It varies slightly from winter to summer season according to the chiller type running. On Monday and Tuesday, the COP is usually higher due to huge cooling loads.

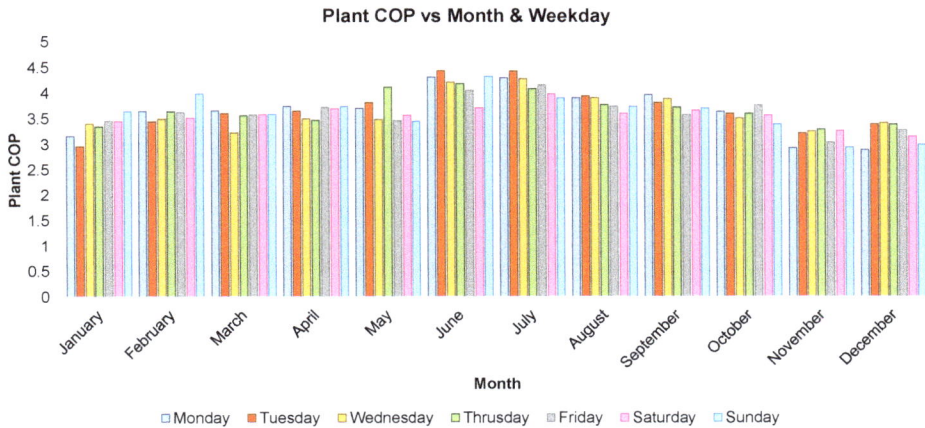

Figure 14. Plant COP each month and weekday after changes.

Our proposal takes advantage of real past data from the chillers and the plant, instead of using simulated data. Moreover, it can be deployed into conventional controllers since the rule-based management system requires low computational resources. Therefore, it is not necessary to substitute the existing hardware in the building. On the other hand, the addition of more data in subsequent years would improve the coverage, providing incremental improvements in efficiency to the plant operation.

One drawback of the proposed methodology is that it is not a completely automatic method to convert the knowledge extracted from the data analysis of the chillers and plant into management rules. Another limitation is that the iterative application of the proposed approach only provides incremental improvements to the system operation but does not guarantee an optimal result. Furthermore, the presence of a data science expert in the process would be beneficial for its correct application.

7. Conclusions

In this paper, a comprehensive methodology for improving the efficiency in multiple-chiller plants has been proposed. This methodology is based on a data analysis of the operation of the chillers and the overall plant, using real data instead of simulations. The proposed data analyses highlight relevant information by applying aggregation, filtering and data projection. Using the knowledge extracted specifically from the plant, control parameters of the chillers can be adjusted and management rules can be defined or tuned. The aim is to achieve an efficient management of the plant, without the need of incorporating cutting-edge controllers, since the management rules obtained through the proposed approach can be easily deployed in existing controllers.

The proposed methodology has been applied on a real chiller plant at the Hospital of León (Spain). Data analyses have helped to understand the operation of each chiller and the plant with regard to a chiller load ratio or outdoor temperature, which are variables that affect the efficiency of cooling production systems. The extracted knowledge about chiller performance has enabled the adjustment of internal control parameters and setpoints, detect faults and inefficient operations and define management rules, whereas the knowledge about the plant has allowed us to redefine and tune some rules. As a result, noteworthy enhancements on efficiency have been obtained after applying that methodology. In this sense, chiller COPs (especially for air-cooled chillers) and the overall plant

COP (12.33% higher) have been increased. This implied an electricity savings of 380,000 KWh during the year studied at the Hospital building.

The main limitations of the proposed methodology have been discussed in the paper. On the one hand, the extracted knowledge is not automatically converted into management rules. On the other hand, our approach does not guarantee an optimal result, so iterative applications of the approach will be required.

As future work, new data from subsequent time periods (one year) will be analyzed in order to redefine or tune the management rules, evaluating the incremental improvement in efficiency provided by the approach. Furthermore, a dynamic global optimization approach will be applied, in order to compare it with our methodology in terms of efficiency, resources and computational cost.

Author Contributions: Conceptualization, S.A. and M.D.; methodology, S.A. and A.M.; software, S.A. and A.M.; validation, M.A.P. and P.R.; formal analysis, S.A.; investigation, S.A., A.M. and M.A.P.; data curation, S.A. and A.M.; writing—original draft preparation, S.A. and A.M.; writing—review and editing, M.A.P. and P.R.; visualization, J.J.F. and P.R.; supervision, J.J.F. and M.D.; project administration, M.D.; funding acquisition, M.D.

Funding: This research was funded by the Spanish *Ministerio de Ciencia e Innovación* and the European FEDER under project CICYT DPI2015-69891-C2-1-R./2-R.

Conflicts of Interest: The authors declare no conflict of interest.

Abbreviations

The following abbreviations are used in this manuscript:

HVAC	Heating, Ventilating and Air Conditioning
COP	Coefficient of Performance
EEI	Energy Efficiency Indicator
EER	Energy Efficiency Ratio
SCOP	Seasonal Coefficient of Performance
SEER	Seasonal Energy Efficiency Ratio
IPLV	Integrated Part Load Value
BMS	Building Management System
EEV	Electronic Expansion Valve
VAV	Variable Air Volume
PMS	Plant Manager Software

References

1. Pérez-Lombard, L.; Ortiz, J.; Pout, C. A review on buildings energy consumption information. *Energy Build.* **2008**, *40*, 394–398. [CrossRef]
2. Perez-Lombard, L.; Ortiz, J.; Maestre, I.R. The map of energy flow in HVAC systems. *Appl. Energy* **2011**, *88*, 5020–5031. [CrossRef]
3. Chung, M.; Park, H.C. Comparison of building energy demand for hotels, hospitals, and offices in Korea. *Energy* **2015**, *92 Pt 3*, 383–393. [CrossRef]
4. Teke, A.; Timur, O. Assessing the energy efficiency improvement potentials of HVAC systems considering economic and environmental aspects at the hospitals. *Renew. Sustain. Energy Rev.* **2014**, *33*, 224–235. [CrossRef]
5. Campuzano-Cervantes, J.; Meléndez-Pertuz, F.; Núñez-Perez, B.; Simancas-García, J. Electronic monitoring system of displacement of extension tubes for the expansion joint. *Revista Iberoamericana Automática Informática Industrial RIAI* **2017**, *14*, 268–278. [CrossRef]
6. Chua, K.; Chou, S.; Yang, W.; Yan, J. Achieving better energy-efficient air conditioning—A review of technologies and strategies. *Appl. Energy* **2013**, *104*, 87–104. [CrossRef]
7. Yu, F.; Chan, K. Improved energy management of chiller systems by multivariate and data envelopment analyses. *Appl. Energy* **2012**, *92*, 168–174. [CrossRef]
8. Do, H.; Cetin, K.S. Data-Driven Evaluation of Residential HVAC System Efficiency Using Energy and Environmental Data. *Energies* **2019**, *12*, 188. [CrossRef]

9. Peña, M.; Biscarri, F.; Guerrero, J.I.; Monedero, I.; León, C. Rule-based system to detect energy efficiency anomalies in smart buildings, a data mining approach. *Expert Syst. Appl.* **2016**, *56*, 242–255. [CrossRef]
10. Kim, H.; Stumpf, A.; Kim, W. Analysis of an energy efficient building design through data mining approach. *Autom. Constr.* **2011**, *20*, 37–43. [CrossRef]
11. Hong, T.; Yang, L.; Hill, D.; Feng, W. Data and analytics to inform energy retrofit of high performance buildings. *Appl. Energy* **2014**, *126*, 90–106. [CrossRef]
12. Ahmad, M.W.; Mourshed, M.; Rezgui, Y. Trees vs Neurons: Comparison between random forest and ANN for high-resolution prediction of building energy consumption. *Energy Build.* **2017**, *147*, 77–89. [CrossRef]
13. Lin, Y.; Zhou, S.; Yang, W.; Shi, L.; Li, C.Q. Development of Building Thermal Load and Discomfort Degree Hour Prediction Models Using Data Mining Approaches. *Energies* **2018**, *11*, 1570. [CrossRef]
14. Zeng, Y.; Zhang, Z.; Kusiak, A. Predictive modeling and optimization of a multi-zone HVAC system with data mining and firefly algorithms. *Energy* **2015**, *86*, 393–402. [CrossRef]
15. Afram, A.; Janabi-Sharifi, F. Review of modeling methods for HVAC systems. *Appl. Therm. Eng.* **2014**, *67*, 507–519. [CrossRef]
16. Alves, O.; Monteiro, E.; Brito, P.; Romano, P. Measurement and classification of energy efficiency in HVAC systems. *Energy Build.* **2016**, *130*, 408–419. [CrossRef]
17. Yu, F.; Chan, K. Experimental determination of the energy efficiency of an air-cooled chiller under part load conditions. *Energy* **2005**, *30*, 1747–1758. [CrossRef]
18. Yu, F.; Chan, K. Part load performance of air-cooled centrifugal chillers with variable speed condenser fan control. *Build. Environ.* **2007**, *42*, 3816–3829, [CrossRef]
19. Kabeel, A.; El-Samadony, Y.; Khiera, M. Performance evaluation of energy efficient evaporatively air-cooled chiller. *Appl. Therm. Eng.* **2017**, *122*, 204–213. [CrossRef]
20. Weng, T.; Agarwal, Y. From Buildings to Smart Buildings—Sensing and Actuation to Improve Energy Efficiency. *IEEE Des. Test Comput.* **2012**, *29*, 36–44. [CrossRef]
21. Oldewurtel, F.; Parisio, A.; Jones, C.N.; Gyalistras, D.; Gwerder, M.; Stauch, V.; Lehmann, B.; Morari, M. Use of model predictive control and weather forecasts for energy efficient building climate control. *Energy Build.* **2012**, *45*, 15–27. [CrossRef]
22. Mahmoud, M.S. Multilevel Systems Control and Applications: A Survey. *IEEE Trans. Syst. Man Cybern.* **1977**, *7*, 125–143. [CrossRef]
23. Findeisen, W.; Bailey, F.N.; Bryds, M.; Malinowski, K.; Tatjewski, P.; Wozniak, A. *Control and Coordination in Hierarchical Systems*, 1st ed.; John Wiley & Sons: Hoboken, NJ, USA, 1980; p. 478.
24. Figueiredo, J.; da Costa, J.S. A SCADA system for energy management in intelligent buildings. *Energy Build.* **2012**, *49*, 85–98. [CrossRef]
25. Li, Z.; Huang, G.; Sun, Y. Stochastic chiller sequencing control. *Energy Build.* **2014**, *84*, 203–213. [CrossRef]
26. Shan, K.; Wang, S.; ce Gao, D.; Xiao, F. Development and validation of an effective and robust chiller sequence control strategy using data-driven models. *Autom. Constr.* **2016**, *65*, 78–85. [CrossRef]
27. Sun, Y.; Wang, S.; Huang, G. Chiller sequencing control with enhanced robustness for energy efficient operation. *Energy Build.* **2009**, *41*, 1246–1255. [CrossRef]
28. Chan, K.; Yu, F. Applying condensing-temperature control in air-cooled reciprocating water chillers for energy efficiency. *Appl. Energy* **2002**, *72*, 565–581. [CrossRef]
29. Yu, F.; Chan, K. Advanced control of condensing temperature for enhancing the operating efficiency of air-cooled chillers. *Build. Environ.* **2005**, *40*, 727–737. [CrossRef]
30. Seem, J.E. Using intelligent data analysis to detect abnormal energy consumption in buildings. *Energy Build.* **2007**, *39*, 52–58. [CrossRef]
31. Thangavelu, S.R.; Myat, A.; Khambadkone, A. Energy optimization methodology of multi-chiller plant in commercial buildings. *Energy* **2017**, *123*, 64–76. [CrossRef]
32. Beghi, A.; Cecchinato, L.; Cosi, G.; Rampazzo, M. A PSO-based algorithm for optimal multiple chiller systems operation. *Appl. Therm. Eng.* **2012**, *32*, 31–40. [CrossRef]
33. Ardakani, A.J.; Ardakani, F.F.; Hosseinian, S. A novel approach for optimal chiller loading using particle swarm optimization. *Energy Build.* **2008**, *40*, 2177–2187. [CrossRef]
34. Hartman, T. Designing Efficient Systems With the Equal Marginal Performance Principle. *ASHRAE J.* **2005**, *47*, 64–70.

35. Mařík, K.; Rojíček, J.; Stluka, P.; Vass, J. Advanced HVAC Control: Theory vs. Reality. *IFAC Proc. Volumes* **2011**, *44*, 3108–3113. [CrossRef]
36. Wang, S.; Ma, Z. Supervisory and Optimal Control of Building HVAC Systems: A Review. *HVAC R Res.* **2008**, *14*, 3–32. [CrossRef]
37. Díaz, I.; Cuadrado, A.A.; Pérez, D.; Domínguez, M.; Alonso, S.; Prada, M.A. Energy analytics in public buildings using interactive histograms. *Energy Build.* **2017**, *134*, 94–104. [CrossRef]
38. Morán, A.; Fuertes, J.J.; Domínguez, M.; Prada, M.A.; Alonso, S.; Barrientos, P. Analysis of electricity bills using visual continuous maps. *Neural Comput. Appl.* **2013**, *23*, 645–655. [CrossRef]
39. Behrooz, F.; Mariun, N.; Marhaban, M.H.; Mohd Radzi, M.A.; Ramli, A.R. Review of Control Techniques for HVAC Systems—Nonlinearity Approaches Based on Fuzzy Cognitive Maps. *Energies* **2018**, *11*, 495. [CrossRef]

Article

Energy and Economics Analyses of Condenser Evaporative Precooling for Various Climates, Buildings and Refrigerants

Bo Shen *, Joshua New and Moonis Ally

Oak Ridge National Laboratory, Oak Ridge, TN 37830, USA; newjr@ornl.gov (J.N.); allymr@ornl.gov (M.A.)
* Correspondence: shenb@ornl.gov; Tel.: +1-865-574-5745

Received: 29 April 2019; Accepted: 27 May 2019; Published: 31 May 2019

Abstract: Condenser evaporative pre-coolers provide a low cost retrofit option for existing packaged rooftop air conditioning application units. This paper aimed to provide a comprehensive study to assess energy savings and peak power reductions of condenser evaporative cooling. Condenser evaporative cooling leads to a lower temperature of the air entering the condenser of a rooftop unit, which results in smaller compressor power consumption. Using EnergyPlus building energy simulations, we mapped the impacts on energy savings and energy reductions at peak ambient temperatures in three building types and 16 locations with levels of pad effectiveness and demonstrated the effects on air conditioner using either R22 or R410A as refrigerants. Economics and control strategy to maximize the cost saving were also investigated. The results demonstrate that energy savings are much greater for HVAC systems with the refrigerant R410A than they are with R22, and evaporative pre-cooling provides the opportunity for annual energy savings and peak demand reductions, with significant potential in hot, dry climates. Additionally, we validated an improved mathematical model for estimating the condenser pre-cooling wet bulb efficiency which shows clear advantage over the current EnergyPlus model.

Keywords: condenser evaporative precooling; rooftop air conditioners; building energy modelling; control strategy

1. Introduction

Evaporative cooling is a process that cools air through the simple evaporation of water. It is best suited for climates where the air is warm and dry (low humidity) because both conditions favor high rates of water evaporation that produce the desired cooling effect. Even in locations with moderate humidity, evaporative cooling may be used effectively, especially in the commercial and industrial sectors. In direct evaporative cooling, water is sprayed on a substrate that is placed in the air stream to be cooled. This technique may be used to cool indoor as well as outdoor air. When applied to outdoor air, the wetted media cools down the air entering the condenser coil and the process is called evaporative pre-cooling. Pre-coolers achieve energy savings because the entering air temperature to the condenser coil is lower, decreasing the lift imposed on the vapor compression cycle. These systems have the potential to be deployed immediately at scale for annual power savings and peak power demand reduction in many parts of the country at an attractive payback.

There have been many literatures addressing the evaporative pre-cooling effect around the world. Wu et al., [1] developed a simplified model to describe heat and mass transfer process in evaporative coolers, and validated the model against a wide range of air temperatures, humidities, velocities, and pad thicknesses, etc. They conducted numerical and analytical modeling practices for evaporative pre-cooling, and applied the model in four different regions in northwest China. Wu et al., [2] applied the same model to optimize the frontal air velocity for a given pad geometry. They demonstrated that the pre-cooling effects were significant for the desert climates in China. Waly et al., [3] evaluated the condenser evaporative pre-cooling option on a 2.8-ton/9.8 kW R-22 split unit in Kuwait, and they identified power savings ranging from 8.1 to 20.5%, as a result of the increase in coefficient of performance (COP) ranging from 36 to 59%. Hajidavalloo [4] investigated a window- size R-22 air conditioner with condenser pre-cooling in Iran. The experimental results demonstrated that the pre-cooling reduced the power consumption by 16% and increased COP by 55%. The author recommended using the pre-cooling technique in very hot climate zones, where the pay-back period for adding the condenser pre-cooling devices can be less than one year.

Goswani et al., [5] performed an experimental investigation of performance of a residential air conditioning system with an evaporatively-cooled condenser, based on a 2.5-ton/8.8 kW residential air conditioning system. For the residential application, they investigated seven U.S. locations (Miami, Orlando, Jacksonville, Key West, Panama City, Fort Meyers and Tampa). It should be noted that the seven locations include a very humid climate zone like Panama City, where the evaporative pre-cooling saving is less significant. At the end, they concluded the payback periods to cover the retrofit cost (pad, pipe, pump, etc.) are less than 2 years. Yu et al., [6] used mist precooling to enhance efficiency of an air-cooled chiller. They concluded that the mist precooling could increase the coefficient of performance up to 10.2%. Liu et al., [7] applied an evaporative-condenser in a gas-engine drive heat pump system, and their experimental results showed that the efficiency increase due to the evaporative condenser was 28.1%, as compared to an air-cooled condenser. Xuan et al., [8] conducted a comprehensive review of evaporative cooling technologies as applied in China. Kim et al., [9] used direct and in-direct evaporative cooling in a 100% outdoor air system. They observed more significant energy savings in the intermediate season than the cooling season. Eidan et al., [10] studied the effects of condenser evaporative pre-cooling on a small window air conditioner in Iraq's climate where the air temperature can reach 55 °C. They reported that the precooling was able to extend the working range to the extreme temperature, and reduce peak power consumption. The evaporative pre-cooling enabled the compressor to work at 16% lower voltage than 220 V.

The above literatures indicate significant saving potentials and short pay-back periods worldwide using condenser evaporative pre-cooling techniques, but these investigations have been limited in equipment, building types, climate zones, etc. Especially, we didn't find much investigation related to U.S. commercial applications. In order to promote the quick and wide application of the pre-cooling technique in the U.S. market, we need to extensively assess this promising technique covering all the variances of equipment, buildings and climates, etc. This work intends to provide a comprehensive reference evaluation, in terms of annual energy savings, peak power reduction, water consumption, payback period, for applying the condenser evaporative pre-cooling technique nationwide.

We introduced a mathematical model to correlate the wet bulb efficiency of an evaporative-precooling pad, which requires less data points for fitting and has better accuracy than the current model used in EnergyPlus [11]. The model comparison was based on a manufacturer's data in a wide range of pad thicknesses and frontal air velocities. Next, EnergyPlus was used to conduct parametric simulation studies covering sixteen US cities, three commercial buildings, three generations (present, before 1980 and post-1980) of rooftop air conditioners using two refrigerants (R-22 and R-410A), respectively. Energy and economics saving potentials were revealed with respect to the multiple influential factors.

2. Mathematical Models

To model the evaporative cooling heat and mass transfer process, the basic assumptions generally include the following: the pad is fully wet, and is exposed to entering air of uniform temperature, humidity and velocity. The boundary layer between the entering air and the wet pad is saturated air at the pad surface temperature. The pad surface has a uniform bulk temperature, i.e., no heat and mass transfer occur within the pad structure. The pad's thermal capacity and transient temperature change are ignored. There is no heat transfer from the surroundings, other than the entering air.

EnergyPlus [11] is capable of simulating direct evaporative cooling for both indoor as well as outdoor components. Described by EnergyPlus [11], the parameter for gauging efficiency of the evaporative pre-cooling process is the wet bulb (WB) efficiency, which is a measure of the drop in the dry bulb temperature due to the evaporative process. The wet bulb efficiency, E_{wb}, is defined in Equation (1) as:

$$E_{wb} = \frac{T_{db,i} - T_{db,o}}{T_{db,i} - T_{wb,i}} \tag{1}$$

where E_{wb} is the wet bulb efficiency; $T_{db,i}$ and $T_{db,o}$ are the entering and leaving air dry bulb temperatures, respectively; and $T_{wb,i}$ is the entering air wet bulb temperature. The maximum possible wet bulb efficiency can be unity when the leaving air dry bulb temperature is cooled to the entering wet bulb temperature. If the wet bulb efficiency is zero, the pad, substrate, or media is dry and evaporative cooling effect is absent.

In EnergyPlus [11], evaporative cooling is treated as a heat transfer process between air and water, following a constant wet bulb temperature line in the psychrometric chart. The media geometry (thickness) and the face velocity (the velocity of air entering the outer surface of the media) are major independent variables that impact the wet bulb efficiency. Other factors like supply water temperature, flow rate, and external heat to the water stream have relatively minor impact. For direct evaporative cooling, EnergyPlus simulates the WB efficiency using a curve-fit polynomial equation with media thickness (Depth) and face velocity (Velocity) as the independent variables, as depicted in Equation (2):

$$
\begin{aligned}
E_{wb} = a_1 + \ & a_2(Depth) + a_3(Velocity) + a_4\left(Depth^2\right) + a_5\left(Velocity^2\right) \\
& + a_6(Depth \cdot Velocity) + a_7\left(Depth^2 \cdot Velocity\right) \\
& + a_8\left(Depth \cdot Velocity^3\right) + a_9\left(Depth^3 \cdot Velocity\right) \\
& + a_{10}\left(Depth^2 \cdot Velocity^3\right) + a_{11}\left(Depth^3 \cdot Velocity^2\right)
\end{aligned} \tag{2}
$$

The EnergyPlus curve-fit form of calculating evaporative cooling efficiency requires a large body of empirical data to obtain the 11 parameters required in the equation and, being a curve fit, is also not physically meaningful.

Braun et al., [12] presented an innovative effectiveness-NTU approach to model a wide range of cooling towers and cooling coils. Through detailed analytical work, Braun et al., [12] treated the simultaneous heat and mass transfer process between sprayed water and entering air as an enthalpy-potential-driven process, characterized by a Lewis Number of unity. Since the Lewis Number is defined as the ratio of thermal diffusivity to mass diffusivity, a value of unity means that the thermal and mass exchanges are of equal significance. Braun et al., [12] proposed three new operational concepts in their model of simultaneous heat and mass transfer process: (1) that the heat and mass transfer rates are proportional to the difference of the enthalpy of saturated air at the entering water temperature and the enthalpy of entering air; (2) introduced a new parameter C_s called the specific heat of saturated air, which can be obtained as in Equation (4); (3) defined the ratio of saturated air specific heat flow rate to the water specific heat flow rate, m^* as in Equation (5):

$$\dot{Q} = E_H \cdot \dot{m}_{air}\left(H_{s,water,i} - H_{air,i}\right) \tag{3}$$

$$C_s \equiv \left[\frac{dH_s}{dT}\right]_{T = T_{water}} \tag{4}$$

$$m^* = \frac{\dot{m}_{air} \cdot C_s}{\dot{m}_{water} \cdot C_{P_{water}}} \tag{5}$$

where:

$\dot{Q}$ = total energy transfer rate,

$H_{s,water,i}$ = enthalpy of saturated air at the entering water temperature,

$H_{air,i}$ = enthalpy of entering air,

E_H = heat and mass transfer effectiveness, defined later in Equation (7).

With these three operational concepts, Braun et al., [12] defined the number of transfer units (NTU) and the effectiveness E_H for evaporative cooling by a simple mathematical form. The NTU and E_H both maintain the same form as that for sensible heat transfer applicable to a wide range of counter, parallel and cross-flow geometries. Assuming constant water temperature on the surface of the wet media, we can get E_H in the form of Equation (7). Thus:

$$NTU = \frac{h_a \cdot A_{surface}}{\dot{m}_{air} \cdot C_{P_{air}}} \tag{6}$$

$$E_H = \frac{T_{db,i} - T_{db,o}}{T_{db,i} - T_{s,water}} = 1 - \exp(-NTU) \tag{7}$$

where:

h_a = heat transfer coefficient between air and water stream,

$A_{surface}$ = heat transfer surface area, usually taken as the pad area,

$\dot{m}_{air}$ = mass flow rate of dry air,

$C_{P_{air}}$ = specific heat of dry air at dry bulb conditions,

$T_{s,water}$ = water surface temperature (constant).

It should be noted that Braun et al., [12] defined the wet bulb efficiency, E_H (Equation (7)) more generally than is described by E_{wb} in Equation (1). In the case of direct evaporative cooling, if the amount of water that is sprayed is much less than the amount of water resident on the pad, i.e., sensible water heat transfer is negligible, then $T_{s,water} \cong T_{wb,i}$ and E_H becomes identical to E_{wb}. In practice, we want the amount of water sprayed on the pad to be small and, in fact, commercial equipment makers control water supply very diligently to apply a fine mist, just enough to wet the pad surface area.

Under these conditions, Wu et al., [1] combined Equation (1) and Equation (7), taking $T_{s,water} = T_{wb,i}$ to yield:

$$E_{wb} = E_H = 1 - \exp(-NTU) \tag{8}$$

Further, Wu et al. [2] formulated the following equations for the air heat transfer coefficient, h_a, the surface area for evaporative cooling, $A_{surface}$, and the air mass flow rate:

$$h_a = a \times V_{air}^m \tag{9}$$

$$A_{surface} = c \times \delta \times A_F \tag{10}$$

$$\dot{m}_{air} = \rho_{air} \times V_{air} \times A_F \tag{11}$$

where a, c, and m in Equations (9) and (10) are constants specific to the pad geometry, δ is the pad thickness, and ρ_{air} is the density of ambient air [kg/m³]. V_{air} is the frontal air velocity [m/s]. Substituting $\dot{m}_{air}$ from Equation (11) in Equation (6) gives the final form for the NTU:

$$NTU = \frac{a \cdot \delta}{\rho_{air} \times C_{P_{air}} \times V_{air}^n} \tag{12}$$

where α and n are empirical constants specific to the pad.

An overall mass balance on water gives the water mass flow rate in terms of the humidity ratios of the entering and leaving air, $\omega_{air,i}$ and $\omega_{air,o}$ [kg water/kg dry air], respectively, given by:

$$\dot{m}_{water} = \dot{m}_{air}(\omega_{air,i} - \omega_{air,o}) \tag{13}$$

Based on the approaches of Braun et al., [12] and Wu et al., [1] our proposed coupled heat and mass transfer model utilizes Equations (3), (7) and Equations (8)–(13) to solve for $\dot{m}_{water}$ and $T_{s,water}$ iteratively.

All the equations above from the references are used for the model validation in the section below. Our approach to model validation is based on a comprehensive set of manufacturer's performance data for a specific type of evaporative pad. The data set consists of 13 levels of pad thicknesses ranging from 1 in to 24 in (0.025 m to 0.61 m), frontal air velocities from 250 fpm to 900 fpm (1.3 to 4.6 m/s) (12 levels) and wet bulb efficiencies ranging from 17.5% to 99.6%. The 12×13 matrix of data points were used to fit the coefficients from a_1 through a_{11} for the polynomial curve fit used by EnergyPlus as depicted by Equation (2), and to obtain the parameters α and n used in Equation (12) by Braun's model.

Our criteria for deciding goodness of a model is to demonstrate how closely the model predicts the measured data gathered by the manufacturer. A good model will show small deviations from actual field measurements. Figure 1 compares the predicted deviations of the EnergyPlus polynomial curve-fit calculations of the wet bulb efficiency and the effectiveness-NTU model versus manufacturer's data. The effectiveness-NTU method is significantly more accurate with prediction errors having a standard deviation of 0.67% with maximum deviation of 2.9%. In contrast, EnergyPlus with the polynomial curve-fitted equation has a standard deviation of 4% with a maximum deviation of 31%. Prediction deviations are defined as (Model Prediction—Manufacturer's data)/Manufacturer's data. We compared the predictions over a wide range of face velocities. With EnergyPlus, big outliers are observed at the upper and lower bounds of the face velocity, whereas for the effectiveness-NTU approach, the predictions are closer and uniformly distributed across a wide range of face velocities even at the upper and lower bounds of face velocity. These simulations are shown in Figure 1.

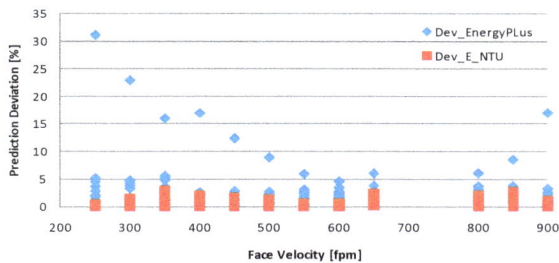

Figure 1. Relative deviations from manufacturer's data predicted by the effectiveness-NTU model and EnergyPlus.

On the basis of these comparisons, we conclude that the effectiveness-NTU model is easier to use, and yet gives much better agreement with manufacturer's data than the approach used in EnergyPlus for evaporative cooling. The current EnergyPlus requires constant wet bulb efficiency for modeling condenser evaporative pre-cooling at each equipment speed. So the effectiveness-NTU model can be an efficient tool to predict the wet bulb efficiency as an input to EnergyPlus, based on the actual pad geometry and condenser frontal air velocity.

We used EnergyPlus to conduct parametric building energy simulations. In EnergyPlus, direct-expansion (DX) air conditioners, i.e., rooftop air conditioners, are modeled in the form of performance curves. The equation forms are given as below:

$$CAPFT_{coil,cooling} = a + b * T_{wb,i} + c * T_{wb,i}^2 + d * T_o + e * T_o^2 + f * T_{wb,i} * T_o \tag{14}$$

$$\dot{Q}_{coil(i),cooling,total} = \dot{Q}_{coil(i),cooling,rated} * CAPFT_{coil,cooling} \tag{15}$$

where $CAPFT_{coil,cooling}$ = Coil Cooling Capacity Correction Factor (function of temperature); $\dot{Q}_{coil(i),cooling,total}$ = unit total (sensible + latent) cooling capacity [W]; $\dot{Q}_{coil(i),cooling,rated}$ = rated total (sensible + latent) cooling capacity [W]; $T_{wb,i}$ = wet-bulb temperature of the air entering the cooling coil (°C); T_o = temperature of the air entering an outdoor heat exchanger (°C); and $a - f$ = bi-quadratic equation coefficients for Cooling Capacity Correction Factor.

The DX cooling coil energy input ratio (*EIRFT*) also depends on the wet-bulb temperature of the air entering the cooling coil and ambient temperature.

$$EIRFT_{cooling} = a + b * T_{wb,i} + c * T_{wb,i}^2 + d * T_o + e * T_o^2 + f * T_{wb,i} * T_o \qquad (16)$$

$$COP_{cooling} = COP_{cooling,rated} / EIRFT_{cooling} \qquad (17)$$

$EIRFT_{cooling}$ = the cooling energy input ratio correction factor (function of temperature). $COP_{cooling,rated}$ is the unit rated COP. The rated conditions for obtaining the capacities, COPs and SHRs are at an indoor dry-bulb temperature of 26.67 °C (80 °F), wet bulb temperature of 19.44 °C (67 °F), and condenser entering air temperature of 35 °C (95 °F).

For simulating a unit with an evaporatively cooled condenser, EnergyPlus simply corrects the condenser entering air temperature $T_{c,i}$ in the form as Equation (18):

$$T_{c,i} = T_{wb,o} + (1 - E_{wb}) * \left(T_{db,o} - T_{wb,o} \right) \qquad (18)$$

where $T_{db,o}$ and $T_{wb,o}$ are the ambient dry bulb and wet bulb temperatures, respectively. Since the condenser air velocity and the evaporative pad thickness are usually fixed for an existing rooftop unit, the E_{wb} can be inputted as a constant value for a building simulation case at one fixed condenser air flow rate.

EnergyPlus auto-sizes rooftop equipment to match building sensible loads under design days—the statistically hottest day in summer. Operating conditions in design days would vary regarding to locations. Working as a safety factor, the equipment sizing factor is set to scale the equipment rated capacity. If the sizing factor is larger than unity, it means the equipment is oversized in comparison to the required building sensible load. Consequently, the building comfort level is increased, but at the expense of the cyclic loss. However, the equipment sizing factor might vary according to building type, since an individual building type might have a specific comfort level regulation.

EnergyPlus is capable of simulating single-speed, two-speed and multiple-speed rooftop equipment. For two-speed equipment, EnergyPlus uses the rated capacity at high speed to match the building peak sensible load in design days, and the rated capacity at low speed would be assumed as 1/3 of the rated value at the high speed. In addition, for two-speed equipment, we can input evaporative pre-cooling wet bulb efficiency specific to each speed and condenser air flow level.

Certainly, for modeling condenser evaporative pre-cooling, we need to consider the expense in exchange of the energy saving, which are water evaporation and pump power consumed. In EnergyPlus, the evaporative condenser pump rated power consumption is modeled as the total cooling capacity times 0.004266 Watts pump power per Watt rated capacity, i.e., 15 W/ton. For two-speed rooftop equipment, at low speed, the pump power is set equal to 1/3 times the total cooling capacity times 0.004266 Watts pump power per Watt capacity. It should be mentioned that we ignored the extra condenser fan power due to adding the precooling pad upstream of the condenser, since the condenser fan power consumption is less than 10% of the total equipment power consumption, and the extra fan power caused by the precooling pad is marginal, i.e., 2 to 3% to the total power.

The water consumption rate is calculated by the difference between the entering and exit air specific humidity multiplied by the condenser air mass flow rate. A standard rated condenser air flow rate is between 0.00004027 m³/s and 0.00006041 m³/s per Watt of rated total cooling capacity, i.e., 300–450 cfm/ton. The water evaporation amount would be integrated along the running time fraction during operation.

3. Parametric Building Energy Simulations

3.1. Performance Curves of Different Refrigerants

For air conditioning application at high ambient temperature, R-410A usually works under near-critical conditions. With increasing ambient temperature, particularly above 35 °C (95 °F), the equipment performance of R-410A can degrade much faster than that of R-22. Payne et al., [13] compared an R-22 and an R-410A air conditioner operating at high ambient temperature. The two air conditioner systems were tested using the same evaporator and condenser heat exchangers. The capacity and COP of the R-410A system was compared to those of the R-22 system in terms of normalized ratios. The capacities of R-22 and R-410A systems matched each other at 35 °C and COPs matched each other at 27 °C. With changing the ambient temperature from 25 °C to 55 °C, the capacity normalized ratio, R-410A versus R-22, degraded from 1.05 to 0.90, and the COP ratio was reduced from 1.05 to 0.80. The results implied that the R-410A system was more sensitive to the increased outdoor temperature. Rice [14] conducted a comprehensive system simulation studies comparing R-410A to R-22, which basically confirmed the same conclusion.

The condenser evaporative pre-cooling achieves energy savings by lowering the air temperature entering the condenser coil. Consequently, the energy saving effect is directly related to how the working refrigerant responds to the entering air temperature. To have a direct comparison, we obtained product literature and reduced the data in the forms of EnergyPlus equipment performance curves (Equations (14)–(17)). The comparisons of the performance curves at the full compressor capacity between a 10-ton (rated cooling capacity of 35.2 kW) R-22 rooftop unit and a 10-ton R-410A rooftop unit can be seen in Figures 2 and 3. Both the units have a rated cooling COP of 3.0 at 35 °C (95 °F) outdoor air temperature and 26.7 °C DB/19.4 °C WB (80 °F DB/67 °F WB) indoor air, With fixing indoor wet bulb temperature at 67 °F (19.4 °C), Figure 2 shows the capacity correction ratio (capacity at various ambient temperatures versus the rated capacity at 35 °C) and Figure 3 shows the EIR correction ratio (EIR at individual temperatures versus the rated EIR), as a function of the air temperature entering the condenser coil. At 35 °C, i.e., the rated condition, the ratios are unity. With lowering the air temperature, the capacity correction fraction increases while the EIR (reverse of cooling COP) correction fraction decreases, because the units deliver higher capacities with lower power consumptions at reduced temperatures.

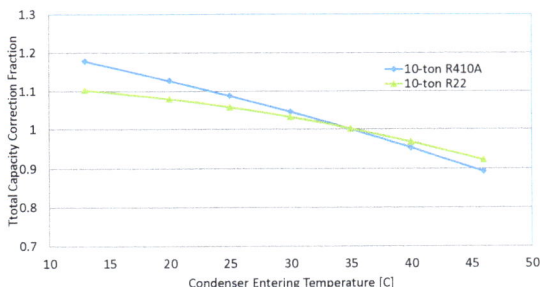

Figure 2. Total cooling capacity correction ratios of R-22 and R-410A varying with outdoor air dry bulb temperature and fixing indoor wet bulb at 67 °F (19.4 °C).

We can see that both the total cooling capacity and EIR curves of R-410A change more drastically than R-22. Changing the outdoor temperature from 115 °F (46.1 °C) to 55 °F (12.8 °C), the relative change in capacity for R-22 (compared to the total capacity at 95 °F/35 °C outdoor temperature) is 52%, however, the change for R-410A is 79%; the relative change in EIR for R-22 (compared to the EIR at 95 °F/35 °C outdoor temperature) is 18%, however, the change for R-410A is 29%. This means, given the same rating capacity and EIR at the design condition, the R-410A equipment would get more capacity and operate more efficiently, with the same drop in outdoor temperature. Consequently, applying the same condenser evaporative precooling load, the R-410A equipment will get more relative power reduction than the R-22 equipment.

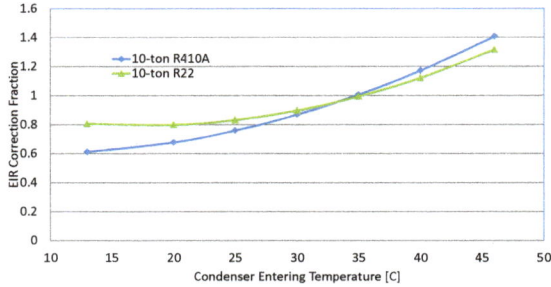

Figure 3. EIR correction ratios of R-22 and R-410A varying with outdoor air temperature and fixing indoor WB at 67 °F (19.4 °C).

3.2. DOE Benchmark Commercial Buildings and Locations

DOE selected 16 building types classified as benchmark buildings that represent most of the commercial building stock, across 16 locations (representing all U.S climate zones, NREL/CP-550-43291 [15]). In this study, we examined the effect of pre-cooling technology in three of the 16 building types in all 16 locations to gain an understanding of the extent to which annual and peak energy reductions are realistically possible. Descriptions of these buildings reside within the supplied EnergyPlus input files. The output from our effectiveness-NTU model was input to EnergyPlus, which performed the simulations for the various buildings in the particular climate zones, as the current EnergyPlus requires constant wet bulb efficiency for modeling condenser evaporative pre-cooling at each equipment speed.

We selected a medium office, a secondary school, and a supermarket for the three building types because they have very different equipment sizes, load profiles, zones, and uses. The building characteristics are shown in Table 1.

Table 1. Three benchmark commercial buildings selected for this study.

Building	Floor Area (m²)	Number of Floors	Zones	DX Cooling Coils	Sizing Factor
Medium office	5017	3	15	3 units, 2-speed	1.33
Secondary school	19,509	2	46	5 units, 1-speed	1.5
Supermarket	4181	1	6	6 units, 1-speed	1.

In addition to the three building types, our simulations covered three building generations (new, pre-1980, and post-1980) and two refrigerants (R-22 and R-410A) in 16 climate zones. For retrofit applications, the pad is directly added to the original rooftop unit sizes selected by EnergyPlus as equipment sizing. It shall be noted that this study only investigates the retrofit applications; possible equipment size reductions (and any equipment first cost savings) due to the precooling effect are not taken into account.

Building types represent different load distributions. For example, in Phoenix, AZ, the load profile in a medium office building is different from that in a supermarket as shown in Figures 4 and 5, respectively. In cold weather, when the ambient temperature is low, the load in the supermarket is practically zero (no cooling needed, only perhaps heating), whereas in the medium office building heating and cooling may be needed depending on the zones in the building because the loads in the exterior portions are different from that in the interior portions. Most of the cooling load in the supermarket is in the hotter months, whereas in the medium office building, the cooling loads are distributed over more temperature bins. The building load distributions lead to utilizations of evaporative precooling at different ambient conditions.

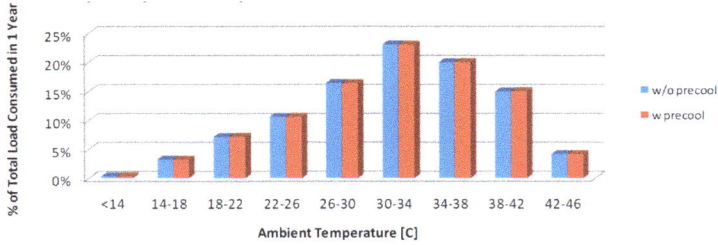

Figure 4. Medium office building in Phoenix, AZ, USA: Percentage of annual building load that is delivered in each temperature bin.

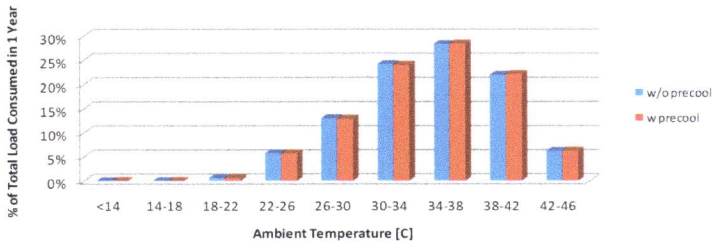

Figure 5. Supermarket in Phoenix, AZ, USA: Percentage of annual building load that is delivered in each temperature bin.

Using EnergyPlus, we estimated the annual power savings for R-410A rooftop equipment in Phoenix, AZ, for the three classes of buildings (present buildings): medium office, secondary school and supermarket, as shown in Figure 6.

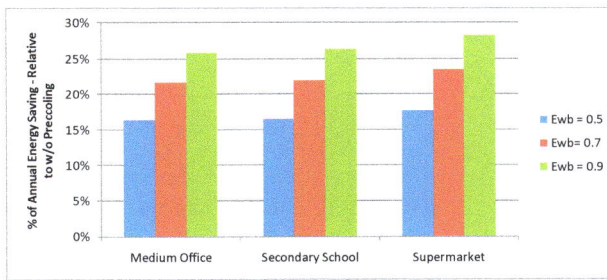

Figure 6. Annual savings depend on the building type (present buildings) and the WB efficiency.

Due to the different load profiles, the annual savings vary according to building type slightly. Figure 6 further exemplifies the benefit of using pre-cooling technology. Relative to no pre-cooling, the percentage of annual energy savings can be as high as 23% with a WB efficiency of 0.7 or as much as 27% with a higher WB efficiency of 0.9 (where many pre-cooling units operate). The decision to operate at a specific WB efficiency depends largely on the pad thickness and face velocity as discussed above and these are operating variables available to the HVAC engineer.

3.3. Effect of Refrigerant Type

Older rooftop units use R-22 as refrigerant while newer models utilize R-410A. We examined the effect of refrigerant type on annual energy and peak power reductions for a medium office building (present) with a two-speed rooftop unit and WB efficiency of 0.7 in 16 climate zones as shown in

Figures 7 and 8 A peak power reduction is calculated via comparing power consumptions with/without the condenser evaporative precooling at the hottest temperature of summer in each city. Clearly, the largest benefits are derived with the R-410A refrigerant in hot and dry climates. Note that even in hot and humid climates such as Houston, TX, the benefits with R-410A are significantly higher than it is for R-22.

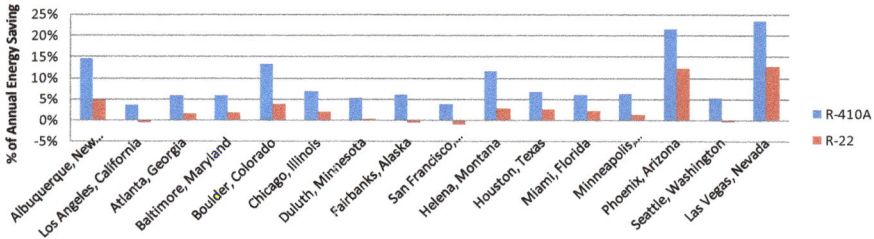

Figure 7. Annual energy savings for a medium office (present) building with two-speed unit and WB efficiency fixed at 0.7 in all 16 climate zones.

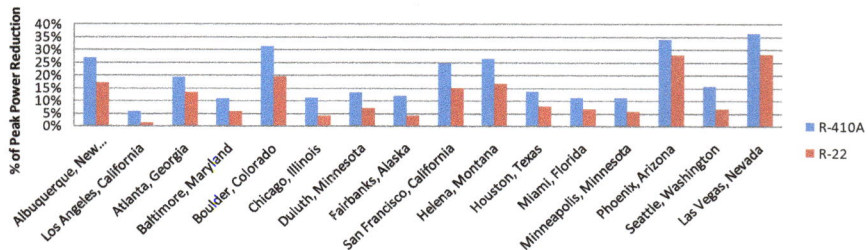

Figure 8. Peak power reductions for a medium office building with two-speed unit and WB efficiency fixed at 0.7, in 16 climate zones.

The effect of refrigerant type can be understood by noting that R-410A condensing pressure operates near the critical region (where the refrigerant liquid and vapor become identical) and any lowering of the condenser temperature (as would be the case of evaporative cooling) causes the two-phase isotherm to move downwards from the apex of the critical region enabling the system to increase its cooling capacity and efficiency. In other words, the refrigerant operates over a larger enthalpy change across the two-phase region, resulting in improved overall condenser heat transfer effectiveness, and hence can reject more heat to the ambient for the same mass flow rate. Consequently, precooling has a more beneficial effect on energy savings for R-410A systems, as compared with R-22 units.

3.4. Annual Energy Savings and Peak Power Reduction of Three Building Types (Present) in 16 Climate Zones

Commercially available evaporative pre-coolers provide a low-cost retrofit for existing packaged rooftop units, commercial unitary split systems, and air cooled chillers. We mapped the impact of energy savings and peak energy reduction in the three building types (present medium office, secondary school, and supermarket) in 16 locations with a fixed pad thickness (giving an effectiveness of 0.816) and show the effect of both refrigerants, R-22 and R-410A. The selected pad wet bulb effectiveness of 0.816 was recommended by a manufacturer for standard 8-inch-thick pad for commercial applications. Simulations are performed with EnergyPlus. In each of the three building types, there is tangible opportunity to reduce annual energy and peak power consumption if pre-cooling is used and if a switch is made from R-22 equipment to R-410A equipment. The air frontal velocity was selected as 250 fpm (1.3 m/s), which is the actual condenser frontal air flow velocity according to the 10-ton

R-410A unit obtained from the manufacturer's literature. Inputting the calculated wet bulb efficiency into EnergyPlus, we obtain annual energy savings and peak power reductions as below. Tables 2–4 illustrate annual energy savings and peak power reductions, respectively for the R-410A and R-22 units in the three builing types of medium office, secondary school and supermarket. It can be seen that the condenser evaporative pre-cooling leads to larger savings and peak power reductions in dry and hot cities, e.g., Phoenix, AZ, etc. The reduction percentages with using R-410A is 10% higher than using R-22. On the other hand, there are no apparent differences between the three building types.

Table 2. Medium office building: Annual energy savings and peak power reductions.

City/Saving	Annual Energy Saving		Peak Power Reduction	
	R410A	R22	R410A	R22
Albuquerque, NM	16.1%	5.1%	29.5%	18.4%
Los Angeles, CA	4.2%	−0.5%	6.7%	1.8%
Atlanta, GA	6.9%	1.9%	22.1%	14.9%
Baltimore, MD	7.0%	2.0%	12.7%	6.7%
Boulder, CO	14.8%	3.9%	34.4%	20.4%
Chicago, IL	8.1%	2.2%	12.6%	4.7%
Duluth, MN	6.1%	0.5%	15.2%	7.9%
Fairbanks, AK	6.8%	−0.5%	13.5%	4.2%
San Francisco, CA	4.6%	−0.9%	27.8%	16.3%
Helena, MT	13.0%	2.8%	29.4%	17.7%
Houston, TX	7.8%	3.0%	15.4%	8.9%
Miami, FL	7.1%	2.5%	13.3%	8.0%
Minneapolis, MN	7.3%	1.6%	13.0%	6.8%
Phoenix, AZ	24.2%	13.3%	38.0%	30.5%
Seattle, WA	6.1%	−0.2%	17.6%	7.5%
Las Vegas, NV	26.0%	13.5%	39.0%	29.7%
Average	10.4%	3.1%	21.3%	12.8%
Max	26.0%	13.5%	39.0%	30.5%
Min	4.2%	−0.9%	6.7%	1.8%

Table 3. Secondary school building: Annual energy savings and peak power reductions.

City/Saving	Annual Power Saving		Peak Power Reduction	
	R410A	R22	R410A	R22
Albuquerque, NM	18.8%	6.7%	29.8%	17.1%
Los Angeles, CA	6.8%	0.7%	12.3%	2.5%
Atlanta, GA	9.3%	3.0%	15.6%	9.5%
Baltimore, MD	9.5%	3.4%	12.3%	7.7%
Boulder, CO	18.7%	6.2%	31.2%	17.9%
Chicago, IL	9.9%	3.2%	11.6%	6.5%
Duluth, MN	9.4%	2.2%	9.4%	3.1%
Fairbanks, AK	11.5%	1.6%	20.2%	6.4%
San Francisco, CA	10.5%	2.0%	22.1%	11.2%
Helena, MT	17.7%	6.0%	25.2%	14.4%
Houston, TX	8.3%	3.0%	11.2%	6.7%
Miami, FL	6.7%	2.1%	9.5%	4.8%
Minneapolis, MN	9.1%	2.6%	10.8%	4.8%
Phoenix, AZ	24.6%	13.5%	31.7%	25.2%
Seattle, WA	10.2%	1.8%	14.8%	5.5%
Las Vegas, NV	26.9%	14.4%	34.2%	25.1%
Average	13.0%	4.5%	18.9%	10.5%
Max	26.9%	14.4%	34.2%	25.2%
Min	6.7%	0.7%	9.4%	2.5%

Table 4. Supermarket: Annual energy savings and peak power reductions.

City/Saving	Annual Power Saving		Peak Power Reduction	
	R410A	**R22**	**R410A**	**R22**
Albuquerque, NM	19.2%	7.3%	26.2%	16.0%
Los Angeles, CA	6.5%	0.7%	8.1%	2.7%
Atlanta, GA	8.9%	3.2%	20.7%	13.4%
Baltimore, MD	9.5%	3.7%	12.2%	6.0%
Boulder, CO	18.7%	6.6%	31.0%	19.1%
Chicago, IL	9.7%	3.5%	18.1%	11.0%
Duluth, MN	6.0%	0.7%	15.2%	7.1%
Fairbanks, AK	6.5%	-0.3%	17.4%	8.0%
San Francisco, CA	9.9%	1.9%	27.2%	15.7%
Helena, MT	17.7%	6.1%	32.1%	18.6%
Houston, TX	9.1%	4.1%	13.0%	7.6%
Miami, FL	7.3%	2.7%	10.4%	4.9%
Minneapolis, MN	8.4%	2.6%	10.7%	5.8%
Phoenix, AZ	26.4%	16.0%	37.0%	29.4%
Seattle, WA	10.7%	2.5%	28.3%	18.4%
Las Vegas, NV	28.8%	16.7%	37.0%	27.6%
Average	12.7%	4.9%	21.5%	13.2%
Max	28.8%	16.7%	37.0%	29.4%
Min	6.0%	-0.3%	8.1%	2.7%

4. Economics and Control Strategy

The economics of condenser evaporative pre-cooling is driven by the climate, refrigerant, equipment vintage, utility rate structure, control strategy, and equipment costs. To assess the economics, we collected electricity and water rates in six cities, as given in Table 5.

Table 5. Electricity and water rates in six cities.

Cost in Cents	Albuquerque	Atlanta	Houston	Minneapolis	Phoenix	Seattle
Water/m^3	49.8	27.7	94.1	107.4	124.0	158.6
Electricity/kwh	8.7	8.9	9.5	8.1	10.1	7.3

Annual Electricity Savings and Cost of Water

In our analyses, savings are relative to the baseline equipment without pre-cooling. The variables in the relative savings are the type of building, the pad wet bulb efficiency, the utility rates, and the equipment rated COP and the type of refrigerant. Percent savings represents the annual operating cost savings with pre-cooling (taking into account the cost of water and pumping costs) relative to the cost of operating the same equipment without pre-cooling. One water conserving strategy is to use water for evaporative cooling only when the dry bulb temperature exceeds 90 °F (32.2 °C) instead of using it at all temperatures during the cooling season. Our findings are summarized below.

If pre-cooling is deployed at all ambient temperatures during the cooling season, then the percent energy savings relative to using no pre-cooling at wet bulb efficiency of 0.7 and equipment rated COP of 3.0 is shown Figure 9. It should be mentioned that COP used here considers total cooling capacity and compressor and outdoor fan power consumptions. The indoor blower power and heat are not taken into consideration. This is in line with EnergyPlus input. The COP of 3.0 represents lower efficiency equipment used in current commercial buildings. We observe that annual operating cost savings are increased due to precooling for all three types of buildings particularly in the hot dry climates (Phoenix, Albuquerque) but less pronounced in wet climates (Seattle), and that cost savings with R-410A are significantly greater than it is for the older R-22 refrigerant. Hot dry climates are amenable to water evaporation and hence such climates benefit from this technology. For the same wet bulb efficiency and COP, the percent savings if pre-cooling is deployed when the wet bulb temperature exceeds 90 °F/32.2 °C is shown in Figure 10 for comparison with Figure 9.

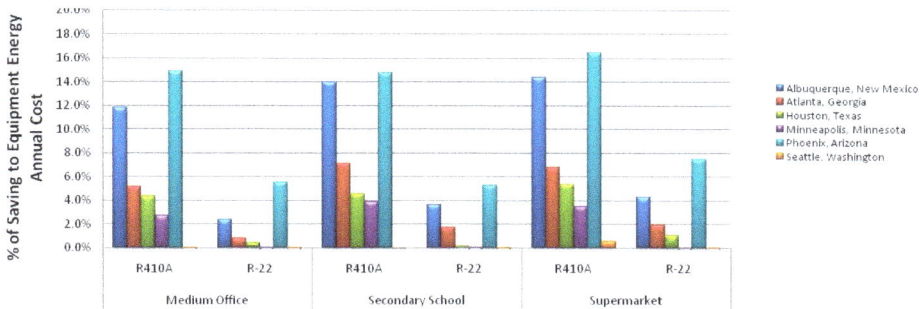

Figure 9. Percentage of energy savings due to pre-cooling relative to no pre-cooling when water is sprayed at all ambient temperatures and wet bulb efficiency of 0.7, and COP = 3.

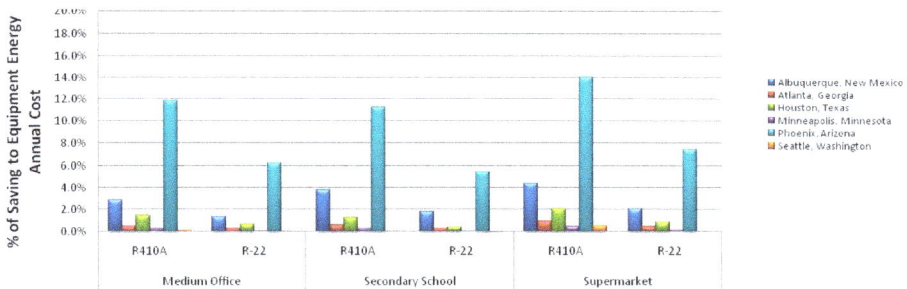

Figure 10. Percentage of energy savings due to pre-cooling relative to no pre-cooling when water is sprayed after ambient temperature equals or exceeds 90 °F/32.2 °C. Wet bulb efficiency = 0.7, COP = 3.

Here we observe that, for the equipment using R-410A, if precooling is deployed only when the ambient air temperature exceeds 90 °F/32.2 °C, the percent operating cost savings are less than if precooling was deployed at all operating temperatures in the respective climates. However, the results are different for the equipment using R-22. For example, for the medium office in Phoenix, AZ, spraying water at all temperatures yields an annual relative cost saving of 5.5%, whereas spraying water only when the DB is above 90 °F/32.2 °C, yields a slightly higher annual cost saving of 6.3%. Similar conclusions are reached for the secondary school and supermarket. Therefore, for R-22, it may be beneficial to use water only when DB temperatures exceed 90 °F/32.2 °C to maximize the annual operating cost savings while simultaneously minimizing water use. However, it must be pointed out clearly that the annual and peak energy reductions using R-410A are far greater than that of using R-22, and for R-410A the savings are greater if the equipment is operated at all DB temperatures rather than only when the DB exceeds 90 °F/32.2 °C.

In our EnergyPlus simulations, we consider an upper wet bulb efficiency limit of 0.9; the results for this case are shown in Figures 11 and 12:

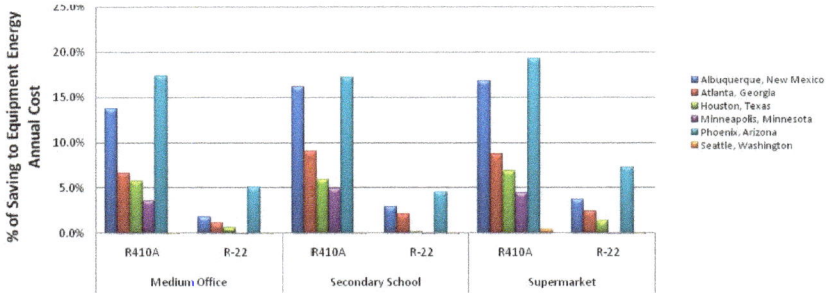

Figure 11. Percentage of energy savings due to pre-cooling relative to no pre-cooling at all ambient temperatures. Wet bulb efficiency = 0.9, equipment COP = 3.

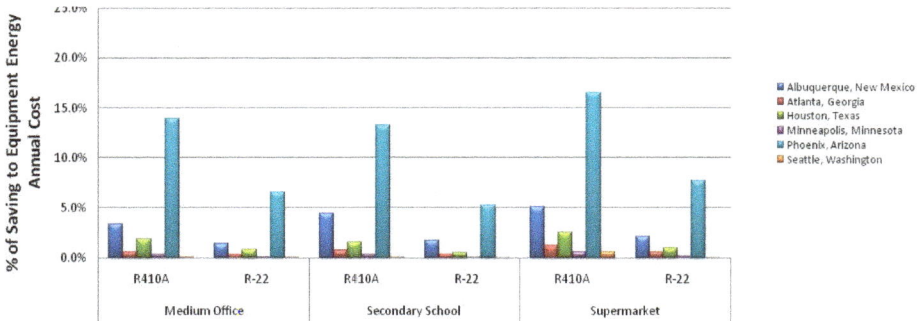

Figure 12. Percentage of energy savings due to pre-cooling relative to no pre-cooling when water is sprayed after ambient temperature is equal to or exceeds 90°F. Wet bulb efficiency = 0.9, equipment COP = 3.

Again we see significantly higher annual operating cost savings using R-410A versus R-22 for the three building types in all six cities except Seattle, for the reasons mentioned above, and that increasing the wet bulb efficiency from 0.7 to 0.9 marginally improves the annual operating cost savings as seen by comparing Figures 10 and 12. If precooling is deployed only when the ambient temperature exceeds 90 °F/32.2 °C, annual percent savings using R-410A equipment drop slightly, but the annual percent savings using R-22 equipment increase slightly.

Next, we examine the effect of installing a more efficient rooftop unit with a rated COP = 5, which represents the higher efficiency level of equipment on the market, to replace an older rooftop unit and make the same comparisons described above. As expected, the impact on annual energy savings with precooling as compared to no precooling in more efficient equipment (COP = 5) is smaller than it would be for a less efficient equipment (COP = 3) as shown by comparison of Figure 9 with Figure 13.

A similar trend is observed at higher wet bulb efficiency of 0.9, as shown by comparing Figure 11 to Figure 14. The impact of pre-cooling on annual savings compared to no pre-cooling is attenuated for equipment that operate at high efficiency (COP = 5 compared to COP = 3), nonetheless, pre-cooling does provide energy savings, especially in hot, dry climates.

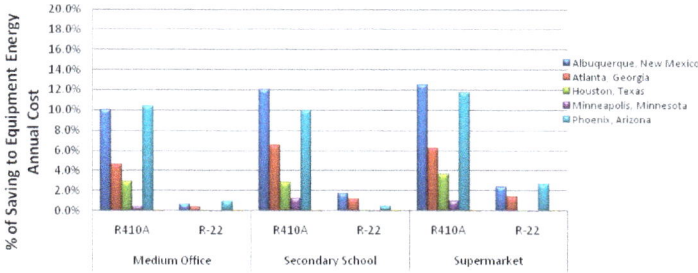

Figure 13. Percentage of savings due to pre-cooling relative to no pre-cooling at all ambient temperatures. Wet bulb efficiency = 0.7, equipment COP = 5.

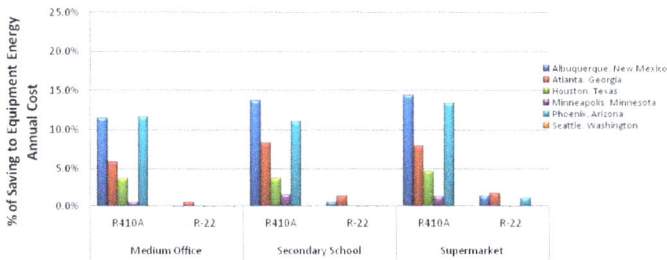

Figure 14. Percentage of energy savings due to pre-cooling relative to no pre-cooling at all ambient temperatures. Wet bulb efficiency = 0.9, equipment COP = 5.

Comparing the annual energy savings estimates for the higher efficient (COP = 5) equipment that utilizes a water control strategy of deploying pre-cooling only when the DB temperature exceeds 90 °F/32.2 °C, we find similar trends in Figure 15 as discussed for the lower efficiency equipment (COP = 3).

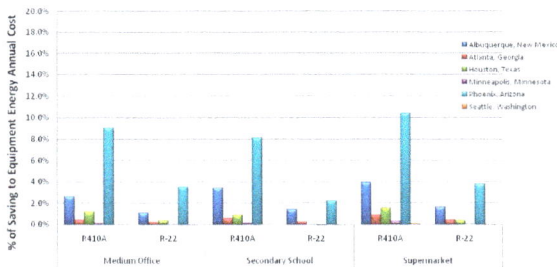

Figure 15. Percentage of energy savings due to pre-cooling relative to no pre-cooling when water is sprayed after ambient temperature is equal to or exceeds 90 °F/32.2 °C. Wet bulb efficiency = 0.9, equipment COP = 5.

For equipment using R-410A, when pre-cooling is deployed at DB temperatures exceeding 90 °F/32.2 °C, annual operating cost savings are somewhat lower than if pre-cooling is deployed at all DB temperatures. The same trend is observed at the higher WB efficiency of 0.9. Therefore, at least for these three building types, the recommended strategy would be to pre-cool at all DB temperatures, for the equipment using R-410A. On the other hand, for the equipment using R-22 in the cities where the water price is high, we recommend applying pre-cooling at higher ambient temperatures (for example when DB exceeds 90 °F/32.2 °C).

5. Conclusions

- Commercially available evaporative pre-coolers provide a low cost retrofit option for many existing packaged rooftop units, commercial unitary split systems, and air cooled chillers.
- Relative energy savings are much greater for HVAC systems with the refrigerant R-410A than they are with R-22. The relative energy savings and peak power reductions of the R-410A unit is 10% higher than the R-22 unit in hot and dry climates.
- For R-410A equipment, operating cost percent savings are greater if pre-cooling is used at all DB temperatures than they are if used only when the DB temperature exceeds 90 °F/32.2 °C. For R-22 equipment in the cities where the water price is high, we would recommend applying pre-cooling only at high ambient temperatures.
- Evaporative pre-cooling provides the opportunity for annual energy savings and peak demand reduction, with significant potential in hot, dry climates. For example, in Las Vegas, NV, use of an 81.6% efficiency pre-cooling pad can achieve annual energy savings up to 26% and peak power reductions up to 39.0%. Even in cold, wet climates, for example, Chicago, IL, it still achieves 10% annual saving and 18% peak power reduction.

Author Contributions: B.S. conducted major analytical studies and writing; M.A. monitored the work progress, reviewed the analytical work, and greatly enhanced the technical writing; J.N. provided indispensable support for the EnergyPlus building energy simulations in multiple climate zones and three template buildings.

Funding: The authors acknowledge the support provided by the US DOE Building Technologies Office and the Technology Manager, Antonio Bouza. The work was funded under Contract No. DE-AC05-00OR22725.

Acronyms

DOE	U.S. Department of Energy.
DB	air dry bulb temperature [C].
COP	coefficient of performance [w/w]
C_s	specific heat of saturated air [J/kg/K].
$CAPFT_{coil,cooling}$	coil cooling capacity correction factor.
$EIRFT_{cooling}$	cooling energy input ratio correction factor.
E_H	heat and mass transfer effectiveness.
E_{wb}	wet bulb efficiency.
EIR	energy input ratio [w/w].
$\dot{m}_{air}$	air mass flow rate [kg/s].
NTU	Number of heat transfer units.
SHR	sensible heat ratio.
$T_{s,water}$	water surface temperature [C].
WB	air wet bulb temperature [C].
$\omega_{air,i}$	air inlet specific humidity [kg water/kg dry air].
$\omega_{air,o}$	air outlet specific humidity [kg water/kg dry air].

References

1. Wu, J.; Huang, X.; Zhang, H. Numerical investigation on the heat and mass transfer in a direct evaporative cooler. *Appl. Therm. Eng.* **2009**, *29*, 195–201. [CrossRef]
2. Wu, J.; Huang, X.; Zhang, H. Theoretical analysis on heat and mass transfer in a direct evaporative cooler. *Appl. Therm. Eng.* **2009**, *29*, 980–984. [CrossRef]

3. Waly, M.; Chakroun, W.; Al-Mutawa, N.K. Effect of pre-cooling of inlet air to condensers of air-conditioning units. *Int. J. Energy Res.* **2005**, *29*, 781–794. [CrossRef]

4. Hajidavalloo, E.; Eghtedari, H. Performance improvement of air-cooled refrigeration system by using evaporatively cooled air condenser. *Int. J. Refrig.* **2010**, *33*, 982–988. [CrossRef]

5. Goswami, D.Y.; Mathur, G.D.; Kulkarni, S.M. Experimental Investigation of Performance of a Residential Air Conditioning System with an Evaporatively Cooled Condenser. *J. Sol. Energy* **1993**, *115*, 206–211. [CrossRef]

6. Yu, F.W.; Ho, W.T.; Chan, K.T.; Sit, R.K.Y. Theoretical and experimental analyses of mist precooling for an air-cooled chiller. *Appl. Therm. Eng.* **2018**, *130*, 112–119. [CrossRef]

7. Liu, H.; Zhou, Q.; Zhao, H. Experimental study on cooling performance and energy saving of gas engine-driven heat pump system with evaporative condenser. *Energy Convers. Manag.* **2016**, *123*, 200–208. [CrossRef]

8. Xuan, Y.; Xiao, F.; Niu, X.; Huang, X.; Wang, S.; Wang, S. Research and application of evaporative cooling in China: A review (I)—Research. *Renew. Sustain. Energy Rev.* **2012**, *16*, 3535–3546. [CrossRef]

9. Kim, M.H.; Jeong, J.W. Cooling performance of a 100% outdoor air system integrated with indirect and direct evaporative coolers. *Energy* **2013**, *52*, 245–257. [CrossRef]

10. Eidan, A.A.; Alwan, K.J.; Alsahlani, A.; Alfahham, M. Enhancement of the Performance Characteristics for Air-Conditioning System by Using Direct Evaporative Cooling in Hot Climates. *Energy Procedia* **2017**, *142*, 3998–4003. [CrossRef]

11. EnergyPlus Engineering Reference, Version 8.7, DOE. 2017. Available online: https://energyplus.net/ (accessed on May 2017).

12. Braun, J.E.; Klein, A.; Mitchell, J.W. Effectiveness models for cooling towers and cooling coils. *ASHRAE Trans.* **1989**, *95*, 164–174.

13. Payne, W.V.; Domanski, P.A. A Comparison of an R22 And an R410A Air Conditioner Operating at High Ambient Temperatures. In Proceedings of the International Refrigeration and Air Conditioning Conference, West Lafayette, IN, USA, July 2002.

14. Rice, C.K. *Investigation of R-410A Air Conditioning System Performance Operating at Extreme Ambient Temperatures up to the Refrigerant Critical Point*; Final Report, ARTI-21CR/605-50015-01, ORNL/TM-2005/277; December 2005.

15. Torcelini, P.; Deru, M.; Griffith, B.; Benne, K.; Halverson, M.; Winiarski, D.; Crawley, D.B. DOE Commercial Building Benchmark Models: preprint. Presented at the 2008 ACEEE Summer Study on Energy Efficiency in Buildings, Pacific Grove, Ca, USA, 17–22 August 2008. NREL/CP-550-43291.

energies

MDPI

Article

Analysis of the Influence Subjective Human Parameters in the Calculation of Thermal Comfort and Energy Consumption of Buildings

Roberto Robledo-Fava [1,*], Mónica C. Hernández-Luna [1], Pedro Fernández-de-Córdoba [1], Humberto Michinel [2], Sonia Zaragoza [3], A Castillo-Guzman [4] and Romeo Selvas-Aguilar [4]

[1] UMPA, Universidad Politécnica de Valencia, Camino de Vera s/n, 46022 Valencia, Spain; moher@doctor.upv.es (M.C.H.-L.); pfernandez@mat.upv.es (P.F.-d.-C.)

[2] Applied Physics Department, Universidade de Vigo, As Lagoas s/n, 32004 Ourense, Spain; hmichinel@uvigo.es

[3] Escuela Politécnica Superior, Universidade da Coruña, Mendizábal s/n, 15403 Ferrol, Spain; szaragoza@udc.es

[4] Universidad Autónoma de Nuevo León, Facultad de Ciencias Físico-Matemáticas, Av. Universidad S/N, Ciudad Universitaria, San Nicolas de los Garza, Nuevo León 66455, Mexico; arturo.castillogz@uanl.edu.mx (A.C.-G.); romeo.selvasag@uanl.edu.mx (R.S.-A.)

* Correspondence: rorobfa@doctor.upv.es; Tel.: +52-818-250-7386

Received: 11 March 2019; Accepted: 16 April 2019; Published: 23 April 2019

Abstract: In the present work, we analyze the influence of the designer's choice of values for the human metabolic index (met) and insulation by clothing (clo) that can be selected within the ISO 7730 for the calculation of the energy demand of buildings. To this aim, we first numerically modeled, using TRNSYS, two buildings in different countries and climatologies. Then, we consistently validated our simulations by predicting indoor temperatures and comparing them with measured data. After that, the energy demand of both buildings was obtained. Subsequently, the variability of the set-point temperature concerning the choice of clo and met, within limits prescribed in ISO 7730, was analyzed using a Monte Carlo method. This variability of the interior comfort conditions has been finally used in the numerical model previously validated, to calculate the changes in the energy demand of the two buildings. Therefore, this work demonstrated that the diversity of possibilities offered by ISO 7730 for the choice of clo and met results, depending on the values chosen by the designer, in significant differences in indoor comfort conditions, leading to non-negligible changes in the calculations of energy consumption, especially in the case of big buildings.

Keywords: Monte Carlo method; ISO 7730; TRNSYS

1. Introduction

The building envelope is the interface between the outdoor environment and the interior of buildings and its primary function is to act as a physical barrier to offer a comfortable place to develop different activities, in exchange for a certain demand for energy [1–6]. 40% of the primary energy consumption in the world and 17% of carbon dioxide emissions are caused by the building sector [7–10]. Thus, power consumption in buildings is one of the leading causes, among other adverse effects, of ozone layer depletion, global warming, and climate change [11,12]. Besides, residential buildings are often mentioned as one of the most profitable areas to decrease global power demand [13,14].

A rapid search on scientific databases shows that comfort in buildings is addressed by different areas, such as Engineering ($\approx$ 39%) energy ($\approx$14%), environmental ($\approx$10%), computer ($\approx$7.5%) and social sciences ($\approx$7%). Most of the analysis emphasizes the fact that adequate levels of indoor comfort (thermal, acoustic, lighting, and air quality) are essential to guarantee the health and welfare of the occupants and to prevent adverse problems such as disease, among others [15].

Some recent examples of the application of comfort analysis concepts in singular environments include a synagogue design [16], strategies for low-cost housing [17], the evaluation of a modern low-energy office [18] or materials for hip-protective pads [19]. Thermal comfort is thus one of the most important elements in the relationship with the estimation of indoor environment quality, and it has been defined in parallel by ASHRAE (American Society of Heating, Refrigerating and Air-Conditioning Engineers) Standard 55 and ISO (International Standard Organization) Standard 7730 as *"the condition of the mind in which satisfaction is expressed with the thermal environment"* [20,21].

Thus, comfort ranges will depend on the physical and human parameters of its occupants and therefore, should be established according to the specificity of each region or country [22]. Well established international standards from countries and regions such as USA [23] or Europe [24], restrict comfort intervals based on numerical indices, such as the predicted mean vote (PMV) and the predicted percentage dissatisfied (PPD) [21,24,25], both defined in the ISO 7730 [26]. In the present work, we precisely focus on ISO 7730, which is the mandatory standard in Europe. In this norm, the designer has some freedom, within a given range, to choose the adequate clo and met parameters, based on his/her observation of the occupants of the building under analysis [27]. Thus, as we will explain in detail below, we have analyzed with a Monte Carlo code, the possible variations in energy demand calculations, due to aleatory designer's choices *within the range provided by ISO 7730*. In this way, we quantify the impact of the choice of met and clo on energy demand, according to the norm. We must stress that other norms such as EN 15251-2007 or ISO 17772-1-2017 have been used in studies related with occupants modeling [28]; however, they are not mandatory in many countries, such as Spain or Mexico, that we analyzed in the present work, where ISO 7730 is the standard.

The above parameters are used to predict the thermal sensation as a function of the four classical thermal environmental parameters (air temperature, mean radiant temperature, air velocity, and humidity) and two subjective human parameters: activity and clothing; meaning activity the intensity of the physical tasks a person is developing (expressed through the metabolic rate index) and clothing standing for the total thermal resistance from the skin to the outer surface of the clothed body [26,29]. This comfort estimate is evidently more significant for buildings located in extreme weather regions, either cold, such as northern Europe, Canada or Asia [30,31] or warm such as Africa, South-East Asia or Latin America [32]. On the other hand, the use of HVAC (Heating, Ventilation and Air Conditioning) systems is strongly related with thermal comfort and energy consumption and it is well known that in countries with extreme weather conditions, HVAC systems may represent more than one half of the total power demand of a single dwelling [33,34].

Building Information Modeling (BIM)-based simulations are commonly used as an effective tool to analyze simultaneously energy consumption and thermal comfort in buildings [35]. Some building simulation programs such as Energy Plus, ESP-r or TRNSYS are software used to calculate comfort levels inside of buildings [36,37]. TRNSYS (TRaNsient SYStems simulation) is one of the most widely established programs, providing a flexible and graphical simulation tool, based on FORTRAN code, originally developed in 1975 by the University of Wisconsin [38].

Over the last few years, multiple energy modeling studies have been carried out, taking into account the level of insulation of clothing and the level of activity [39–41]. In these studies, the relationship between clo and met with energy consumption has been analyzed under different perspectives [42–44]. There have also been many papers relating the parameters of thermal comfort with the most widely used international standard as is the case of the ASHRAE 55 [45–47] or ISO 7730 [48,49].

A factor of considerable uncertainty is the modeling behavior of the occupant [50]. Adaptive thermal comfort, which relies on changing clothing levels to gain a wider comfortable temperature range has been also widely researched [51]. Clothing levels cannot be sensed electronically and therefore surveys and observations have been commonly used [41] and in recent years methodological advances have been developed that make dynamic simulations of building-occupant systems more tractable, such as visualization systems [52] agent-based models [53] or adaptive building simulations [54] among others. However, a significant problem when simulating the performance of a building is still the precise determination of the complete requirements of the models of the behavior of the occupants [55,56].

While most of the available simulators are focused on the performance of electrical and thermal networks, TRNSYS can similarly well be used to model other dynamic systems such as traffic flow, or even biological processes [57,58]. Studies using TRNSYS are reported in the literature as comparing cooling strategies for an office in different European climates [59], thermal comfort in an indoor swimming pool [60], or operating performance in cooling mode of a nearly zero energy building in China [61].

In this paper, we use TRNSYS for analyzing two different buildings: one in Spain and another one in Mexico. The numerical simulations have been carried out by multi-zone analysis, including technical/physical properties of the buildings, focusing on the cold (winter) and warm (summer) conditions in these countries, as there is more influence on thermal comfort. The novelty of our approach mostly lies in the combination of such BIM-based computer simulations with a sensitivity analysis (SA), to determine which input parameters have the maximum impact on the output [29].

The SA can be done by different popular methods such as Morris, regression, variance-based, or Monte Carlo, among others [62,63]. In our case, the SA based on Monte Carlo method [49] offers the possibility of prioritizing and fixating inputs when considering multiple outputs, in comparison with other methods [64,65]. *Several studies can be found in the literature with SA performed on the effects of technical and physical parameters on the energy consumption of buildings and simulation programs* [49,66,67]. However, to our knowledge, less attention has been paid to the effect of subjective human parameters [51] and thus, this aspect will be our main focus for the present work. Therefore, as a final step, we will show the significant impact of the designer's subjective choices on the calculation of the energy demand of buildings, depending on the election of the amounts of activity or clothing, based on the environmental ergonomic standard ISO 7730. We consider that our SA of the above-mentioned parameters using the Monte Carlo method will help the designers to define the most appropriate parameters to be considered, depending on the aim and type of the numerical modeling performed.

2. Case Study

In the present work, we analyze two non-residential buildings localized in different continents (America and Europe), with particular characteristics in their structure, use, and surrounding climate. The first building is a recent one in Cambre, in north-western Spain (latitude 43°18′22.55″ N, longitude 8°17′36.13″ W) located at a height of 99 m above sea level. The second is a building situated in San Nicolás de los Garza, in north-eastern Mexico (latitude 27°32′02″ N, longitude 99°58′33″), at a height of 239 m above sea level.

2.1. Climate Conditions

The Map of K'oppen-Geiger Map [68] define different regions classifications in function of the central values of temperature and humidity, the Spanish location is classified in the Cfb region (C—temperate, f—without dry season, b—warm summer), where Oceanic Climate has a soft summer with a minimum rainfall and does not exceed of 22 °C on average in the warmest month, and at least four months averaging above 10 °C. No significant precipitation variation between seasons.

With respect to Mexico, according to the same classification, San Nicolás de los Garza can be considered to be BSh (BS—semi-arid, h—hot). A semi-arid (or steppe) region receives precipitation below potential evapotranspiration, but not extremely. Semi-arid climates tend to support short or scrubby vegetation, with large zones usually dominated by either grasses or shrubs. Also, the average annual temperature is higher than 18 °C. From the databases available in TRANSYS, the web platform

Equs [69], and Meteonorm [70], we obtain the values of different climatic variables such as air temperature, relative humidity, direct irradiance (solar) and diffuse horizontal irradiance.

2.2. Building Structure And Use

The base of the first building is of rectangular shape with an area of 218.14 m^2 (see Figure 1) and its main wall faces southwest (216° N). The building is distributed on two floors (defined as ground and first), being the lower level used for storage and the upper zone for offices, meeting rooms, and kitchen, among others uses. This building is the headquarters of a private company specialized in air-conditioning systems. The working hours, considered from Monday to Friday throughout the year, are from 08:30 a.m. to 06:00 p.m.

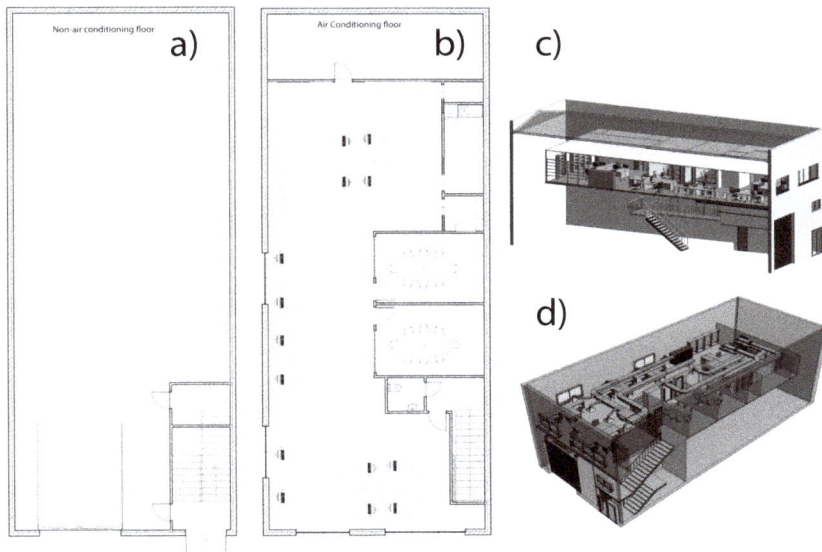

Figure 1. Architectural blueprints of the first case study building (Spain): (**a**) ground floor; (**b**) first floor. Numerical model geometry: (**c**) Northeast view and (**d**) Southwest view.

For the second case, the plant of the building has an "L" shape, covering an area of 1214 m^2 in each of its three heights that we define as: ground floor (GF), first floor (FF) and second floor (SF). The GF corresponds to a parking lot without vertical enclosures and a reception area, while the other two floors are classrooms, meeting rooms, laboratories, offices, etc. as is shown in Figure 2. The building hosts a research center on physics and mathematics (CICFIM), which is located behind of the School of Physics and Mathematics of the Universidad Autónoma de Nuevo León (Mexico). Its principal access is on the facade facing east, next to the campus stadium, to the west. This building starts the activities at 7:00 a.m., ending at 9:00 p.m. from Monday to Friday. Saturday's schedule is from 10:00 a.m. to 2:00 p.m.

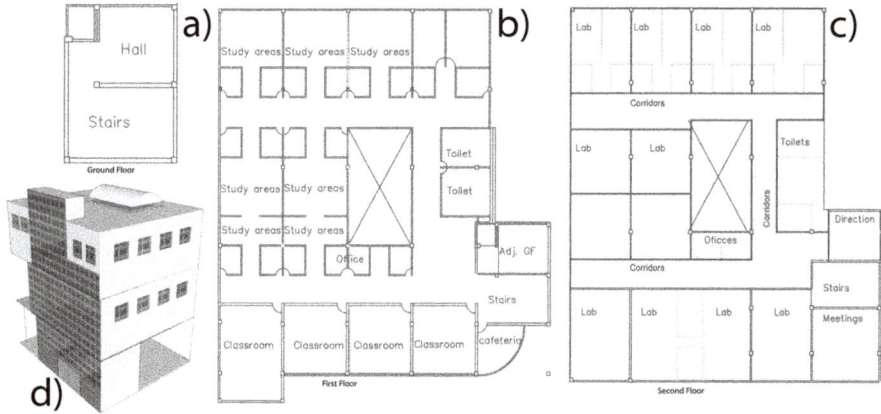

Figure 2. Architectural blueprints of the second case study building (Mexico): (**a**) ground floor and (**b**) first floor (**c**) second floor and (**d**) Numerical model geometry: Northeast view.

The building in Spain is made with three different types of envelopes: the vertical facing walls, the floor, and the roof. In Table 1 we indicate that their respective thicknesses (t) are 46.5 cm, 30.4 cm and 185 cm. The building in Mexico Table 2 has four types of envelopes: internal (9 cm) and external (35 cm) walls, as well as 10 cm-thick divisions. In this case, the same materials were used in the floor and the roof, with a total thickness of 40 cm.

In both Tables 1 and 2, the physical characteristics of each layer of material are specified, such as thermal transmittance ($U - value$), thermal conductivity (κ), heat capacity (C_p), density (ρ) and resistance (R) in specific cases.

Table 1. Thermal and structural properties of adopted constructive solutions for case 1 (Spain).

Wall Types	Structure	Layer	t (cm)	κ (kJ/mK)	C_p (kJ/kgK)	ρ (g/cm³)	R (hm²K/kJ)
External Walls		Cement roughcast	1.5	5.040	1.10	2.00	--
		Concrete block	20	1.764	1.10	1.20	--
		Air chamber	5.0	--	--	--	0.05
		Plasterboard	20	0.900	1.00	0.90	--
		$U - value = 0.637$ W/m²K					
Floor		Ceramic Brick	25	4.104	0.90	1.25	--
		Compressed concrete	5.0	1.00	1.10	1.50	--
		Air chamber	0.4	0.612	1.40	1.20	--
		$U - value = 2.402$ W/m²K					
Roof		Galvanized Metal	2.5	180.1	1.50	7.85	--
		Air chamber	180	--	--	--	0.05
		Perlite ceiling	2.5	0.187	1.50	0.12	--
		$U - value = 0.934$ W/m²K					

For both buildings, the properties of the windows are described in Table A1. In the first building (Spain) there are two types of zones, depending on whether there is air conditioning or not (see Table A2) in Appendix A. The second edifice (Mexico) is distributed in three partitions, because of the different uses and temperatures inside. Among its main characteristics are frame percentage, thermal transmittance (U) and solar gain ($G - Value$).

In Spain, the HVAC system consists on an inverter heat pump (2 × 1) and a split air–air unit, with a total refrigeration capacity of 17.2 kW, a heat power of 19.5 kW and an installed power of 6.5 kW. The diffusion of air is carried out by ducts and impulse/return grids. For the building in Mexico which a conventional technology, based on air-air heat pumps with 26 exterior units and indoor cassette units and a total cooling capacity of 349.4 kW. The heat output is 447.7 kW and the installed power is 166.4 kW.

Table 2. Thermal and structural properties of adopted constructive solutions for case 2 (Mexico).

Wall Types	Structure	Layer	t (cm)	κ (kJ/mK)	C_p (kJ/kgK)	ρ (g/cm³)	R (hm²K/kJ)
Internal Walls		Plasterboard	1.5	0.900	1.00	0.90	--
		Mineral wool	6.0	0.130	1.03	0.05	--
		Plasterboard	1.5	0.900	1.00	0.90	--
		U − value = 0.512 W/m²K					
External Walls		Plasterboard	1.0				
		Perforated brick	9.0	1.080	1.00	0.80	--
		Extruded polystyrene	5.0	2.736	1.00	1.60	--
		Air Chamber	5.0	0.122	1.45	0.03	--
		Hollow brick	14	--	--	--	0.05
		Cement mortar	1.0	1.764	0.92	1.20	--
				5.040	1.05	2.00	--
		U − value = 0.441 W/m²K					
Floor & Roof		Concrete	3.0	4.140	1.00	1.80	--
		Cement mortar	1.0	5.040	1.05	2.00	--
		Extruded polystyrene	6.0	0.122	1.45	0.03	--
		Reinfor. concrete	15	8.280	1.00	2.30	--
		Pebble	15	2.916	0.92	1.70	--
		U − value = 0.450 W/m²K					
Divisions		Aluminum	3.0	575	0.90	2.80	--
		Mineral wool	4.0	0.11	1.00	0.05	--
		Aluminum	3.0	575	0.90	2.80	--
		U − value = 3.296 W/m²K					

3. Methodology

The main objective of this research is to show the consequences on the results of BIM-based simulations of subjective decisions made by the designers when using the environmental ergonomic standard ISO 7730 to calculate thermal comfort ranges and its impact on the energy demand of buildings, as well as its economic repercussion. *The thermal sensation experienced by a human being is mainly due to the overall thermal balance of the body, which depends on the physical activity (met) and clothing (clo) of the subject, as well as by the environmental parameters: air temperature, radiant temperature, air speed, and air humidity* [71]. We must stress that the definition of the parameters of human clothing and activity is done subjectively by the designer and therefore we refer to these quantities as Subjective Human Parameters (SHPs). The use of biometric data from wearable devices has been recently proposed as an interesting option for metabolic rate estimation in thermal comfort analysis [72].

The PMV and PPD indices, according to Fanger's method [71] are commonly used to express warm and cool discomfort for the body as a whole, although thermal dissatisfaction may also be caused by unwanted heating or cooling of one particular part of the body (local discomfort). Therefore, when the previous factors are known, the thermal sensation for the occupants can be predicted, using a seven-point scale ranging from cold (−3) to hot (3), being 0 the value for "neutral sensation" [73]. The PMV may be computed from tables or may be measured by an instrument [74]. The methodology proposed in this paper consists of obtaining first the thermal performances of the buildings and then performing the SA for the SHPs used in the thermal comfort calculation, with the purpose of identifying which quantity has more influence on the BIM-based simulation outputs. Figure 3 shows a scheme of the procedure applied, consisting of the following steps, for both cases studied:

1. Real-time tracking of the main variables using Equs web platform.
2. TRNSYS modeling, getting the indoor temperature.
3. Verification of the thermal simulation with measured temperature from Equs database.
4. Mathematical estimation of PMV and PPD indices, according to ISO 7730.

5. SA (Monte Carlo) to determine which parameter (activity and clothing) values yield significant variation on PMV and PPD outputs.
6. New set-point temperatures are established, depending on the changes in the thermal categories in the previous step.
7. TRNSYS recalculation of the energy demand for each set-point temperature from the Monte Carlo method.
8. Evaluation of the variations in the energy demand for the different options.

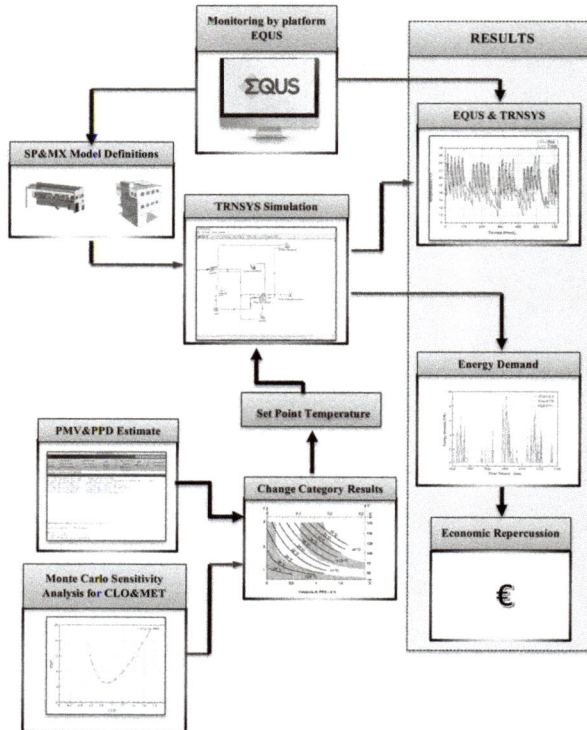

Figure 3. Scheme of the methodology followed.

3.1. Monitoring by Web Platform

A web platform for energy management of buildings named "Equs", was developed with the purpose of control, maintenance and monitoring of energy consumptions and environmental variables as temperature, humidity and CO_2 levels. This tool also includes the possibility of remotely controlling any device by SCADA (Supervisory Control and Data Acquisition) system, ensuring the quality 6 of the statistical results. Equs collects in its database the information sent by the different sensors, allowing us to know the variation of the physical variables remotely, as access to these data can be obtained with any computer or other devices such as smartphones, which are connected to the Internet, making it very simple the treatment of the data, which can be done in the same device accessing to the values, in real time [69].

3.2. Trnsys Modeling of The Buildings

Through dynamic simulation tools, such as TRNSYS, we can define and evaluate the comfort and energy demand of the two buildings accurately; in both cases, we work with multi-zone building models. Notice that TRNSYS is integrated into two parts: TRNBuild, which reads and processes an input file, solving and plotting the system; and TRNSYS Studio, which is a library of components that can be used to model different systems. An advantage of using TRNSYS is the possibility to convert complex problems as multi-zone building models into smaller components, to allow the designer to establish the connections between the building and numerous other subsystems/components in the simulation environment [75].

To verify the energy demands of the structures, after analyzing the sensitivity with the Monte Carlo method of the set-point temperatures of the air-conditioning systems with the met and clo parameters, it has opted for its simulation in the program TRNSYS. To simulate the building, both in Spain and in Mexico, all the necessary inputs, dimensions, location, construction characteristics, etc. have been introduced into the program. This simulation will be verified later comparing the results that the TRNSYS throws in certain indoor temperature conditions, with the actual measurements of both the environmental conditions and the interior temperatures. Once the model is validated, we can calculate the variation of the energy demand in both buildings as indicated by the Monte Carlo SA to the parameters studied in ISO 7730.

Entering the buildings models in TRNSYS was done in two steps: first, in TRNBuild we implemented the different zones and all its elements as location, orientations, walls, layers of materials, gains, ventilation, infiltration, schedules, and comfort in TRNBuild. The second step corresponds to include in TRNSYS models as weather, inputs such as the building previously described and outputs that will show through graphics the thermal performance and energy demands of the buildings. Thus, first, we obtained the database of the buildings, which includes information such as orientations of the facades, materials for the walls and windows, the energy demand of the installed equipment, the working hours, the occupation, etc. The process of the definition of the buildings in TRNBuild includes the following information:

- Location and orientations. This information was described in Section 2.2
- Structure of the building: the materials detailed in Tables 1, 2 and A1 were assigned from the software database.
- The zones described in Table A2, were created as WAC and AC for the first case, GF-FF, SF, and SL for the second case in the TRNBuild Manager.
- In Table A3 we show the different types of schedules created for each building, we set a value of 1 in the periods where there is activity in the building, in any other case, including the weekend, the value is 0.
- In Table A4 we summarize the infiltrations for summer and winter in both cases. This information was calculated based on ASHRAE Standard 55.
- For calculating the ventilation we use standard EN 13779, ASHRAE 62 R. It was estimated a value for each zone which is shown in Table A2.
- The internal gains of the buildings are shown in Tables 3 and 4.
- The values of clothing factor, metabolic rate, relative air velocity were introduced in TRNSYS based on the ISO Standard 7730.

For estimation of comfort, it is important to note that thermal sensations are different between persons sharing the same environment. Perceptions of human beings about thermal comfort are very different because even people who stay in similar places have different opinions, because there exist many elements that impact the perception of human beings.

Thermal comfort is the main goal and has a strong influence in the thermal behavior of a building, for this reason, the values of clothing factor, metabolic rate, relative air velocity were introduced in TRNSYS, based on the ISO Standard 7730. In Table 5, we show the numbers for both cases in summer and winter. We decided to set the same values of clothing and activity, because the characteristics of the occupants are very similar. It should be noted that these values are subjective decisions on the part of the designer. The analysis of the impact of these decisions is one of the main goals of this work.

Table 3. Input data of internal gains in the AC zone on Spain case.

Gains	Description	Schedule	Total Energy Rate (W)
Occupancy	15 Adults seated doing office work	Weekly 1	2250
Computers	16 PC with monitor	Weekly 1	2240
Artificial Lighting	37 fluorescent lamps in 218.12 m^2	Weekly 1	3600
Other Gains	HVAC	Weekly 1	3332
	Coffee machine (10 cups)	Weekly 2	1500
	Copy Machine (office type)	Weekly 3	1060
	Microwave	Weekly 2	600
	Refrigerator	All time	322
	Plotter	Weekly 1	250
	TV	Weekly 1	90

Table 4. Input data of internal gains in the different zones on Mexico case.

Gains	Zones		Description	Schedule	Total Energy Rate (W)	
	GF-FF	SF			GF-FF	SF
Occupancy	63	51	Adults seated doing office work	Weekly A	9450	7650
Computers	54	50	PC with monitor	Weekly A	7650	7000
Artificial	37	9	Fluorescent lamps (64 W)	Weekly A	2304	576
Lighting	60	46	Fluorescent lamps (80 W)	Weekly A	4800	3680
	64	62	Fluorescent lamps (40 W)	Weekly A	2560	2480
Other			HVAC	Weekly B	21,320	
Gains	5	3	Coffee machine (10 cups)	Weekly A	1500	
	2	1	Water Cooler	Weekly B	1060	
	1	— —	Microwave	Weekly A	600	
	2	1	Refrigerator	All time	322	
	10	12	Plotter	Weekly A	250	
	2	— —	TV	Weekly A	90	

Table 5. Input data of comfort in the two buildings, according to ISO 7730.

Parameters	Description	Value
Clothing factor	Summer 1: panties, t-shirt, shorts, thin socks, sandals	0.30 clo/0.050 m^{2}K/W
	Summer 2: underpants, t-shirt, light pants, thin socks, shoes	0.50 clo/0.080 m^{2}K/W
	Summer 3: underwear, shirt, pants, socks, shoes	0.70 clo/0.110 m^{2}K/W
	Winter: underwear, shirt, pants, thermal jacket, socks, shoes	1.20 clo/0.185 m^{2}K/W
Metabolic Rate	Rest, seated	1.0 met/58 W/m^2
	Seated, light work (office, home, school, laboratory)	1.2 met/70 W/m^2
External Work	In general, the external work is around 0	0 met
Relative air velocity	The air velocity relative to the person	0.3 m/s

4. Simulation of Building Models

In this section, it is presented the dynamic thermal simulation model in TRNSYS Studio of the building previously described in TRNBuild, for this purpose we use multi-zone building model "TYPE 56" that can be connected to a large number of other components, including weather data, HVAC systems, occupancy schedules, controllers, output functions, etc. Other models used were Type 62, Type 109-TMY2, and Type 65.

Under ISO Standard 7730, the evaluation of the levels of comfort was calculated using a mathematical MATLAB software package for the interactive numerical study of dynamical systems.

The change of PMV and PPD in base of its parameters such as metabolic rate, clothing insulation, and air temperature, is easier of visualizing in MATLAB than in TRNSYS because you can modify its parameters independently and realize studies more specific such SA.

Monte Carlo Sensitivity Analysis

The principal aim of the SA using a Monte Carlo code is to obtain the variation of the PPD and PMV concerning the possible choices of the clo and met that are indicated in the annexes of the 7730 for thermal comfort calculations. In this way, by changing the values of clo and met according to what is reported in the tables of ISO 7730, we try to find out how the environmental category varies and, consequently, the set-point temperature. The SA is extensively suitable for associating input parameters with the overall building performance (for example, the temperature or demand for heating or cooling) [64]. The individual contribution from specific input can be determined by SA to prepare for future optimization of energy, climate, and economic performance [76,77]. SA methods can typically be classified to the local SA where input parameters are varied one at a time and the global SA where all inputs are changed simultaneously.

Sampling-based methods for uncertainty and SA have become very popular. The sampling-based Monte Carlo method is a procedure that performs a model repeated times with random samples generated from input distributions. It provides approximate solutions to both uncertainty and sensitivity analyses by making statistical modeling investigations [78]. The method can manage complex black box models irrespective of the linearity and continuity, and generate a probability distribution for each output depending on input distribution types [62].

The analysis under consideration can be represented by a function $y(x)$ where:

$$\mathbf{y}(x) = [y_1(x), y_2(x), \ldots, y_{n_Y}(x)] \tag{1}$$

and

$$\mathbf{x} = [x_1, x_2, \ldots, x_{n_X}] \tag{2}$$

designates the outputs and inputs of the analysis, respectively. Successively, sensitivity in x results in a corresponding sensitivity in $y(x)$. Furthermore, the distributions

$$\mathbf{D} = [D_1, D_2, \ldots, D_{n_X}] \tag{3}$$

must be well-defined to characterize the uncertainty associated with the elements of $\mathbf{x}$, where D_i is the distribution associated with x_i for $i = 1, 2 \ldots n_X$.

The most common way for applying SA in building performance analysis consists on the following steps: determine input variations, create building energy models, run energy models, collect simulation results, run SA and presentation of SA results [64]. Determining the probability distributions of input parameters is the first step in any SA, which depends on the election of sensitivity method. In the present study, Monte Carlo SA is used, considering N input parameters defining the system:

$$O_i(P_1, P_2, \ldots P_N) \qquad i = 1, 2, \ldots, M. \tag{4}$$

Suppose that the parameter P_j can change within a range given by ΔP_j. To analyze the influence of the parameter P_j in the objective O_i, we proceed first generating a uniform random number distribution $r \in [0,1]$, thus:

$$P_j^1 = (P_j - \Delta P_j) + 2r\Delta P_j, \qquad P_j^1 \in [P_j - \Delta P_j, P_j + \Delta P_j], \tag{5}$$

gives us the first random value P_j^1. With the value of N randomly generated parameters $P_j^1 (j = 1, 2, \ldots, N)$, we obtain a value of O_i^1, which is the first evaluation of the objective O_i. We repeat this procedure up to the desired number of random evaluations. We can randomly change each one of the parameters or we can apply random variations to one of the parameters keeping constant the value of the rest.

For this work, the objectives O_i are the PMV and PPD indices, which are a function of different parameters mentioned in the previous sections. To analyze the influence on the thermal performance of the buildings, the P_j parameters are CLO and MET. Depending on the admissible ranges for PMV and PPD, three kinds of comfort zones or categories of thermal requirements are defined by UNE-EN ISO 7730 as: category I (or class A) (PPD < 6%, i.e., −0.2 < PMV < 0.2), category II (or class B) (PPD < 10%, i.e., −0.5 < PMV < 0.5) and category III (or class C) (PPD < 15%, i.e., −0.7 < PMV < 0.7). The ranges of recommended air temperatures for different types of buildings depending on the previous categories are shown in Table 6 [26]. Based on the results of the SA, we can establish the set-point temperature for the cooling and heating seasons.

Table 6. The range of recommended air temperatures for offices and classrooms, according to ISO 7730.

Type of Building	Activity (W/m²)	Category	Temperature (°C)	
			Summer	Winter
Classrooms		A	24.5 ± 1.0	22.0 ± 1.0
Main Hall	70	B	24.5 ± 1.5	22.0 ± 2.0
Offices		C	24.5 ± 2.5	22.0 ± 3.0
Conferences room				

We use the SA based on the Monte Carlo method, randomly generating a uniform distribution for clo from 0 to 2 and met from 0.8 to 4, since in this range the PMV index lies between −2 and +2, taking into account that this is the range suggested by the standard. A uniform distribution corresponds to the case of a random variable that can only have values between two extremes a and b, so that all the intervals of the same length (within (a, b)) have the equal probability.

The above definition shows that the density function must be the same value for all points within the interval (a, b) (and zero outside the range), i.e.,

$$f_X(x) = \left\{ \frac{1}{b-a} \ si \ \chi \in (a,b) \ y \ 0 \ si \ \chi \notin (a,b) \right\} \tag{6}$$

The results of the SA on the comfort index (PMV and PPD) for the clothing and metabolic rate are shown in Figures 4 and 5, respectively. The thermal category changes according to PMV and PPD values, which in turn depend on the subjective decisions of clo and met. Figure 4 shows that for category A, the clo values vary from 0.46 to 0.66, for B of 0.34 to 0.46 clo and 0.66 to 0.87 clo. Category C is between 0.27 to 0.34 clo and 0.87 to 1.04 clo. For the case of met and PPD, in Figure 5 we show that category A lies between met values from 0.83 to 1.04. For the case of category B, it remains from 0.8 to 0.83 met and 1.04 to 1.16 clo. Finally, category C corresponds to the range from 1.16 to 1.31 met.

An attractive alternative in this type of SA, it is to use a normal distribution, which is one of the theoretical distributions best studied in statistical texts and more in practice, also called Gaussian distribution. Its importance is mainly due to the frequency with the variables other than natural and everyday phenomena, approximately, this distribution.

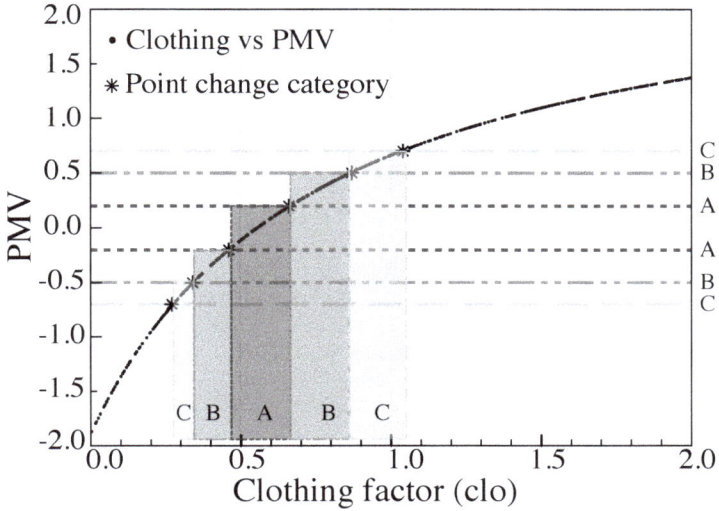

Figure 4. Change of level of thermal comfort in base of sensitivity analysis of clothing factor on PMV.

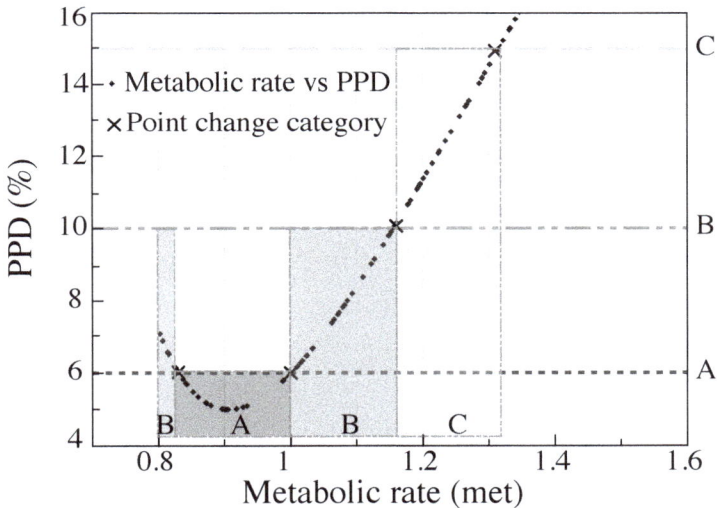

Figure 5. Change of level of thermal comfort in base of sensitivity analysis of metabolic rate on PPD.

The normal distribution is determined by two parameters, their mean and their standard deviation, usually denoted by μ y σ [79]. With this notation, the density of the distribution Gaussian is given by the equation:

$$f(x) = \frac{1}{\sigma\sqrt{2\pi}} exp\left\{\frac{-1}{2}\left(\frac{x-\mu}{\sigma}\right)^2\right\}; \quad -\infty < x < \infty \tag{7}$$

For the case of a normal distribution, we have randomly generated values for clo with $\mu = 1.0$, setting the metabolic factor to 1.2 met and, varying the standard deviation $\sigma_{clo} = 0.025, 0.05, 0.075, 0.1$ and 0.125 . For these values, we obtain the distribution of the PMV and PPD values and, by adjusting to a normal distribution, we obtain the values of μ and σ for the PMV and PPD. Similarly, for the case of the met (metabolic factor) a $\mu = 1.2$ has been established, setting the clothing factor to 1.0 clo and varying the standard deviation $\sigma_{met} = 0.025, 0.05, 0.075, 0.1$ and 0.125.

That is to say, we see easily as variations in the values of clo and met, they are transferred through the model to variations of PMV and PPD, and therefore, we can see the sensitivity of the model to these variations of the clo and met parameters.

That is, we have obtained the relation of the variations of the clothing (clo) and activity (met) with the widths (σ) of the PMV and PPD, with the objective of observe which of the two parameters that the designer enters in a subjective manner, produces a more significant impact on the output of the model.

In the Figure 6a, the relationship between the widths of the distributions of clo and met and the value of σ_{PMV} is presented. For example, for a width of 5 % with respect to the central value of clo $\mu = 1$ and $(\sigma_{clo} = 0.10)$, a value of σ_{PMV} of 0.150 is obtained and for a width of 5 % with respect to the central value $\mu = 1.2$ of met $(\sigma_{met} = 0.10)$, a value of σ_{PMV} of 0.184 is obtained. Therefore, the impact of the metabolic factor (met) is higher than that of the clothing factor (clo) for the PMV case.

In the same way, for each percentage of variation in the values of clo and met we have a value of σ_{PPD}, which is shown in the Figure 6b, for example: for a width of 12.5 % the value of σ_{PPD} is 1.990 for a variation in clo and 2.594 if the parameter met varies. Also, the values predict that the impact of the metabolic factor (met) is greater than the clothing factor (clo) in the PPD.

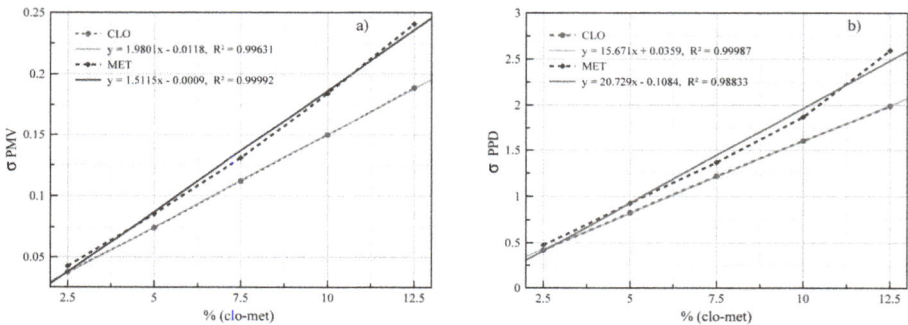

Figure 6. Association between variations of clothing and activity with the widths of (a) σ_{PMV} (b) σ_{PPD}.

5. Results

In this section, we discuss the results of the simulations of both buildings. First, we will show the verification of the monitoring using the data from Equs and the TRNSYS simulation of each building. Then, after a second simulation, we will get the energy demand using the calculated results for the category change, depending on the estimation of the PMV and PPD values as well as the sensitivity analysis of the CLO and MET variables. Finally, we analyze the economic impact of energy demand savings.

As noted before, this work joins BIM-based energy simulations and SA of two non-residential buildings in two seasons (summer and winter) under the international standard ISO 7730. This norm has been studied in previous papers from different perspectives: the effect of personal factors in comfort assessment [80], energy simulation [27], thermal comfort in outdoor urban spaces [81] or the validity of ISO-PMV for predicting comfort votes in everyday thermal environments [82], among others. However, to our knowledge, the effect on the energy demand of the diversity of possibilities offered by ISO 7730 for the choice of clo and met results, depending on the values chosen by the designer, was not previously calculated.

5.1. Real-Time Monitoring and Validation of The Models

Real-time tracking made online with Equs demonstrates the reliability of the measurements that validate the computational simulations. In Figure 7 we show the temperatures measured from 1 January to 31 October 2015 of the air-conditioning zone of the building in Spain, located 43 °18′ north and 8°17′ west. In the first month (January, winter season) the lowest temperature of the period is 12.53 °C registered on Saturday 17. The highest temperature (30.48 °C), was achieved on Monday, 15 June (summer). The mean temperature was 22.43 °C. For the study of the outside temperature, we observe that the cold and hot temperatures are between 2 °C and 41 °C, approximately. The mean temperature outside the building was 15.49 °C.

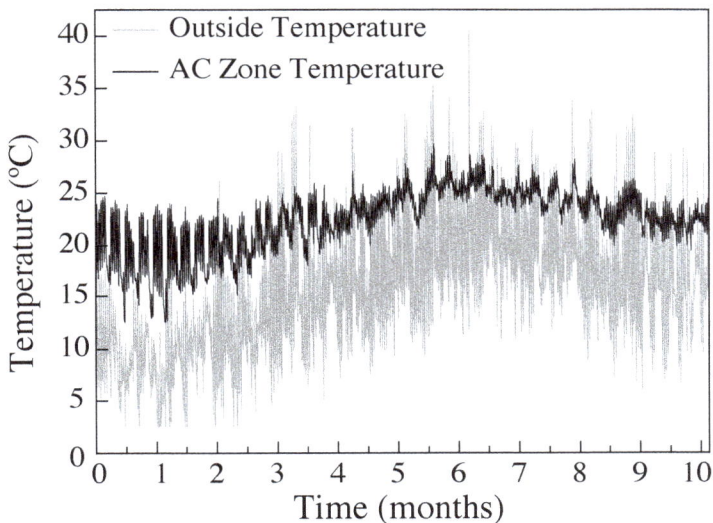

Figure 7. Internal and external temperatures measured by the web platform Equs. In the case of the outside temperature, we observe that the cold and hot temperatures are between 2 °C and 41 °C, approximately. The mean temperature was 15.49 °C.

The comparison between the real data and the simulations is required to have reliable simulations of the buildings. Currently, a simulation of a building's energy consumption is usually considered "calibrated" if the criteria set out by ASHRAE Guideline 14 is met [83]. Accordingly, the model can be considered "calibrated," according to current international criteria to accept BES models, provided that considerable accordance is reached between the measured and simulated data [84]. We must stress that there is no unique model that meets the previous criteria and therefore, several alternative models of the same building can be considered to be "calibrated". Moreover, it should be noted that our calibration is very accurate for the simulated environment (e.g., temperature profiles). As an example, the parallelism among experimental data and the modeling for internal air temperature for the month of June from the hour 3624 to 4344, is shown in Figure 8a. For the period of June, the maximum difference between the simulated and measured temperatures is 1.47 °C, being the mean deviation 0.49 °C. These values have no influence with respect to indoor comfort, showing that the results of our model are basically correct.

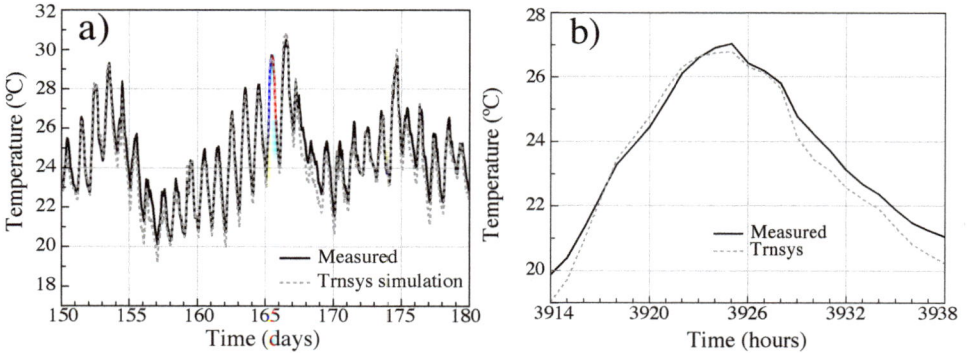

Figure 8. Comparison between the real data obtained by Equs and theoretical simulations results of the first case of study for the month of June (**a**) and a single day of June (**b**).

In Figure 8b we show the temperature of a single day of June (24 h from hour 3914 to 3938 of the simulation). We can observe in more detail the difference between the calculated and the measured temperature for the period, where the minimum variation is of 0.04 °C (obtained in working hours) and the mean difference is of 0.35 °C. The maximum variation is 0.72 °C. These results were obtained within the situation described in the previous sections. In particular, during the occupancy periods, the metabolic rate (met) of the occupiers was selected as to be "seated light work", which corresponds to 70 W/m^2 or 1.2 met. The clothing factor (clo) was established to 0.7 for the summer period and 1.2 for winter. Notice that a clothing with factor 1.2 clo corresponds to: underwear, shirt, pants, thermal jacket, socks, shoes. Summer clothing of 0.7 clo corresponds to underwear, shirt, pants, socks, shoes [26]. Table 5 shows the input that was used for the simulation for the base case scenario. In Figure 9 we can observe the energy demand in the initial configuration, where the set point was established in 23.8 °C.

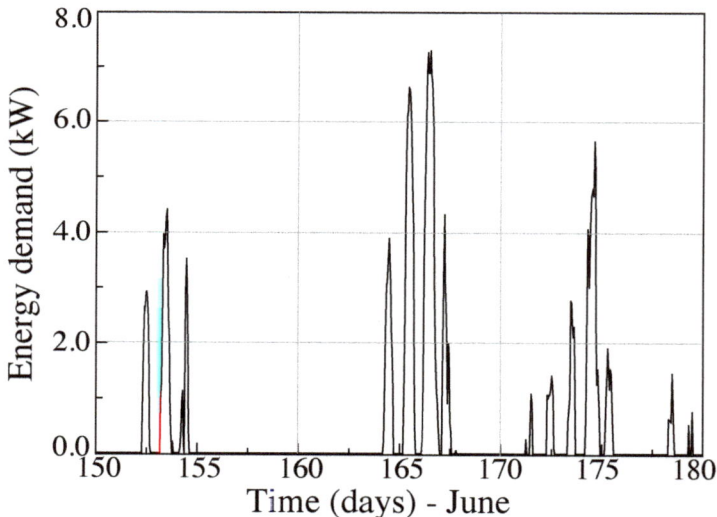

Figure 9. Energy demand of the building in initial configuration of thermal comfort parameters.

5.2. Evaluation of the Variations in the Energy Demand

The main goal of the application of a SA in this work is to visualize in a simpler and more detailed way than in TRNSYS, the impact of subjective decisions made by the designer on the values of insulation of clothes and activity for thermal comfort, which are defined in the TRNSYS simulation program. Based on ISO 7730, the insulation of clothes (clo) can be estimated directly from the combination of clothes or by adding the partial insulation values for each clothes. For example, typical value of clothes insulation for summer (cooling) and winter (heating) are 0.5 and 1 clo, respectively. Therefore, these quantities can be determinant for the different types of clothes that the occupants use and thus, the designers have a greater margin of error when determining by subjective decisions the values of clothing parameters.

Once we have made the variation of clo and met, we obtain variations in the PMV and PPD indices, these values of the indices allow us to define the thermal category based on the definition of the ISO 7730 standard. This category establishes the range of recommended temperatures for these input parameters of comfort. In this permissible temperature range, we can select a value of set-point temperature, to introduce it to our validated BIM model and thus obtain the new energy demand.

For instance: summer light clothing insulation of 0.5 clo includes items such as underpants, short–sleeved shirt, light pants, thin socks, or shoes. In our case, we choose for both cases of study a mean value of 0.7 clo that includes underwear, shirt, pants, socks, shoes, which are also considered light clothing for the same season in this type of buildings. Analogously, the estimation of the value of activity or metabolic rate (met), which represents a heat production depending on the activity level, also depends on the subjective decision made by the designer. According to ISO 7730 for an activity corresponding to this type of buildings such as office, school, and laboratory, the activity parameter can vary from 1.0 to 1.6 met. For this work we assumed a value of 1.2 met for the energy simulations on TRNSYS.

Thus, it is of fundamental importance to determine the impact of the mentioned subjective choices of clothing and metabolic factor, on the values of the PMV and PPD indexes, which show the level of thermal comfort and local thermal comfort criteria, reporting the environmental requirements that are considered acceptable for global thermal comfort, including the local discomfort.

We must notice that a variation in the clo values of summer clothing, from 0.3 to 0.8, causes a change in category from A to C. In Figure 10, the impact on the calculation of energy demand is clearly shown that has the change of environmental category.

According to the SA results and the results of Table 6, we established in TRNSYS the new values for clo and set-point temperature. The first configuration was set for a value of clo of 0.7 with a set-point temperature of 23.5 °C, which belongs to category A. For category B, we used a temperature of 24 °C and clo of 0.5. This configuration was initially defined for the validation of the TRNSYS simulation previously shown. Finally, a set-point temperature of 25 °C for category C and clo of 0.3 was considered. As an example, for the month of June, from Figure 10 for clo values of 0.3, 0.5 and 0.7, we obtain a total energy demand of 347.19 kW, 565.05 kW, and 885.2 kW, respectively.

In Table 7 it can be observed that changes in the energy demand due to variations in the designer's criteria in climates such as Mexico, can reach up to 22% depending on the choice of clo and up to 23% in energy demand, depending on the choice of the activity rate made by the designer. As it can be seen after the analysis of the Monte Carlo method, the BIM calculation is very sensitive to both the appropriate choice of the met and the clo choice. In the case of the clo we have studied variations of daily clothing in summer with clo 0.3 (panties, t-shirt, shorts, thin socks, sandals), clo 0.5 (underpants, t-shirt, light pants, thin socks, shoes), clo 0.7 (underwear, shirt, pants, socks, shoes).

All previous values of clo are assignable to the attire of the occupants of both buildings studied both in the case of Spain and in Mexico, in summer. However, after analyzing it by the Monte Carlo method and calculating the environmental category by the TRNSYS program, we see that in both buildings for a room 0.3 the comfort in the buildings studied is in the category A ranges; for a clo of 0.5 in category B and if the clo is 0.3, in category C (Figure 4). This variation in the category, due to the uncertainty in the choice of the clo on the part of the designer, can have, consequently, alterations in the energy demand.

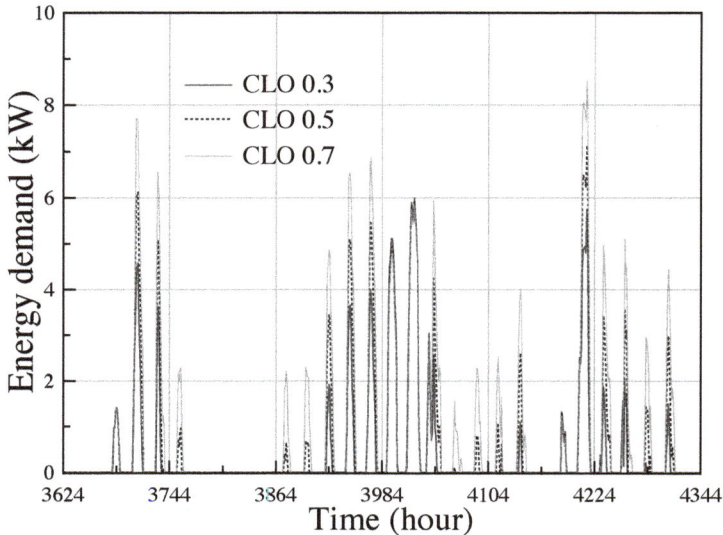

Figure 10. Energy demand for several values of clothing factor.

Table 7. Cumulated energy demand between June and September, with varying values of clo and met, for different set temperatures based on the categories of thermal comfort.

Category (Case)	Set temp	Cumulated Energy Demand (kWh/m²) for clo & met (June to September)					
		0.3 clo	0.5 clo	0.7 clo	0.8 met	1.0 met	1.2 met
A (Spain)	23 °C	46.01	46.32	46.51	48.02	48.20	48.85
B (Spain)	24 °C	44.67	44.69	44.91	46.61	46.72	46.84
C (Spain)	25 °C	43.86	43.93	43.99	45.84	45.89	45.95
A (Mexico)	23 °C	119.53	120.76	122.03	150.35	152.85	155.36
B (Mexico)	24 °C	108.42	109.42	110.44	126.43	128.71	131.02
C (Mexico)	25 °C	99.67	100.44	101.23	106.09	106.85	108.93

As it can be appreciated from the values of Table 7 the change is more than 20%, for the building in Mexico. For the building in Spain, the variation in the calculation of the energy demand, despite having a higher sensitivity, is much lower (only 6%). This discrepancy is due to the different volumes of the two buildings. The higher the volume, the greater the energy demand and therefore the variations of the environmental categories that determine the thermal comfort inside have a much greater impact.

With respect to the met, according to ISO 7730, the cases of a technical office (main activity of the building of Spain) and a University (main activity of the building of Mexico) correspond to met 1 (sitting rest), and met 1.2 (office activity, school, home, laboratory). As it can be appreciated from the values of Table 7, the higher variations in the calculation of energy demand in buildings have been again in the case of Mexico (more than 20%). In the case of the building in Spain the oscillation is only of 6.6%.

6. Conclusions

In the present work, we have analyzed, within the mandatory standard ISO 7730, the influence of the designer's choice of values for human metabolic index and insulation by clothing that can be selected for the estimation of the energy demand of structures. We have demonstrated that the diversity of possibilities offered by ISO 7730 for the choice of clo and met results in significant differences in indoor comfort conditions, leading to non-negligible changes in the calculations of energy consumption.

It has been clearly outlined that the higher the cooling demands, the greater the sensitivity, generating greater differences according to the designer's choice. Since ISO 7730 is an international norm, given the great variety of existing climates, it is vital to take into account this peculiarity of the system at the time of its application and in the choice of the designer, who must be familiar with the procedure. The international norm ISO 7730 is a reference in most of the countries of the world and nevertheless requires of an experienced designer for an appropriate application. Therefore, we have shown in the present work the considerable impact of the designer's subjective choices on the calculation of the energy demand of buildings, as well as the extra costs cumulated, depending on the election of the amounts of activity or clothing, based on the environmental ergonomic standard ISO 7730. We consider that our SA of the parameters mentioned above using the Monte Carlo method will help to define the most appropriate metrics to be admitted, depending on the aim and type of the modeling performed.

We have calculated that the effect of to the different possibilities of election within the ISO 7730 can reach a quarter of the energy demand, showing that this standard is susceptible to the designer's choice of the clo and met. As the ISO 7730 is the only standard which applies to the building regulations in many parts of the world, we conclude that the norm could be revisited from the perspective of energy efficiency, maybe including recommendations from other standards such as EN 15251-2007 or ISO 17772-1-2017, designed explicitly for providing environmental input parameters for design and assessment of energy performance of buildings.

Author Contributions: Conceptualization, R.R.F., M.C.H.-L. and S.Z.; Methodology, M.C.H.-L. and H.M.; Software, R.R.F. and P.F.-d.-C.; Validation, S.Z., P.F.-d.-C. and H.M.; Formal Analysis, M.C.H.-L. and R.R.F.; Investigation, R.R.F., S.Z., A.C.-G. and M.C.H.-L.; Resources, P.F.-d.-C., A.C.-G. and R.S.-A. ; Data Curation, R.R.F., R.S.-A. and M.C.H.-L.; Writing-Original Draft Preparation, S.Z., R.R.F., M.C.H.-L. and H.M.; Writing-Review & Editing, A.C.-G., R.S.-A., H.M., S.Z., M.C.H.-L., R.R.F. and P.F.-d.-C.; Visualization, A.C.-G., M.C.H.-L and R.S.-A.; Supervision, P.F.-d.-C., H.M. and S.Z.; Project Administration, S.Z. and P.F.-d.-C.; Funding Acquisition, P.F.-d.-C., A.C.-G. and R.S.-A.

Funding: This work was partially funded by grants OHMERA MAT2017-86453-R, FIS2017-83762-P and ENE2015-71333-R from MINECO (Spain). R. Robledo and M. Hernández were supported by CONACYT grants 298503 and 296471, respectively. We also thanks to supporting given by the project number INFRA-187906 from the Mexican National Council of Science and Technology-CONACYT.

Acknowledgments: The authors acknowledge Fridama Instalaciones, S.L. and Facultad de Ciencias Físico-Matemáticas, U.A.N.L., for the access to the data used in the present work.

Conflicts of Interest: The authors declare no conflict of interest.

Appendix A

In this appendix section, some tables used to complete the information of the manuscript are shown.

Table A1. Thermal and structural properties in windows.

	Window Types	Glass	Thickness (mm)	% Frame	U (Wm^{-2}K^{-1})	G-Value
Case 1 (Spain)	Type A	double low emissivity glass	4/6/4	15	3.44	0.76
Case 2 (Mexico)	Type 1	Single solar control glass	6	5	5.73	0.482
	Type 2	Single clear glass	6	15	5.73	0.837
	Type 3	Double clear glass	6/12/6	10	3.21	0.722
	Type 4	Single clear glass	6	35	5.73	0.837
	Type 5	Single clear glass	2	5	5.87	0.888
	Type 6	Single solar control glass	6	25	5.73	0.482

Table A2. Distribution of the cases of study.

	Zones	Partitions	Characteristics
Case 1 (Spain)	Without air-conditioning (WAC)	Storage and air chamber of the roof.	Area (253.36 m^2) Volume (1910 m^3) Capacitance (2292.71 kJ/°K)
	Air-conditioned (AC)	2 Offices, 2 meeting rooms, small storage kitchen, rack, access and cleaning room.	Area (218.13 m^2) Volume (610.78 m^3) Capacitance (732.94 kJ/°K)
Case 2 (Mexico)	Zone GF-FF	GF: the reception of the building. FF: labs, offices, toilets, study areas, dining room.	Area (1142.68 m^2) Volume (3366.99 m^3) Capacitance (4040.39 kJ/°K)
	Zone SF	Labs, offices, toilets computer center, cleaning room, meeting room.	Area (1142.68 m^2) Volume (3225.0 m^3) Capacitance (3870.0 kJ/°K)
	Zone SL	Unused open space (Skylight)	Area (84.0 m^2) Volume (462.0 m^3) Capacitance (554.4 kJ/°K)

Table A3. Types of schedules in the buildings.

	Schedule Type		Hours	Use Factor
Case 1 (Spain)	Daily	Daily 1	08:00 to 18:30	1
		Daily 2	13:00 to 16:00	1
		Daily 3	08:00 to 13:00 16:00 to 18:30	1 1
	Weekly	Weekly 1	Monday to Friday	Daily 1
		Weekly 2	Monday to Friday	Daily 2
		Weekly 3	Monday to Friday	Daily 3
Case 2 (Mexico)	Daily	Daily A	09:00 to 21:00	1
		Daily B	07:00 to 21:00	1
		Daily C	10:00 to 14:00	1
	Weekly	Weekly A	Monday to Friday Saturday	Daily A Daily C
		Weekly B	Monday to Sunday	Daily B

Table A4. Infiltrations and ventilation for each zone in summer and winter.

Case of Study	Zones	Infiltrations Air Change per Hour		Ventilation Air Change per Hour
		Summer	Winter	
Case 1: Spain	WAC	0.143	0.031	5.00
	AC	0.066	0.067	1.65
Case 2: Mexico	GF-FF	3.39	1.80	8.00
	SF	3.41	1.81	8.00
	SL	25.27	13.41	10.0

References

1. Hemsath, T.L.; Bandhosseini, K.A. Sensitivity analysis evaluating basic building geometry's effect on energy use. *Renew. Energy* **2015**, *76*, 526–538. [CrossRef]
2. Griego, D.; Krarti, M.; Hernandez-Guerrero, A. Energy efficiency optimization of new and existing office buildings in Guanajuato, Mexico. *Sustain. Cities Soc.* **2015**, *17*, 132–140. [CrossRef]
3. Lin, H.W.; Hong, T. On variations of space-heating energy use in office buildings. *Appl. Energy* **2013**, *111*, 515–528. [CrossRef]
4. Pikas, E.; Thalfeldt, M.; Kurnitski, J. Cost optimal and nearly zero energy building solutions for office buildings. *Energy Build.* **2014**, *74*, 30–42. [CrossRef]
5. Terés-Zubiaga, J.; Campos-Celador, A.; González-Pino, I.; Escudero-Revilla, C. Energy and economic assessment of the envelope retrofitting in residential buildings in Northern Spain. *Energy Build.* **2015**, *86*, 194–202. [CrossRef]
6. Lee, J.; Kim, J.; Song, D.; Kim, J.; Jang, C. Impact of external insulation and internal thermal density upon energy consumption of buildings in a temperate climate with four distinct seasons. *Renew. Sustain. Energy Rev.* **2017**, *75*, 1081–1088. [CrossRef]
7. IEA. *World Energy Outlook 2017*; Organisation for Economic Co-Operation and Development, OECD: Paris, France, 2017.
8. Anderson, J.E.; Wulfhorst, G.; Lang, W. Energy analysis of the built environment—A review and outlook. *Renew. Sustain. Energy Rev.* **2015**, *44*, 149–158. [CrossRef]
9. Abdelaziz, E.; Saidur, R.; Mekhilef, S. A review on energy saving strategies in industrial sector. *Renew. Sustain. Energy Rev.* **2011**, *15*, 150–168. [CrossRef]
10. Nejat, P.; Jomehzadeh, F.; Taheri, M.M.; Gohari, M.; Majid, M.Z.A. A global review of energy consumption, CO_2 emissions and policy in the residential sector (with an overview of the top ten CO_2 emitting countries). *Renew. Sustain. Energy Rev.* **2015**, *43*, 843–862. [CrossRef]
11. Balaras, C.A.; Droutsa, K.; Dascalaki, E.; Kontoyiannidis, S. Heating energy consumption and resulting environmental impact of European apartment buildings. *Energy Build.* **2005**, *37*, 429–442. [CrossRef]
12. Pérez-Lombard, L.; Ortiz, J.; Pout, C. A review on buildings energy consumption information. *Energy Build.* **2008**, *40*, 394–398. [CrossRef]
13. Galvin, R. Thermal upgrades of existing homes in Germany: The building code, subsidies, and economic efficiency. *Energy Build.* **2010**, *42*, 834–844. [CrossRef]
14. Creyts, J.C. Reducing US Greenhouse Gas Emissions: How Much at What Cost?: US Greenhouse Gas Abatement Mapping Initiative. *McKinsey & Co.* 2007. Available online: https://www.mckinsey.com/business-functions/sustainability/our-insights/reducing-us-greenhouse-gas-emissions (accessed on 1 March 2018).
15. Chappells, H.; Shove, E. Debating the future of comfort: Environmental sustainability, energy consumption and the indoor environment. *Build. Res. Inf.* **2005**, *33*, 32–40. [CrossRef]
16. Geva, A.; Saaroni, H.; Morris, J. Measurements and simulations of thermal comfort: A synagogue in Tel Aviv, Israel. *J. Build. Perform. Simul.* **2014**, *7*, 233–250. [CrossRef]
17. Nguyen, A.T.; Reiter, S. Passive designs and strategies for low-cost housing using simulation-based optimization and different thermal comfort criteria. *J. Build. Perform. Simul.* **2014**, *7*, 68–81. [CrossRef]

18. Rey Martínez, F.J.; Chicote, M.A.; Peñalver, A.V.; Gónzalez, A.T.; Gómez, E.V. Indoor air quality and thermal comfort evaluation in a Spanish modern low-energy office with thermally activated building systems. *Sci. Technol. Built Environ.* **2015**, *21*, 1091–1099. [CrossRef]

19. Wardiningsih, W.; Troynikov, O. Force attenuation capacity and thermophysiological wear comfort of vertically lapped nonwoven fabric. *J. Text. Inst.* **2017**, *109*, 1–9. [CrossRef]

20. Ashrae, A.S. *Standard 62-1989, Ventilation for Acceptable Indoor Air Quality, Atlanta, GA*; American Society of Heating, Refrigerating, and Air Conditioning Engineers, Inc.: New York, NY, USA, 1989.

21. MacArthur, J.; Arens, E.; Gonzalez, R.; Berglund, L.; Spain, S.; Madsen, T.; Oleson, B.; Reid, K. HVAC is for people. *Ashrae Trans.* **1986**, *92*, 5–64.

22. Oropeza-Perez, I.; Petzold-Rodriguez, A.H.; Bonilla-Lopez, C. Adaptive thermal comfort in the main Mexican climate conditions with and without passive cooling. *Energy Build.* **2017**, *145*, 251–258. [CrossRef]

23. *Standard 55 Thermal Environmental Conditions For Human Occupancy*; ASHRAE: New York, NY, USA, 2010.

24. CEN. *15251-Criteria for the Indoor Environment, Including Thermal, Indoor Air Quality (Ventilation), Light And Noise, 2006*; CEN: Brussels, Belgium, 2006.

25. Matzarakis, A.; Mayer, H.; Iziomon, M.G. Applications of a universal thermal index: Physiological equivalent temperature. *Int. J. Biometeorol.* **1999**, *43*, 76–84. [CrossRef]

26. *Standard 7730. Ergonomics of the Thermal Environment—Analytical Determination and Interpretation of Thermal Comfort Using Calculation of the Pmv and Ppd Indices and Local Thermal Comfort Criteria*; International Organization for Standardization: Geneva, Switzerland, 2005.

27. Peeters, L.; De Dear, R.; Hensen, J.; D'haeseleer, W. Thermal comfort in residential buildings: Comfort values and scales for building energy simulation. *Appl. Energy* **2009**, *86*, 772–780. [CrossRef]

28. Ahmed, K.; Akhondzada, A.; Kurnitski, J.; Olesen, B. Occupancy schedules for energy simulation in new prEN16798-1 and ISO/FDIS 17772-1 standards. *Sustain. Cities Soc.* **2017**, *35*, 134–144. [CrossRef]

29. Antoniadou, P.; Papadopoulos, A.M. Occupants' thermal comfort: State of the art and the prospects of personalized assessment in office buildings. *Energy Build.* **2017**, *153*, 136–149. [CrossRef]

30. Oropeza-Perez, I.; Østergaard, P.A. Potential of natural ventilation in temperate countries–a case study of Denmark. *Appl. Energy* **2014**, *114*, 520–530. [CrossRef]

31. Zhang, L.; Zhang, L.; Wang, Y. Shape optimization of free-form buildings based on solar radiation gain and space efficiency using a multi-objective genetic algorithm in the severe cold zones of China. *Sol. Energy* **2016**, *132*, 38–50. [CrossRef]

32. Lei, J.; Yang, J.; Yang, E.H. Energy performance of building envelopes integrated with phase change materials for cooling load reduction in tropical Singapore. *Appl. Energy* **2016**, *162*, 207–217. [CrossRef]

33. Chen, C.W.; Lee, C.W.; Lin, Y.W. Air Conditioning—Optimizing Performance by Reducing Energy Consumption. *Energy Environ.* **2014**, *25*, 1019–1024. [CrossRef]

34. Sivak, M. Potential energy demand for cooling in the 50 largest metropolitan areas of the world: Implications for developing countries. *Energy Policy* **2009**, *37*, 1382–1384. [CrossRef]

35. Attia, S.; Hensen, J.L.; Beltrán, L.; De Herde, A. Selection criteria for building performance simulation tools: contrasting architects' and engineers' needs. *J. Build. Perform. Simul.* **2012**, *5*, 155–169. [CrossRef]

36. Crawley, D.B.; Lawrie, L.K.; Winkelmann, F.C.; Buhl, W.F.; Huang, Y.J.; Pedersen, C.O.; Strand, R.K.; Liesen, R.J.; Fisher, D.E.; Witte, M.J.; et al. EnergyPlus: Creating a new-generation building energy simulation program. *Energy Build.* **2001**, *33–34*, 319–331. [CrossRef]

37. Attia, S.; De Herde, A. Early design simulation tools for net zero energy buildings: A comparison of ten tools. In Proceedings of the Conference 12th International Building Performance Simulation Association, Sydney, Australia, 14–16 November 2011.

38. Klein, S.A. *TRNSYS-A Transient System Simulation Program*; University of Wisconsin-Madison, Engineering Experiment Station Report; University of Wisconsin-Madison: Madison, WI, USA, 1988; pp. 38–12.

39. Newsham, G.R. Clothing as a thermal comfort moderator and the effect on energy consumption. *Energy Build.* **1997**, *26*, 283–291. [CrossRef]

40. Hensen, J.L.; Lamberts, R. *Introduction to Building Performance Simulation*; Building Performance Simulation for Design and Operation: London, UK, 2011; pp. 365–401.

41. Schiavon, S.; Lee, K.H. Dynamic predictive clothing insulation models based on outdoor air and indoor operative temperatures. *Build. Environ.* **2013**, *59*, 250–260. [CrossRef]

42. Lee, Y.S.; Malkawi, A.M. Simulating multiple occupant behaviors in buildings: An agent-based modeling approach. *Energy Build.* **2014**, *69*, 407–416. [CrossRef]

43. Kang, D.H.; Mo, P.H.; Choi, D.H.; Song, S.Y.; Yeo, M.S.; Kim, K.W. Effect of MRT variation on the energy consumption in a PMV-controlled office. *Build. Environ.* **2010**, *45*, 1914–1922. [CrossRef]

44. Luo, M.; Cao, B.; Zhou, X.; Li, M.; Zhang, J.; Ouyang, Q.; Zhu, Y. Can personal control influence human thermal comfort? A field study in residential buildings in China in winter. *Energy Build.* **2014**, *72*, 411–418. [CrossRef]

45. De Dear, R.; Brager, G.S. *Developing an Adaptive Model of Thermal Comfort And Preference*; UC Berkeley: Berkeley, CA, USA, 1998.

46. De Dear, R.J. A global database of thermal comfort field experiments. *ASHRAE Trans.* **1998**, *104*, 1141.

47. Manu, S.; Shukla, Y.; Rawal, R.; Thomas, L.E.; de Dear, R. Field studies of thermal comfort across multiple climate zones for the subcontinent: India Model for Adaptive Comfort (IMAC). *Build. Environ.* **2016**, *98*, 55–70. [CrossRef]

48. Hwang, R.L.; Shu, S.Y. Building envelope regulations on thermal comfort in glass facade buildings and energy-saving potential for PMV-based comfort control. *Build. Environ.* **2011**, *46*, 824–834. [CrossRef]

49. Ioannou, A.; Itard, L. Energy performance and comfort in residential buildings: Sensitivity for building parameters and occupancy. *Energy Build.* **2015**, *92*, 216–233. [CrossRef]

50. Hong, T.; Taylor-Lange, S.C.; D'Oca, S.; Yan, D.; Corgnati, S.P. Advances in research and applications of energy-related occupant behavior in buildings. *Energy Build.* **2016**, *116*, 694–702. [CrossRef]

51. Yan, D.; O'Brien, W.; Hong, T.; Feng, X.; Gunay, H.B.; Tahmasebi, F.; Mahdavi, A. Occupant behavior modeling for building performance simulation: Current state and future challenges. *Energy Build.* **2015**, *107*, 264–278. [CrossRef]

52. Chen, Y.; Liang, X.; Hong, T.; Luo, X. Simulation and visualization of energy-related occupant behavior in office buildings. *Build. Simul.* **2017**, *10*, 785–798. [CrossRef]

53. Putra, H.C.; Andrews, C.J.; Senick, J.A. An agent-based model of building occupant behavior during load shedding. *Build. Simul.* **2017**, *10*, 845–859. [CrossRef]

54. Thomas, A.; Menassa, C.C.; Kamat, V.R. Lightweight and adaptive building simulation (LABS) framework for integrated building energy and thermal comfort analysis. *Build. Simul.* **2017**, *10*, 1023–1044. [CrossRef]

55. Lindner, A.J.; Park, S.; Mitterhofer, M. Determination of requirements on occupant behavior models for the use in building performance simulations. *Build. Simul.* **2017**, *10*, 861–874. [CrossRef]

56. Laurent, J.G.C.; Samuelson, H.W.; Chen, Y. The impact of window opening and other occupant behavior on simulated energy performance in residence halls. *Build. Simul.* **2017**, *10*, 963–976. [CrossRef]

57. Kuznik, F.; Virgone, J.; Johannes, K. Development and validation of a new TRNSYS type for the simulation of external building walls containing PCM. *Energy Build.* **2010**, *42*, 1004–1009. [CrossRef]

58. Klein, S.; Beckman, W.; Mitchell, J.; Duffie, J.; Duffie, N.; Freeman, T.; Mitchell, J.; Braun, J.; Evans, B.; Kummer, J.; et al. *TRNSYS 16—A TRaNsient System Simulation Program, User Manual*; Solar Energy Laboratory, University of Wisconsin-Madison: Madison, WI, USA, 2004.

59. Salvalai, G.; Pfafferott, J.; Sesana, M.M. Assessing energy and thermal comfort of different low-energy cooling concepts for non-residential buildings. *Energy Convers. Manag.* **2013**, *76*, 332–341. [CrossRef]

60. Lebon, M.; Fellouah, H.; Galanis, N.; Limane, A.; Guerfala, N. Numerical analysis and field measurements of the airflow patterns and thermal comfort in an indoor swimming pool: A case study. *Energy Effic.* **2017**, *10*, 527–548. [CrossRef]

61. Zhang, S.; Jiang, Y.; Xu, W.; Li, H.; Yu, Z. Operating performance in cooling mode of a ground source heat pump of a nearly-zero energy building in the cold region of China. *Renew. Energy* **2016**, *87*, 1045–1052. [CrossRef]

62. Rodríguez, G.C.; Andrés, A.C.; Muñoz, F.D.; López, J.M.C.; Zhang, Y. Uncertainties and sensitivity analysis in building energy simulation using macroparameters. *Energy Build.* **2013**, *67*, 79–87. [CrossRef]

63. Basinska, M.; Koczyk, H.; Szczechowiak, E. Sensitivity analysis in determining the optimum energy for residential buildings in Polish conditions. *Energy Build.* **2015**, *107*, 307–318. [CrossRef]

64. Tian, W. A review of sensitivity analysis methods in building energy analysis. *Renew. Sustain. Energy Rev.* **2013**, *20*, 411–419. [CrossRef]

65. Lomas, K.J.; Eppel, H. Sensitivity analysis techniques for building thermal simulation programs. *Energy Build.* **1992**, *19*, 21–44. [CrossRef]

66. Breesch, H.; Janssens, A. Performance evaluation of passive cooling in office buildings based on uncertainty and sensitivity analysis. *Sol. Energy* **2010**, *84*, 1453–1467. [CrossRef]

67. Ruiz Flores, R.; Bertagnolio, S.; Lemort, V. Global sensitivity analysis applied to total energy use in buildings. In Proceedings of the 2nd International High Performance Buildings Conference, Purdue, West Lafayette, IN, USA, 16–19 July 2012.

68. Peel, M.C.; Finlayson, B.L.; McMahon, T.A. Updated world map of the Köppen-Geiger climate classification. *Hydrol. Earth Syst. Sci. Discuss.* **2007**, *4*, 439–473. [CrossRef]

69. Balvís, E.; Sampedro, Ó.; Zaragoza, S.; Paredes, A.; Michinel, H. A simple model for automatic analysis and diagnosis of environmental thermal comfort in energy efficient buildings. *Appl. Energy* **2016**, *177*, 60–70. [CrossRef]

70. Remund, J. Meteonorm: Irradiation data for every place on Earth. Bern2014: Switzerlan. 2014. Available online: https://meteonorm.com (accessed on 20 September 2018).

71. Fanger, P.O. Thermal environment—Human requirements. *Environmentalist* **1986**, *6*, 275–278. [CrossRef]

72. Hasan, M.H.; Alsaleem, F.; Rafaie, M. Sensitivity study for the PMV thermal comfort model and the use of wearable devices biometric data for metabolic rate estimation. *Build. Environ.* **2016**, *110*, 173–183. [CrossRef]

73. Fanger, P.O. Thermal comfort. Analysis and applications in environmental engineering. In *Thermal Comfort. Analysis and Applications in Environmental Engineering*; Danish Technical Press: Copenhagen, Denmark, 1970.

74. Madsen, T. *Description of Thermal Manikin for Measuring Thermal Insulation Values of Clothing*; Thermal Insulation Report; Technical University of Denmark: Lyngby, Denmark, 1976.

75. DIN. *13779: Ventilation For Non-Residential Buildings-Performance Requirements for Ventilation and Room-Conditioning Systems*; DIN: Berlin, Germany, 2007.

76. Ascione, F.; Bianco, N.; De Stasio, C.; Mauro, G.M.; Vanoli, G.P. Multi-stage and multi-objective optimization for energy retrofitting a developed hospital reference building: A new approach to assess cost-optimality. *Appl. Energy* **2016**, *174*, 37–68. [CrossRef]

77. Echenagucia, T.M.; Capozzoli, A.; Cascone, Y.; Sassone, M. The early design stage of a building envelope: Multi-objective search through heating, cooling and lighting energy performance analysis. *Appl. Energy* **2015**, *154*, 577–591. [CrossRef]

78. Lam, J.C.; Hui, S.C. Sensitivity analysis of energy performance of office buildings. *Build. Environ.* **1996**, *31*, 27–39. [CrossRef]

79. Pértegas Díaz, S.; Pita Fernández, S. La distribución normal. *Cad Aten Primaria* **2001**, *8*, 268–274.

80. Havenith, G.; Holmér, I.; Parsons, K. Personal factors in thermal comfort assessment: Clothing properties and metabolic heat production. *Energy Build.* **2002**, *34*, 581–591. [CrossRef]

81. Nikolopoulou, M.; Baker, N.; Steemers, K. Thermal comfort in outdoor urban spaces: Understanding the human parameter. *Sol. Energy* **2001**, *70*, 227–235. [CrossRef]

82. Humphreys, M.A.; Nicol, J.F. The validity of ISO-PMV for predicting comfort votes in every-day thermal environments. *Energy Build.* **2002**, *34*, 667–684. [CrossRef]

83. ASHRAE. Guideline 14-2002. *Meas. Energy Demand Sav.* **2002**, *22*, 32–43.

84. Coakley, D.; Raftery, P.; Keane, M. A review of methods to match building energy simulation models to measured data. *Renew. Sustain. Energy Rev.* **2014**, *37*, 123–141. [CrossRef]

energies

MDPI

Article

A Study of the Effects of Enhanced Uniformity Control of Greenhouse Environment Variables on Crop Growth

Chan Kyu Lee [1], Mo Chung [2,*], Ki-Yeol Shin [2], Yong-Hoon Im [3,*] and Si-Won Yoon [3]

[1] Department of Mechanical Engineering, Graduate School of Yeungnam University, 280, Daehak-Ro, Gyeongsan 38541, Korea; cklee@kier.re.kr
[2] Department of Mechanical Engineering, Yeungnam University, 280, Daehak-Ro, Gyeongsan 38541, Korea; shinky@yu.ac.kr
[3] Korea Institute of Energy Research, 152, Gajeong-ro, Yuseong-gu, Daejeon 34129, Korea; siwon@kier.re.kr
* Correspondence: mchung@yu.ac.kr (M.C.); iyh@kier.re.kr (Y.-H.I.); Tel.: +82-53-810-2459 (M.C.); +82-860-3327 (Y.-H.I.)

Received: 4 April 2019; Accepted: 6 May 2019; Published: 9 May 2019

Abstract: In order to ensure high crop yield and good quality in greenhouse horticulture, the major environment control variables, such as temperature, humidity, and CO_2 concentration, etc., need to be controlled properly, in order to reduce harmful effects on crop growth by minimizing the fluctuation of the thermal condition. Even though a hot water-based heating system is evidently superior to a hot air-based heating system, in terms of the thermally stable condition or energy saving, a hot air-based heating system has occupied the domestic market due to its economic efficiency from an initial investment cost saving. However, the intrinsic drawbacks of a hot air-based heating system, being more frequent variation of thermal variables and an inordinate disturbance on crops due to its convective heat delivery nature, are believed to be the main reasons for the insufficient crop yield and/or the quality deterioration. In addition, the current thermal environment monitoring system in a greenhouse, in which a sole sensor node usually covers a large part of cultivating area, seems to have a profound need of improvement in order to resolve those problems, in that the assumption of thermal uniform condition, which is adequate for a sole sensor node system, cannot be ensured in some cases. In this study, the qualitative concept of the new control variable—the degree of uniformity—is suggested as an indicator to seek ways of enhancing the crop yield and its quality based on the multiple sensor nodes system with a wireless sensor network. In contrast to a conventional monitoring system, for which a newly suggested concept of qualitative variable cannot be estimated at all, the multiple sensor nodes-based thermal monitoring system can provide more accurate and precise sensing, which enables the degree of uniformity to be checked in real-time and thus more precise control becomes possible as a consequence. From the analysis of the results of the experiment and simulation, it is found that the crops in plastic vinyl houses can be exposed to a serious level of non-uniform thermal condition. For instance, the temperature difference in the longitudinal and widthwise direction is 3.0 °C and 6.5 °C, respectively for the case of 75 × 8 m dimension greenhouse during a typical winter season, and it can be hypothesized that this level of non-uniformity might cause considerable damage to crop growth. In this paper, several variants of control systems, within the framework of the multiple sensor nodes system, is proposed to provide a more thermally-stable cultivating environment and the experimental verification is carried out for different scales of test greenhouses. The results showed that a simple change of heating mode (i.e., from a hot air- to a hot water-based heating system) can bring about a significant improvement for the non-uniformity of temperature (more or less 80%), and an additional countermeasure, with local heat flux control, can lead to a supplementary cut of non-uniformity up to 90%. Among the several variants of local heat flux control systems, the hydraulic proportional mass flow control valve system was proven to represent the best performance, and it can be hypothesized that the newly suggested qualitative variable—the degree of uniformity—with the multiple sensor nodes system can be a good alternative for seeking enhanced cultivating performance, being higher crop yield and better quality along with energy cost saving.

Keywords: greenhouse; indoor temperature uniformity; multiple sensor nodes; qualitative control

---

1. Introduction

As the effects of climate change are becoming persistently serious, it is reported that we will need 50% more energy, 40% more water, and 35% more food, which is in reference to the statistics that a population of about 8.3 billion will be attained by 2030 [1,2] and that the food system will consume more or less 30% of the final energy use, along with the current trend of natural resource exploitation and the corresponding greenhouse gas (GHG) emission increase by the same rate [3]. However, overcoming the shortage of water and securing a fat supply of food will come down to an energy problem, since production of fresh water and food also require a substantial amount of energy; and the consequent additional production of heat and power with fossil fuels to cope with the aforementioned energy problems can lead to the extra emission of greenhouse gases, thus making the situation of climate change worse. In the long run, the conventional mass production-based outdoor culture becomes inappropriate to deal with the huge risk of the security of food supply caused by climate change, and a transition from the conventional outdoor culture into greenhouse horticulture, in the form of smart or sustainable farms, is needed for ensuring food security [4–6].

As greenhouse horticulture is getting more attention as one of the appropriate measures to resolve food security problems in the era of climate change, the market is expected to expand rapidly around the globe at an impressive CAGR (compound annual growth rate) of approximately 19% by between 2017 and 2022 [7,8]. This can be understood as being a distinctive indicator of the relevant market reflecting the awareness about the crisis created by climate change, and, furthermore, that greenhouse horticulture is the right endeavor to overcome the potential risk of reliable food supply capability. In South Korea, it is also not difficult to see the effects of climate change in the cultivation environment, not to mention in ordinary life, in that the suitable region for crop growth, or fruits, has been rapidly changing according to the change of the annual average temperature in the region, and tropical foods even are starting to substitute the conventional ones out of necessity. In addition, greenhouse horticulture, including the concept of smart farming, is rapidly expanding in the market, with the recognition of the potential threat in terms of food supply security as mentioned above [9]; however, small-scale versions of greenhouse the horticulture model, like a plastic vinyl house, still hold a 99% majority in the market [10–12].

Although a large-scale greenhouse with a glass skin is expected to enjoy the advantages of economy of scale, to some extent, the strong market share of the traditional small-scale plastic vinyl house model will be managed for the time being, due to the advantage of the relatively cheap facility capital costs and public acceptability in the agricultural area. It is interesting to note that the majority of the greenhouse horticulture market is still comprised of the plastic vinyl house type model (99%), whereas the glass greenhouse type model has been stagnant or has only had some increase in their market share, but remain within 1% at the moment in South Korea [12–14].

Along with the rapid annual growth rate of greenhouse horticulture in the era of climate change, the relevant market for the multi-variable control of the environment system for greenhouse horticulture is expected to repeat its high growth rate, and one can notice, without difficulty, that slightly different features have been manifested in each product, of the various cultivating control systems, in the market of greenhouse horticulture. However, what all systems have in common is the quantitative control of the variables of temperature, humidity, CO_2 concentration, and light intensity, via auxiliary light source, etc., in order to provide comfortable growth conditions for the crops inside the greenhouse, and, so far, a quite satisfactory accomplishment in terms of improved productivity and enhanced quality has been achieved [15–19]. However, there is still large room for improvement, especially to cope with increasingly harsh environments from climate change.

In conventional greenhouse environment control systems, regardless of the types of greenhouse or the scale of it, the energy cost represents the major share of total operating costs. For example, at present in South Korea, it usually forms about 30% to 35% among the whole operating cost annual, in the case of plastic vinyl houses, and a little bit higher portion, about 40%, in the case of a large-scale glass-type greenhouse [20–22].

In addition, a variety of researches and developments for improving crop yield and its quality, or reducing the energy consumption, have been carried out consistently to ensure cost competitiveness in the market. The optimal value of the thermal variables for successful greenhouse cultivation of tomato has been reviewed [23], and it has been pointed out that sustainable greenhouse production requires the integration of information and management strategies, as well as excellent understanding of the influencing microclimate parameters, which does not seem to be accomplished properly with the current sensing platform of a limited number of sensor nodes to cover a wide range of cultivating areas. Various types of controller techniques, such as proportional integral (PI) control, fuzzy logic control, artificial neural network control, and adaptive neuro-fuzzy control have also been developed and tested in the field [24–27]. Numerous simulation models and tools for a greenhouse such as TRNSYS, HORTICERN, MICGREEN, and HORTITRANS have been developed to describe and analyze the microclimate of a greenhouse in which a static, dynamic, or intermediate model are included [28]. For example, greenhouse building energy simulation and relevant sub-model development have been extensively carried out to estimate the energy load and corresponding cost management using a versatile dynamic building energy simulation software, of TRNSYS, where the annual time-varying energy load prediction for a greenhouse, the renewable energy resources, and various energy saving techniques can be simulated for single- or multi-zone buildings [29–31]. Furthermore, a computational fluid dynamics technique, for which a detailed spatial distribution of the fluid dynamic and thermal variables are provided, is being applied to greenhouse microclimate simulation to increase understanding of the detailed mechanisms, causes, or effects of the applied systems [32,33].

It is also quite surprising to recognize that it occurs even though the lowest electricity price rate is applied to the energy facilities, which is operated by using electricity such as a heat pump, electricity boiler, etc., in the agricultural sector in South Korea. It can be supposed that, in some ways, such a low rate electricity price system could lead to the thoughtless waste of energy, or, on the other hand, that there is a large opportunity for energy consumption reduction by adopting the proper energy efficient measures that are commonly used in other sectors, such as residential or industrial sectors. One thing we know it will do is that the energy cost will continue its upward movement in the forthcoming years of climate change, because the persistent pressure to reduce GHG emission at the national scale, in order to implement the Paris Agreement on climate change, will be reflected in the prices of energy, in the form of taxation, etc., to achieve the final goals of GHG emission reduction in each sector. The portion of electricity usage in the agricultural area of South Korea is about 4% [34] at present, but it is indicated that the transition from fossil fuel to electricity to produce the energy for cultivation has been rapidly realized recently. Moreover, the rapid expansion of the greenhouse horticulture market along with the new concept of smart farming, which is estimated to grow at an impressive compound annual growth rate (CAGR) of approximately 19% by 2020 [7], will certainly work to change or withdraw the current policy of the low rate electricity price policy in the agricultural sector in due time. In that sense, the development of the proper measures or solutions to enable the reduction of energy consumption in operating the greenhouse has become increasingly critical, in order to survive in the forthcoming era of climate change and to secure sustainable business circumstances, and, subsequently, to have a connected ring between energy savings and crop productivity improvement by providing more comfortable thermal conditions.

In this study, a qualitative concept of a new control variable—the degree of uniformity—is proposed as a main control variable for seeking ways to save energy, to enhance the productivity and the quality of crops simultaneously. In order to accomplish the above goals, an increased number of sensors are installed, based on the wireless sensor network, to monitor the status of uniformity more accurately compared to the conventional monitoring system, whereby a sole sensor node is used for wide range of cultivating areas. It was fully verified that the hot water system is superior to the hot air supplying system (which is dominant in the market due to its low level of initial capital cost and ease of control), in that it is more stable, has little fluctuation of thermal variables, and provides comfortable thermal conditions, due to the natural convection or radiative heat transfer mode that can be attained via the hot water pipe system. From the elaborate analysis of the experiment data, it was deduced that the simple change of heating mode, from the convective heat transfer to the radiative one, provides a serious enhancement of the degree of uniformity inside a greenhouse, about more than 80%. Additionally, three systems of local heat flux variation control measures were tested, along with the precise sensor nodes network of multiple wireless sensors based on the hot water system. Among the counter measures of local heat flux variation, based on the hot water-based heating system, the one with hydraulic proportional mass flow control valve showed the best performance and was evidenced to be beneficial by managing the degree of uniformity in many ways. Furthermore, the newly suggested qualitative variable of the degree of uniformity was believed to be a good indicator, and deserved to be a good control variable to provide a better greenhouse thermal environment for ensuring the enhanced productivity and quality of the crops. This new measure to exert active control over the local heat flux, with the hydraulic proportional mass flow control valve, was able to aid a supplementary reduction of non-uniformity up to 90% effectively. In the following, the experiment conditions and corresponding control system configurations are described in detail and various technical aspects of enhanced degree of uniformity, for the main environment variables on energy saving and the productivity improvement, is assessed in detail.

2. Experimental Verification

2.1. Specification of test Greenhouses and Control Measures

In this study, the experimental verification for the newly proposed control concept, the degree of uniformity, and the several corresponding control measures based on the hot water-based heating system were carried out for two different scales of test greenhouses. In order to seek the best performing measure for thermal environment control, several variants of control measures were implemented and tested for each greenhouse. The several variants of adopted control measures based on hot water supply are summarized in Table 1, and the different dynamic features for the applied measures are described in the following. The experimental verification for the all variants of control measures were not carried out in the same test beds of a greenhouse due to several reasons, and the variant of pump on–off by the beds, which was designed to seek a degree of freedom on heat supply by a bed in contrast to the simultaneous supply to all beds, was applied on both test greenhouse. The dimension of the first greenhouse test bed (greenhouse A) was 70×8.3 m of area and the inside volume was about 2146 m^3. The other experiment greenhouse test bed (greenhouse B) had the dimensions of heating areas of 25×6 m, and volume of 400 m^3, which was relatively smaller to the first one. The thermal and air conditioning environment of the greenhouses were controlled mainly with the side window and skylight operation, and an auxiliary roof fan and hot air-based heating system, etc., was equipped, as used for an original greenhouse. The hot water-based heating system was newly installed for this experimental verification and somewhat different heat supplying configurations were adopted for each test greenhouse in terms of hot water temperature management, which is to be mentioned briefly later.

Figures 1 and 2 show the view of both greenhouses and one can easily recognize the difference of scales between each test greenhouse. The greenhouse for Case 1 is located in the coastal area of the east sea with the values for longitude and latitude of 37°50′56.2″ N 128°51′10.5″ E and the annual average temperature was about 13.1 °C and the highest and lowest annual average temperature was 17.5 and 9.2 °C. The greenhouse for Case 2 is located in the central part of the Korean peninsula with the values for longitude and latitude of 36°34′19.1″ N 127°19′16.9″ E. The annual average temperature was 13 °C, but the highest and lowest one is 25.6 °C in August and –1 °C in January and the orientation is shown in Figures 1 and 2, respectively. It is necessary to note that the experiment verification for the former greenhouse was performed in real cultivating operation conditions, but the latter one was done in test operating conditions. The strawberry was grown in the former greenhouse and the indoor temperature is managed to be kept more or less 10 °C on average, not less than 4~5 °C in the worst case during winter, to prevent abrupt deterioration of crop quality, not to mention of its productivity [35,36]. In the conventional heating system, a hot air-based heating system, automatic heating fan control, and the on-and-off control algorithm had been in operation based on a sensor, which is located in the center of the greenhouse. In the newly installed hot water-based heating system, the oil boiler-based heat accumulator was installed so as to supply hot water via the main pipes of the supply and return headers, for which the sub-pipe system was connected to each bed in order to enable the same inlet temperature to be supplied to each bed, as shown in Figure 3, which is definitely beneficial to enhance the degree of temperature uniformity inside the greenhouse, as compared to other possible variants of pipe systems, for which one pipe system is connected to the accumulator directly without the headers, experiencing temperature drop along the pipe as it is passes through several beds, resulting in a different overall heat flux for each bed.

Figure 1. The view of experimental test greenhouse A (Case 1).

Figure 2. The view of experimental test greenhouse B (Case 2).

The mass flow rate into each sub-pipe system, from the header, was controlled by a large pump to provide the same flow rate condition along with the same inlet temperature, denoted by the simultaneous supply method, A2 in Table 1, and settled to be controlled proportional to the temperature difference between the set point and the real-time monitored averaged temperature value of the greenhouse. Along with the variable mass flow rate control, the hot water temperature of the heat accumulator is also managed to be varied according to the outdoor temperature change. The colder the outdoor temperature, the higher the heat accumulator hot water temperature becomes.

At the design stage, of the heating pipe network inside of the greenhouse, it is important to determine the proper size of the pipe to ensure enough surface area to cover the maximum heating load (e.g., on the coldest day during winter) [37]. In this case, the size of pipe was set up as shown in Table 2, with the aid of auxiliary calculation results by the greenhouse heat load simulator, securing 60,000 Kcal/h heating capability to prevent crops from freezing during the experiment in winter. From the experiment, it was shown that the simple substitution from a hot air-based heating system to a hot water-based heating system, including the extra sub-pipe network, can lead to an outstanding enhancement of the degree of uniformity inside of the greenhouse, which will be mentioned in detail later. In addition to this control measure, an additional one, A3 in Table 1, for which each bed mass flow rate can be controlled independently by adopting a solenoid valve, open and closed, at the inlet of the auxiliary hot water supplying system in each bed. As one can conceive with ease, the thermal condition for each bed is not in the same situation, for example, the bed located near the side wall is to be exposed to more severe thermal conditions, by the infiltration of cold outdoor air and the thermal diffusion, with a higher temperature gradient, near the greenhouse skin, than the bed in center. In that sense, the same amount of hot water with the same inlet temperature for each bed, like A2 in Table 1, may not be an optimal control measure from the view point of managing the degree of uniformity close to a desired level, and more enhancement can be achieved by increasing the level of freedom for each bed in supplying heat flux. In this second control measure of hot water supply in the greenhouse, the on-and-off type of mass flow control to each bed, the execution of it is designed to be determined by the information from the multiple sensor nodes, which is also newly installed in the test greenhouse. In the experiment, the representative value for each bed, the averaged value calculated with several sensing data near the bed, is calculated in each time step of control, by several minutes. Additionally, the current level of temperature difference between the target and the representative value on each bed is referred to in order to determine whether to open or close the solenoid valve, to supply more heat or not, in each bed. This control measure, A3, with the aid of the multiple sensor nodes monitoring system has been proven to be quite effective in enhancing the degree of uniformity inside of the greenhouse against the counter measure A2.

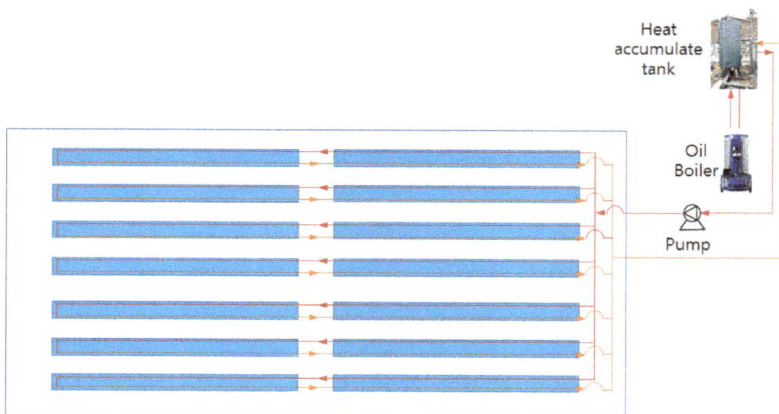

Figure 3. Schematic diagram of hot water supply heating system for Case 1 (greenhouse A).

Figure 4 shows another pipe system configuration for experimental test greenhouse B in this study. It is noted that three individual greenhouses are connected via the main hot water supply and return pipe networks, for which a compact heat exchanger was installed to control the inlet temperature supplied to the hot water header for each building. The main test facility of the wireless multiple sensor nodes system and the physical control one, such as a hydraulic proportional mass flow control valve, was installed in greenhouse A, as shown in Figure 4. An additional control measure of a hydraulic proportional mass flow control valve was adopted to seek an enhanced result of control precision, and, as a consequence, the reactive time to reach to target point, not to mention of the degree of uniformity. At the inlet point of the supply sub-pipe for each bed, a hydraulic proportional mass flow control valve plays the role of controlling the rate of mass flow into each bed according to the monitored data from multiple sensor nodes. It means that the heat release in each bed can be managed independently, unlike the case of the former greenhouse, A2 and A3 in Table 1, in which the same mass flow rate is induced simultaneously for each bed, with the same inlet temperature of hot water, or independently for each bed (i.e., the mass flow variation is not applicable in A2 and A3). In real operating condition, it is observed that the bed near the side wall of the greenhouse needs more heat flux to make the local temperature of the bed the same with that of the bed in the center (i.e., securing the degree of uniformity of the thermal environment). In other words, the same amount of heat flux in each bed, as in the case of the former greenhouse experiment A2 and A3, might cause a heat imbalance again, which would hinder the degree of uniformity from being enhanced above a certain level. Along with this heat release independent control measure, with variable mass flow control for test greenhouse (Case 2 B2, B3), the on-and-off control measure was tested too in test greenhouse B, B1, in order to make an assessment on the effects of greenhouse scale on the adopted control measures, and it will be described in detail in the following. In the case of B2, it had linear logic for opening heat supply control valve. Logic will control the valve opening by a value between 4 and 20 mA with a gradient of 4. The temperature is adjusted with a deviation of ±2 °C from the set temperature. B3 has the same control logic as B2, but B3 calculated the valve opening value with current temperature measured data and the previous one. This averaging logic control of the values was a role for the valve in order to prevent sudden changes. The specification of installed pump and proportional valve is given in Table 3.

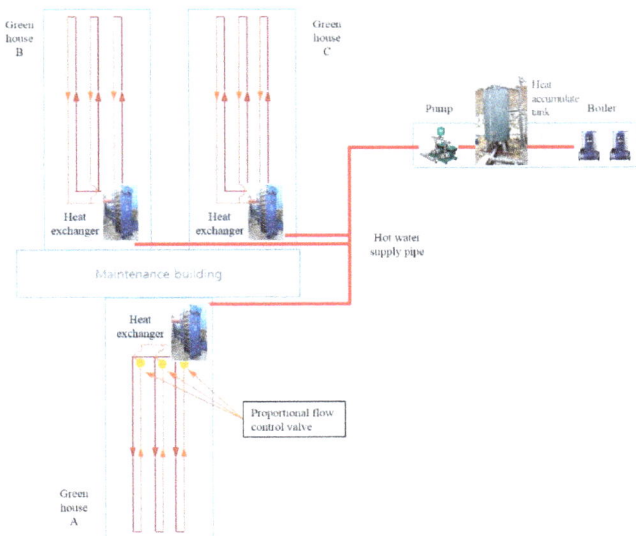

Figure 4. Schematic diagram of hot water supply heating system for Case 2.

Table 1. Clarification of the variants of control measures for test cases.

Case 1 (Greenhouse A)		Case 2 (Greenhouse B)	
A1	Hot air	B1	Hot water (pump on–off by bed)
A2	Hot water (simultaneous supply)	B2	Hot water (valve con. 1, linear control)
A3	Hot water (pump on–off by bed)	B3	Hot water (valve con. 2, two step avg. control)

Table 2. Sub-pipe specification of Case 1 and 2.

	Case 1	Case 2
Pipe outer diameter and thickness	60.5/3.91 mm	42.7/3.58 mm
Total pipe length	70 m × 7 line ×2 supply and return = 980 m	23 m × 3 line × 2 supply and return = 138 m
Total area of heat radiation	186 m^2	19 m^2

Table 3. Specification of installed pump and proportional valve.

Item	Specification
Pump	AC 3 phase motor pump
Wilo PBI-D 803 MA	1.85 kW, 6.9 A, 60 Hz, 220 V, 220 L/min
Proportional Valve	DC 24 V, 15 W
HSH–FLO DN32 2 Way SS304	Control signal: 4–20 mA (0–10 v/0–20 mA)

2.2. Specification for the Monitoring System of Multiple Sensor Nodes

Unlike the conventional multi-variable control of the greenhouse environment system, for which the operating condition monitoring for the large volume of a greenhouse tends to be covered by a sensor node, a multiple sensor nodes system is introduced in this study to overcome the drawback of the current cost optimized system, in order to seek for a precise monitoring system [38] in which the operating environment inside a greenhouse can be monitored more accurately, resulting in more precise environment control measures becoming applicable, as introduced in this study. The conceptual diagram for the multi-sensor nodes system adopted in this study is shown in Figure 5. From the view point of precise thermal environment control, with which one can expect a significant enhancement crop productivity and quality, a precise data monitoring system with multiple nodes, a sort of volumetric data, could be said to be a prerequisite for accomplishing it. Although the additional cost for installing a multiple sensor nodes system, instead of the previous cheap and simple monitoring system, might cause an economic matter in terms of a proper pay-back period, the expected ripple effects of primary energy saving, productivity increase, and quality enhancement, etc., by introducing precise control measures supported by the volumetric data for thermal variables, is worth considering properly. As shown in Figure 5, the sensor module can measure relative humidity with ±2% Relative humidity (RH) accuracy, temperature range from −40 °C to 80 °C with ±0.1 °C accuracy and CO_2 range from 400 to 2000 ppm with ±50 ppm accuracy [39]. To ensure reliability, five random sensors were placed in a box and temperature data were measured during every minute of a half hour. As shown in Figure 6, more than 86% of data is distributed in the 0.1 °C range from the average temperature. The precision of the sensor can be regarded as being reliable enough for temperature measurement in the greenhouse. As for the error analysis for the measurement data, it is assumed that the measurement data is within the precision of the sensor provided by the manufacturer, and the data with abnormal behavior was excluded in the analysis process.

A total of twenty sensor modules were installed for the experiment and the arrangement of the sensor nodes inside of the test greenhouse were reshuffled and adapted for the experiment purpose. For example, as shown in Figures 7 and 8, the sensor modules are arranged to cover the one sectional area of the greenhouse to monitor the variation of the thermal variables in real-time, which was applied in the experiment of test greenhouse A, or they are re-arranged in the longitudinal direction to investigate the variation in that direction, for test greenhouse B. In case of a test greenhouse A, the whole control volume of the test building, about 2000 m^3, is covered by a sensor node in the center under the assumption that the thermal variables were in uniform condition or had a minor deviation from the value of monitored data.

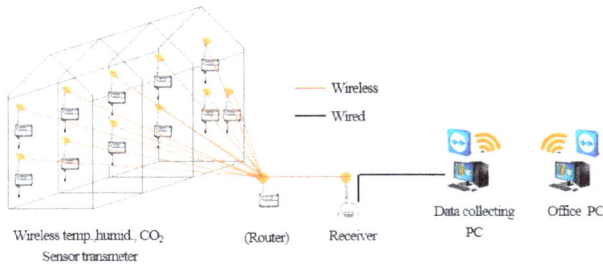

Figure 5. Schematic for greenhouse indoor environmental data collecting system.

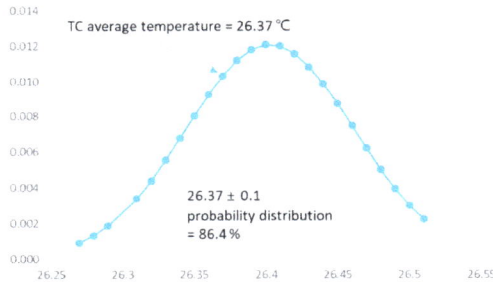

Figure 6. Normal distribution of sensor data graph.

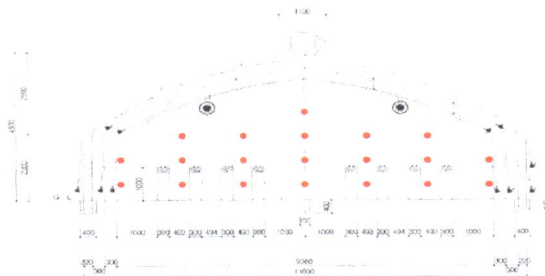

Figure 7. Twenty sensing points in one plane (temperature/humidity/CO$_2$).

However, with the multiple sensor nodes system, it can be reduced up to 100 m^3, a twentieth of its original volume so as to substantially improve the level of preciseness for the control. Besides, the multi-points information from the sensor nodes enables the newly proposed control measures or algorithms to be realized successfully. Another benefit with the introduction of multiple sensor nodes system, as in this study, is the reduced risk of malfunction of the thermal control system with the error value from a sensor node, due to the breakdown of the sensor itself, or of data missing during the communications, which is supposed to give rise to substantial damage in crop growth. More detailed description of the monitoring or sensor system in terms of telecommunication aspects is omitted here for the sake of simplicity. The time frequency for measuring the data can be adjusted according to the situation, by seconds, or by minutes, but in this study the measured data were collected every 10 min, as it was noticed that the thermal environments did not seriously vary within the scale of several minutes.

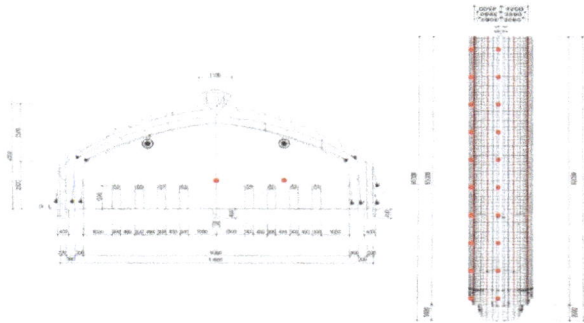

Figure 8. Twenty sensing points in longitudinal direction (temperature/humidity/CO_2).

3. Results and Discussion

3.1. Comparison of Thermal Environment by Heat Supplying Systems (Hot Air vs. Hot Water)

As for the thermal environment control, the representative independent variable for the greenhouse was definitely the inner temperature, since the relative humidity was also a function of the temperature. The qualitative behavior of the temperature and the humidity according to the heat supply systems could be hypothesized to be similar, although the behavior of the humidity according to the thermal condition variation inside of the greenhouse did not exactly match with that of temperature, due to several aspects that should be considered further, such as different diffusivity, the effect of plant respiration, etc. The effects of heat supply systems on other main variables, such as humidity or gas concentrations, will be analyzed in a follow-up study later, thus the present analysis focused on the temperature distribution and variation primarily.

Figure 9 shows the comparison of the profiles of the temperature with different heat supply systems (i.e., hot air and hot water) in test greenhouse A. One can observe that quite different thermal behaviors were created by the different heat supply systems during the night and the dawn, in that a more comfortable thermal condition was provided by the hot water supply system, in contrast to the hot air one, for which frequent oscillating patterns during the period appeared, and the temperature difference was approximately 3~4 °C. It was evident that a large variation or disturbance of the thermal variable can exert a bad influence on crop growth, which could eventually lead to a productivity decrease and potentially lower the quality of the crop. The data for a hot air system was monitored for another building of test greenhouse A, for which a sensor node was implemented in the center of the greenhouse as usual. On the contrary, the data for a hot water-based heating system was monitored from multiple sensor nodes, 20 sensor nodes in this case, and the data in Figure 9 denote the average value of those data from multiple sensor nodes. For Case 1, experimentation was carried out on the same day, and the average ambient air temperature was −5.8 °C, the highest and lowest temperature was −9.5 and 0 °C, respectively. Average wind speed was 3.8 m/s. Average humidity of air was 36.9%, as shown in Figure 10. The raw data of hot air- and hot water-based heating supply are shown in Figure 11. For the case of hot air, the temperature changed drastically compared to the hot water system, since the hot air blower frequently repeats the on-and-off operation. On the other hand, with the hot water system, the fluctuation of it was not distinctively observed. However, the maximum temperature gap between the data for different sensing positions was nearly 5 °C, which means that there is still room for improvement in terms of the degree of uniformity by adopting additional measures, such as a varying local heat flux for each bed.

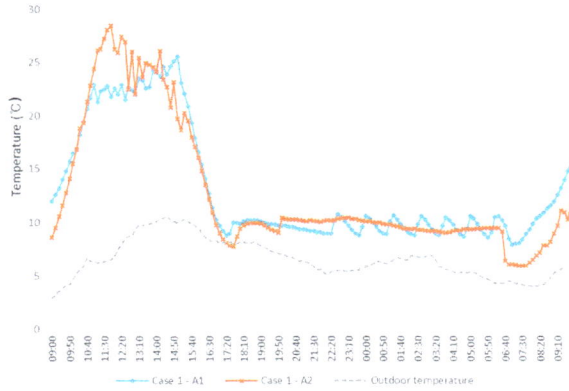

Figure 9. Temperature profiles of greenhouse A for different heat supply systems, A1 and A2 in Table 2.

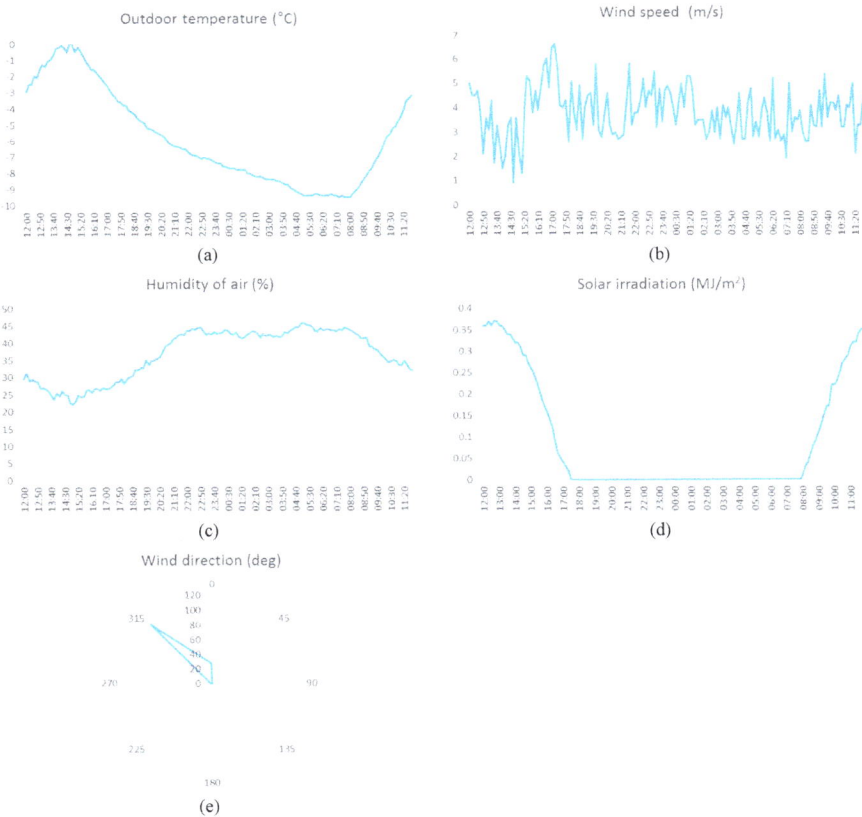

Figure 10. The climatic data from the experimental day of Case 1. (**a**) Outdoor temperature; (**b**) Wind speed; (**c**) Humidity of air; (**d**) Solar irradiation; (**e**) Wind direction.

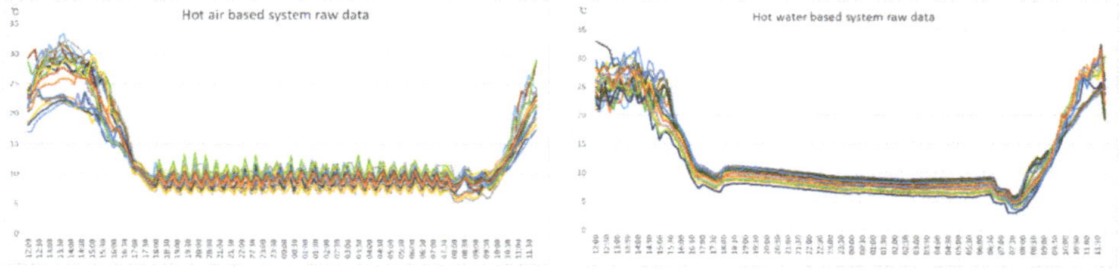

(a) (b)

Figure 11. Raw temperature data of Case 1. (**a**) Hot air based system temperature; (**b**) Hot water based system temperature.

In Figure 12, the temperature distribution on a specific plane of the greenhouses, in the center position of the building, by the heat supplying systems is compared for the assessment of the degree of uniformity. In the case of the conventional monitoring system with a sensor node in the center, it is shown that the representative temperature of 11.2 °C and the maximum temperature difference in the plane, with the lower right corner one of 17.4 °C and left corner one of 14.1 °C, is about 6 and 3 °C, respectively. In contrast, the multi sensor node-based average temperature in the hot water system is about 13.2 °C and the maximum temperature difference is estimated to be less than 1.33 °C at the lower center of the plane, coinciding with the position of the heat source, and the heating pipe under a bed. This certainly does have a variety of implications on greenhouse thermal condition management for providing a more favorable environment that seeks a high productivity and quality of the crop. Firstly, from the view point of securing a uniform crop quality, to enable increased profit on the market, the large difference of thermal variability in a control volume, regardless of whether it is in the longitudinal or widthwise direction, is certainly not a desired operating condition, thus it needs to be avoided as far as possible. In that sense, the conventional, and widely installed and operated, hot air-based heating system is, then, not the best, or optimal, heating solution at all for ensuring, or seeking, better economic profits.

			11.63			
	11.37	11.33	11.53	11.18	11.48	
11.61	11.23	11.38	11.20	11.18	11.33	11.84
14.14	11.81	11.72	11.70	11.51	11.57	17.36

			13.36			
	13.05	13.35	13.52	13.33	13.15	
12.82	12.71	13.04	12.93	13.33	12.51	12.01
13.20	13.96	14.55	13.56	14.35	13.12	12.39

Figure 12. Temperature distribution of Case 1 (A1) and Case 1 (A2) (heating stage).

In Figure 13, the patterns of temperature change for different types of heating systems based on the experiment data are shown as time passed. The hot air ejected from the successive hole on the plastic vinyl tube in the longitudinal direction, which is located on both sides of the greenhouse, is convected upward, mixing with the relatively lower temperature air in order to heat up the whole volume of the greenhouse eventually. The hot air supply system has the benefit of a quick response in relation to reaching the target temperature of the internal air of the greenhouse, but the area of a high temperature spot is unavoidable due to the intrinsic feature of the forced convection-based heat delivery mechanism. Additionally, when the hot air supply stops due to the signal that the center temperature from a sensor has reached a set temperature, the heat flux into the greenhouse is not sustained further any more by the stop of hot air supply, which means that the continuous heat loss via the greenhouse skin and the infiltration of cold outdoor air via the gap make the inside temperature drop rapidly again, as shown in Figure 9. In contrast, the natural convection, or somewhat radiative heat transfer mechanism of a hot water-based heating system can provide a continuous or unwearying heating, which can be executed moderately, as compared to the forced convective mode of a hot air system, via the large area of heat ejection pipe under the bed. Unlike the hot air-based heating system, the hot water one can be said to be beneficial in that the hot water that remains inside of the heating pipe can emit the heat continuously to compensate for the heat loss from a greenhouse, even after the pump operation is stopped by the controller. Secondly, it can also be conceived that the current thermal management control system, that relies on a single sensor node, can cause serious results in terms of productivity or crop quality decrease and, more seriously, a substantial damage of the whole cultivation due to disease by mold or bacteria, which can be initiated by condensation on a product. For example, when the greenhouse is controlled by referring to only the sensor data of a sole sensor node with the hot air system, the multi-variable control system cannot recognize the observed large temperature difference in a target control volume—the building of a greenhouse—at all (specifically during the night in winter), and thus does not take any action to alleviate the large temperature difference under the assumption that the inside temperature is uniform or within the range of minor differences. The possibility of this situation of danger can be significantly reduced with the simple substitution of the heating system into a hot water-based heating system. For example, in the ripening stage of thermal management for a strawberry, the temperature inside of a greenhouse is desired to be in a chilled temperature range, around 5~7 °C, in order to secure higher sugar content; the heating system is often not triggered until drops below 4 °C. It is expected that, in this case of operation, the local temperature difference would become large, especially between the beds located in the center and those by the side walls, so in a certain cold day of a winter, the crops in a side bed would likely be suffering from being exposed to a cold air condition below 4 °C without being recognized by a control system. If this type of exposure or condition occurs often or for a certain period of time, it would surely lead to undesirable consequences (e.g., damage by diseases).

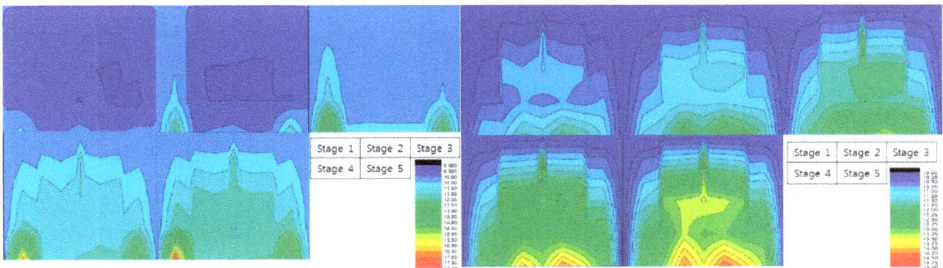

Figure 13. Temperature changes due to time variation in Case 1 (A1) and Case 1 (A2).

During the day, the thermal condition in the greenhouse can be highly asymmetric due to the effects of solar irradiation. However, during the night, when the experiment was performed, the non-uniformity due to the solar irradiation is greatly reduced. Moreover, since the windows are almost closed during the experiment at night for heating, the effect of wind blowing outside of the greenhouse is assumed to be marginal, even though a strong and continuous wind condition might cause an asymmetric thermal condition inside a greenhouse. In this study, the experiment with sensors in the longitudinal direction were configured to set all sensors located on one side of the greenhouse in order to increase the measurement resolution, under the assumption that the effect due to wind is too marginal to suppose a symmetric thermal condition. Figure 13 shows the symmetrical result, to some extent, of temperature in the greenhouse during the heating experiment period, except for the values for both side corners. However, it was considered that the level of asymmetry is not significant to sacrifice the high resolution of data in the longitudinal direction by installing the sensors on both side.

From the analysis of the results of the experimental verification and the promoted understanding of the different features of the heating systems with the volumetric data for the thermal variables, it can be concluded that the introduction of a new concept of control variable, of a thermal condition for a greenhouse, might be beneficial in various aspects, such as productivity and quality enhancement, or risk management to prevent disease outbreak preemptively, by providing a more uniform thermal environment. In this study, the variable of the degree of uniformity of temperature is newly introduced as a main control variable to a greenhouse, and the effects of the adopted heating systems or the counter control measures on the variable is assessed as follows.

The degree of uniformity of the temperature inside of a greenhouse can be defined by the following equation.

$$U = \frac{\left(\sum_{i=1}^{n} \frac{(T_i - T_a)^2}{n}\right)}{T_a} \times 100,\tag{1}$$

U denotes the degree of uniformity, T_i is a local temperature of a sensor at a certain measuring point, and T_a is the average temperature of the whole measured points. The definition of the degree of uniformity of the thermal variable, temperature in this case, means the degree of dispersion of the temperature from the average value, and it is only available when the multiple sensor nodes-based monitoring system is implemented to be able to provide lots of data in real-time. From the view point of thermal environment control in the real field, this new concept of a thermal condition control variable—the degree of uniformity—can bring about substantial benefits for several aspects that have been described above. The estimation, or calculation, of the variable based on the monitored data from lots of sensor nodes in real-time is to be done with ease and it can be estimated that an evolution from a pointwise, one-dimensional control to a volumetric, three-dimensional control can be attained with more precise control algorithms, which can be realized with this multiple sensor nodes platform, in other words, more rigorous or precise real-time monitoring about the status of crop growth and more precise control of thermal environment can be accomplished, with which a substantial improvement of crop production and quality can also be pursued.

Figure 14 shows the comparison of the degree of uniformity according to different heating systems and the control measures of hot water supply on the beds as in Table 1 (A1, A2, and A3). It represents the values of the degree of uniformity in the longitudinal direction of a greenhouse, as shown in Figure 8, and it is shown in Figure 14 that the degree of uniformity for a hot air-based heating system is worse than the others as expected, and it can be substantially improved by the simple substitution into a hot water system, about 80% decrease of its non-uniformity. The additional control measure of independent heat supply by the bed can also improve another 10%, resulting in the value decrease of a tenth. It is also interesting to note that the temperature difference in the longitudinal direction for a typical winter day at night is more or less 3.0 °C, between the center and the end of the greenhouse, in the conventional hot air-based heating system, on the other hand, within more or less 0.5 and 0.35 in case of the hot water-based heating system based on control measures of A2 and A3 as shown in Table 4. This aspect of serious non-uniformity that is supposed to be prevailing in the current thermal environment control system relying on a sensor node monitoring system has long been ignored or overlooked implicitly, and there is substantial room for enhancement of greenhouse horticulture with the adoption of this newly proposed concept of precise control measures based on volumetric sensing data; however, the level of the improvement potential needs to be investigated further for a variety of crops' thermal environments, not to mention of the types of greenhouse and the scale, etc.

Table 4. Greenhouse temperature difference by location by the variants of control measures (Case 1).

Temperature Difference (°C)	Case 1 – A1	Case 1 – A2	Case 1 – A3
Front-Rear	0.22	0.52	0.53
Mid-Front	2.97	0.55	0.36
Mid-Rear	2.75	0.03	−0.17

Figure 15 shows the comparison for the fuel consumption of the heating systems; hot air- and hot water-based heating systems. The measurement of fuel consumption was performed for 11 days in a typical winter period, which was from 21 December to 31 December 2017. The comparison shows that the fuel consumption of A2 hot water-based heating decreased by about 30% compared to A1 hot air-based heating.

Although the accurate measurement for the crop yield for the different buildings of a test greenhouse with different heating systems was not performed, it was approximately estimated that around a 30% crop yield increase was accomplished by providing a more stable thermal condition, with the concept of qualitative control in the real crop cultivating condition of the test greenhouse A (Case 1).

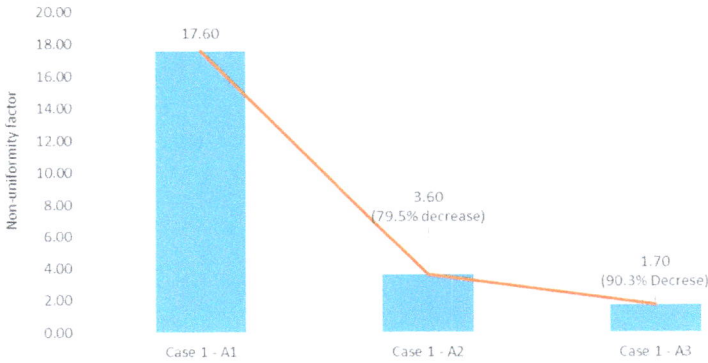

Figure 14. Comparison of the degree of uniformity by the variants of control measures (Case 1).

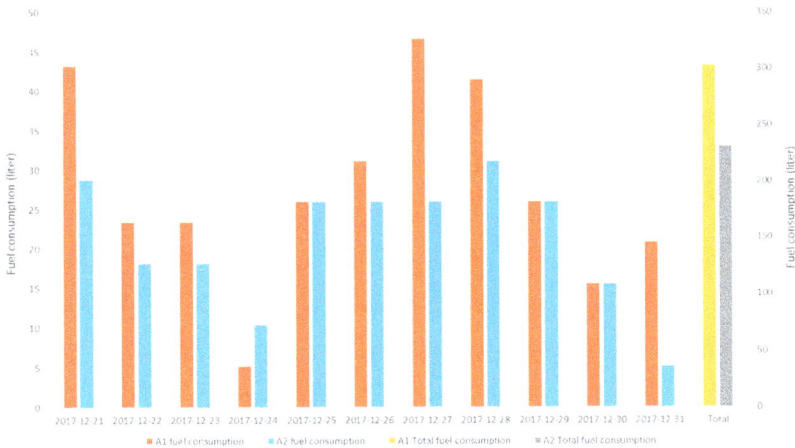

Figure 15. Comparison of the fuel consumption of Case 1 (A1 and A2).

3.2. Comparison of Thermal Environment by Variants of Control Measures (Hydraulic Mass Flow Control)

In the previous experiment for test greenhouse A, the concept of a new control variable with the multiple sensor nodes system—the degree of uniformity—was introduced and the effects on it by the heating systems and, further, the enhancement of the variable along with the additional mass flow rate control by a bed was verified experimentally. In this section, a new measure to be able to increase the degree of freedom in controlling heat flux in each bed is proposed and its effects on the degree of uniformity is assessed, and the experimental verification of it is described in detail. In comparison with test greenhouse A, where the inlet temperature into the inner pipe network of the greenhouse was determined from the operating status of a heat accumulator connected to the oil boiler, the inlet temperature into each building of the test greenhouse B could be controlled with a more reliable manner via the heat exchanger located in each building, in that the secondary loop of hot water was decoupled from the main loop for which the heat accumulator was connected, as shown in Figure 4 (i.e., the inlet temperature into each building's heat supply pipe could be adjusted and controlled with relatively less variation compared to the test greenhouse A (Case 1). In addition to that, a hydraulic proportional mass flow control valve is also adopted at the inlet of each bed of the building, by which the mass flow rate into the heat emitting pipe for each bed can be adjusted in an appropriate manner, in order to be able to control the amount of heat flux for each bed, as shown in Figure 16. Since the drastic effects of the hot water-based heating system, compared to the hot air one, on the degree of uniformity is quite straightforward, as proven in the previous experiment, it is not to be considered hear any more. Instead, several control measures, as shown in Table 1, for improving the resolution of uniformity, that can be adopted to the hot water-based heating system, was investigated for test greenhouse B for the comparison of the performances. In this experimental verification for greenhouse B, along with the previously mentioned control measure A3, two new control algorithms are applied with the adoption of hydraulic proportional mass flow control valve. Furthermore, it differs in that the opening rate of the valve is proportional to the temperature difference between the set value of temperature, the target value for each bed, and the averaged value of temperature for each bed in the case of B2. As for the case of B3, the opening rate of the valve is not determined with the data for a current time step, and the opening rate of the valve for the previous step is also referred to in determining the new value of the opening rate of the valve (i.e., the arithmetic mean value is finally applied for the current step in this study even though more various weight average methods can be considered, which might be studied in further study).

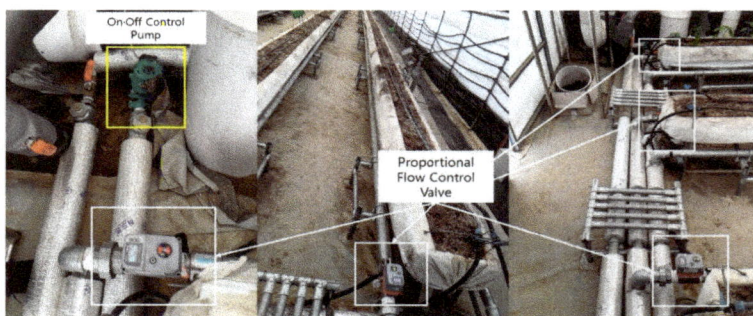

Figure 16. Images of the pump and valves installation for heating control in greenhouse B (Case 2).

Figure 17 shows the experimental test results of temperature profiles inside of test greenhouse B according to the applied variants of control measures. The experiments were conducted for three days, and the ambient air data are shown in Figure 18. In the results, at first it is observed that the temperature profile over time is oscillating, to some extent, regardless of the applied measures, which differs from that of it in the case of test greenhouse A. It is supposedly caused by the small heat capacity of greenhouse B (i.e., a relatively large surface to volume ratio) due to the small-scale of the greenhouse. The performance of each variants of control measures shows that B3 in Table 1 provided the best temperature control feature, in that the deviation from the set temperature, 10 °C, is the smallest, among the other variants, to be able to provide a more comfortable thermal environment to the crop. In contrast, the simple on-and-off pump control for each bed without regulating the rate of mass flow, B1 in Table 1, tends not to be able to meet the target temperature, although the fluctuation is smaller than the case with the variant of B2, linear mass flow rate control algorithm.

The measured temperature data of different control measures, B1–B3, for Case 2 are shown in Figure 19. It is clearly observed that more stable thermal conditions can be managed with B3 control algorithm, in that the temperature at various measured points remained within a range of ±0.5 °C (i.e., the degree of uniformity was highly improved), although the inflection point of temperature over time was adversely increased due to the more rapid response for the temperature gradient variation.

The measurement data for the temperature variation in case of on-and-off pump control, B1, and the variants using proportional valve control, B2 and B3, is given in Table 5. It can be highlighted that the variant of B3, with referring to the data of the previous time step, induced more peaks but the fluctuation from its mean value is half to that of simple on-and-off control, B1, and one third to that of another variant of B2. From the result, it can be understood that the abrupt change of mass flow rate into the heat ejection pipe is not favorable for securing a more comfortable thermal environment, in terms of meeting the set value or the fluctuation, etc. In that sense, the adjusted control algorithm of B3, with gradual change of hot water supply into the heat ejection pipe, can be hypothesized to be the best or optimal control algorithm at present. However, more extensive or rigorous investigation for a variety of test cases is certainly required in order to draw a general conclusion, and will be carried out in further study. With regard to the enhancement of the degree of uniformity by the variants of the control measures, it is summarized in Figure 20. As mentioned above, the value for the hot air-based heating system is excluded for simplicity, and the comparison reveals that an unattractive improvement of the degree of uniformity can be attained just by introducing hydraulic mass flow valve control. Therefore, more elaborate control algorithms need to be adopted, as in the case of B3, where the previous time step opening rate was also referred to in order to determine the current time step of the rate having an effect, to aid an abrupt change of heat flux being avoided. The new control algorithm of B3, referring to the previous value of the opening rate of the valve, is able to accomplish a certain level of enhancement of the degree of uniformity inside the greenhouse, more or less 55% compared to the value of B1 or B2. From the analysis of the experimental verifications, it can be concluded that B3 shows the best performance in terms of tracing the target temperature and providing a more favorable thermal condition with a higher degree of uniformity.

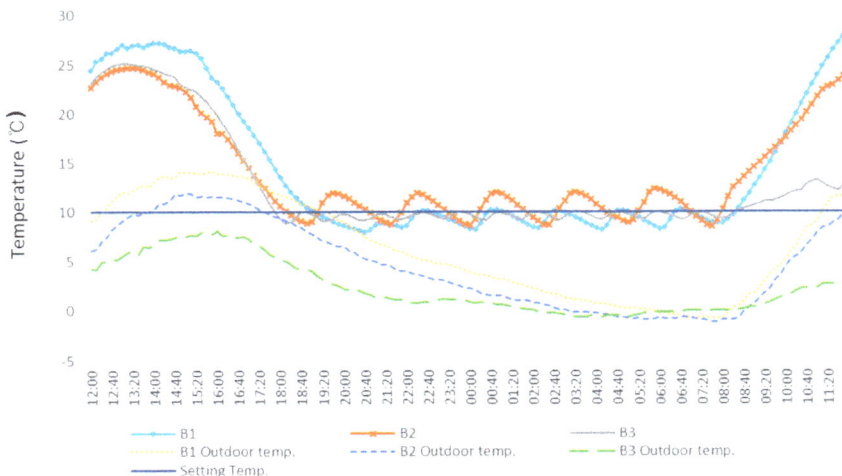

Figure 17. Profiles of temperature variation by the variants of control measures (Case 2).

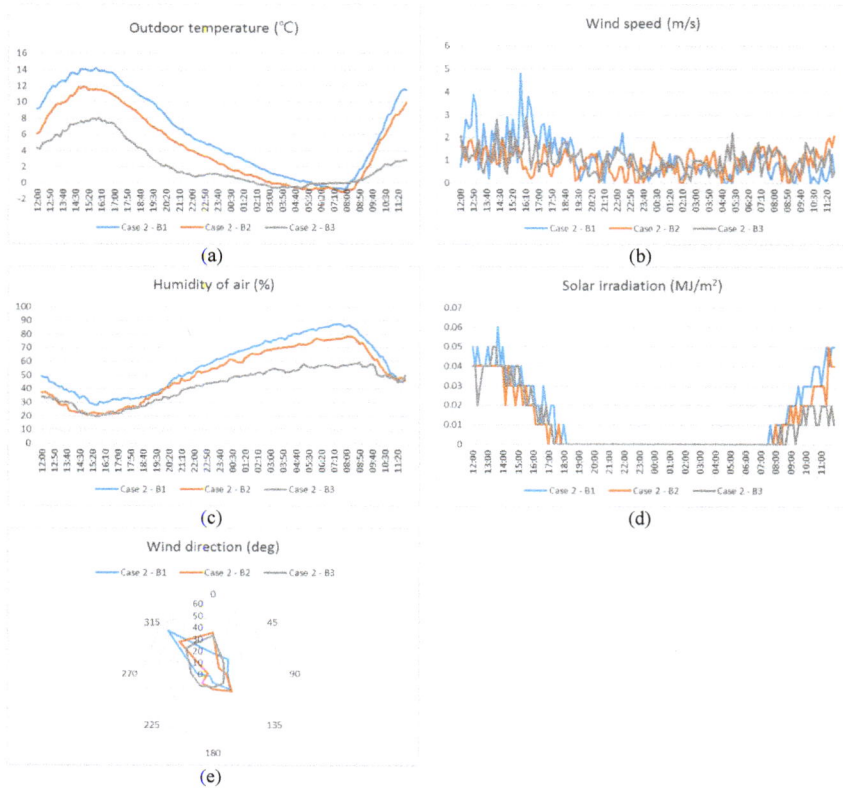

Figure 18. The climatic data of the experimental day of Case 1. (**a**) Outdoor temperature; (**b**) Wind speed; (**c**) Humidity of air; (**d**) Solar irradiation; (**e**) Wind direction.

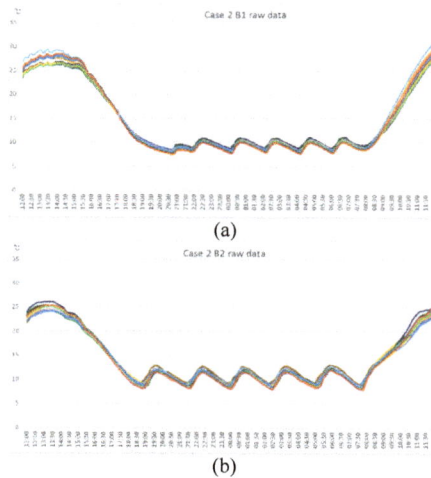

Figure 19. *Cont.*

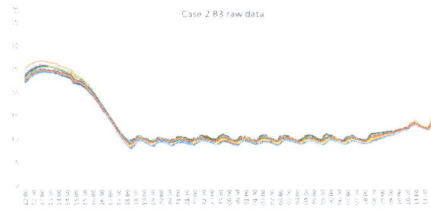

(c)

Figure 19. Temperature measured data of different control measures of Case 2. (**a**) B1; (**b**) B2; (**c**) B3.

Table 5. Maximum and minimum temperature data by the variants of control measures (Case 2).

	Case 2 – B1			Case 2 – B2			Case 2 – B3		
	Max. Temp	Min. Temp	Temp. diff	Max. Temp	Min. Temp	Temp. diff	Max. Temp	Min. Temp	Temp. diff
1st	10.4	7.9	2.5	11.9	9.0	2.9	10.2	8.7	1.5
2nd	10.4	8.1	2.3	11.9	8.8	3.1	10.1	9.0	1.1
3rd	10.3	8.3	2.0	12.0	8.7	3.3	9.9	9.1	0.8
4th	10.4	8.1	2.3	12.0	8.7	3.3	10.0	9.2	0.8
5th	10.5	8.2	2.3	12.3	8.9	3.4	10.1	9.2	0.9
6th	10.3	8.8	1.5				10.1	9.2	0.9
7th							10.0	9.2	0.8
8th							10.2	9.1	1.1
9th							10.2	9.2	1.0
10th							10.1	9.2	0.9
11th							10.1	9.2	0.9
12th							10.0	9.3	0.7
Avg.	10.4	8.2	2.2	12.0	8.8	3.2	10.1	9.1	1.0

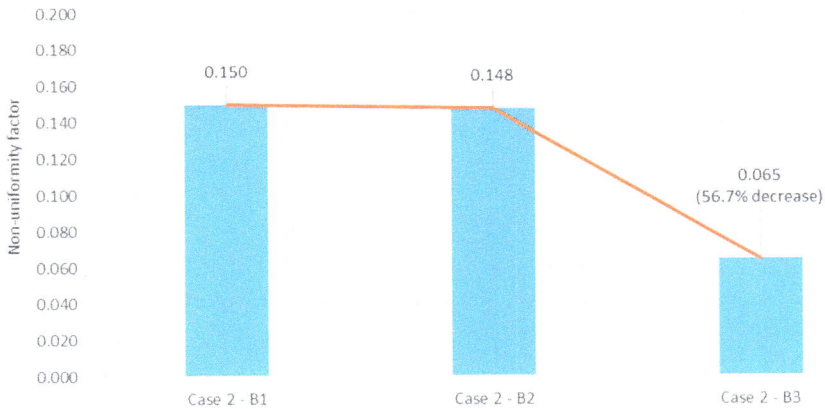

Figure 20. Comparison of the degree of uniformity by the variants of control measures (Case 2).

In Figure 21, the time variation of the degree of uniformity for two different control measures (i.e., B1 and B3) is compared and it shows that the instantaneous difference of it can be larger than that for the averaged value. It also implies that the newly proposed control variable—the degree of uniformity—is quite adequate for securing a more comfortable thermal environment for greenhouse horticulture, and it becomes more powerful when it is utilized as a real control variable in managing the transient thermal operating conditions for which the support of volumetric thermal variables data from a multiple sensor nodes system is positively necessary, as shown in this study.

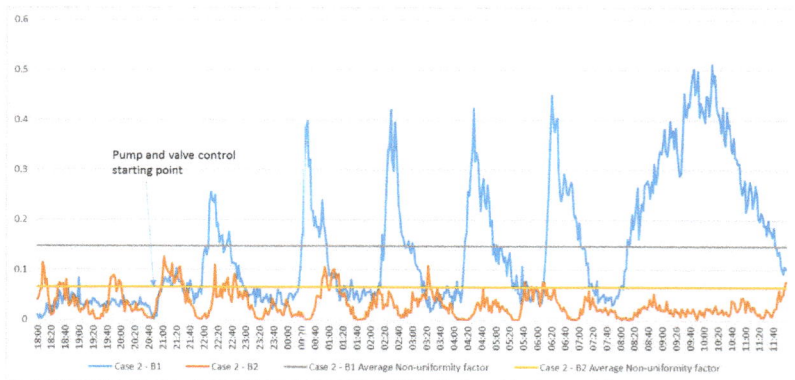

Figure 21. Comparison of the variation for the degree of uniformity for the variants of control measures of B1 and B3 (Case 2).

In order to make an assessment for the energy saving mechanism with the adoption of varying heat flux control for each bed, a computational fluid dynamics (CFD) simulation has been performed using commercial software of FLUENT, and the results are compared in Figures 22 and 23. The simulation was conducted for the condition of the same amount of heat flux supply for both cases. It is worthy to note that the total amount of heat supplied to a greenhouse is the same, but the thermodynamic or fluid dynamic phenomena, including the degree of uniformity, are shown to be quite different according to the variation of the local heat flux to each bed. As for the varying heat flux for each bed, it was adjusted to be increasing as it approached the side walls. It is interesting to note that the varying heat flux operating control can shorten the time to reach target temperature inside a greenhouse, as shown in Figure 23, by about 600 s, and the air flow pattern differs in that different heat flux controls lead to a counter clockwise tumble flow pattern, with an additional clockwise one at the side wall. On the other hand, for the case of the same heat flux control for each bed, a wide and single clockwise tumble flow pattern is observed. From the comparison, it can be hypothesized that the temperature non-uniformity is mainly caused by the chilled air due to the heat loss by cold outside air infiltration at the side wall, which is convected into the central part of the greenhouse. However, when more heat flux was imposed on the bed near the side wall, a kind of thermal barrier forms to prevent the cold air from being penetrated easily into the central part. However, more rigorous CFD simulations for a variety of operating conditions, including the three-dimensional effect, are highly desired to be performed in further study.

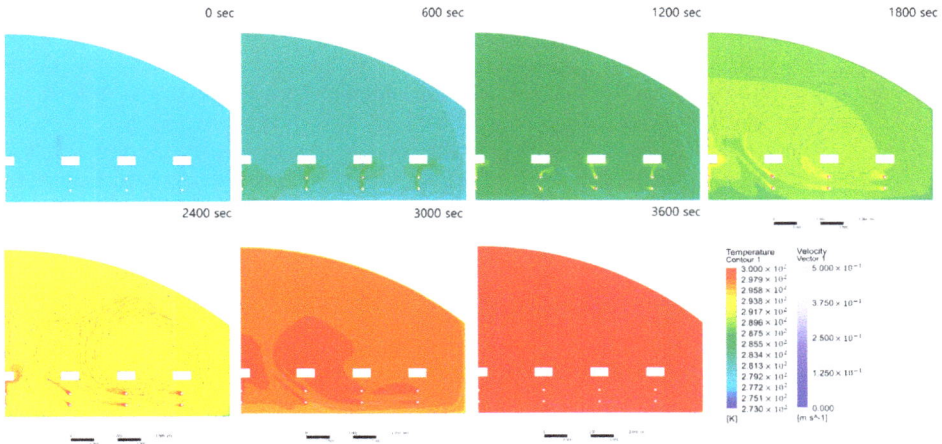

Figure 22. Temperature and air velocity distribution for the case of same heat flux for all beds.

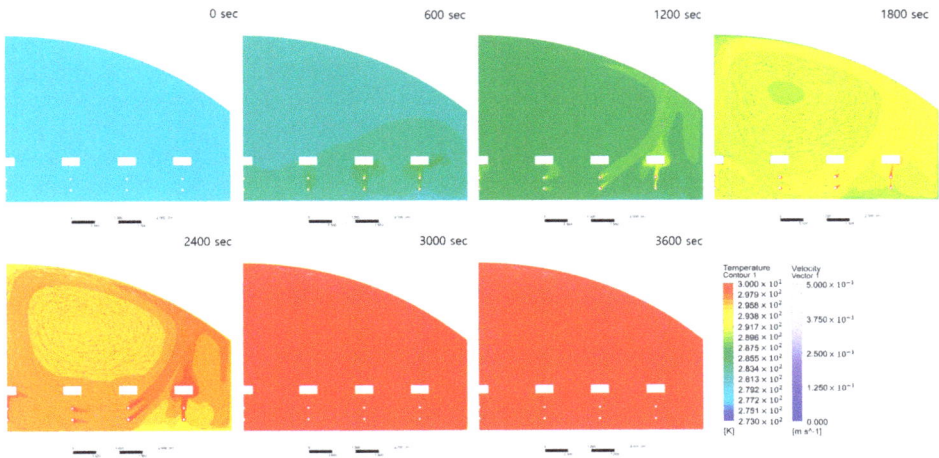

Figure 23. Temperature and air velocity distribution for the case of varying heat flux for each bed.

4. Conclusions

In this study, the new concept of qualitative control variable, the degree of uniformity, is proposed for seeking enhanced cultivating performance, being higher crop yield and better quality. A more precise sensing or monitoring system with higher spatial resolution is introduced with the multiple sensor nodes system based on a wireless sensor network, instead of the conventional monitoring system with a sole sensor node to cover a large cultivating area, for which the proposed qualitative control variable—the degree of uniformity—cannot be estimated appropriately.

It is found out that there is a big difference in the degree of uniformity by the different heating systems—a hot air-based heating system and a hot water-based heating system—for the test cases of a greenhouse. A simple substitution for a hot water-based heating system can make a significant improvement to the degree of uniformity, more or less 80%, by providing more thermally stable conditions with less temperature fluctuation. This improved level of uniformity with the hot water-based heating system is due largely to the change of the dominant heat transfer mode, from the forced convective heat transfer to the radiative or natural convective one, and the continuous supply of local heat flux by the hot water, that remains in the pipe even after the set temperature is achieved, seems to contribute to form a thermally-stable condition to some extent. The additional counter measure to control the local heat flux independently for each bed is able provide a supplementary reduction of non-uniformity up to 90%, effectively, as compared to the equal heat flux for each bed. It is worthy to note that this local heat flux control can be effectively executed only when the state of local thermal condition—the degree of uniformity in this study—is informed properly by the multiple sensor nodes system. Among the variants of counter measures of local heat flux variation for the hot water-based heating system, the hydraulic proportional mass flow control valve, with two-step average value, showed the best performance, with the additional reduction of its non-uniformity to half of the level for other variants of local heat flux variation. It can be highlighted that it was also verified that the thermally stable condition inside a greenhouse, with the concept of a qualitative variable and corresponding control—the degree of uniformity—can also bring about the energy cost reduction along with the crop yield improvement, about 30%, simultaneously, in this study. It means that the provision of a more stable thermal condition (i.e., higher level of thermal uniformity) is quite crucial in order to secure economic benefits in greenhouse horticulture, by attaining an enhanced crop yield and reduced energy costs simultaneously.

In summary, the new concept of qualitative control variable—the degree of uniformity—is proposed and its behavior and dynamic features according to the control of the variants of counter measures in a greenhouse has been investigated experimentally with the auxiliary simulation approach, using precise sensing data from a multiple sensor nodes system. The improved stable thermal environment in a greenhouse is proven to be beneficial, in that one can attain both crop yield improvement and energy cost reduction simultaneously, to be able to compensate for the initial investment cost increase from installing a hot water-based heating pipe system and multiple sensors, etc. In further study, the behavior and characteristics for other main control variables, such as a humidity and CO_2 concentration, is to be investigated, and the effects of this approach on real crop yield improvement, including quality of the crop, will also be verified from a demonstration in real cultivating conditions.

Author Contributions: Conceptualization, M.C. and Y.-H.I.; Methodology, K.-Y.S.; Validation, C.K.L., S.-W.Y. and Y.-H.I.; Formal Analysis, C.K.L. and S.-W.Y.; Writing-Original Draft Preparation, C.K.L.; Writing-Review & Editing, K.-Y.S. and Y.-H.I.; Supervision, M.C. and Y.-H.I.

Funding: This work was supported by the National Research Council of Science & Technology (NST) grant by the Korea government (MSIP) (No. CRC-15-01-KIST).

References

1. Kim, Y.J. Agricultural and Rural Energy Policies in Major Countries. *World Agric.* **2015**, *182*, 1–13.
2. National Intelligence Council. *Global Trends 2030: Alternative Worlds*; National Intelligence Council: Washington, DC, USA, 2012; pp. 31–35.
3. Laso, J.; Hoehn, D.; Margallo, M.; García-Herrero, I.; Batlle-Bayer, L.; Bala, A.; Fullana-i-Palmer, P.; Vázquez-Rowe, I.; Irabien, A.; Aldaco, R. Assessing Energy and Environmental Efficiency of the Spanish Agri-Food System Using the LCA/DEA Methodology. *Energies* **2018**, *11*, 3395. [CrossRef]
4. Djojodihardjo, H.; Ahmad, D. Opportunities and Challenges for Climate-Smart Agriculture. In Proceedings of the 3rd International Conference on Natural Resource Management (NRM '15), New Delhi, India, 10–13 February 2015.
5. Jeong, E.M.; Lee, U.Y. *Research Report R614-2 Energy Utilization in Agricultural Sector*; Korea Rural Economic Institute: Seoul, Korea, 2010; pp. 5–12.
6. Kim, Y.J.; Park, S.H.; Han, H.S.; Park, Y.K. *Research Report R740 Status of Energy Utilization in Agriculture, Rural, and Policies*; Korea Rural Economic Institute: Seoul, Korea, 2014; pp. 19–29, 59–74.

7. Marketing Report—Global Smart Greenhouse Market 2016–2020. Available online: https://www.businesswire. com/news/home/20161007005274/en/Smart-City-Market-Grow-Tremendously-CAGR-Close (accessed on 3 August 2018).

8. Kim, Y.H. *Smart Farm & Smart Water Management*; Korea Rural Community Corporation-Rural Research Institute: Seoul, Korea, 2018.

9. Kim, B.R. *Smart Farm—The Convergence of Agriculture and ICT, Convergence Weekly TIP*; Convergence Research Policy Center: Seoul, Korea, 2016; pp. 6–9.

10. Kang, C.Y.; Seo, D.S. *Research Report—A Study on the Development of the Horticultural Production Materials Industry*; Korea Rural Economic Institute: Seoul, Korea, 2015; pp. 13–14.

11. Choi, B.O.; Jeong, E.M.; Kim, D.K.; Shin, Y.S.; Kim, H.S. *Research Report—A Study on the Improvement of the System and Development Plan for the Horticultural Industry*; Korea Rural Economic Institute: Seoul, Korea, 2017; pp. 11–16.

12. Stallen, M.; van Ulffen, R. *Greenhouse Sector Study South Korea*; Final Report; Ministry of Agriculture, Nature and Food Quality: Wageningen, The Netherland, 2006.

13. Choi, S.; Hinkle, A.F. *Korea's Controlled Horticulture*; USDA Foreign Agricultural Service Gain Report; USDA: Washington, DC, USA, 2018.

14. Park, H.M. *The Present Situation of Horticultural Industry and Policy Implications*; Korea Society for Horticulture Science: Seoul, Korea, 2016; Volume 34, p. 37.

15. Jiaqiang, Y.; Yulong, J.; Jian, G. An Intelligent Greenhouse Control System. *Telkomnika* **2013**, *11*, 4627–4632. [CrossRef]

16. Nicolosi, G.; Volpe, R.; Messineo, A. An Innovative Adaptive Control System to Regulate Microclimatic Conditions in a Greenhouse. *Energies* **2017**, *10*, 722. [CrossRef]

17. Su, Y.; Xu, L.; Goodman, E.D. Greenhouse Climate Fuzzy Adaptive Control Considering Energy Saving. *Int. J. Control Autom. Syst.* **2017**, *15*, 1936–1948. [CrossRef]

18. Castilla, N.; Montero, J.I. Environmental Control and Crop Production in Mediterranean Greenhouses. *Acta Hortic.* **2008**, *797*, 25–36. [CrossRef]

19. Baeza, E.J.; Stanghellini, C.; Castilla, N. Protected Cultivation in Europe. *Acta Hortic.* **2013**, *987*, 11–27. [CrossRef]

20. Kang, S.; Kang, Y.; Paek, Y.; Kim, Y.; Kim, Y. Analysis of Energy Consumption in Agricultural Facilities. In Proceedings of the Korean Society for New and Renewable Energy Conference, Seoul, Korea, 23 February 2017; p. 293.

21. Choi, C.K.; Park, J.S.; Park, S.Y.; Bae, H.H.; Lee, N.R. *Research Report—An Analysis of Management Situation for the Major Greenhouse Horticultural Crops in Types of Heating Facilities*; Rural Development Administration: Suwon, Korea, 2014; pp. 8–28.

22. Korea Energy Economics Institute. *2014 Energy Consumption Survey*; Korea Energy Economics Institute: Seoul, Korea, 2015; pp. 125–130.

23. Man, H.C.; Thorp, K.R.; Shamshiri, R.R.; Ahmad, D.; Taheri, S.; Jones, J.W. Review of optimum temperature, humidity, and vapour pressure deficit for microclimate evaluation and control in greenhouse cultivation of tomato: A review. *Int. Agrophys.* **2018**, *32*, 287–302. [CrossRef]

24. Atia, D.M.; El-Madany, H.T. Analysis and design of greenhouse temperature control using adaptive neuro-fuzzy inference system-NC-ND license (http://creativecommons.org/licenses/by-nc-nd/4.0/). *J. Electr. Syst. Inf. Technol.* **2017**, *4*, 34–48. [CrossRef]

25. Trigui, M.; Barrington, S.; Gauthier, L. A Strategy for Greenhouse Climate Control, Part I: Model Development. *J. Agric. Eng. Res.* **2001**, *78*, 407–413. [CrossRef]

26. Javadi Kia, P.; Tabatabaee Far, A.; Omid, M.; Alimardani, R.; Naderloo, L. Intelligent Control Based Fuzzy Logic for Automation of Greenhouse Irrigation System and Evaluation in Relation to Conventional Systems. *World Appl. Sci. J.* **2009**, *6*, 16–23.

27. Radojević, N.; Kostadinović, D.; Vlajković, H.; Veg, E. Microclimate control in greenhouses. *FME Trans.* **2014**, *42*, 167–171. [CrossRef]

28. Sethi, V.P.; Sumathy, K.; Lee, C.; Pal, D.S. Thermal modeling aspects of solar greenhouse microclimate control: A review on heating technologies. *Sol. Energy* **2013**, *96*, 56–82. [CrossRef]

29. Vadiee, A.; Martin, V. Energy analysis and thermoeconomic assessment of the closed greenhouse - The largest commercial solar building. *Appl. Energy* **2013**, *102*, 1256–1266. [CrossRef]

30. Rasheed, A.; Lee, J.W.; Lee, H.W. A Review of Greenhouse Energy Management by Using Building Energy Simulation. *Prot. Hortic. Plant Fact.* **2016**, *24*, 317–325. [CrossRef]

31. Ha, T.; Lee, I.B.; Kwon, K.S.; Hong, S.W. Computation and field experiment validation of greenhouse energy load using building energy simulation model. *Int. J. Agric. Biol. Eng.* **2016**, *8*, 116–127. [CrossRef]

32. Tadj, N.; Nahal, M.A.; Draoui, B.; Constantinos, K. CFD simulation of heating greenhouse using a perforated polyethylene ducts. *Int. J. Eng. Syst. Model. Simul.* **2017**, *9*, 3. [CrossRef]

33. Couto, N.; Rouboa, A.; Monteiro, E.; Viera, J. Computational Fluid Dynamics Analysis of Greenhouses with Artificial Heat Tube. *World J. Mech.* **2012**, *02*, 181–187. [CrossRef]

34. Korean Electric Power Corporation. *Statics of Electric Power in Korea*; Korean Electric Power Corporation: Seoul, Korea, 2018.

35. Gupta, M.J.; Chandra, P. Effect of greenhouse design parameters on conservation of energy for greenhouse environmental control. *Energy* **2002**, *27*, 777–794. [CrossRef]

36. Spanomitsios, G.K. Temperature Control and Energy Conservation in a Plastic Greenhouse. *J. Agric. Eng. Res.* **2001**, *80*, 251–259. [CrossRef]

37. Shen, Y.; Wei, R.; Xu, L. Energy Consumption Prediction of a Greenhouse and Optimization of Daily Average Temperature. *Energies* **2018**, *11*, 65. [CrossRef]

38. Pawlowski, A.; Guzman, J.L.; Rodríguez, F.; Berenguel, M.; Sánchez, J.; Dormido, S. Simulation of Greenhouse Climate Monitoring and Control with Wireless Sensor Network and Event-Based Control. *Sensors* **2009**, *9*, 232–252. [CrossRef] [PubMed]

39. E+E Elektronik. Available online: https://www.epluse.com/ko/products/humidity-instruments/wireless-sensors-1/ee244/ (accessed on 27 March 2017).

MDPI

St. Alban-Anlage 66

4052 Basel

Switzerland

Tel. +41 61 683 77 34

Fax +41 61 302 89 18

www.mdpi.com

Energies Editorial Office

E-mail: energies@mdpi.com

www.mdpi.com/journal/energies

www.ingramcontent.com/pod-product-compliance
Lightning Source LLC
Chambersburg PA
CBHW051709210326
41597CB00032B/5424